IFIP Advances in Information and Communication Technology 393

IFIP – The International Federation for Information Processing

IFIP was founded in 1960 under the auspices of UNESCO, following the First World Computer Congress held in Paris the previous year. An umbrella organization for societies working in information processing, IFIP's aim is two-fold: to support information processing within its member countries and to encourage technology transfer to developing nations. As its mission statement clearly states,

> *IFIP's mission is to be the leading, truly international, apolitical organization which encourages and assists in the development, exploitation and application of information technology for the benefit of all people.*

IFIP is a non-profitmaking organization, run almost solely by 2500 volunteers. It operates through a number of technical committees, which organize events and publications. IFIP's events range from an international congress to local seminars, but the most important are:

- The IFIP World Computer Congress, held every second year;
- Open conferences;
- Working conferences.

The flagship event is the IFIP World Computer Congress, at which both invited and contributed papers are presented. Contributed papers are rigorously refereed and the rejection rate is high.

As with the Congress, participation in the open conferences is open to all and papers may be invited or submitted. Again, submitted papers are stringently refereed.

The working conferences are structured differently. They are usually run by a working group and attendance is small and by invitation only. Their purpose is to create an atmosphere conducive to innovation and development. Refereeing is also rigorous and papers are subjected to extensive group discussion.

Publications arising from IFIP events vary. The papers presented at the IFIP World Computer Congress and at open conferences are published as conference proceedings, while the results of the working conferences are often published as collections of selected and edited papers.

Any national society whose primary activity is about information processing may apply to become a full member of IFIP, although full membership is restricted to one society per country. Full members are entitled to vote at the annual General Assembly, National societies preferring a less committed involvement may apply for associate or corresponding membership. Associate members enjoy the same benefits as full members, but without voting rights. Corresponding members are not represented in IFIP bodies. Affiliated membership is open to non-national societies, and individual and honorary membership schemes are also offered.

Daoliang Li Yingyi Chen (Eds.)

Computer and Computing Technologies in Agriculture VI

6th IFIP WG 5.14 International Conference, CCTA 2012
Zhangjiajie, China, October 19-21, 2012
Revised Selected Papers, Part II

Springer

Volume Editors

Daoliang Li
Yingyi Chen
China Agricultural University
China-EU Center for Information
and Communication Technologies in Agriculture (CICTA)
17 Tsinghua East Road, Beijing, 100083, P.R. China
E-mail:{dliangl, chenyingyi}@cau.edu.cn

ISSN 1868-4238
e-ISSN 1868-422X
ISBN 978-3-642-44526-2
ISBN 978-3-642-36137-1 (eBook)
DOI 10.1007/978-3-642-36137-1
Springer Heidelberg Dordrecht London New York

CR Subject Classification (1998): J.0, I.2.9-11, I.2.6, H.4.2-3, I.4.8-9, I.5.1, I.5.4, C.2.1, C.2.m, H.2.8, I.2.4, C.3

Typesetting: Camera-ready by author, data conversion by Scientific Publishing Services, Chennai, India

Printed on acid-free paper

Springer is part of Springer Science+Business Media (www.springer.com)

Preface

First of all, I must express my sincere thanks to all authors who submitted research papers to support the Sixth International Conference on Computer and Computing Technologies in Agriculture (CCTA 2012), held in Zhangjiajie, China, 19–21 October 2012.

This conference was hosted by China Agricultural University, the IFIP TC5 Working Group (WG) on Advanced Information Processing for Agriculture (AIPA), and the Agricultural Engineering Information Committee of the Chinese Society of Agricultural Engineering, and was organized by the China-EU Centre for Information and Communication Technologies (CICTA).

Proper scale management is not only the necessary approach to agro-modernization and agro-industrialization but is also required by the growth in agricultural productivity, so, the application of different technologies in agriculture is becoming especially important. 'Informatized Agriculture' and the 'Internet of Things' have been chased by many countries recently in order to scientifically manage agriculture to achieve low costs and high income. CICTA aims to promote research and development in advanced and practical technologies applied in agriculture and to encourage international communication and cooperation, and has successfully held six international conferences on Computer and Computing Technologies in Agriculture since 2007.

The topics of CCTA 2012 cover a wide range of interesting theory and applications related to kinds of technology in agriculture, including the Internet of Things and cloud computing; simulation models and decision-support systems for agricultural production; smart sensors, monitoring and control technology; traceability and e-commerce technology; computer vision, computer graphics and virtual reality; the application of information and communication technology in agriculture; and universal information service technology and service systems development in rural areas.

We selected 108 best papers among all the papers submitted to CCTA 2012 for these proceedings. The papers have divided into two thematic sections. Creative thoughts and inspirations have been discovered, discussed and disseminated. It is always exciting when creative experts, professionals and scholars get together to share inspiring ideas and hopefully accomplish great developments in the technologies in high demand.

Finally, I would like to express my sincere thanks to all authors, speakers, session chairs and attendees both coming from abroad and from mainland China for their active participation and support of this conference.

December 2012 Daoliang Li

Conference Organization

Organizer

China-EU Center for Information and Communication Technologies in Agriculture (CICTA)

Chairman

Daoliang Li

Conference Secretariat

Lihong Shen

SPONSORS

China Agricultural University

The IFIP TC5 Working Group (WG) on Advanced Information Processing for Agriculture (AIPA)

Agricultural Engineering Information Committee, Chinese Society of Agricultural Engineering

Table of Contents – Part II

Table of Contents – Part I

Remote Sensing of Forest LAI from Multitemporal Optical Satellite Images over Mountain Area

Yuechan Shi[1,2], Guijun Yang[1,*], Haikuan Feng[1], and Renli Wang[2]

[1] Beijing Research Center for Information Technology in Agriculture, Beijing 100097, China

[2] Geomatics College, Shandong University of Science and Technology, Qiangdao 266510, China

Abstract. Leaf area index (LAI) is one of the most important structural parameters in terrestrial ecosystems. In this study, multitemporal Landsat TM images covering experiment station in the northwest mountain areas of Beijing were acquired and in-situ forest LAI was measured. By correlation analysis of three vegetation indexes (NDVI, EVI and TGDVI) and LAI, it is found that the correlation between LAI and NDVI in exponential form behaved a good relativity. This model was applied in mapping multitemporal forest LAI. Further, the sensitivity of the models between vegetation indices and LAI were tested for the broadleaf forest, coniferous and mixed forests, respectively. The results show that the accuracy was improved both in broadleaf and mixed forests compared with the advanced findings which didn't take care of different vegetation types, except for conifer stands with little accuracy decreasing. However, the accuracy of all models has reached a significant level.

Keywords: remote sensing, landsat TM, forest, LAI, seasonal variability.

1 Introduction

Leaf area index (LAI) is an important ecological parameter used for characterizing the interface between a vegetation canopy and the atmosphere. And it is also crucial for quantifying the energy and mass exchange of terrestrial ecosystems (Sellers et al., 1986; Running et al., 1989; Bonan et al., 1993; Turner et al., 1999). Chen and Black (1992) defined LAI was a half of the total leaf area per unit ground surface area.

All the methods of obtaining LAI values can be classified into two categories: direct or indirect way (Xiang, 2004; Zhang, 2010). The traditional direct way by destructive sampling and litter collection are time- and strain- consuming. Both the optical instruments and remote-sensing method are classified as the indirect way. The optical instruments (e.g. the LAI-2000, TRAC, and hemispherical photography) were used to measure LAI through the gap fraction method (Inge, 2004). But, this method is limited due to the extent of observation only representing small-scale area. On the contrary, the remote sensing method could acquire the images at a moment covering large area.

[*] Corresponding author.

D. Li and Y. Chen (Eds.): CCTA 2012, Part II, IFIP AICT 393, pp. 1–9, 2013.

Most of previous researches on estimating LAI from optical satellite data are based on two approaches: empirical models and physical models (Weiss, 1999; Liang, 2003; Xing, 2010; Yang, 2010). The empirical model involved the relationship between LAI and vegetation indices, while physical model uses the inversion of canopy reflectance models. Due to the quantitative methods usually faced the problem of ill-posed inversion because too many parameters should be providing before inverting the model. Therefore, many scientists preferred to use the empirical model. The empirical model method, as a simple and practical way, is widely used (Chen, 1999; Turner, 1999; Luo, 2005; Pauline, 2008).

In this study, forest LAI was measured using Hemispherical Photography (Cambridge, UK) at experiment station and multitemporal Landsat5 TM images are acquired simultaneously. All image data were geometrically and radiometrically corrected with the DEM data. Three typical vegetation indices (NDVI, EVI and TGDVI) were calculated and selected through the correlative analysis method. This model was applied in mapping multitemporal forest LAI.

2 Study Areas

2.1 In-Situ LAI Measurements

The study area (Figure 1) is located in the northwest mountain areas of Beijing, China (116°0'38"E–116°8'43"E, 40°0'48"N–40°4'59"N) with an elevation of 150-1265 m. With the semi-humid continental monsoon climate in warm temperate zone, the woodland area is 823.04 hectares and the rate of forest coverage reached 96.2%. Broadleaf and broadleaf-coniferous mixed forests, shrub and meadow extensively distribute in the study area in relatively homogenous spatial patterns.

Five ground-based forest LAI measurements were conducted using Hemispherical Photography over study areas from April to October, 2011.The histogram of LAI seasonal variability was given in Figure 2. Each plot was designed 30 m×30m, which is the same size with the pixel of Landsat5 TM data. The geographic coordinates of the centre place of plots were measured with GPS (Trimble Geo XT). The forest type

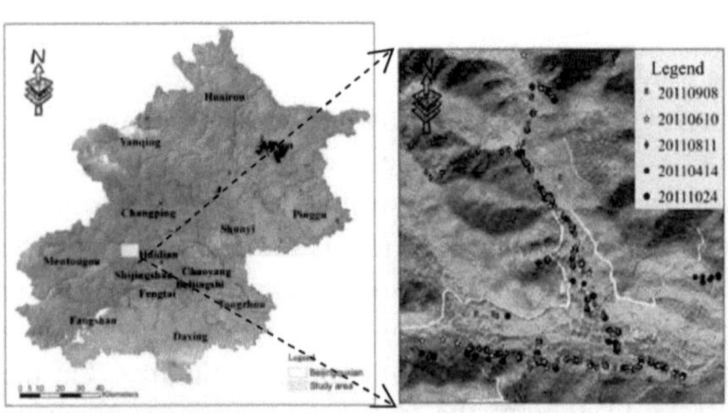

Fig. 1. The location of study areas and in-situ Measurements

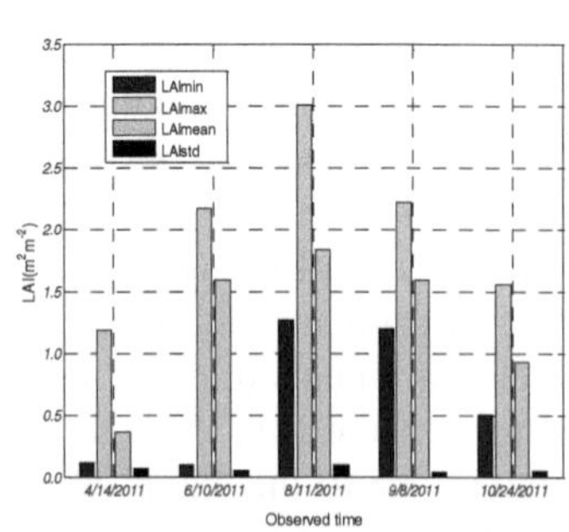

Fig. 2. The histogram of observed LAI with the seasonal variability

was also recorded. Hemispherical photography is a technique for studying plant canopies via photographs acquired through a hemispherical (fisheye) lens with 180° field of view from beneath the canopy (oriented towards zenith) or placed above the canopy looking downward.

2.2 Multitemporal Landsat5 TM Data and DEM Data

Multitemporal Landsat5 TM data were acquired coincident with the ground measurements. They are acquired respectively in April 15, 2009, June 5, 2010, August 8, 2010, September 22, 2009 and October 24, 2009. All image were geometrically corrected to a subpixel accuracy (<10 m) and GPS measurements. They were projected to UTM coordinates (WGS84 datum). Topographic and atmospheric calibration was performed combined with the DEM data.

The DEM data covering the study area with a resolution of 30 m were received from the NASA Land Processes Distributed Active Archive Center (LP DAAC), which were produced from optical stereo data of the Advanced Space borne Thermal Emission and Reflection Radiometer (ASTER). ASTER GDEM is in geotiff format with geographic lat/long coordinates (WGS84 datum). According to the zenith angle, azimuth angle and imaging time, the DEM data was used to calculate the slope, aspect, skyview and shadow factors which are the parameters of terrain radiometric correction.

3 Analyze and Result

3.1 Inversion LAI Model

Three vegetation indices (NDVI, EVI and TGDVI) (Table 2) were employed for correlative analysis to LAI (Chen, 2011; Zhu, 2010; Robert, 2003). Normalized Difference Vegetation Index (NDVI).Compared to NDVI, Enhanced Vegetation Index (EVI) adds blue band, which used to further correction for aerosol residue and the impact of soil background on vegetation index (Li, 2010). According to the vegetation and soil spectral characteristics, Tang et al (2003) advanced Three Gradient Difference Vegetation Index (TGDVI) and indicated this index has advantages in a high saturation point and eliminating background effects.

In our study, vegetation indices, which extracted from matched spectral reflectance image using the GPS points, are used for corrective analysis in two cases: individual observed period and the whole observations. The accuracy in Table 3 indicated that the vegetation indices were little sensitive to LAI especially in August and September. Nerveless, when all the LAI observations are used to established model, the significance was increasing largely and they reached significant level (Table 4). The LAI inversion model based on NDVI showed the most performance (Fig. 3), the equation as follows.

$$LAI = 0.042e^{4.6*NDVI}\ (R^2=0.487) \tag{1}$$

In Figure 4, the retrieved results were compared with the available LAI measurements with R^2 of 0.44 and RMSE of 0.38. The significance of retrieved LAI models improved considerably using all the LAI, which can be explained by: (1) the error caused by the signal LAI value makes stronger influence on the trend of NDVI with LAI changes in individual observed period than that in all observations. This situation can be detected in September in Figure 5 that the sample located in small NDVI value but large LAI value. (2) In contrast with individual observed period, the total dataset located in 2D space of NDVI-LAI uniformly, showing the trend of NDVI with LAI changes clearly. For example, 80% of LAI values ranged from $1.3m^2m^{-2}$ to$1.8m^2m^{-2}$ while NDVI from 0.7 to 0.78 in June. 68% of LAI values distributed from $0.7m^2m^{-2}$ to$1.1m^2m^{-2}$ while NDVI from 0.64 and 0.7 in October. A similar situation also existed in other months. But, the multitemporal samples extend the dynamic range obviously which LAI values change from $0.12m^2m^{-2}$ to $3.01m^2m^{-2}$ and NDVI from 0.46 to 0.92. Further, showing in Figure 5, the April values are relatively in low value areas of 2D space of NDVI-LAI while the June, August and September values are relatively in high value areas, resulting in the trend of NDVI with LAI changes.

Table 1. Three selected vegetation indices

Vegetation Index	VI formulate
NDVI (Normalized Difference Vegetation Index)	$\mathrm{NDVI} = \frac{\mathrm{NIR} - \mathrm{R}}{\mathrm{NIR} + \mathrm{R}}$
EVI (Enhanced Vegetation Index)	$\mathrm{EVI} = 2.5\frac{\mathrm{NIR} - \mathrm{R}}{\mathrm{NIR} + 6\mathrm{R} - 7.5\mathrm{B} + 1}$
TGDVI (Three Gradient Difference Vegetation Index)	$\mathrm{TGDVI} = \frac{\mathrm{NIR} - \mathrm{R}}{\lambda_{nir} - \lambda_{nir}} - \frac{\mathrm{R} - \mathrm{G}}{\lambda_r - \lambda_g}$

Table 2. The accuracy of the relationship between all LAI values and vegetation index

Vegetation Index	Liner(R^2)	Exponent(R^2)	Logarithm(R^2)
NDVI	0.456**	0.487**	0.454**
TGDVI	0.164**	0.102**	0.177**
EVI	0.175**	0.130 **	0.176**

Note: **a significant level in 0.001

Table 3. The accuracy of the relationship between LAI and vegetation index in different observed time

Observed Time	NDVI Lin.	NDVI Exp.	NDVI Log.	TGDVI Lin.	TGDVI Exp.	TGDVI Log.	EVI Lin.	EVI Exp.	EVI Log.
20110414	0.062	0.047	0.049	0.195	0.05	0.166	0.272*	0.133	0.223
20110610	0.123	0.154*	0.128	0.143	0.143	0.140	0.141	0.154*	0.149*
20110811	0.001	0.002	0.001	0.001	0.002	0.001	0.003	0.002	0.002
20110908	0.038	0.031	0.030	0.002	0.001	0.001	0.002	0.001	0.002
20111024	0.001	0.003	0.001	0.022	0.021	0.031	0.049	0.043	0.037

Note: * a significant level in 0.05. Lin., Exp. and Log. indicate Liner, Exponent and Logarithm model.

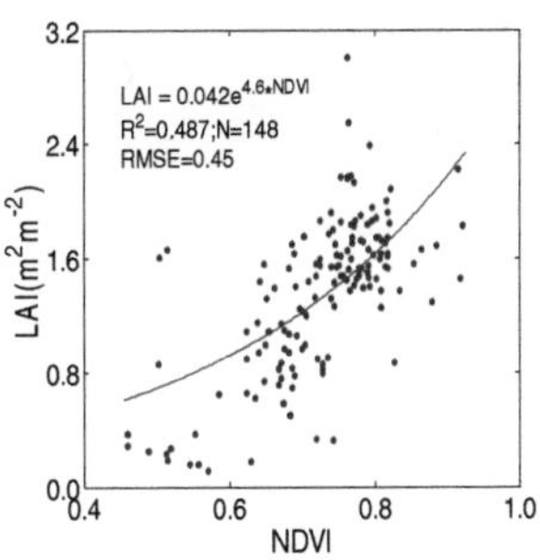

Fig. 3. Exponential model Between NDVI and LAI

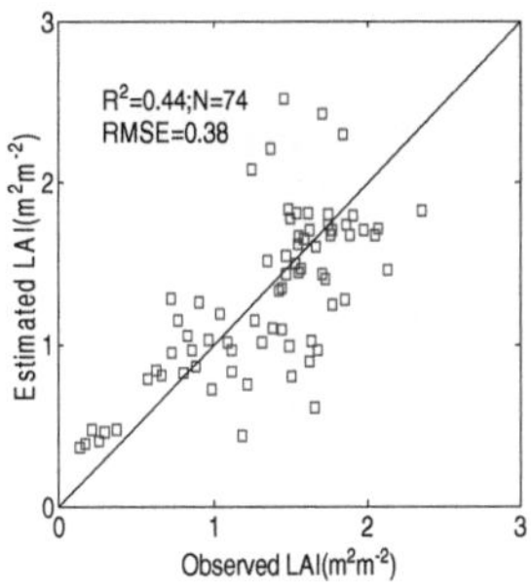

Fig. 4. Estimated and observed LAI

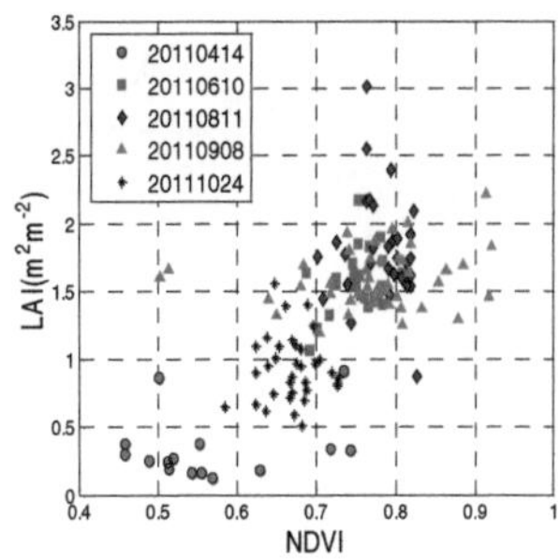

Fig. 5. The scatter of multitemporal LAI and NDVI

3.2 Inversion of Forest LAI and Analysis of Seasonal Variability

The retrieved forest LAI (Figure 6) show the seasonal variability from April to October. The estimated LAI value changes from 0-3.1m^2m^{-2} and the April and October values(less than 1.5m^2m^{-2}) are lower than those in June, August and September (0.5m^2m^{-2} to 3.1m^2m^{-2}). Both the dynamic range and seasonal variation of the retrieval LAI are consistent with the situation of ground-based LAI measurements (Figure 2). Those proved that LAI model based on multitemporal LAI values can be used for estimating individual seasonal map.

To further confirm applicability of this model, 180 samples are extracted from each estimated LAI map and used to explore the seasonal variability of estimated LAI values (Figure.7). We can see that LAI increased from April to August and decreased from August to October. The values reached the peak in August which is similar to the trend of ground-based LAI observations in Figure 2.

From figure 6, the most spatial distribution variation areas of forest LAI mainly located in Mentougou, Fangshan, Miyun, Huairou, Yanqing and Changping. In April, the value is obviously lower in Mentougou and Fangshan (less than 0.5m^2m^{-2}) than other regions (less than 1.5m^2m^{-2}). Through investigation, the difference is caused by the differences in vegetation type: Mentougou and Fangshan are mainly covered by bush while Huairou dominated covered by shelterbelts. With the growth of the vegetation, this difference becomes smaller especially in August.

In this part, the sensitivity of the models between vegetation indices and LAI were tested for the broadleaf forest, coniferous and mixed forests, respectively.Whatever the vegetation types is, the LAI value increased with NDVI and the values exceed in certain rang in LAI-NDVI 2D space (NDVI: 0.6-0.9； LAI: 0.5 m^2m^{-2}-2.5 m^2m^{-2}) in Figure 8.We found that LAI model based on NDVI in exponential form is more sensitive than other models: broadleaf forest R^2=0.601, coniferous R^2=0.432 and mixed stands R^2=0.667(Table 5). Compared with aforementioned results which didn't take into account the effect of vegetation type, the significance of the relationship increased slightly both in broadleaf and mixed stands and decreased a little in conifer stand. However, the accuracy of all models has reached a significant level.

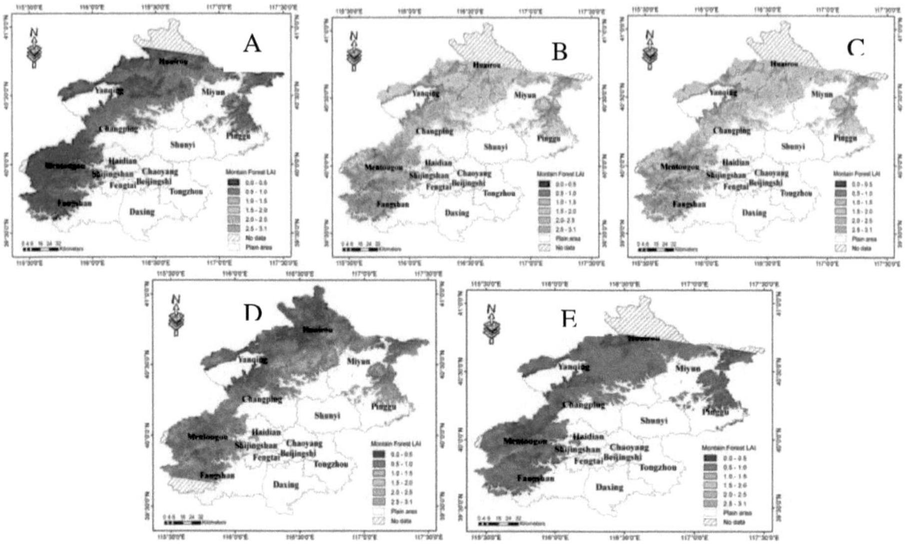

Fig. 6. Multitemporal Maps of retrieved forest LAI (A) Forest LAI map in April ; (B) Forest LAI map in June ; (C) Forest LAI map in August ; (D) Forest LAI map in September ; (E) Forest LAI map in October

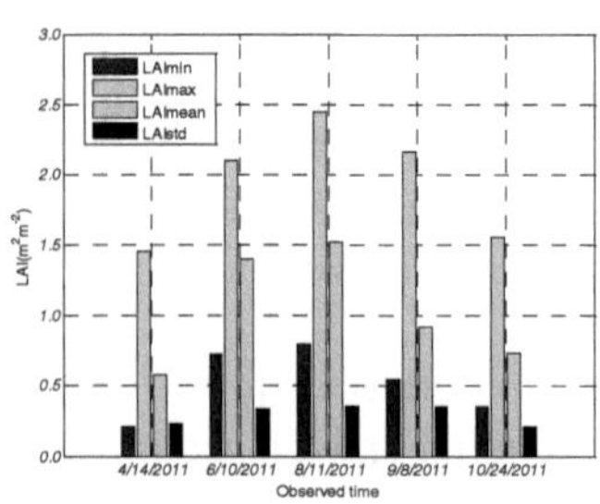

Fig. 7. The histogram of estimated LAI with the seasonal variability

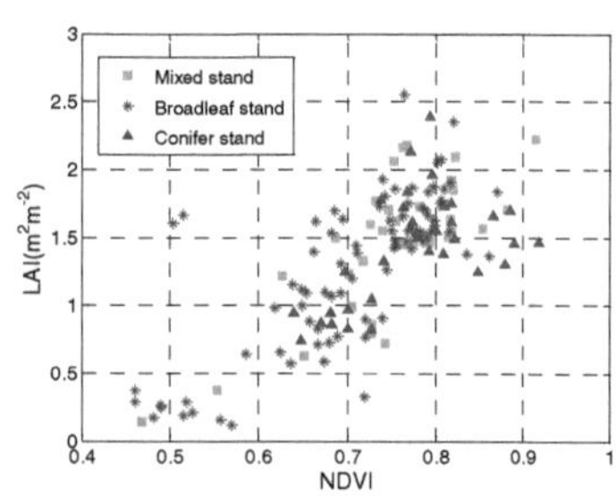

Fig. 8. The scatter of LAI and NDVI in different vegetation types

Table 4. The accuracy of the relationship between LAI and vegetation index in different vegetation types (R^2)

Vegetation index	Mixed forest			Broadleaf forest			Conifer forest		
	Lin.	Exp.	Log.	Lin.	Exp.	Log.	Lin.	Exp.	Log.
NDVI	0.578**	0.667**	0.585**	0.579**	0.601**	0.566**	0.402**	0.432**	0.418**
TGDVI	0.143*	0.074*	0.136*	0.250**	0.110**	0.261**	0.103*	0.112*	0.111*
EVI	0.122*	0.076*	0.106*	0.300 **	0.193**	0.303**	0.128*	0.141*	0.145*

Note: * a significant level in 0.05; * * a significant level in 0.001.

4 Conclusions

The purpose of this paper is to retrieve forest LAI from multitemporal optical satellite images over mountain areas in Beijing. In-situ Forest LAI were measured using Hemispherical Photography. Three typical vegetation indices were employed, and the best index was selected through the correlative analysis method. The selected best LAI model is used to estimate forest LAI in this study. In the preprocessing of images, both terrain and atmospheric correction are conducted using DEM data. The major conclusions were drawn from this study as follows: the significance of the relationship between NDVI and LAI using all observations is better than others (R^2=0.487). The retrieved LAI maps presented seasonal variability, forest LAI increased from April to August and decreased from August to October, which was similar to the trend of observed LAI. These situations proved that LAI model based on multitemporal LAI measurement could more suitable be used in the inversion of individual seasonal LAI.

The sensitivity of the models between vegetation indices and LAI were tested for the broadleaf forest, coniferous and mixed forests, respectively. The results show that the accuracy was improved both in broadleaf and mixed forests if we take care of different vegetation types, except for conifer stands with little accuracy decreasing. However, the accuracy of all models has reached a significant level.

Acknowledgment. This work was supported in part by the State Key Basic Research and Development Program (2011CB311806), the National Natural Science Foundation of China (41071228, 41271345), the Beijing Municipal Natural Science Foundation (4102021), the Beijing Municipal Talents Training Funded Project (2012D002020000007), the Special Funds for Technology innovation capacity building sponsored by the Beijing Academy of Agriculture and Forestry Sciences (KJCX201104012), and the State Key Laboratory of Remote Sensing Science sponsored by the Institute of Remote Sensing Applications of Chinese Academy of Sciences, through its open funds (OFSLRSS201109).

References

1. Bonan, G.B.: Importance of leaf area index and forest type when estimating photosynthesis in boreal forests. Remote Sensing of Environment 43(3), 303–314 (1993)
2. Chen, J.M., Rich, P.M., Gower, S.T., Norman, J.M., Plummer, S.: Leaf area index of boreal forests: Theory, techniques and measurements. Geophys. Res. 102, 29429–29443 (1997)
3. Chen, C., Zhu, Y., Li, X.: The Applicability of Different Methods for Retrieving Leaf Area Index of Forests in Red Soil Hilly Regions. Acta Agriculturae Universitatis Jiangxiensis 33(3), 508–513 (2011)
4. Yang, G., Huang, W., Wang, J., Xing, Z.: Inversion of Forest Leaf Area Index Calculated from Multi-source and Multi-angle Remote Sensing Data. Chinese Bulletin of Botany 45(5), 566–578 (2010)

5. Huete, A.R., Liu, H.Q., Batchily, K., Van Leeuwen, W.: A comparison of vegetation indices over a global set of TM images for EOS-MODIS. Remote sensing of Environment 59(3), 440–451 (1997)
6. Jonckheere, I., Fleck, S., Nackaerts, K., Muys, B., Coppin, P., Weiss, M., Bare, F.: Review of methods for in situ leaf area index determination Part I. Theories, sensors and hemispherical photography. Agricultural and Forest Meteorology 121, 19–35 (2004)
7. Chen, J.M., Leblanc, S.G., Miller, J.R., Freemantle, J., Loechel, S.E., Walthall, C.L., Innanen, K.A., Peter White, H.: Compact Airborne Spectrographic Imager (CASI) used for mapping biophysical parameters of boreal forests. Journal of Geophysical Research 104(22), 27945–27958 (1999)
8. Liang, S.L., Fang, H., Thorp, L., Kaul, M., Niel, T.G.V., McVicar, T.R.: Estimation and validation of land surface broadband albedos and leaf area index from EO-1 ALI data. IEEE Transactions on Geosciences and Remote Sensing 41(6), 1260–1267 (2003)
9. Li, W.-M., Qin, Z.-H., Li, W.-J., Yang, Q.: Comparison and Analysis of MODIS NDVI and MODIS EVI. Remote Sensing Information 6, 73–78 (2010)
10. Luo, Z.-M., Tian, Q.-J., Hui, F.-M.: Retrieving Leaf Area Indices for Coniferous Forest in Xingguo County, Jiangxi Province, in Use of Landsat ETM + Images. Journal of Nanjing University (Natural Sciences) 41(3), 253–258 (2005)
11. Mao, R.: Study on LAI Extraction Based on Landsat Image. Nanjing Normal University (2004)
12. Stenberg, P., Rautiainen, M., Manninen, T., Voipio, P., Mottus, M.: Boreal forest leaf areas index from optical satellite images: model simulations and empirical analyses using data from central Finland. Boreal Environment Research 13, 433–443 (2008)
13. Rouse, J.W., Haas, R.H., Schell, J.A., Deering, D.W.: Monitoring vegetation systems in the Great Plains with ERTS. In: Proceedings of the Third ERTS Symposium, vol. 1, pp. 48–62 (1973)
14. Rich, P.M., Clark, D.B., Clark, D.A., Oberbauer, S.F.: Long-term study of solar radiation regimes in a tropical wet forest using quantum sensors and hemispherical photography. Agric. For. Meteorol. 65, 107–127 (1993)
15. Colombo, R., Bellingeri, D., Fasolini, D., Marino, C.M.: Retrieval of leaf area index in different vegetation types using high resolution satellite data. Remote Sensing of Environment 86(1), 120–131 (2003)
16. Running, S.W., Nemani, R.R., Peterson, D.L., Band, L.E., Potts, D.F., Pierce, L.L., Spanner, M.A.: Mapping regional forest evapotranspiration and photosynthesis by coupling satellite data with ecosystem simulation. Ecology 70(4), 1090–1101 (1989)
17. Sellers, P.J., Mintz, Y., Sud, Y.C., et al.: A simple biosphere model (SiB) for use within general circulation models. Journal of the Atmospheric Sciences 43, 505–531 (1986)
18. Turner, D.P., Cohen, W.B., Kennedy, R.E., Fassnacht, K.S., Briggs, J.M.: Relationships between leaf area index, FPAR, and net primary production of terrestrial ecosystems. Remote Sensing of Environment 70, 52–68 (1999)
19. Turner, D.P., Cohen, W.B., Kennedy, R.E., Fassnacht, K.S., Briggs, J.M.: Relationship between leaf area index and Landsat TM Spectral Vegetation Indices across three temperate zone sites. Remote Sens. Environ. 70, 52–68 (1999)
20. Tang, S.-H., Zhu, Q.-J.: Theoretical basis and application of Three Gradient Difference Vegetation Index. Science China (Series D) 11(33), 1094–1102 (2003)
21. Weiss, M., Baret, F.: Evaluation of canopy biophysical variable retrieval performances from the accumulation of large swath satellite data. Remote Sens. Environ. 70(3), 293–306 (1999)

22. Xiang, H., Guo, Z., Zhao, Z., Wang, J.: Estimating Method of Forest Leaf Areas Index on Different Space Scales. Scientia Silvae Sinicae 45(6), 139–144 (2004)
23. Xing, Z., Feng, Y., Li, W., Wang, P., Yang, G.: The research status of inversion of leaf area index with hyperspectra l remote sensing. Science of Surveying and Mapping 35(S1), 162–164 (2010)
24. Zhang, J., Zhang, G., Wang, P.: Vegetation and Ecological Remote Sensing. Science Press, Beijing (2010)
25. Zhao, Y.: Principles and Analysis of Remote Sensing Application. Science Press, Beijing (2003)
26. Zhu, G.-L., Ju, W.-M., Chen, J.M., Fan, W., Zhou, Y., Li, X., Li, M.: Forest canopy leaf area index in Maoershan Mountain: Ground measurement and remote sensing. Chinese Journal of Applied Ecology 21(8), 2117–2124 (2010)

A Sampling Design for Monitoring of the Cultivated Areas of Main Crops at National Scale Based 3S Technologies in China

Quan Wu*, Zhiyan Pei, Fei Wang, Hu Zhao, Lin Guo, Juanying Sun, and Lijuan Jia

CAAE, Chinese Academy of Agricultural Engineering, Remote Sensing Application Centre.
41, Maizidian Street Chaoyang District, Beijing, China
wuquan95@tom.com, {peizhiyuan,wangfei}@agri.gov.cn

Abstract. Remote Sensing Application Centre (RSAC) has been working on monitoring the main crop cultivated areas as operational task and research project for several years.The main problems is that the time and features of background data used in sampling maybe produce system errors. The sampling method called stratified sampling adopted by RSAC is also used in this study. But there is an important difference that the background data used to stratify is the same year's cultivated land data derived from an analysis result to MODIS data, containing crop cultivated area proportion information, instead of cultivated land data. The Area Proportion Enlargement Estimation (APEE) model designed in this study is different with Area Enlargement Estimation (AEE) adpoted by RSAC. The new APEE model was set up through calculating with twice weighted averaging. A case study limitedly presents the process of the sampling design and testifies feasibility of the method.

Keywords: Sampling, Stratified sampling, Sampling error, Cultivated land, Crop cultivated area proportion, RS, GIS.

1 Introduction

Sampling methods are widely used to estimate main crop cultivated areas by many countries, such as US, China, EU, etc., in the world. But there is much difference among these countries about technologic content introduced below[1][2][3][4][5][6].

1.1 In European Union (EU)

Monitoring Agriculture with Remote Sensing (MARS) is a project facing Europe in order to obtain crop yield information constituted by European Union Committee[7]. It is a kind of three-stage sampling based on unsupervised classification using multitemporal RS data[8]. The first-stage sampling unit consisted of 60 sites is designed to square with the side length of 40 kilometer. There are 16 component parts

* Corresponding author.

D. Li and Y. Chen (Eds.): CCTA 2012, Part II, IFIP AICT 393, pp. 10–19, 2013.

in each site and each part has 40 sampling points. Because unsupervised classification method is used with RS data to cover the third-stage sampling points 5-6 times every year, all crops in sampling units are recognized and then the acreage and yield will be worked out after statistical calculation from the third-stage sampling points to the first-stage sampling units.

1.2 In America

In America, the prediction of total crop yield is acquired from crop acreage and crop yield per unit. The crop acreage data is gotten by June Agricultural Survey (JAS) [3][7]. Two different sampling units used by JAS are area frame covering America and name list frame. The name list consists of registered farmers. Every year about 2400 investigators contact more than 120 thousand farmers in the first two weeks of June in order to get crop acreage data [3].

In NASS, these parcels are usually about 6 to 8 square miles in size and are called PSUs, or Primary Sampling Units. These are further divided into 1 square mile parcels, called segments. During data collection, enumerators determine who operates the land within the randomly selected segment and what kinds of agriculture occur on that land. Notice that the subject PSU is yellow and the selected segment is red.

1.3 In China

The operating prediction of crop cultivated area is mainly provided by RSAC. RSAC adopts two methods to obtain the acreage of main crops such as wheat, corn, cotton, soybean, rice, etc.. One method is stratified sampling using RS and the other is ground random sampling using GPS [2][3][4]. The Stratified Sampling Method (SSM) using RS is the major one.

To use SSM, the first step is to select a sampling unit according to the surveyed crop and to order appropriate RS images covering all sampling units in the region. The second step is to discriminate the crop on RS images and get the area of the crop distributed in the sampling units. The last step is calculating the result with AEE (Area Enlargement Estimation) model [3][4].

2 Problems

It is evident that the US' sampling method does not adapt to China for lack of base data while the one of EU also does not meet China for lack of money [4]. However, one problem coming from the monitoring system is that the background data used to stratify is too old to reflect the survey year's ground truth [9]. The time intervals vary from three years to ten years based surveyed crops. So, as a sampling error the systematic error is theoretically produced. Another problem is that the background data is cultivated land data while the monitoring object is one kind of crop among many crops, such as cotton, corn, soybean, etc., distributed on the cultivated land. So, in the abstract, another systematic error also probably exists [1][4][10][11].

3 Objectives

It is the objective of the research to theoretically set up a new sampling system, including the APEE model, based cultivated land data with crop cultivated area proportion information, for RSAC.

4 Methods

In order to achieve the goal RSAC made up a workgroup to research the above problems. At the same time, designing a new sampling method is also a research project authorized by CAAE (Chinese Academy of Agricultural Engineering), launched by RSAC in 2012.

4.1 General and Basic Points

Because RSAC's operational system of crop cultivated area monitoring has been working well for many years while the technologic and working systematic structure has come into being [7][12], the research key point should be firstly to theoretically modify the technologic system at some technologic points according to the actual demand produced by change of the background data from cultivated land data layer to crop data layer, not throw it away. It is feasible and reasonable to do so based RSAC's work condition and experience. So, a new sampling method may be set up through rebuilding the old one.

The main task is theoretically to set up a new sampling systematic structure in former half year of 2012 and experiment using the new sampling method in next half year. So, this paper has only given detailed account of the theoretical research for the new sampling method and a case based simulate data. The technologic flow chart is below show in figure 1.

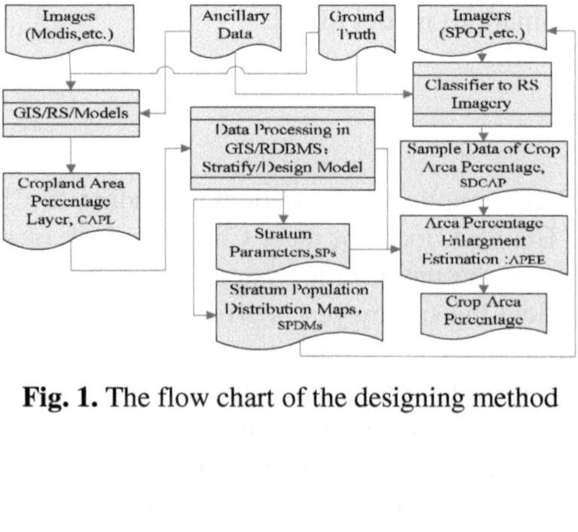

Fig. 1. The flow chart of the designing method

4.2 Sampling Units

Sampling units are basic to a sampling system. It is directly related to sampling accuracy and sampling efficiency. At the same time, sampling methods are also involved with sampling units. County administrative boundaries, as a sampling unit, was adopted by RSAC before 2004. In order to decrease the sampling error derived from the different area and shape among sampling units and increase sample size, it was changed to quadrangle frames of relief map at national scale from that year. So, the sampling unit has been used for many years although it also holds systematic errors[4]. In the research, the quadrangle frame of relief map was acted as a sampling unit.

4.3 The Stratified Sampling

Adapting to spatial distribution of Chinese main crops and surveying condition, RSAC has selected a sampling method named SSM acted as main technical method to monitor crop cultivated area for many years. The SSM was also adopted in the research, but the stratified index was changed.

Stratified Indexes. A stratified index should describe main characters of surveyed objects. Optional stratified indexes in the research are mainly crop cultivated area, crop type amount and objective crop cultivated area percentage, etc., which are derived from the analysis result of low spatial resolution RS images, such as MODIS, spatially distributed on the surveyed geographic regions[14][15]. The research selects objective crop cultivated area percentage to act as a stratified index for an experiment which will be performed in back half year of 2012.

Confirming an Objective Crop. Because different crops distribute on different geographic regions with different cultivated habits and area proportion, it is necessary to stratify crop cultivated area proportion only aiming at one kind of crop. So, the stratum parameters vary with crop types. The investigated regions and imaging time of RS images can be confirmed when a kind of investigated crops is acted as an objected crop.

Resampling. As basic data used to stratify is produced of low spatial resolution RS images, crop cultivated area proportion is derived from estimation to pixels and that pixel area is generally not equal to sampling unit area. So, it is necessary to resample crop cultivated area from pixel level to sampling unit level.

Based one important hypothesis that is the objective crop evenly distributed in pixel space, the resampling can be performed. The resampling includes two area transferred steps which orientation is opposite. The first step is to transfer objective crop area proportion to objective crop area according to polygons derived from pixel geographic boundaries. The second step is transfer objective crop area to objective crop area proportion according to polygons derived from sampling unit geographic boundaries.

Stratification. It is simple to stratify directly based crop cultivated area proportion within sampling units. The stratum threshold values can be designed to 30%, 50% and 100%. In this article Frequency Accumulation Stratification (FAS) was used, with which RSAC mainly stratify sampling units in crop cultivated area survey. The essence of the method includes dividing data into groups according to a certain step length such as 0.1%, between two groups in ascending or descending data array, calculating the amount of sampling units of each group, which is called frequency, accumulating frequency and square root of frequency of each group, getting the total accumulating value of square root, obtaining another step length between two layers through dividing the total accumulating value of square root by layer amount such as six, which is an equal step length method. The stratification process is shown figure 2.

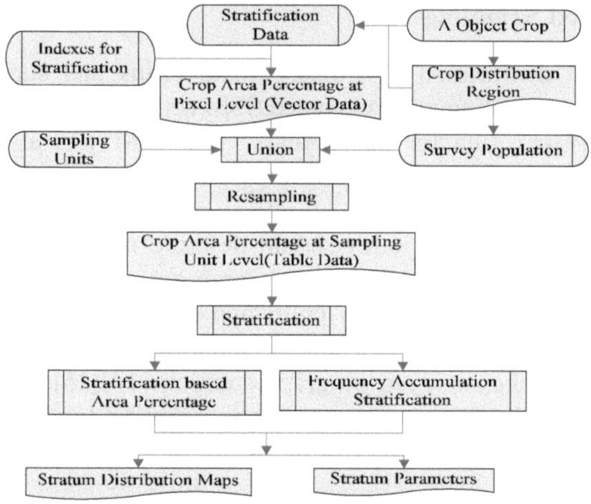

Fig. 2. The flow chart of the stratification process

4.4 Parameters and Equations

When stratification is finished, the following task is to calculate stratum sample average and variance based stratum population average and variance. Sample size is a very important parameter. The parameter should be calculated before performing a crop cultivated area sampling investigating.

Stratum Sample Average and Variance. Stratum sample average and variance can be calculated by following equations[3].

$$\overline{y_h} = \frac{1}{n} \sum_{i=1}^{n_h} y_{hi} \tag{1}$$

$$s_h^2 = \frac{1}{n_h - 1} \sum_{i=1}^{n_h} (y_{hi} - \overline{y_h})^2 \tag{2}$$

Where $\overline{y_h}$ = the sample average of the h layer

$n =$ the sample size

$n_h =$ the sample size of the h layer;

$y_{hi} =$ the SOV of the i unit of the h layer

$s_h^2 =$ the sample variance the h layer

Sample Size. So, the least sample size is determined by the follow equations[3].

$$n = \frac{\sum_{h=1}^{L} \frac{N_h s_h^2}{W_h}}{V + \sum_{h=1}^{L} N_h s_h^2} \tag{3}$$

Where $n =$ the sample size

$W_h = N_h / N$

$L =$ the amount of layers

$V =$ the variance of estimated value

The others are as same as the above.

Distribution according to proportion:

$$n = \frac{n_0}{1 + n_0 / N} \tag{4}$$

Where $n_0 = \frac{N \sum_{h=1}^{L} N_h s_h^2}{V}$

The others are as same as the above.

The surveying accuracy signed δ is set to a certain value such as 0.95, and the reliability signed $(1-\alpha)$ is also set to 0.95 or other value, the value of V should be calculated according to the follow equation:

$$V = (d / t)^2 \tag{5}$$

Where $d = (1-\delta)Y$, $Y =$ the total value of the population

$t = \mu_{\alpha/2}$ When the sample size is bigger than 45

4.5 Area Proportion Enlargement Estimation (APEE)

The stratification outcome is used to design a model for estimating crop cultivated area proportion based SOV derived from surveying year. The model is below shown with equations (12).

$$\hat{Y} = \sum_{h=1}^{L} (W_h \times \overline{y_h}) \tag{6-1}$$

$$\overline{y_h} = \sum_{i=1}^{n_h} (p_{hi} \times y_{hi}) \qquad (6\text{-}2)$$

$$p_{hi} = a_{hi} / \sum_{i=1}^{n_h} a_{hi} \qquad (6\text{-}3)$$

Where $\hat{Y}$ = the population average of crop cultivated area proportion

$\overline{y_h}$ = the sample average of crop cultivated area proportion of the h layer

$W_h = N_h / N$

p_{hi} = the cultivated area weight of the i unit of the h layer

y_{hi} = the crop cultivated area proportion of the i unit of the h layer

a_{hi} = the cultivated area of the i unit of the h layer

The others are as same as the above.

The process of designing the model APEE is presented by figure 3.

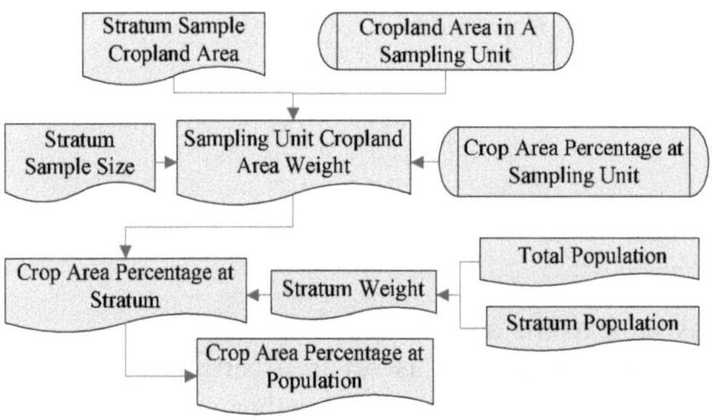

Fig. 3. The model of APEE

5 A Case Presentation

The case presents the working process including SSM and APEE model using simulative data. According to the process of the sampling designed above, the FAS of SSM was performed on the assumption that the used background data was derived from resampling to low spatial resolution RS images. Figure 4 below shows the result of stratified sampling, in which sampling units are the quadrangle frames of relief map with the scale 1 : 50000 and the stratum number is 6. There are totally 3083 sampling units mostly covering 9 provinces' early rice planting region of China in this map.

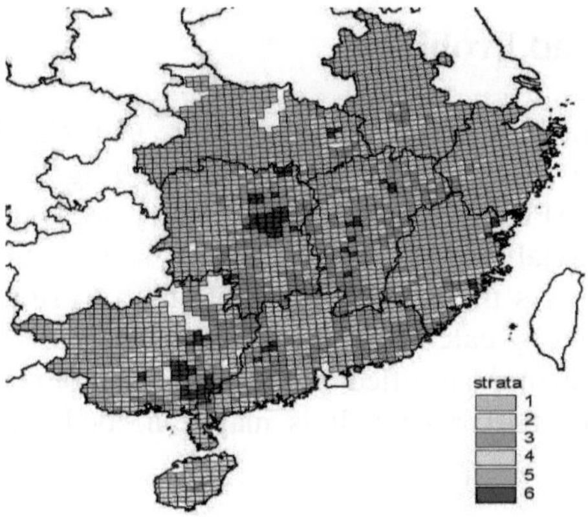

Fig. 4. Stratum distribution of early rice area

The table 1 below shows main parameters produced by FAS of SSM using early rice cultivated area proportion , produced by the latest land-use data and 2009's early rice cultivated area data which is derived from MOA, acted as a stratified index when the population is 9 provinces.

It is evident that the early rice cultivated area in main paddy land is relatively small while early rice plots separately distribute on paddy fields with other crops such as peanut, potato and vegetable, etc..

Table 1. The parameters of sampling

Layers	Population	Minimum sample size	Area percentage means (%)	Sampling rate
1	1483	42	2.99	0.0285
2	532	15	15.83	
3	404	12	24.99	
4	339	10	33.29	
5	232	7	41.32	
6	93	3	57.61	

Using the APEE model with the above parameters and early rice cultivated area proportion data produced by RSAC' monitoring result to early rice cultivated area in 2011 and cropland area data derived from the latest land-use data, the 2011's early rice cultivated area percentage is below shown in table 2.

Table 2. The outcome derived from APEE

Region	Minimum sample size	Practical sample size	Area percentage means (%)	Sampling rate
9 Provinces	88	212	1.80	0.0285
Central China	69	160	1.63	0.0379
South China	84	52	1.82	0.0655
Main productive provinces	48	134	2.86	0.0289

6 Conclusion and Problems

Through the case based theoretically calculation and simulate data, a primary conclusion is that the sampling design for monitoring a crop cultivated area, consisting of SSM and APEE, is feasible. But it is very necessary to experiment using practical data in operational productive work.

A problem in the case is that the simulate data used to stratify and estimate crop area is directly produced by calculating with other data. The resampling designed in the sampling for producing stratified parameters was omitted because it is too complicated to simulate the process. It is maybe needed further discussing based another topic.

Acknowledgement. This study is supported by the project named Preceding Study to Agricultural Remote Sensing Monitoring and Evaluating Subsystem which number is E0201/1112-1.

References

1. Wu, Q., Yang, B.J., Pei, Z.Y., Wang, F.: Influence of small features on crop area estimation at a nation scale using remote sensing and a double sampling method. Chinese Journal of Agricultural Resources and Regional Planning 20(3), 130–133 (2004) (in Chinese)
2. Jiao, X.F., Yang, B.J., Pei, Z.Y.: Design of sampling method for cotton field area estimation using remote sensing at a national level. Transactions of the Chinese Society of Agricultural Engineering 18(4), 159–162 (2002) (in Chinese)
3. Wu, Q., Sun, L.: Sampling methods using RS and GPS in crops acreage monitoring at a national scale in China. In: Chen, J., Jiang, J., Alain, B. (eds.) The International Archives of the Photogrammetry, Remote Sensing and Spatial Information Sciences, Beijing, vol. XXXVII, Part B7, WG VII/7, pp. 1337–1342 (2008)
4. Wu, Q., Sun, L., Wang, F., Jia, S.: Theory of Double Sampling Applied to Main Crops Acreage Monitoring at National Scale Based on 3S in China—CT316. In: Li, D., Liu, Y., Chen, Y. (eds.) CCTA 2010, Part III. IFIP AICT, vol. 346, pp. 198–211. Springer, Heidelberg (2011)
5. Zhao, T., Pravesh, A.S.: Geochemical sampling scheme optimization on mine wasters based on hyperspectral data. In: Chen, J., Jiang, J., Alain, B. (eds.) The International Archives of the Photogrammetry, Remote Sensing and Spatial Information Sciences, Beijing, vol. XXXVII, Part B7, WG VII/7, pp. 1529–1532 (2008)
6. Xie, H., Tong, X., Jiang, Z.: The quality assessment and sampling model for the geological spatial data in China. In: Chen, J., Jiang, J., Kainz, W. (eds.) The International Archives of the Photogrammetry, Remote Sensing and Spatial Information Sciences, Beijing, vol. XXXVII, Part B2, WG VII/7, pp. 819–824 (2008)
7. Yang, B.J.: Monitoring the agricultural condition using RS, pp. 19–29. China Agriculture Press, Beijing (2010) (in Chinese)
8. Duda, T.: Unsupervised classification of satellite imagery: choosing a good algorithm. Remote Sensing 23(11), 2193–2212 (2002)

9. Wu, Q., Sun, L., Wang, F.: The applications of 3S in operational monitoring system of main crops acreage in China. In: Li, D., Yang, S.X. (eds.) Computer and Computing Technologies in Agriculture -II, vol. 24, pp. 319–324. TSI Press, USA (2010)
10. Sun, L., Wu, Q., Zhang, S.L., Wang, F.: Estimation of area ratios of small features in crop area obtained by remote sensing technique in Hebei Province. Transactions of the Chinese Society of Agricultural Enginerring 22(2), 79–82 (2006) (in Chinese)
11. Huang, H.P., Wu, B.F.: Analysis to the relationship of feature size, objects scales, image resolution. Remote Sensing Technology and Application 21(3), 243–248 (2006) (in Chinese)
12. Wu, Q., Pei, Z.Y., Zhang, S.L., Wang, F., Wang, Q.F.: The methods for monitoring land-use change with RS at non-large scale, pp. 33–40. China Agriculture Press, Beijing (2010) (in Chinese)
13. Yang, X.: Using a time series of satellite imagery to detect land use and land cover changes in the Atlanta, Georgia metropolitan area. Remote Sensing 23(9), 1175–1798 (2002)
14. Zeng, Y., Li, G.: Using Data Grid Technology to Build MODIS Data Management System in Agriculture Application. In: Li, D., Liu, Y., Chen, Y. (eds.) CCTA 2010, Part III. IFIP AICT, vol. 346, pp. 241–248. Springer, Heidelberg (2011)
15. Lillesand, T.M.: Remote sensing and imagery interpretation, pp. 576–586. John Wiley & Sons, Inc., New York (2002)

A Discussion on Spatial Distribution Differentiation Law of Peanut Quality in China Based on GIS Technology

Liping Yang[1], Honghai Guo[2], Xinhua Li[1], and Shubo Wan[3]

[1] Shan Dong Institute of Agriculture Sustainable Development, Jinan, Shandong, China
[2] Agricultural Resources and Environment Institute of Shandong Provincial Academy of Agricultural Sciences, Jinan, Shandong, China
[3] Shan Dong Academy of Agricultural Sciences, Jinan, Shandong China
{yangping96,honghaig,xinhuali_2008}@163.com, wansb@saas.ac.cn

Abstract. In order to realize spatial differentiation law of peanut quality in China by quantitative analysis method based on GIS technology, a spatial analysis model for peanut quality was established based on 9 ecological factors influencing peanut quality (light, temperature, water and soil, etc). The results showed that it can be achieved in the analysis of peanut quality differentiation law in space with GIS technology. Taking peanut protein content as an example, the spatial distribution of peanut protein content gradually increases from north to south, high-content region distributes mainly in south of middle-lower Yangtze region; The results have instructive significance for production and breeding department to improve peanut protein quality, establish peanut special production base, optimize regional production layout, and provide a reference for quality rating and planting of other crops.

Keywords: Peanut quality, Protein content, GIS technology, Spatial distribution.

1 Introduction

Peanuts are major oil and food crops in the world, also major oil and cash crops in China. Since the 1990s, China's peanut production has ranked first around the world, accounting for about 1/3 of global production. Peanuts are widely cultivated all over China, except for Qinghai and Ningxia [1,2].

The differences in peanut quality spatial distribution are formed under some ecological conditions, resulted from the combined effects of temperature, light, water, soil and other ecological factors which vary among regions, so geographical differences in peanut quality exist. Scholars at home and abroad mostly concern about species, farming systems and fertilizers related to peanut quality, however, there are contradictions such as regional imbalances, general purpose and special varieties, production and processing, so it is necessary to clarify the spatial distribution of peanut quality. As a result, Arcgis software is used to explore the spatial distribution of peanut quality based on ecological and geographical principles. In this paper, Peanut protein content, an important index of peanut quality, was used. A spatial analysis model for protein content is established based on 9 ecological factors influencing protein content, and spatial distribution of peanut protein

D. Li and Y. Chen (Eds.): CCTA 2012, Part II, IFIP AICT 393, pp. 20–26, 2013.

content in China is analyzed by GIS technology. The results have instructive significance for production and breeding department to improve peanut protein quality, establish peanut special production base and optimize regional production layout.

2 Spatial Analysis Model for Peanut Protein Content

2.1 Index System of Spatial Analysis Model for Peanut Protein Content

Academic research suggests that temperature is the dominant factor influencing protein content whose accumulation is significantly positively correlated with temperature; as temperature rises, protein content increases [3]. During seedling stage, mild and moderate drought stresses can increase seed protein content, but severe drought conditions have little impact[4]; during pegging stage, drought has a larger impact on peanut protein: drought increases fat content but decreases protein content [5,6]; pod maturity, which determines peanut quality largely, is affected by accumulated temperature above 15°C[7]; shading processing can reduce peanut crude protein content [8]; under similar climatic factors, soil type, texture and fertility levels become important factors determining peanut quality [9-11]. The spatial analysis model for peanut protein content is established based on 9 ecological factors influencing protein content, in which the meteorological factor only covers the whole growth period.

Table 1. Index system of peanut protein content analysis model

Factors		Interpretation
Meteorological factors	X_1:>15°C accumulated temperature(the whole growth period of peanut)	>15°C daily average temperature accumulated during the whole growth period of peanut
	X_2:Precipitation (the whole growth period of peanut)	Precipitation accumulated during the whole growth period of peanut
	X_3:Sunshine hours(the whole growth period of peanut)	Sunshine hours accumulated during the whole growth period of peanut
	X_4:Diurnal variation of temperature(the whole growth period of peanut)	An average daily diurnal variation of temperature during the whole growth period of peanut
Soil factors	X_5: Soil organic matter	The data of sampling points of 2008
	X_6: Soil pH	
	X_7:Soil available phosphorus	
	X_8:Soil available potassium	
	X_9:Soil clay content	National 2KM×2KM grid data

2.2 Data Source of Index System

140 sampling points (sampling points are as shown in Fig. 1) were set in major peanut production areas of China, the pods and corresponding soil samples of dominant peanut cultivars were collected, and then pod protein content, soil organic matter, available phosphorus, available potassium, PH, and other indexes were determined in accordance with national standards. The soil clay content is 2KM×2KM raster data from the Institute of Soil Science, Chinese Academy of Sciences[12,13]. Use extraction command in Arcgis9.2 software to extract the value of the sampling points.

Fig. 1. Sampling points distribution map of 2008

Meteorological factors are provided by China Meteorological Data Sharing Service System, they are log data of 756 stations during peanut growth period. Process meteorological factors in Geostatistical Analyst module of Arcgis9.2 software, use ordinary Kriging method to calculate the spatial interpolation of meteorological factors at 756 points, and then use Extraction command to extract the data.

2.3 Analysis Model

In spss13.0 software, Regression was used to establish the linear analytical model of peanut protein content, in which regression uses the enter method to include all factors into the equation. According to goodness-of-fit test of the model (Table 2) and ANOVA (Table 3), R=0.408, R^2=0.166, sig=0.046, the model passes the significance testing.

Table 2. Summary table of peanut protein content analysis model

Model	R	R Square	Adjusted R Square	Std. Error of the Estimate
1	0.408	0.166	0.084	2.311

Table 3. ANOVA table of peanut protein content analysis model

Model		Sum of Squares	df	Mean Square	F	Sig.
1	Regression	97.051	9	10.783	2.02	0.046
	Residual	485.862	91	5.339		
	Total	582.912	100			

Linear regression model of peanut protein content is as shown below:

$$Y=32.432+0.0001X_1+0.00014X_2-0.00104X_3-0.30594X_4-0.04503X_5-0.33025X_6-0.02715X_7-0.00351X_8+0.01884X_9 \quad (1)$$

3 Spatial Analysis for Peanut Protein Content

3.1 Determination of Analysis Unit

The analysis unit is required to indicate the spatial difference of peanut protein content. Each unit shall have its unique characteristics, including both variability and comparability. In the principle of quality consistency, this paper uses 2KM×2KM raster as the unit, and to enhance accuracy, all data were masked by 1:1000000 countrywide farmland vector layer, so only the farmland part was analyzed.

3.2 Factor Processing

Soil Factor Processing. Soil indexes are 2KM×2KM raster data from Institute of Soil Science of Chinese Academy of Sciences.

Meteorological Factor Processing. The meteorological indexes are processed based on 30-year log data of peanut growth period during 1978-2007 (from 756 stations in China, provided by China Meteorological Data Sharing Service System). The 2KM×2KM raster data are acquired through Kriging interpolation calculation of the 756 stations' meteorological indexes under the Geostatistical Analyst module of Arcgis9.2.

Meteorological and soil factors are both masked by 1:1000000 countrywide farmland vector layer to obtain the farmland raster data.

Peanut Quality Spatial Distribution Analysis (based on protein content). Upon the Raster Calculator command of Spatial Analyst of Arcgis9.2, apply the model to carry out raster calculation, gain the value of each unit, which is protein content data for 2KM×2KM raster of countrywide farmland, thus the spatial distribution map for countrywide farmland is gained.

4 Spatial Distribution Analysis for Peanut Quality

Analyze the spatial distribution of peanut protein content for countrywide farmland with the classification function of Arcgis and the result shows a significant spatial distribution trend of peanut protein content: the minimum value which appears in Jilin Province is 13.3%; the maximum value which appears in Taiwan Province is 32.9%; the average value is 26.9%. As shown in Fig. 2, the countrywide peanut protein content increases from south to north, with an average value of 28.3%, higher value mainly in south of middle-lower Yangtze region, and slightly lower value in Inner Mongolia, Ningxia, Shaanxi, Gansu and Xinjiang, which is consistent with the research conclusion drawn by Xiao Changzhen etc. that peanut protein content is inversely related to the latitude [14].

5 Discussion

A spatial analysis model for peanut quality based on 9 ecological factors influencing protein content was established to analyze the spatial distribution of peanut quality in China (farmland), which breaks the old way of analyzing the relationship between peanut quality and environment on field scale and qualitatively describing spatial distribution of peanut quality.

It realizes quantitative analysis and point-to-dimension spatial analysis with GIS technology. As this research shows that the high-protein-content region mainly distributes in south of middle-lower Yangtze region, so production and breeding authorities shall give play to such areas, especially in southeast coastal areas, to make full use of the environment advantages, change the usage pattern of peanuts, extend the industrial chain of peanut protein, build peanut production bases and promote peanut industry development. The research result of this paper will provide a scientific theoretical basis for peanut quality classifying, strong breed selecting and planting,

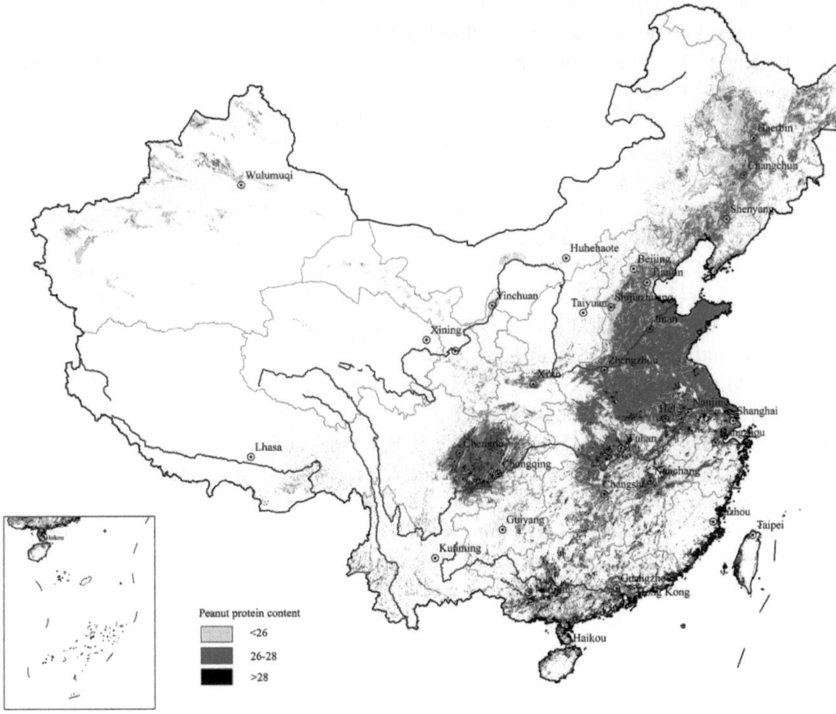

Fig. 2. Spatial distribution map of peanut protein content in China (cultivated land)

which will facilitate the implementation of the national policies of "setting up the sources of editable oils, vegetable proteins, raw materials, cash crops, and farmland biodiversity", to ensure oil crop safety and stable supply of high quality agricultural products.

The method of quantitatively analyzing spatial distribution of peanut quality with GIS technology discussed in this paper is scientific and feasible, but the accuracy of this model and the result needs to be further improved.

References

1. Yang, J.: Present Status and Outlook of Peanut Production and Trade in China. Journal of Peanut Science 38(1), 27–31 (2009)
2. Wan, S.: Peanut Quality Science. China Agricultural Science and Technology Publishing House, Beijing (2007)
3. Zhen, Z., Wang, X., Duan, Y., et al.: The Effect of Climatic Factors on Protein and Oil Content of Peanut Seeds. Journal of Peanut Science 33(3), 22–24 (2004)
4. Yan, M., Li, X., Lin, Y., et al.: Effects of Drought during Seedling Stage on Physiological Traits, Yield and Quality of Different Peanut Cultivars. Acta Agronomica Sinica 33(1), 113–119 (2007)

5. Yao, J., Luo, Y., Yang, X., et al.: Study on peanut early variety critical moisture in different growth stages. Peanut Science and Technology 4, 23–27 (1982)
6. Yao, J., Yang, X.: Study on appropriate water of Summer Peanut. Peanut Science and Technology 2, 31–32 (1986)
7. Shi, K., Xue, X., Song, J., et al.: Preliminary study on meteorological condition of peanut quality. Peanut Science and Technology 1, 5–7 (1994)
8. Lin, C.H., McGraw, R.L., George, M.F., Garrett, H.E.: Nutritive Quality and Morphological Development under Partial Shade of Some Forage Species with Agroforestry Potential. Agroforesty Systems 53(3), 269–287 (2001)
9. Zhao, W., Jin, C.: Study on the low between peanut and soil texture. Peanut Science and Technology 3, 6–11 (1983)
10. Hu, W., Qiu, Q., Li, Z., et al.: Studies of the Effect Factors on Peanut Qualities II. Cultural Factors. Journal of Peanut Science 31(4), 14–18 (2002)
11. Zhang, J., Miao, H., Wu, L., et al.: The Impact of Different Soils and Fertilizers on Peanut Quality Traits. Journal of Peanut Science 32, 372–374 (2003)
12. Shi, X.Z., Yu, D.S., Warner, E.D., Sun, W.X., Petersen, G.W., Gong, Z.T.: Cross-Reference System for Translating Between Genetic Soil Classification of China and Soil Taxonomy. Soil Sci. Soc. Am. J. 70(1), 78–83 (2006)
13. Shi, X.Z., Yu, D.S., Warner, E.D., Pan, X.Z., Petersen, G.W., Gong, Z.G., Weindorf, D.C.: Soil Database of 1:1,000,000 Digital Soil Survey and Reference System of the Chinese Genetic Soil Classification System. Soil Survey Horizons 43(4), 129–136 (2004)
14. Xiao, C., Wu, Y., Gan, D., et al.: Analysis on Main Biochemical Quality of Peanut in China, pp. 161–166 (1999)

Research and Implementation of Measurement Data Wavelet De-noising and 3D Visualization of Farmland

Weidong Zhuang, Chun Wang, and Xi Wang

College of Engineering, Heilongjiang Bayi Agriculture University, 163319, Daqing, Heilongjiang Province, China
81nd@163.com

Abstract. The precision agriculture decision-making needs of geographic information visualization analysis for farmland. In order to get 3D visualization of farmland, the Matlab software and wavelet transform were used for 3D-terrain survey data processing. Haar, DB6, Sym8, coif3 wavelet were used on the 3 layer decomposition of elevation data de-noising test, DB6 wavelet was used on the two to seven layer decomposition de-noising. The measurement plane scatter plots, contour map, 3D surface chart of the farmland was drew. The research shows that using a db6 wavelet five level decomposition de-noising for elevation data measured by DGPS continuously was suitable for the 3D-terrain farmland surveying and mapping in which farmland have the more moderate wavy. This may well reflect the farmland terrain features.

Keywords: Precision agriculture, farmland, Wavelet de-noising, 3D visualization.

1 Introduction

The goal of precision agriculture (PA) is the management of crop and soil variability to increase profitability and reduce environmental impact. The PA decision-making needs of geographic information visualization analysis for farmland. The measurement of elevation data for sloping farmland's 3D topography uses real-time differential GPS system in order to reduce the cost of measuring equipment and make it convenient and easy. The positioned GPS observation is a set of sequence related to time and the discretization of time functions. GPS measurement error contains the ionosphere delay, troposphere delay, multi-path effect, receiver software, hardware and so on. The difference technique can eliminate most of the system error except for random measurement error that belongs to the high-frequency measurement noise. If the high frequency measurement noise was removed, it can reduce the disturbance of positioning precision, which is caused by measurement noise, as a result of improving the accuracy and reliability of the GPS.

The real-time differential GPS system was used to gather 3D topography information for farmland in the field of No. 4-3-1-4, Sinograin northern agriculture

D. Li and Y. Chen (Eds.): CCTA 2012, Part II, IFIP AICT 393, pp. 27–35, 2013.

development Co, LTD. Nenjiang County, Heilongjiang Province, China on April 27th, 2011. The farmland is the typical sloping field and its square is 61.4hm^2. The measuring speed is between 20 to 30 kilometers per hour. It is equipped with AgGPS332 receiver that can receive the signal of Satellite Based Augmentation Systems (SBAS). The GPS position data was recorded by computer once per second, 1302 points were collected. The thesis researches the measurement data processing and visualization of farmland's 3D topography.

2 Measurement Data Wavelet De-Noising

Wavelet is the wave of small area, limited length and zero mean value. Wavelet transformation is the space frequency of localization analysis. It takes signal to Multi-scale Analysis through the idea of dilation and translation in order to achieve the applied time in both high and low frequency and get the analysis in both time and frequency domain. It also can be focused to the signal in any detail, thus extracting information effectively. It is called "Math microscope". At present the analysis of wavelet is widely used in Mathematics study, signal analysis, image processing, classification and recognition, synthetic music and language, medical imaging and diagnosis and so on.

2.1 Principle of Wavelet Noise Elimination

It usually contains non-use signal gathering Data signals from the computer, which is called noise signal. The signal noise elimination prevents non-use part of signal Elimination and reserves useful part of signal enhancement. A d signal model that contains noise shows in this form:

$$s(k)=f(k)+\sigma\cdot e(k) \quad k=0,1,\ldots,n-1 \tag{1}$$

In the formula, f(k) is useful signal, e(k)is noise, s(k) is signal including noise. In fact, f(k) is usually low frequency or some stable signal, e(k) is usually high frequency. So there are three procedures to deal with noise elimination for a d signal. (1): It resolve Original signal into wavelet so that the noise usually belongs to high frequency coefficient. (2): It takes quantization process for high frequency coefficient of Wavelet decomposition in some form, such as threshold. (3): It takes reconstruction of a d wavelet for the bottom of low frequency coefficient and high frequency coefficients of every layer of Wavelet decomposition in order to eliminate noise.

2.2 Choice of Wavelet Basis Function

The process of reducing the noise of signal is a operation that the signal is decomposed and restructured. We should search for a set of functional form that stands for the features of the signal.

Wavelet function is an infinite number as well as wavelet basis function. That is because different wavelet basis function has different characteristics of the time frequency. And different wavelet basis function may come out different results when using the same method to analysis problem. We choose the proper wavelet basis according to the characteristics and application of wavelet function in practice.

Choosing wavelet basis we should consider its Mathematics characteristics including support, symmetry, orthogonality, regularity, vanishing moments and wavelet itself. There are Common wavelet basis in the Matlab toolkit including haar, dbN, symN, CoifN.

The premises of GPS data processing of changing wavelet are discrete values observed quantity. So wavelet basis function has the characteristics of Discrete wavelet transformation and orthogonality, symmetry. In order to position GPS data processing fast and precisely we could choose wavelet basis function, such as Haar, dbN, symN, CoifN.

2.3 Wavelet De-noising

In order to study the effect of GPS data noise elimination of changing wavelet, wavelet de-noising is analyzed.

Using elevation data to the variable z1 with assignment, then Haar, DB6, Sym8, coif3 wavelet were used on the three layer decomposition. In addition, choosing threshold model of sure and scale noise to eliminate noise. The original elevation curve data and wavelet de-noising curve of elevation show in Fig.1.

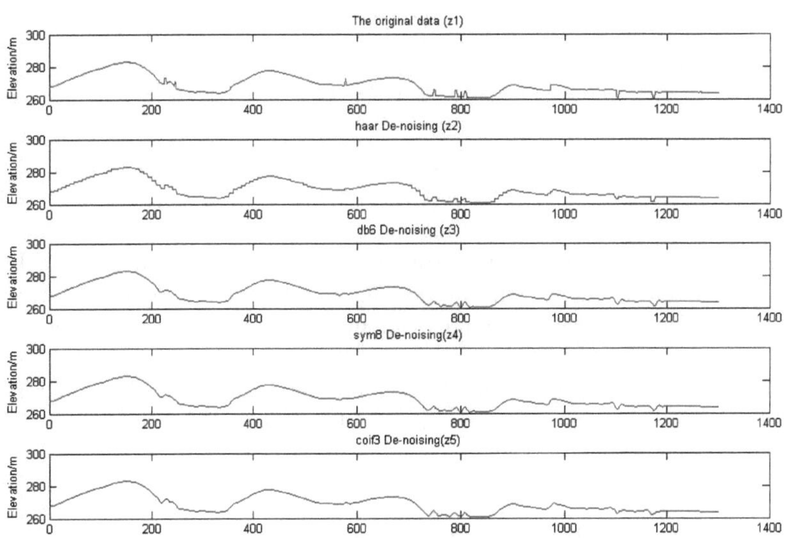

Fig. 1. Wavelet de-noising curve of elevation

It can be seen from the Fig.1 that the wavelet including haar, db6, sym8, coif3 has an effect on noise elimination and filter out outliers. Wavelet Db6 and sym8 for noise elimination is more smooth, so is the effect.

In order to study the GPS elevation curve data de-noising effect of the layering wavelet, using elevation data to the variable z1 with assignment, then the wavelet db6 is used from the two layers to seven layers for de-noising. The original elevation data and the wavelet db6 de-noising curve of elevation show in Fig.2.

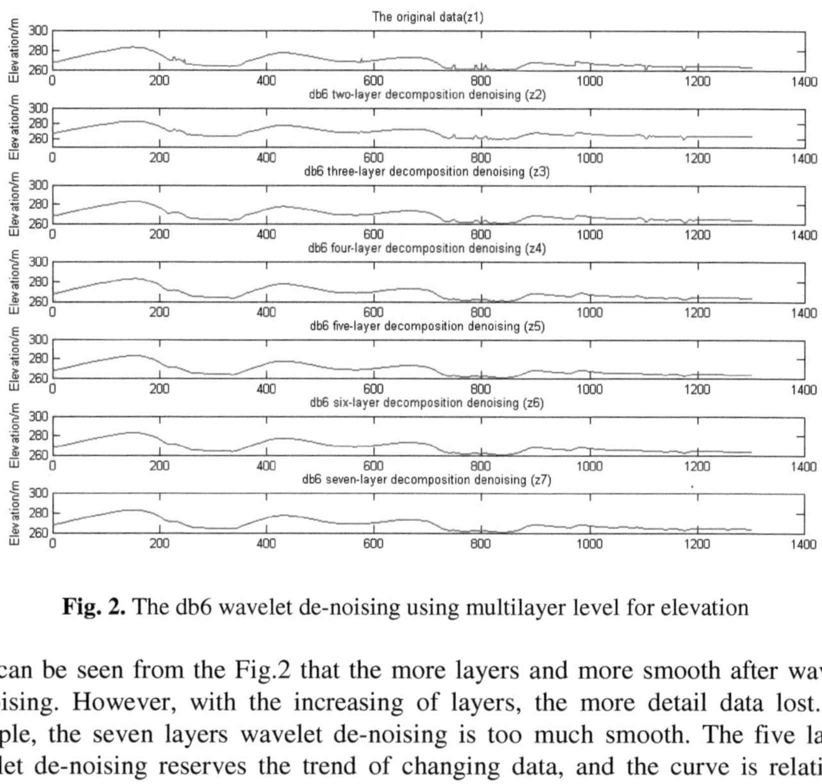

Fig. 2. The db6 wavelet de-noising using multilayer level for elevation

It can be seen from the Fig.2 that the more layers and more smooth after wavelet de-noising. However, with the increasing of layers, the more detail data lost. For example, the seven layers wavelet de-noising is too much smooth. The five layers wavelet de-noising reserves the trend of changing data, and the curve is relatively smooth.

Because slop farmland generally have more moderate undulating and field 3D topographic survey is continuous sampling, the five layers db6 wavelet de-noising can be used for farmland elevation data.

3 Drawing Farmland's 3D Topography

The latitude and longitude coordinates of the GPS positioning acquisition of data points in the program using the UTM projection, the transformation to the plane rectangular coordinates.

In this study, using Matlab as data processing and drawing tools. First, projection coordinates and elevation values of the collected data points were assigned to three variables to x, y, and z. In the operation, export Access data to Excel, then to Matlab. In another way, assign the variable x, y, z with Matlab, then copy and paste the corresponding data series to the variable value table of Matlab. Compared with the former, the latter is that a large amount of column values fast assign the variable value of Matlab.

3.1 Plane Scatter Plots

In order to show each sampling point location, need to draw a plane scatter plot of the sampling points. Draw blue scatter diagram, Matlab command sentences are as follows.

```
plot(x,y,'*b');
xlabel('x coordinate');
ylabel('y coordinate');
```

Plane scatter diagram that is made up of 1302 data points shows in Fig.3.

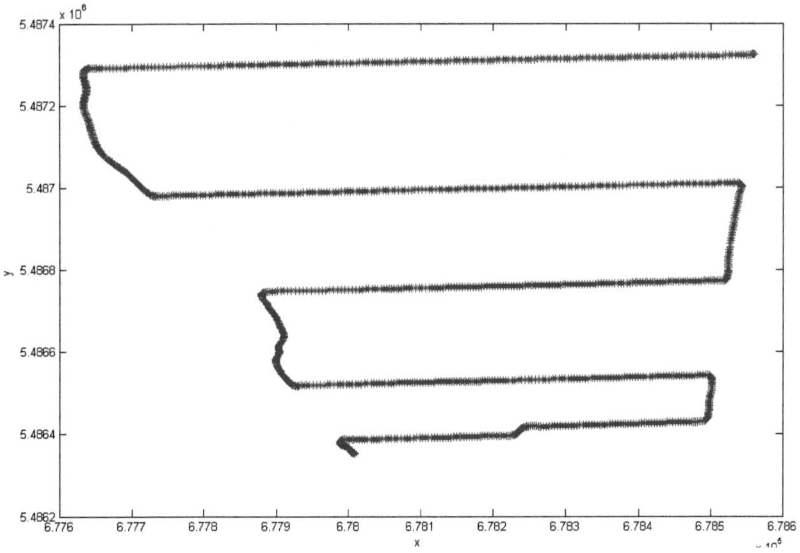

Fig. 3. The sampling points scatter diagram

3.2 Data Interpolation

Sampling points is an irregular distribution practically in farmland. In order to generate isoline and isosurface with Matlab, we should conduct data interpolation and gridding. Matlab command sentences of data interpolation are as follows.

```
X=linspace(min(x),max(x),100);
Y=linspace(min(y),max(y),100);
```

Linspace is one of the instructions that are used to produce Line vector between two points. Take 100 points from minimum to maximum in the experiments. Matlab command sentences of data gridding are as follows.

```
[X,Y]=meshgrid(X,Y);
Z=griddata(x,y,z,X,Y,'linear');
```

The effect of meshigrid (X,Y) is that conducting a matrix for line vector X and column vector Y, generating a cartesian product space and drawing the frame through two points on the axis.

We use designated algorithm. There are four kinds. (1) ‘linear’: Base on the linear (2)‘cubic’: triangle-based cubic interpolation; (3)‘nearest’: the nearest interpolation; (4)‘v4’:Lattice point spline function interpolation algorithm. Through gathering 3D topography information for farmland, the curve that is made up of ‘v4’ interpolation is more smooth. But the interpolation of the four corners will be larger deviation, the curve that is made up of 'linear' interpolation is more ideal.

3.3 Drawing Contour Map

Drawing contour map is an important branch of visualization technology, we can achieve contour map fast with Matlab. The contour map shows in Fig.4.

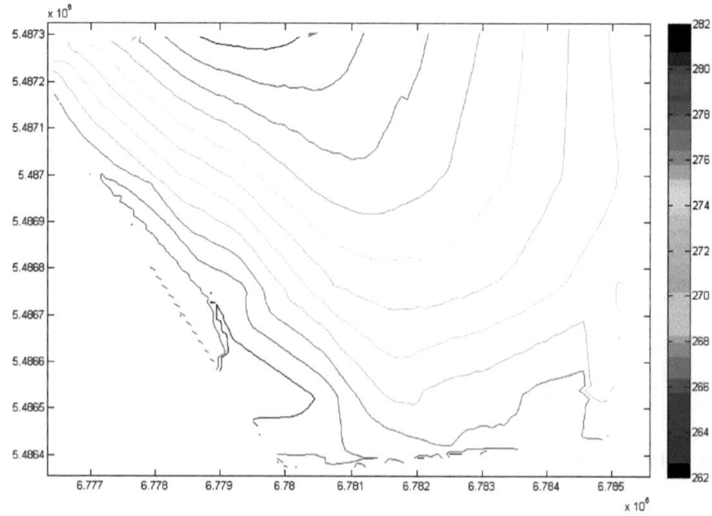

Fig. 4. The contour map of elevation

DB6 wavelet was used on the fifth layer decomposition de-noising. Then make an interpolation. The contour map of elevation denoised shows in Fig.5. Matlab command sentences are as follows.

```
contour(X,Y,Z);
colorbar;
z5=wden(z,'heursure','s','one',5,'db6');
Z5=griddata(x,y,z5,X,Y,'linear');
contour(X,Y,Z5);
colorbar;
Colorbar is used for generating illustrations.
```

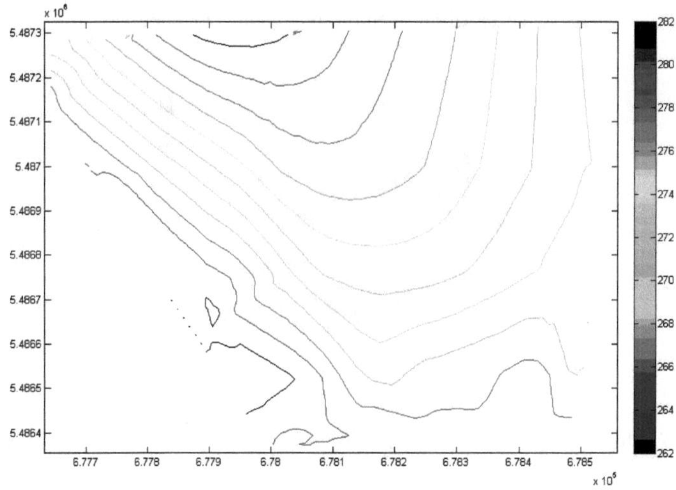

Fig. 5. The contour map of elevation denoising

It can be seen from the Fig.4 and Fig.5 that the contour map of elevation data de-noising is better than before.

3.4 Drawing 3-D Surface Map

3-D surface chart can be fast achieved with Matlab. Matlab command sentences of the3-D surface chart, which is made up of interpolation, are as follows.

```
surf(X,Y,Z);
surf(X,Y,Z5);
```

Three-dimensional topographic surface map and three-dimensional topographic surface map of elevation de-noising show in Fig.6 and Fig.7.

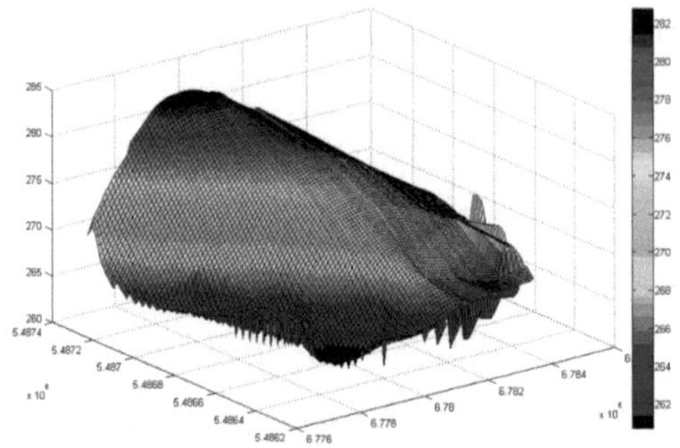

Fig. 6. Three-dimensional topographic surface map

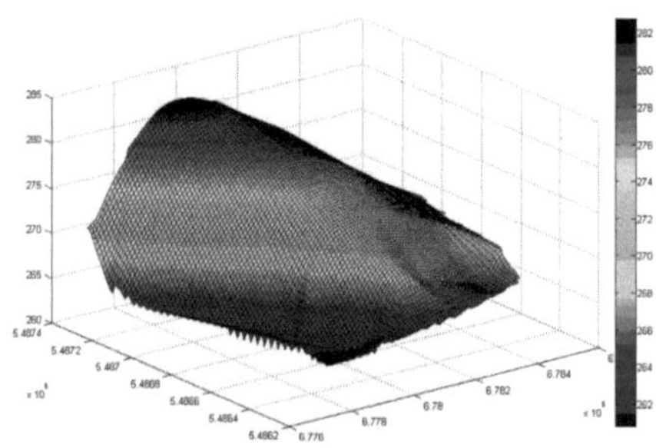

Fig. 7. Three-dimensional topographic surface map of elevation de-noising

It can be seen from the Fig.6 and Fig.7 that three-dimensional topographic surface map of elevation de-noising is better than before. This method was suitable for the sloping farmland, which has the more moderate wavy. The wavelet db6 is used on the fifth layer for de-noising by farmland's 3D topography, and then draw 3-D surface chart. It has a good effect on drawing Geomorphology.

4 Conclusions

It takes the measures of wavelet transformation to process the data of farmland's 3D topography with Matlab. Haar, DB6, Sym8, coif3 wavelet were used on the 3 layer decomposition of elevation data de-noising test, DB6 wavelet was used on the 2 ~ 7 layer decomposition de-noising. The research shows that the wavelet db6 was used on the 5 layer decomposition for de-noising, which could achieve elevation data of farmland.

The precision agriculture decision-making is widely needed in geographic information visualization analysis for farmland. Matlab has the function of Data calculation and drawing. By the technology, plane scatter plots, contour map, 3-D surface map of the farmland are drawn.

The research shows that using the wavelet db6 5 level decomposition de-noising for elevation data measured by DGPS continuously was suitable for the 3D-terrain farmland surveying and mapping in which farmland have the more moderate wavy. This may well reflect the farmland 3D topography features.

Acknowledgments. We wish to thank twelve five years science and technology plan for rural area project No.2011BAD20B03, No.2011BAD20B06 and No.2012BAD04B01, China, for funding this research. We also wish to thank the anonymous reviewers for their useful suggestions that greatly improved this paper.

References

1. Kumhálová, J., Kumhála, F., Kroulík, M., Matějková, Š.: The impact of topography on soil properties and yield and the effects of weather conditions. Precision Agriculture 12(6), 813–830 (2011)
2. Fountas, S., Blackmore, S., Ess, D., Hawkins, S., Blumhoff, G., Lowenberg-Deboer, J., Sorensen, C.G.: Farmer Experience with Precision Agriculture in Denmark and the US Eastern Corn Belt. Precision Agriculture 6, 121–141 (2005)
3. Asaduzzaman, K., Reaz, M.B.I., Mohd-Yasin, F., Sim, K.S., Hussain, M.S.: A Study on Discrete Wavelet-Based Noise Removal from EEG Signals. Advances in Experimental Medicine and Biology 680(pt. 6), 593–599 (2010)
4. Larson, J.A., Roberts, R.K., English, B.C., Larkin, S.L., Marra, M.C., Martin, S.W., et al.: Factors affecting farmer adoption of remotely sensed imagery for precision management in cotton production. Precision Agric. 9, 195–208 (2008)
5. Chung, G., Vese, L.A.: Energy Minimization Based Segmentation and Denoising Using a Multilayer Level Set Approach. In: Rangarajan, A., Vemuri, B.C., Yuille, A.L. (eds.) EMMCVPR 2005. LNCS, vol. 3757, pp. 439–455. Springer, Heidelberg (2005)
6. Zhuang, W.: Study on Soybean Variable Rate Fertilizing and Seeding in the Black Soil Gentle Hilly Regions of the Northeast China. Heilongjiang Bayi Agriculture University, Daqing (2011)
7. Zhuang, W.: Application of GPS &GIS for precision agriculture. Guangming Daily Publishing Company, Beijing (2009)
8. Meng, Z., Fu, W., Liu, H.: Design and implementation of 3D topgraphic surveying system in vehicle for field precision leveling. Transactions of the CSAE 25(S2), 255–259 (2009)
9. Zhang, M., Chen, Y., Jia, W.T.: Design of three dimensional topographic information measuring system. Journal of Jilin University (Engineering and Technology Edition) 37(6), 1451–1454 (2007)
10. Zhang, Y.-L., Bai, W.-B., Jiao, X.-Y., Wang, L.-G., Wang, J.-S.: Study on 3D Visualization of Farmland Soil Nutrients Spatial Variability in Loess Area of Mining Subsidence. Journal of Shanxi Agricultural Sciences 39(8), 894–896 (2011)
11. Fu, Y., Zhang, C.: A New Seismic Data De-Noising Method Based on Wavelet Transform. In: Proceedings of the 2008 International Conference on Computer Science and Information Technology, CCSIT 2008, pp. 92–96 (2008)

Remote Sensing Recognition of Paddy Waterlogging Using Change Vector Analysis Model

Xiaohe Gu, Jingcheng Zhang, Peng Xu, Yingying Dong, and Yansheng Dong*

Beijing Research Center for Information Technology in Agriculture, Beijing 100097, China
dongys@nercita.org.cn

Abstract. It is significant to monitor crop waterlogging range timely and correctly for later field management, agricultural insurance and yield prediction. The change regularity of paddy injured growth was analyzed and the sensitive parameters of growth stress were screened. The model of identifying the paddy waterlogging range based on change vector analysis (CVA) was developed by using the HJ-1/2 CCD images around waterlogging. At last the accuracy of the developed model was evaluated by in-situ sample data. Results showed that the waterlogged paddy mainly scattered around the Huaihe River system. The spatial distribution pattern was in conformity with the occurrence tendency of paddy waterlogging provided by local agricultural department as a whole. By evaluating the accuracy of the model with in-situ samples, the overall accuracy of the model developed in the study reached 87.5%, while the Kappa coefficient reached 0.737. The change vector analysis model could identify the waterlogged paddy and normal paddy correctly and efficient in Huaihe River Basin.

Keywords: Paddy Waterlogging, Change Vector Analysis, Decision Tree Classification, Remote Sensing.

1 Introduction

Flood and waterlogging present such features as high frequency and wide range in China, which have greatly affected national economic development. The flood and waterlogging mainly locate around major rivers with rich soil and developed agriculture. According the statistics, every year more than 7.33 million hectares farmland suffered floods and waterlogging in China. It is significant to monitor crop waterlogging range timely and correctly for later field management, agricultural insurance and yield prediction.

Most researches on loss assessment of flood and waterlogging mostly focused on statistical data such as population and economy, but lack of special loss assessment aiming at crop using remote sensing technology. The loss assessment of crop waterlogging not only needs to monitor the characteristics information of flood, but

* Corresponding author.

D. Li and Y. Chen (Eds.): CCTA 2012, Part II, IFIP AICT 393, pp. 36–43, 2013.

also needs to consider the factors such as crop variety, water-fertilizer, growing stage and soil texture. Li et al (2003) achieved damage evaluation of flood by using remote sensing and spatial economy database, including affected farmland area and affected population. Zhang et al (2007) developed the climate identifying index of drought and flood, analyzed the regularity of outbreak and its influence of crop yield. Recent research on evaluating waterlogging in relation to crop loss mostly focused on experience models, but rarely on mechanistic model (Profeti, et al., 1997 & Srikantha, et al., 1998;).

Because paddy field is usually flat and long-term in water, it is different to distinguish waterlogging region and natural water covering region through water spectrum change with the influence of paddy coronal. In addition, remote sensing images with appropriate resolution are hardly acquired for the cloudy weather when flood and waterlogging happens. So it is different to identify the paddy waterlogging directly using up-to-date images within the flood period. In this study the change regularity of paddy injured growth was analyzed and the sensitive parameters of growth stress were screened. By using the HJ-1/2 CCD images in the former and later periods of waterlogging, the model of monitoring the paddy waterlogging range based on change vector analysis (CVA) was developed. At last the accuracy of the developed model was evaluated by in-situ sample data.

2 Study Area and Materials

2.1 Study Area

The Huaihe River Basin is chosen as study area, where locates at transverse zone of semi-humid area to humid area in the east of China. The weather caused flood in the Huaihe River Basin is very changeable and the space-time location of the precipitation is not well balanced. The average annual precipitation is 883 mm and mainly concentrates on the period from June to September (Bi, et al, 2004). By analyzing the interannual variations, the precipitation of high flow years reaches five times more than that of dry years. The Huaihe River Basin is one of the highest frequency basins of flood in China. The variation law of summer precipitation with 2-3a cycles makes the basin suffering flood once every two or three years. In addition, because of low-lying land and inadequate ability of draining waterlogged fields, the Huaihe River Basin often holds out the situation of outside flood and inside waterlogging. For example, the flood happened in 2007 caused 1.25 million hectares of crop waterlogged in Anhui Province, including 360000 hectares of no harvest land and 44 hundred million yuan of agricultural loss.

The paddy waterlogging is directly related with the weather condition, especially short-term precipitation. Near Huoqiu and Shouxian county of Huaihe River Basin suffered continuous heavy precipitation from 26 to 30 August 2002 on the occasion of mid filling stage of local paddy. According to the records of local weather stations, the precipitation reached 74.3 mm in a short period of time. The observations of local department of agriculture discovered the severe waterlogging locally for hysteretic drainage management.

2.2 Data Acquisition and Preprocessing

Three images of HJ-1/2 CCD with mid resolution were used in the study, of which the temporal included July 13, August 26, September 3 and September 5, 2009. The resolution and response feature of HJ-1/2 CCD image were very close to these of LandSat5 TM image, which could meet accuracy requirement of paddy identification, growth retrieval and yield loss evaluation. The covering cycle is two days with a width of 360 kilometers, of which the advantages of multi-temporal and large scale is help to monitor the crop growth.

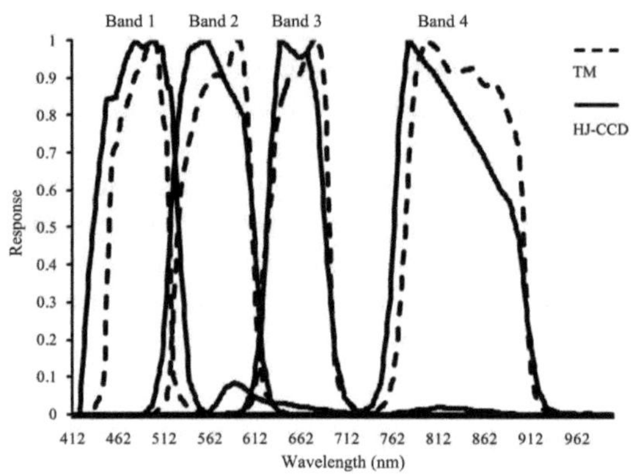

Fig. 1. Response function of former four band between HJ CCD and TM image

The in-situ data used in the study included 42 samples surrounding Fuyang, Huainan and Bengbu city in Anhui Province, which collected from September 17 to 20 in 2009. The collecting indexes of sample parcels included GPS position, planting density and waterlogged level. The representative plant samples of three square meters were gathered, of which the yield were determined in the lab.

In addition, phenomena data in Anhui Province was used to analyze the spectrum feature of fall crops and provide prior knowledge for decision tree classification stat.

The preprocess steps of remote sensing image included geometric correction, atmospheric calibration and projection transformation. The method of atmospheric calibration used in the paper was FLAASH model. The geometric correction used the polynomial model of bilinear interpolation, which could insure the error less than 0.5 pixels. The projection type of all spatial data was Universal Transverse Mercator (UTM), while the geography coordinate system was WGS 84.

3 Method and Technical Process

3.1 Identifying Spatial Distribution of Paddy

The decision tree classification was used to identify paddy by analyzing spectrum feature of multi-temporal HJ CCD images with in-situ sample point data. The confused crops in the study area mainly includes peanut.

By consulting the former studies, the spectrum information and vegetation indexes of two temporal HJ CCD images were used to identify paddy, including the NDVI of HJ-CCD image of August 26 (named T_1), the reflectivity of the fourth band of August 26 image (named T_2), the reflectivity of the fifth band of August 26 image (named T_3) and the NDVI of July 13 image. The classification was carried in the Ecognition7.0.

3.2 Physiological Mechanism of Growth Loss of Paddy Waterlogging

Water management of paddy field is of the utmost importance for growth and maturity of paddy. Paddy growing can not live without water during most growing periods, especially flowering stage and filling stage.

Maintain reasonable water level from heading to mid filling stage is helpful for paddy field. The water caused by heavy rains is apt to flow into paddy field for the restraints of topography and terrain. On account of the limited ability of flood discharge, the paddy field is apt to suffer waterlogging, even the flag leaf and ear is submerged. When the inversion second and third leaf are submerged, paddy waterlogging is considered to be suffered. Because of the depressed photosynthesis and damaged rice embryos, the waterlogging suffered at the filling stage will reduce grain plumpness and affect the yield evidently. If the ears of paddy are submerged more than 48 hours, the grain will stop filling and make for aborted kernels probably.

The spatial distribution of paddy waterlogging in the study was identified by the change vector analysis model with the two HJ CCD images around flood. According the information provided by local agricultural department, the leaves of some submerged rice plants turned white and were apt to lodge after a week of waterlogging. These variations provided the basis for identifying spatial distribution of waterlogging paddy.

3.3 CVA Model for Identifying Waterlogged Paddy

The method of change vector analysis is widely applied for processing multi-bands images. The change vector could be used to describe the direction and amount of spectrum variety from one temporal to another temporal. Assuming the vector of pixel spectrum of temporal t_1 and t_2 represent $G = (g_1, g_2, \ldots, g_k)^T$ and $H = (h_1, h_2, \ldots, h_k)^T$ respectively, the change vector between t_1 and t_2 will be described as follows.

$$\Delta G = G - H = \begin{pmatrix} g_1 - h_1 \\ g_2 - h_2 \\ \\ g_k - h_k \end{pmatrix} \quad (1)$$

The ΔG includes the all change information of two temporal image, of which the change intensity depends on the module of ΔG.

$$\|\Delta G\| = \sqrt{(g_1 - h_2)^2 + (g_2 - h_2)^2 + \cdots + (g_k - h_k)^2} \quad (2)$$

According to the definition of change intensity, the value of $\|\Delta G\|$ discloses the difference of two images and the change possibility. By setting proper threshold values of $\|\Delta G\|$, the changed and unchanged pixels could be distinguished. The method of change vector analysis is usually applied for identifying the pixels of land use and coverage change. The translated types of land use are depended on the direction of ΔG.

In the study, the difference of growth between waterlogged paddy and normal paddy coheres with the spectrum variation caused by waterlogging. For this reason the method of change vector analysis is introduced to identify the spatial distribution of waterlogged paddy.

3.4 Accuracy Evaluation

Through the spatial overlay and analysis between in-situ samples and waterlogged paddy monitoring by remote sensing, the error matrix was established. The accuracy of change vector analysis model was evaluated with overall accuracy and Kappa coefficient.

4 Results Analysis

On account of the degraded chlorophyll level and enlarged water information derived from submerged paddy, normalized difference vegetation index (NDVI), reflectivity of near infrared band (NIR) and greenness normal difference vegetation index (GNDVI) were input the model. The NDVI represented crop growth, which was declined with the waterlogging. The NIR and GNDVI were used to indicate water content of plant, which ascended for submerging plants. So the CVA model was transformed as followed function.

$$\|\Delta G\| = \sqrt{(ST\Delta NDVI - ST\Delta Nir - ST\Delta GNDVI)^2} \quad . \tag{3}$$

The STΔNDVI, STΔNir and STΔGNDV indicated the differences of three normalized eigenvectors. The normalization was intended to eliminate the difference from dimension of eigenvectors and improved the comparability. The normalization function was as follows.

$$ST\Delta F^2 = \frac{x_1 - \overline{x}}{\sigma} \quad . \tag{4}$$

Here, the x_1 means the difference values of eigenvectors of every pixel. The $\overline{x}$ means the average difference values of eigenvectors of every pixel. The σ means standard deviation. The differences of three eigenvectors from two temporal images were calculated by followed function.

$$\Delta NDVI = NDVI_{0826} - NDVI_{0903\text{-}05} \; . \tag{5}$$

$$\Delta Nir = Nir_{0826} - Nir_{0903\text{-}05} \, . \tag{6}$$

$$\Delta GNDVI = GNDVI_{0826} - GNDVI_{0903\text{-}05} \, . \tag{7}$$

The spatial distribution of $\| \Delta G \|$ was mapped as figure 2.

Because two temporal images were used in the study, the change vector had only two directions, including magnifying and reducing. According to the magnifying modulus, the waterlogged paddy and normal paddy were identified by the threshold. In the study, the threshold was determined with statistical analysis of pixel values of two temporal images, which attributed to improve the generality of the method with different region and images. Because paddy submerged in small scale in study area, the mean and standard deviation value (SD) were used to set threshold through statistical analysis of paddy pixels.

$$\text{Threshold} = \text{Mean}_{\|\Delta G\|} + 2 * SD_{\|\Delta G\|} \, . \tag{8}$$

If the modulus of change vector was above the threshold, the change extent of waterlogged paddy pixels exceeded that of normal paddy pixels, that is $\text{Mean}_{\|\Delta G\|} \pm SD_{\|\Delta G\|}$. Because the direction of change vector in the study was fixed and conservative, the threshold was determined as Mean ± SD. The threshold setting was 5.76, which was used to map the waterlogged paddy as figure 3.

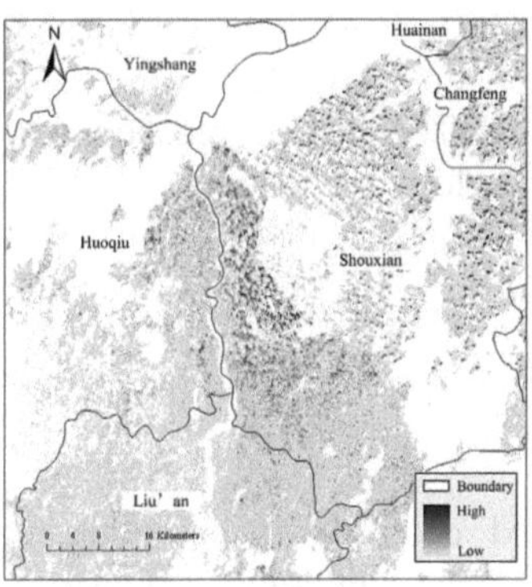

Fig. 2. Spatial distribution of $\|\Delta G\|$

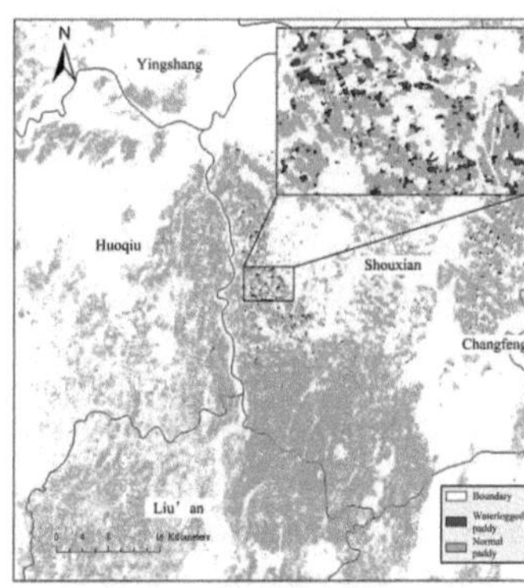

Fig. 3. Spatial distribution of waterlogged paddy

The figure 3 showed that the waterlogged paddy (colored red) scattered everywhere in the study, mainly around the Huaihe River system. The waterlogging in Shouxian County was comparatively serious. The spatial distribution pattern was in conformity with the occurrence tendency of paddy waterlogging provided by local agricultural department as a whole.

By analyzing the attribute of 42 in-situ sample points, except for 2 point of non-paddy, 25 points located in normal paddy range, while 15 points in waterlogged paddy range. Through spatial overlay and analysis of the waterlogged paddy distribution with sample data, the conversion matrix was established according to waterlogging and normal data as follows.

Table 1. Conversion matrix derived from in-situ samples and RS identification

In-situ samples	Remote sensing identification		
	Waterlogged paddy	Normal paddy	total
Waterlogged paddy	13	3	16
Normal paddy	2	22	24
total	15	25	40

From the table 1, the overall accuracy of the model developed in the study reached 87.5%, while the Kappa coefficient reached 0.737.

Table 2. Yield analysis of in-situ samples of waterlogged paddy and normal paddy

Sample type	Sample number	Average yield (kg/hm^2)	standard deviation
Waterlogged paddy	15	5812	384
Normal paddy	25	10662	562

By analyzing the actual output of 40 samples, the average value of normal paddy was 10662 kg/hm^2, 45 percent over that of waterlogged paddy. According the t-test of independent samples, the *p*-value was below 0.0001, which reached extremely significant level. Results showed that paddy waterlogging reduced the local yield considerably. It also indicated that the change vector analysis model could identify the waterlogged paddy and normal paddy correctly.

5 Conclusion and Discussion

Identifying crop waterlogging range timely and correctly is essential for field management afterwards, yield prediction and agricultural insurance. Paddy submerged at filling stage will reduce grain plumpness and affect yield evidently on account of depressed photosynthesis and damaged rice embryos. In this study the change regularity of paddy injured growth was analyzed and the sensitive parameters of growth stress were screened. By using the HJ-1/2 CCD images in the former and later periods of waterlogging, the model of identifying the paddy waterlogging range based on change vector analysis (CVA) was developed. At last the accuracy of the developed model was evaluated by in-situ sample data. Results showed that the waterlogged paddy scattered everywhere in the study, mainly around the Huaihe River system. The spatial distribution pattern was in conformity with the occurrence tendency of paddy waterlogging provided by local agricultural department as a whole. By evaluating the accuracy of the model with in-situ samples, the overall accuracy of the model developed in the study reached 87.5%, while the Kappa coefficient reached

0.737. According the t-test of independent samples, the p-value reached the extremely significant level. Results showed that paddy waterlogging reduced the local yield considerably. These results indicated that the change vector analysis model could identify the waterlogged paddy and normal paddy correctly and efficient in Huaihe River Basin.

The current study requires some further improvements. First, the model of change vector analysis need to test the efficiency of identifying waterlogged paddy in the other region. Second, because the waterlogging scale was relatively small in the study. The accuracy of the method developed in the study needs to be test in great scale of waterlogging.

Acknowledgements. This work is financially supported by the National Natural Science Foundation of China (41001199).

References

1. Profeti, G., Macintosh, H.: Flood Management through Landsat TM and ERS SAR data: a case study. Hudrological Process 11, 1397–1408 (1997)
2. Herath, S., Dutta, D.: Flood inundation modeling and loss estimation using distributed hydrologic model. GIS and RS. In: Proceedings of International Workshop on the Utilization of Remote Sensing Technology to Natural Disaster Reduction, Tsukuba, Japan, October 26-28, pp. 239–250 (1998)
3. Western, I., Messing, I., Liner, H., et al.: Controlled drainage-effects on drain outflow and water quality. Agricultural Water Management 47, 85–100 (2001)
4. Nakasone, H., Abbes, M.K., Kuroda, H.: Nitrogen transport and transformation in packed soil columns from paddy fields. Paddy and Water Environment 23, 115–124 (2004)
5. Reddy, B.B., Ghosh, B.C., Panda, M.M.: Flood tolerance of rice at different crop growth stages as affected by fertilizer application. Plant and Soil 83, 255–263 (1985)
6. Bi, B., Jiao, M., Liao, Y.: Analysis of precipitation and river f low in the Huaihe River basins during the summer of 2003. Journal of Applied Meteorological Science 15(6), 681–687 (2004) (in Chinese)
7. Zhao, L.-N., Yang, X.-D., Qi, D., et al.: The analysis of precipitation and flooding features in the Huaihe River Basin during the summer of 2007. Climatic and Environmental Research 12(6), 728–737 (2007)
8. Zhang, A., Ma, X., Yang, T., et al.: The influence of drought and waterlogging disasters on crop yields in Anhui province. Journal of Applied Meteorological Science 18(5), 69–626 (2007)

The Simulation Models of Nitrogen Accumulation and Partitioning in Plant for Protected Cultivated Tomato

Yuli Chen[1,2], Hongxin Cao[1,*], Yan Zhu[2,*], Yan Liu[1], and Weixin Zhang[1,2]

[1] Institute of Agricultural Economics and Information /Engineering Research Center for Digital Agriculture, Jiangsu Academy of Agricultural Sciences, Nanjing 210014, Jiangsu Province, P.R. China
2009101038@njau.edu.cn, caohongxin@hotmail.com, liuyan0203@yahoo.com.cn, 2011801168@njau.edu.cn
[2] College of Agronomy, Nanjing Agricultural University, Nanjing 210095, Jiangsu Province, P.R. China
yanzhu@njau.edu.cn

Abstract. 【Objective】 In order to quantify the nitrogen accumulation and partitioning of protected cultivated tomato plant. 【Method】 The field experiments on cultivars and fertilizer were conducted in 2009 and 2010, and cultivars: B1, B2, and B3 were adopted. 【Result】 The simulation models of nitrogen accumulation and partitioning for protected cultivated tomato plant were built by analyzing the relationship between partitioning index and product of temperature and light in accordance with the tomato growth characteristics in this paper. The models were validated using independent experiment dataset, the results showed that RMSE, X_{de}, and R^2 between simulated and measured values of nitrogen accumulation in above-ground plant and root were 0.546 $g \cdot m^{-2}$, 0.447 $g \cdot m^{-2}$, and 0.956; 0.041 $g \cdot m^{-2}$, 0.033 $g \cdot m^{-2}$, and 0.942, respectively. That of nitrogen accumulation in stem, leaf, and fruit were 0.165 $g \cdot m^{-2}$, 0.132 $g \cdot m^{-2}$, and 0.839; 0.146 $g \cdot m^{-2}$, 0.099 $g \cdot m^{-2}$, and 0.945; 0.246 $g \cdot m^{-2}$, 0.166 $g \cdot m^{-2}$, and 0.845, respectively. 【Conclusion】 This study indicated that the nitrogen accumulation and partitioning of different cultivars and fertilizer levels in the protected cultivated tomato plant could be well simulated by these models.

Keywords: protected cultivated tomato, nitrogen in plant, accumulation and partitioning, simulation models.

Foundation Information: Jiangsu Province Agricultural Scientific Technology Support Program of China (BE2011342); Jiangsu Province Agricultural Scientific Technology Innovation Fund of China (CX (10)221).

Yuli Chen (1986-), male, born in Zhucheng city of Shandong province, Main research direction: crops simulation models. 2009101038@njau.edu.cn
* Corresponding author.

D. Li and Y. Chen (Eds.): CCTA 2012, Part II, IFIP AICT 393, pp. 44–52, 2013.

1 Introduction

Nitrogen is one of the necessary nutrient elements for crop growth and development. The accumulation of nitrogen have an important effects on the chlorophyll content, leaf area, photosynthetic rate, photorespiration, solar energy use efficiency, yield, quality, and so on[1]. Therefore, the systematical study for nitrogen accumulation and partitioning has an important practical significance for guiding the nitrogen application in high yield cultivation [2]. Nowadays, the studies about nitrogen accumulation and partitioning mainly were involved in the field crop, and as a basis research for protein predicting model. Chen et al. [4], Xu et al. [5], and Shi et al. [6] studied the nitrogen accumulation and partitioning of rice, barely, and wheat, respectively, and built their predicting models for protein content, respectively. However, little research on the model of nitrogen accumulation and partitioning in plant in horticultural crops were reported. The objective of this study was to build the models of nitrogen accumulation and partitioning in plant for protected cultivated tomato based on various cultivars, and fertilizer levels field experiments.

2 Materials and Methods

2.1 Materials

This study used 3 tomato cultivars representing wide variation in maturing characteristics, and they are: (B1) American mole 1(100 ~ 110days), (B2) Chaoshijifanqiedawang (109~119days), and (B3) American 903(106~116days).

2.2 Methods

The experiments were conducted in plastic shed with 80 m long, 9.8 m wide, and horse liver soil (the total nitrogen, 0.239 g/kg; total phosphorus, 1.297 g/kg; available phosphorus, 202 mg/kg; and pH, 6.344 in pre-planting in soils) at Suoshi village in Nanjing from July to October of 2009 (autumn crop) and from March to June of 2010 (spring crop). The experiments were a split plot design with three whole-plot treatments arranged in a randomized complete block design with three blocks and three sub-plot treatments. The whole-plot factors were the fertilizer levels: A1 (Low nitrogen: 1/2 F_N), A2 (Normal: F_N), and A3 (High nitrogen: 3/2 F_N), and the sub-plot factors were varieties (B1, B2, and B3), with 3 replications and 27 plots ($2.96 \times 4.6 m^2$). FN was the normal nitrogen level: compound fertilizer (N:P:K=16%:16%:17%) 750 $kg \cdot hm^{-2}$. The planting density was 5.88 plants/m^2, the pruning method of continuous decapitation was adopted in the experiments of spring crop and autumn crop, and 3~4 and 2~3 clusters were left in spring crop and autumn crop, respectively.

2.2.1 Data Collection

After planting, the representative samples were taken every 7d from seedling to flowering and every 14d during fruit period. Three representative plants selected in each treatment were separated into parts after purification and dried by the filter paper, dried in 30 min. at 105°C, then, at 80°C until reaching a constant weight, measured using a 0.001 g electro-level. Then, the nitrogen contents of various organs were measured by using the micro-kjeldahl method. HOBO-H8 was placed in three different positions in the plastic shed to collect environment elements automatically every 10 s, including air temperature, relative humidity and dew point, the absolute humidity, and light intensity, etc. In that average values every 15 minutes were recorded.

2.2.2 Model Verification

In this study, Excel.2007 and SPSS V 16.0 were used to analysis experimental data. The experiment data in 2009 were applied to model establishment and parameter determination, and the experiment data in 2010 were applied to model verification. The model precision was verified using root mean squared error (RMSE), mean absolute error (X_{de}), the determined coefficient (R^2), and 1:1 plotting between measured values and simulated values. The calculation formula of RMSE and Xde can be expressed as follows:

$$RMSE = \sqrt{\frac{\sum_{i=1}^{n}(OBS_i - SIM_i)^2}{n}}$$

$$de = |OBS_i - SIM_i|$$

$$X_{de} = \frac{\sum de}{n}$$

where OBS_i is measured values, SIM_i is simulated values, de is absolute error, and n is sample numbers.

3 Results and Analysis

3.1 Models Description

3.1.1 The Determination of Product of Temperature and Light (E)

According to the multi factor effect principle [7], the effect of temperature and light could be expressed by the product of temperature and light (E):

$$E = T \times P \tag{1}$$

where E is product of temperature and light, T is effective accumulative temperature, and P is the accumulated light intensity which higher than light compensation point for tomato. T could be expressed as follows:

$$T = \sum_{i=1}^{n}(T_{mean} - T_b) \tag{2}$$

$$T_{mean} = \frac{T_1+T_2+T_3+\cdots+T_{24}}{24} \tag{3}$$

where T_{mean} is day mean temperature, T_b is minimum temperature for tomato growth, and the value is 15°C [8] in all development stages, besides the seedling stage is 10°C [7], and T_1, T_2, T_3, ···, and T_{24} is the plastic shed temperature from 01:00 to 24:00, respectively.

Referring the definition of effective accumulative temperature, P could be expressed as follows:

$$P = \sum_{i=1}^{n}(P_{mean} - P_b) \tag{4}$$

$$P_{mean} = \frac{P_1+P_2+P_3+\cdots+P_m}{m} \tag{5}$$

where P_{mean} is day mean light intensity, P_b is light compensation point for tomato, and the value is 2klx [9], and P_1, P_2, P_3, ···, and Pm is the plastic shed light intensity of integer point time from sunrise to sunset, m is the number of integer point time, and the time of sunrise and sunset were computed by the method of Jing et al. [10]. The other symbols were the same as above.

3.1.2 The Models of Nitrogen Accumulation for Protected Cultivated Tomato Plant

The changes in nitrogen accumulation with the dry matter accumulation in protected cultivated tomato plant were a Logistic curve in accordance with the data of 2009 (Fig.1).

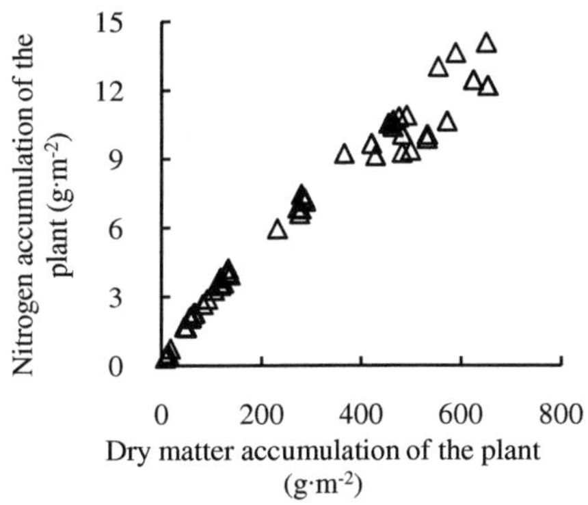

Fig. 1. The relationship between the plant nitrogen accumulation and dry matter

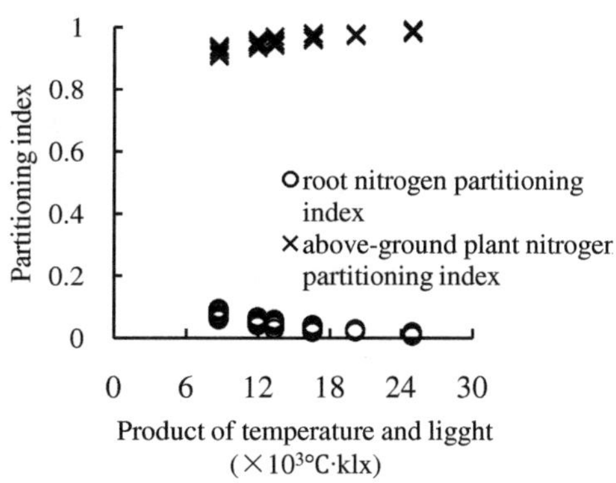

Fig. 2. The relationship between nitrogen partitioning index of above-ground plant and root and E for protected cultivated tomato plant

Therefore, it could be expressed as follow:

$$NB_E = \frac{NB_{max}}{(1+A\times e^{(-k\times DW_E)})^B} \tag{6}$$

where NB_E is the total nitrogen accumulation in protected cultivated tomato plant at E in g·m^{-2}; NB_{max} is the latent total nitrogen accumulation in protected cultivated tomato plant in g·m^{-2}, and it is a model parameter; DW_E is the total dry matter accumulation in protected cultivated tomato plant at E in g·m^{-2}; and A, k, and B are model parameters (table 1).

Table 1. Analysis of variance for all models and its coefficient test

Models	Correlation coefficient	F	A	k	B
Nitrogen contents	0.880**	96.847**	0.008**	0.005**	322.688**
Above-ground plant nitrogen	0.881**	120.973**	3.3396**	0.146**	0.112**
Stem nitrogen partitioning	0.887**	42.985**	1.787×10^{5**}	0.894**	0.106**
Stem nitrogen partitioning	0.967**	64.083**	0.523**	-5.046**	5.198**
Leaf nitrogen partitioning index	0.952**	30.859**	3.661×10^{-7**}	-1.389**	0.066**

Note: The asterisk "*", and "**" denoted the significant level at P<0.05, and P<0.01, respectively. The same below.

3.1.3 The Models of Nitrogen Partitioning in Protected Cultivated Tomato Plant

3.1.3.1 The Simulation of Nitrogen Partitioning of Above-ground Plant and Root in Protected Cultivated Tomato Plant

The nitrogen partitioning index to above-ground plant (root) in protected cultivated tomato plant could be expressed as follows:

$$NPIS = NWSH/NB \tag{7}$$

$$NPIR = 1 - NPIS \tag{8}$$

where NPIS is the nitrogen partitioning index in above-ground plant, NPIR is the nitrogen partitioning index in root, NWSH is the nitrogen accumulation of above-ground plant in g·m^{-2}, and NB is the total nitrogen accumulation in plant in g·m^{-2}.

According to the relationship between the nitrogen partitioning index in above-ground plant or root and E (Fig.2), the trend could be expressed as follows:

$$NPIS_E = \frac{NPIS_{max}}{(1+A\times e^{(-k\times E)})^B} \tag{9}$$

$$NPIR_E = 1 - NPIS_E \tag{10}$$

where $NPIS_E$ and $NPIR_E$ is the nitrogen partitioning index in above-ground plant and root at E, respectively. $NPIS_{max}$ is the latent nitrogen partitioning index in above-ground plant, and it is a model parameter. A, k, and B are model parameters (table 1).

3.1.3.2 *The Simulation of Nitrogen Partitioning for Different Organs of Above-ground Plant*

The partitioning index of nitrogen in various organs (leaf, stem, and fruit) could be expressed as follows:

$$NPIL = NWL/NWSH \tag{11}$$

$$NPIST = NWST/NWSH \tag{12}$$

$$NPIF = 1 - NPIL - NPIST \tag{13}$$

where NPIL, NPIST, and NPIF is the nitrogen partitioning index leaf, stem, and fruit, respectively. NWL, and NWST is the nitrogen accumulation of leaf, and stem in $g\cdot m^{-2}$, respectively. NWSH is the total nitrogen accumulation of above-ground plant in $g\cdot m^{-2}$.

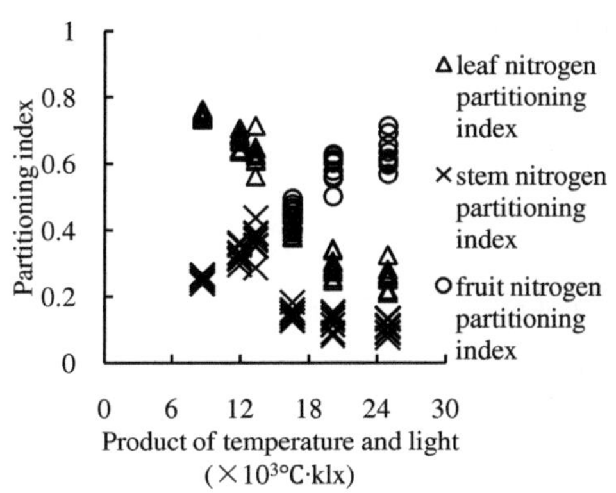

Fig. 3. The relationship between nitrogen partitioning index of stem, leaf, and fruit and E

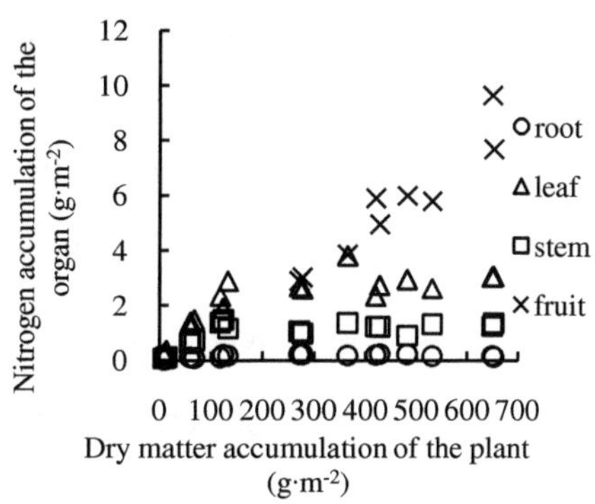

Fig. 6. The relationship between the plant nitrogen accumulation of different parts and dry matter in protected cultivated tomato plant

The data of 2009 showed: the leaf nitrogen partitioning index decreased with the adding of E; the stem nitrogen partitioning index increased firstly, and then decreased with the adding of E; and the fruit nitrogen partitioning index increased with the adding of E (Fig.3).

According to the relationship between nitrogen partitioning index of leaf, stem, and fruit and E, and the trends could be expressed as follows:

$$NPIL_E = \frac{NPIL_{max}}{(1+A\times e^{(-k\times E)})^B} \tag{14}$$

$$NPIST_E = \begin{cases} \frac{NPIST_{max}}{(1+A\times e^{(-k\times E)})^B} & (E \leq 1.332\times 10^4) \\ \frac{NPIST_{max}}{(1+A\times e^{(-k\times E)})^B} & (1.332\times 10^4 \leq E \leq 2.494\times 10^4) \end{cases} \tag{15}$$

$$NPIF_E = 1 - NPIL_E - NPIST_E \tag{16}$$

where $NPIL_E$ is the leaf nitrogen partitioning index at E, $NPIST_E$ is the stem nitrogen partitioning index at E, $NPIF_E$ is the fruit nitrogen partitioning index at E, $NPIL_{max}$ and $NPIST_{max}$ is the latent nitrogen partitioning index of leaf and stem, respectively, and it is a model parameter. A, k, and B are model parameters (table 1).

3.1.3.3 *The Simulation of Nitrogen Accumulation for Different Parts in Protected Cultivated Tomato Plant*

The nitrogen accumulation of different parts of above-ground plant could be expressed as follows:

$$NWSH_E = NB_E \times NPIS_E \tag{17}$$

$$NWR_E = NB_E \times NPIR_E \tag{18}$$

$$NWL_E = NWSH_E \times NPIL_E \tag{19}$$

$$NWST_E = NWSH_E \times NPIST_E \tag{20}$$

$$NWF_E = NWSH_E \times NPIF_E \tag{21}$$

where NB_E, $NWSH_E$, NWR_E, NWL_E, $NWST_E$, and NWF_E are the nitrogen accumulation of total plant, above-ground plant, root, leaf, stem, and fruit in $g{\cdot}m^{-2}$, respectively. $NPISH_E$, $NPIR_E$, $NPIL_E$, $NPIST_E$, and $NPIF_E$ are the nitrogen partitioning index of above-ground plant, root, leaf, stem, and fruit, respectively.

3.2 Model Verification

3.2.1 The Model Verification of Nitrogen Accumulation in Above-ground Plant and Root for Protected Cultivated Tomato

The model of nitrogen accumulation in above-ground plant and root for protected cultivated tomato were verified by independent data in 2010 (Fig.4). The figure 4 showed: RMSE, X_{de}, and R^2 of between measured and simulated values of nitrogen accumulation in above-ground plant and root for protected cultivated tomato were 0.546g·m^{-2} (n=36), 0.447g·m^{-2}, and 0.956; 0.0421g·m^{-2} (n=36), 0.033g·m^{-2}, and 0.942, respectively. The correlation coefficients (r) were 0.978 ($r_{0.01,34}$=0.424) and 0.971 ($r_{0.01,34}$=0.424) with 0.01 significant level, respectively.

3.2.2 The Model Verification of Nitrogen Accumulation in Leaf, Stem, and Fruit for Protected Cultivated Tomato

The model of nitrogen accumulation in leaf, stem, and fruit for protected cultivated tomato were verified by independent data in 2010 (Fig.5). The figure 5 showed: RMSE, X_{de}, and R^2 of between measured and simulated values for protected cultivated tomato leaf, stem, and fruit nitrogen accumulation were 0.165g·m^{-2} (n=36), 0.132g·m^{-2}, and 0.839; 0.146g·m^{-2} (n=36), 0.099g·m^{-2}, and 0.945; 0.246g·m^{-2} (n=36), 0.166g·m^{-2}, and 0.845, respectively. The r values were 0.916 ($r_{0.01,34}$=0.424), 0.972 ($r_{0.01,34}$=0.424) and 0.919 ($r_{0.01,34}$=0.424) with 0.01 significant level, respectively.

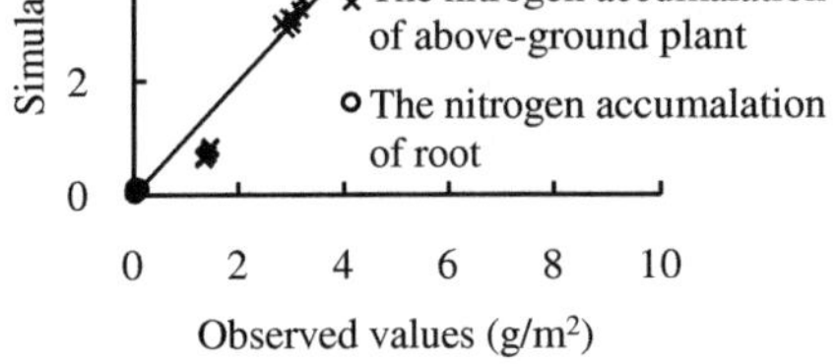

Fig. 4. Comparison of simulated with observed nitrogen accumulation values of above-ground plant and root

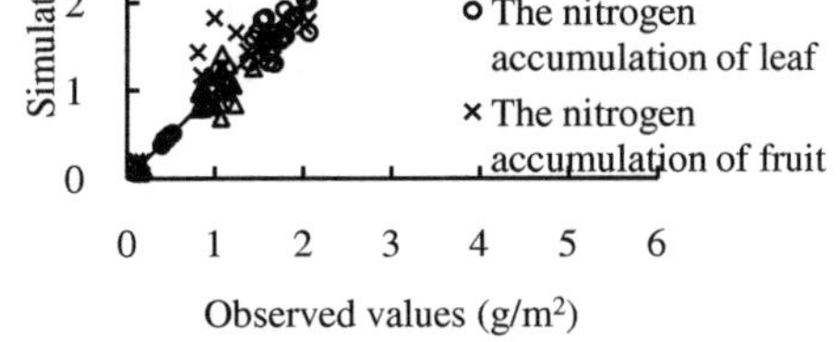

Fig. 5. Comparison of simulated with observed nitrogen accumulation values of stem, leaf and fruit

4 Discussion

The amounts of nitrogen accumulation in leaf, and stem of protected cultivated tomato had a tendency to increase before topping. After topping, the amounts of nitrogen accumulation in leaf, and stem changed into little, but it increased rapidly in fruit, and roots absorbing nitrogen had little changes in the amounts of nitrogen accumulation during the total growth and development processes (Fig.6). Therefore, we could not solve the nitrogen transfer in different parts of protected cultivated tomato plant by using the thought that applied to the field crop. The nitrogen contents measured in different organs were used as the nitrogen accumulation, and the nitrogen transfer in different organs for protected cultivated tomato wasn't be studied in detail in the paper. In order to improve the applicability, accuracy, and precision of the model, the water impact factor should be considered in the model in the further studies.

5 Conclusions

The simulation models of nitrogen accumulation and partitioning for protected cultivated tomato were built by analyzing the relationship between nitrogen accumulation, partitioning, and biomass and E in accordance with the field experiment data of protected cultivated tomato in 2009 and 2010. These models were verified by using independent experiment data, the statistical analysis (RMSE, X_{de}, R^2, and r) and 1:1 diagram.

References

1. Marcelis, L.F.M.: Simulation of biomass allocation in greenhouse-A review. Acta Hort. 328, 49–65 (1993)
2. Liu, K.L., Gao, J.L., Liu, R.X., et al.: Rules of N assimilation, accumulation and distribution of spring wheat. J. Triticeae Crop. 23(3), 97–102 (2003) (in Chinese with English abstract)
3. Zhuang, H.Y., Cao, W.X., Jiang, S.X., et al.: Simulation on nitrogen uptake and partitioning in crops. Syst. Sci. Compr. Stud. Agric. 20(1), 5–8 (2004) (in Chinese with English abstract)
4. Chen, J., Tang, L., Liu, X.J., et al.: Modeling plant nitrogen uptake and grain protein accumulation in rice. Sci. Agric. Sin. 44(10), 1997–2004 (2011) (in Chinese with English abstract)
5. Xu, S.J., Yang, H.S., Fan, F., et al.: A model for predicting grain protein content in barley. Sci. Agric. Sin. 42(11), 3863–3870 (2009) (in Chinese with English abstract)
6. Shi, X.Y., Zhu, Y., Tang, L., et al.: Modeling dynamic grain protein composition in wheat. Sci. Agric. Sin. 42(7), 2326–2331 (2009) (in Chinese with English abstract)
7. Chen, X.J.: Research on tomato fruit development and yield formation model under facility condition. Northwest Agric. For. University, Yangling (2010) (in Chinese with English abstract)
8. Ni, J.H., Luo, W.H., Li, Y.X., et al.: Simulation of leaf area and dry matter production in greenhouse tomato. Sci. Agric. Sin. 38(8), 1629–1635 (2005) (in Chinese with English abstract)
9. Shandong Agricultural University. Vegetable cultivation to learn each theory (north this), 3rd edn. China Agric. Press, Beijing (2000) (in Chinese with English abstract)
10. Jing, C.G., Shu, D.M., Gu, D.Y.: Implementation of the times for sunrise and sunset algorithm in urban streetlight supervisory control system. Mod. Comput. 163, 84–86 (2003) (in Chinese with English abstract)

Systematic Random Deployment for Wireless Sensor Network in Agricultural Sampling-Interpolation Applications

Hui Liu[1,2], Zhijun Meng[*], Hua Wang[2], and Min Xu[2]

[1] National Engineering Research Center for Information Technology in Agriculture, Beijing Nongke Mansion Room A-516,No 11, ShuGuangHuaYuan Middle Road, Banjing, Haidian District, Beijing, 100097, P.R. China
Hui Liu,Zhijun Meng,mengzj@nercita.org.cn
[2] Information Engineering College, Capital Normal University, No.56 North Road of Western 3rd-Ring P.O.Box 801, Beijing, 100048, P.R. China

Abstract. Wireless sensor network can act as a distributed sampling system in precision farming applications which aim at estimating the spatiotemporal variability. Sensor nodes should be deployed referring to the communication range r_c and the sampling range r_s instead of the sensing range. In this paper, a kind of systematic random deployment based on three popular regular patterns-equilateral triangle, square and hexagon is explored to achieve both coverage and connectivity. The efficiency of different regular patterns of systematic random deployment is also compared, thus enabling to make a better deployment decision for different values of r_c/r_s.

Keywords: Precision farming, Wireless sensor network, Sensor nodes deployment.

1 Introduction

Precision farming demands soil, microclimate, crop and other field data for spatial and temporal variability analysis [1]. But limited samples could not explain exactly the variability and support farming decision. The application of Wireless Sensor Network (WSN) technology could provide the optimal and integrated solution for real-time data acquisition, reliable transport, aggregation, in-network processing and statistics [2]. The potential benefits of using the WSN technology resulted in the appearance of a large number of academic and industrial R&D projects in agriculture. Lofar Agro project resulted in an experimental WSN installation in a potato field. The purpose of the WSN is to measure the microclimate factors essential to resist the phytophtora disease [3]. Camalie vineyard currently had an advanced soil moisture monitoring system named Camalie Net. It used twenty Mica2dot motes of Crossbow Inc. to establish the network. These motes sent data of temperature, soil moisture and soil temperature to a central computer once every 10 minutes [4]. It is obvious that wireless

* Corresponding author.

D. Li and Y. Chen (Eds.): CCTA 2012, Part II, IFIP AICT 393, pp. 53–59, 2013.

sensor networks could act as a distributed sampling system in precision farming applications, which aim at estimating the spatiotemporal variability.

Sensor nodes deployment is one of the primary issues for any WSN application. Most researches on deployment concentrate on event detection applications, which aim to detect when and where particular event takes place such as the start of a fire. The communication range and the sensing range are introduced to design the deployment in such applications. However, WSNs in precision farming are mostly kind of continuous sampling-interpolation applications different from event detection. The goal of the network is to carry out continuous spatiotemporal sampling and build field data distribution map by interpolating the values of the field where no sensor reading exist. The sensing range is largely meaningless in this case unlike in the event detection class of sensor networks. In this paper, the sampling range is proposed for WSN deployment instead of the sensing range. A kind of systematic random deployment based on regular patterns to achieve both coverage and connectivity is discussed.

2 Systematic Random Deployment

Wireless sensor nodes could be generally placed in three methods including random deployment, regular deployment and customized deployment [5]. Taking the agricultural scenario, sensors could be mostly possible installed during terrain preparation. It is a good choice for sensor node deployment referring to traditional sampling methods.

The most commonly used sampling design for spatial data is systematic sampling in agricultural experiments because it is easy to implement. A Square grid, an equilateral triangle grid or a regular hexagonal grid is commonly chosen in systematic sampling. Liu et al. show how to deploy sensor nodes in regular patterns in farmland environmental monitoring applications [6]. But Systematic sampling does have a few disadvantages for spatial correlation analysis. There is a danger that the systematic grid might coincide with a systematic feature of the landscape, especially in man made landscapes. Systematic random sampling is another type of sampling spreading the data in space in a similar fashion as systematic sampling. One sampling unit is selected at random in each grid cell. It is more efficient for spatially correlated data than systematic sampling and simple random sampling.

Consequently we present systematic random deployment approach for wireless sensor network in agricultural sampling-interpolation applications. However, it is different from pure sampling that the network connectivity has to be maintained in the sensor nodes deployment at the same time.

Firstly, some definitions and the model description in this paper are as follows.

Definition 1: Let $\{x_0, \cdots, x_{n-1}\}$ be the locations of n sensor nodes spread over a 2-D monitoring area.

Definition 2: Let there be m environmental variables to be sampled by each sensor node in the monitoring network.

Definition 3: Each environmental variable has the spatial correlation. It has a sampling range r^i_s, which could be determined according to historical data, experience results or some literature. r_s is defined as the sampling range for the whole network. Then

$$r_s = \min(r_s^i) \quad i = 1, \cdots, m \tag{1}$$

Definition 4: Each active node has a communication range of r_c, and two active nodes can communicate reliably at a distance of r_c or less. In addition, all nodes have the same communication ranges.

Systematic random deployment is designed based on three popular regular patterns including an equilateral triangle pattern, a square pattern and a regular hexagonal pattern (shown in Fig.1). We need to address the following two questions. (1) How to determine the optimal grid size for these systematic random deployment patterns? (2) Which is the best pattern on the efficiency among these systematic random deployment patterns? We explained the first question in Section 3 and the second in Section 4.

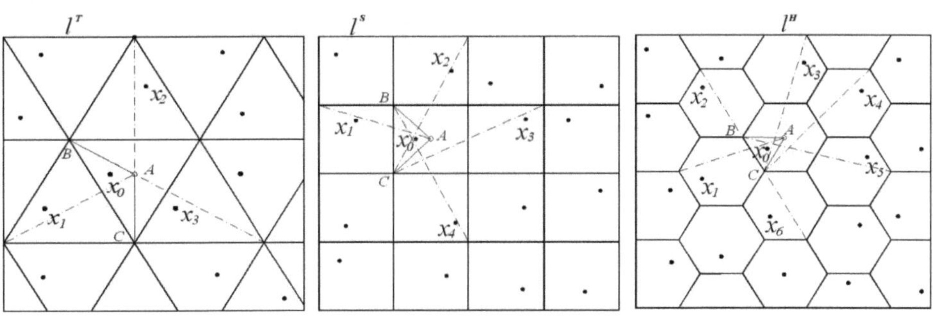

(a) An equilateral triangle pattern (b) A square pattern (c) A regular hexagonal pattern

Fig. 1. Systematic random deployment

3 Optimal Grid Size of Different Deployment Patterns

Let l^T, l^S, l^H denote the grid side length of regular deployment patterns-triangular, square and hexagon. In terms of sampling requirements, the feasible grid size could be determined by the criterion of system sampling patterns-equilateral triangle, square and hexagon. Then,

$$l^T \leq \sqrt{3} r_s \tag{2}$$

$$l^S \leq r_s \tag{3}$$

$$l^H \leq \frac{\sqrt{3}}{3} r_s \tag{4}$$

To achieve connectivity, the communication range r_c must be also considered. In Fig.1a, x_0 is the sensor node with three neighboring nodes x_1, x_2 ,x_3. Let x_0 locate in $\triangle ABC$. Considering connectivity requirement, at least one of the respective distances from x_0 to x_1 ,x_2 ,x_3 is less than r_c. Hence, we must have

$$\begin{cases} \frac{\sqrt{3}}{3} r_c < l^T \le \frac{\sqrt{3}}{2} r_c & \text{at least 1 - } \textit{connectivity} \\ l^T \le \frac{\sqrt{3}}{3} r_c & \text{at least 3 - } \textit{connectivity} \end{cases} \tag{5}$$

Let l^T_{max} denote the maximum grid side length of the equilateral triangle pattern. Using (2) and (5), then

$$l^T_{\max} = \begin{cases} \sqrt{3} \min\left(r_s, \frac{1}{2} r_c \right) & \text{at least 1 - } \textit{connectivity} \\ \sqrt{3} \min\left(r_s, \frac{1}{3} r_c \right) & \text{at least 3 - } \textit{connectivity} \end{cases} \tag{6}$$

Similar calculation could be used for the other two patterns. Let l^S_{max}, l^H_{max} denote the maximum grid side length of the square and hexagon pattern respectively. Hence, we obtain

$$l^S_{\max} = \begin{cases} \min\left(r_s, \frac{\sqrt{10}}{5} r_c \right) & \text{at least 1 - } \textit{connectivity} \\ \min\left(r_s, \frac{\sqrt{5}}{5} r_c \right) & \text{at least 4 - } \textit{connectivity} \end{cases} \tag{7}$$

$$l^H_{\max} = \begin{cases} \min\left(\frac{\sqrt{3}}{3} r_s, \frac{\sqrt{7}}{7} r_c \right) & \text{at least 1 - } \textit{connectivity} \\ \frac{1}{3} \min\left(\sqrt{3} r_s, r_c \right) & \text{at least 3 - } \textit{connectivity} \\ \min\left(\frac{\sqrt{3}}{3} r_s, \frac{\sqrt{13}}{13} r_c \right) & \text{at least 6 - } \textit{connectivity} \end{cases} \tag{8}$$

4 Efficiency Comparison of Different Deployment Patterns

Which is the best pattern on the efficiency among these systematic random deployment patterns? For the same field, the number of sensor nodes could be calculated using the area of grid cell. The bigger the grid cell area, the fewer required nodes.

$A^T_{max}, A^S_{max}, A^H_{max}$ are denoted the maximum area of grid cell in equilateral triangle, square and hexagonal patterns respectively. Then

$$A^T_{\max} = \frac{\sqrt{3}}{4}\left(l^T_{\max}\right)^2 \tag{9}$$

$$A^S_{\max} = \left(l^S_{\max}\right)^2 \tag{10}$$

$$A^H_{\max} = \frac{3\sqrt{3}}{2}\left(l^H_{\max}\right)^2 \tag{11}$$

Using (6) and (9), then

$$A^T_{\max} = \begin{cases} \frac{3\sqrt{3}}{16}(r_s)^2\left[\min(2, \frac{r_c}{r_s})\right]^2 & \text{at least 1 - } \textit{connectivity} \\ \frac{\sqrt{3}}{12}(r_s)^2\left[\min(3, \frac{r_c}{r_s})\right]^2 & \text{at least 3 - } \textit{connectivity} \end{cases} \tag{12}$$

Similar calculation could be used for the other two patterns. Then, we obtain

$$A^S_{\max} = \begin{cases} \frac{2}{5}(r_s)^2\left[\min(\frac{\sqrt{10}}{2}, \frac{r_c}{r_s})\right]^2 & \text{at least 1 - } \textit{connectivity} \\ \frac{1}{5}(r_s)^2\left[\min(\sqrt{5}, \frac{r_c}{r_s})\right]^2 & \text{at least 4 - } \textit{connectivity} \end{cases} \tag{13}$$

$$A^H_{\max} = \begin{cases} \frac{3\sqrt{3}}{14}(r_s)^2\left[\min(\frac{\sqrt{21}}{3}, \frac{r_c}{r_s})\right]^2 & \text{at least 1 - } \textit{connectivity} \\ \frac{\sqrt{3}}{6}(r_s)^2\left[\min(\sqrt{3}, \frac{r_c}{r_s})\right]^2 & \text{at least 3 - } \textit{connectivity} \\ \frac{3\sqrt{3}}{26}(r_s)^2\left[\min(\frac{\sqrt{39}}{3}, \frac{r_c}{r_s})\right]^2 & \text{at least 6 - } \textit{connectivity} \end{cases} \tag{14}$$

Comparing (12), (13) and (14), the maximum area A_{max} of the corresponding patterns is computed for all values of r_c/r_s. Fig.2 shows the results from which we make the following observations.

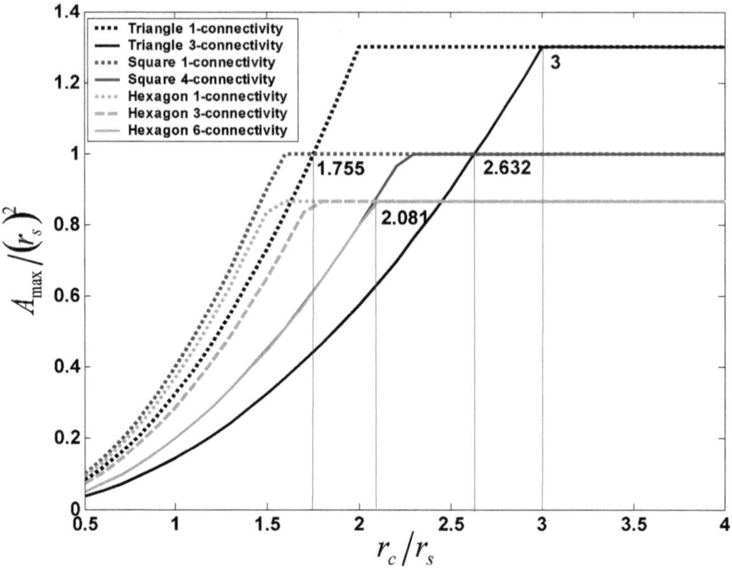

Fig. 2. Efficiency Comparison of different deployment patterns

1. When $r_c/r_s \leq 1.755$, the square based pattern to achieve both coverage and 1-connectivity is optimal because it has the maximum grid area to ensure the best efficiency. While requiring more connectivity, the hexagon pattern which provides coverage and 3-connectivity should be chosen for this range of r_c/r_s.
2. When $1.755 \leq r_c/r_s \leq 2.081$, the triangle based pattern to achieve both coverage and 1-connectivity is optimal because it has the maximum grid area to ensure the best efficiency. If requiring more connectivity, the hexagon pattern which provides coverage and 3-connectivity should be chosen for this range of r_c/r_s.
3. When $2.081 \leq r_c/r_s \leq 2.632$, the triangle based pattern to achieve both coverage and 1-connectivity is optimal because it has the maximum grid area to ensure the best efficiency. If requiring more connectivity, the square pattern which provides coverage and 4-connectivity should be chosen for this range of r_c/r_s.
4. For $r_c/r_s \geq 3$, the triangle based pattern to achieve both coverage and 3-connectivity is optimal because it has the maximum grid area to ensure the best efficiency.

5 Conclusion

In this paper we discussed the problem on the sensor nodes deployment in agricultural sampling-interpolation applications. We proposed systematic random deployment approach to achieve coverage and k-connectivity based on three popular regular patterns-equilateral triangle, square and hexagon. We also compared the efficiency of different regular patterns of systematic random deployment, which enabling to make a better deployment decision for different values of r_c/r_s.

Acknowledgement. This project is supported by National Nature Science Foundation (31101078), Beijing Natural Science Foundation (Connectivity and coverage for wireless sensor network in agricultural applications, 4112013) and Scientific Research Common Program of Beijing Municipal Commission of Education (KM201110028018).

References

1. Zhang, N., Wang, M., Wang, N.: Precision Agriculture-a worldwide overview. Computer and Electronics in Agriculture 36, 113–132 (2002)
2. Wang, N., Zhang, N., Wang, M.: Wireless sensors in agriculture and food industry-Resent development and future perspective. Computer and Electronics in Agriculture 50, 1–14 (2006)
3. Baggio, A.: Wireless sensor networks in precision agriculture. In: Proc. Workshop on Real-World Wireless Sensor Networks (REALWSN 2005), Stockholm, Sweden (June 2005)
4. Camalie Net: Wireless sensor network at Camalie Vineyard. Camalie Vineyards (2006), `http://www.camalie.com`
5. Tilak, S., Abu-Ghazaleh, N., Heinzelman, W.: Infrastructure tradeoffs for sensor networks. In: Proc. of 1st Int'l Workshop on Wireless Sensor Networks and Applications (WSNA 2002), pp. 49–57 (2002)
6. Liu, H., Meng, Z., Shang, Y.: Sensor Nodes Placement for Farmland Envionment Monitoring Applications. In: The 5th Interational Conference Wireless Communications, Networking and Mobile Computing, Beijing, China, September 24-26 (2009)

Using EPIC Model to Determine a Sustainable Potato/Cereal Cropping System in the Arid Region of the Loess Plateau of China

Xuechun Wang[1,*], Jun Li[2], and Shishun Tao[1]

[1] School of Life Science and Technology, Southwest University of Science and Technology, Mianyang, Sichuan, 621010, China;
xuechunwang@swust.edu.cn
[2] College of Agronomy, Northwest A & F University, Yangling, Shaanxi 712100, China

Abstract. Potato has become one of the primary cash crops in the arid region of the Loess Plateau. However, what and how cereal crops to rotate with potato is better for the farmer's income and environment is unclear now. 11 potato based rotation systems were simulated in Yulin, using EPIC model based on local information. Results showed 1) Though mean potato yield decreased from 6.84 to 5.38 t/hm^2, the mean value of farmer's gross income increased from 12807 to 20185 Yuan/hm^2, when the frequency of potato increased from 0.25 to 0.50 for potato based rotation sequence. 2) By comparing, spring maize was better to increase potato yield, while it used more water than foxtail millet and broomcorn millet; Foxtail millet was one of the better preceding crops for potato, and its effects on potato yield was depended on its sequence in potato based rotation system. 3) Comparing with spring maize/potato system, foxtail millet/potato and spring maize/potato/foxtail millet/potato were two better rotation systems to use soil water sustainably and to increase farmer's gross income for a long period in the arid region of the Loess Plateau.

Keywords: The Loess Plateau, Arid region, Potato, Rotation, EPIC model.

Introduction

Potato has become one of the primary cash crops in the arid regions of the Loess Plateau. As a shallow root crop, potato uses less water than other deep root crops [1]. This character increased farmer's interest to plant potato in the arid areas (annual rainfall was not more than 450mm) of the Loess Plateau. Another reason is the great difference of temperature (10-16°C) between day and night is better to accumulate more dry material for potato in the north region of the Loess Plateau. According statistical data [2], potato has been planted more than 2.4×10^5 hm^2 (33% of the grain crop planting area), and contributed about 26.7% of total agricultural cash receipts in Shaaxi Province of China.

* Corresponding author.

D. Li and Y. Chen (Eds.): CCTA 2012, Part II, IFIP AICT 393, pp. 60–68, 2013.

However agricultural technology of potato production still needs to improve in the arid region of the Loess Plateau. Liu [3] and Li [4] reported that potential yield of potato in northern Shaanxi was more than 15 t/hm^2), however actual potato yield in the field was less than 5.9 t/hm^2 [5]. Most reports believed that rotating with cereal crops was one of the better agricultural technologies to increase potato yield [6]. Over the years, increasing costs of acquiring suitable agricultural land and irrigation equipment promoted many farmers to limit rotation of other crops with potatoes. Now, spring maize/potato and broomcorn millet/potato have become the primary potato based agricultural systems in the arid region of the Loess Plateau. However If the present potato-grain rotation system was reasonable, what crop and what sequence of such crop is reasonable and sustainable are still not clear in the arid region of the Loess Plateau.

Field experiment is a better and direct method to determine a sustainable potato-cereal rotation sequence. However, it is a long-term endeavor that is both expensive and time-consuming. With the development of crop model, using computer model to simulate soil water content and crop yield based on local information has become another better approach to answer this question. One such model is the EPIC model which simulates soil water and crop yield simultaneously under the help of its two sub models (growth model and hydrology model). Wang and Li reported that EPIC model can estimate soil water and crop yield well with the new database for the Loess Plateau[1].

Objectives of this paper were to determine a sustainable potato/cereal cropping system in the arid region of the Loess Plateau, and to point out some advices for farmers in the homogeneous areas of the world.

1 Material and Method

1.1 Site Description

Yulin (107.5~111.3° E, 36.9~39.6° N), as a northern city of Shaanxi Province in China, is located at the junction of the Loess Plateau and the Mu Us Desert. It is an arid continental monsoon climate zone and is a representative rain-fed area of China. Its mean annual precipitation is 400mm, with the mean annual temperature of 10°C, and the mean frost-free period of 152 d. Its predominant soil type, used for potato, is Miansha soil.

1.2 Rotation Sequence

11 potato based cropping systems (tale 1) were established based on the following information: 1) Potato was one of the primary cash crops for farmers in Yulin. 2) Foxtail millet and broomcorn millet were the main crops used to build rotation sequence in the arid region of the Loess Plateau. 3) Spring maize was the primary crop to produce food for farmers. Where spring maize/potato and broomcorn millet/potato were two widely practiced potato based agricultural system in Yulin.

Table 1. Information for the simulated potato based rotation system in Yulin

group	Rotation system	Rotation sequence
A	M/M/M/P	spring maize / spring maize / spring maize /potato
	M/B/M/P	spring maize / broomcorn millet/ spring maize /potato
	F/M/B/P	foxtail millet/ spring maize/ broomcorn millet/ potato
	M/F/M/P	spring maize/foxtail millet/ spring maize/potato
B	M/M/P	spring maize / spring maize /potato
	M/F/P	spring maize / foxtail millet /potato
	M/B/P	spring maize /broomcorn millet/potato
	F/B/P	foxtail millet/broomcorn millet/ potato
C	M/P	spring maize /potato
	F/P	foxtail millet /potato
	M/P/F/P	spring maize/ potato / foxtail millet /potato

1.3 Method

1.3.1 Soil Moisture Grade

In this research, based on Soil Desiccation Index (SDI)[7], soil moisture were classified into 6 grades (table 2). To build up a decision table for soil moisture grade, the range of soil water content for each soil moisture grade was calculated using equation 1.

$$MSW = SDI \times (SSM - SWW) + SWW \quad (1)$$

Where MSW is Soil Water Content calculated from SDI, SSM and SWW. SDI is Soil Desiccation Index, SWW is Soil Water content at Wilting point, and SSM is Stable Soil Water content.

Table 2. Decision table of soil moisture grade for Miansha soil in Yulin

Soil moisture grade	Soil water content by weight(SM) (%)	Soil Desiccation Index (SDI)
I	$SM \geq 14.0$	$SDI \geq 1.00$
II	$14.0 > SM \geq 11.5$	$1.00 > SDI \geq 0.75$
III	$12.5 > SM \geq 11.0$	$0.75 > SDI \geq 0.50$
IV	$11.0 > SM \geq 9.5$	$0.50 > SDI \geq 0.25$
V	$9.5 > SM \geq 8.0$	$0.25 > SDI \geq 0.00$
VI	$SM < 8.0$	$SDI < 0.00$

1.3.2 Potato Yield and Gross Income

According calculating method used by State Statistics Bureau of China [2], potato yield was calculated into grain yield using equation 2.

$$Y_{pc}=0.2\times Y_p \tag{2}$$

Where Y_{pc} was the potato yield calculated from fresh tuber yield of potato, Y_p was the fresh tuber yield of potato.

Gross income for different cropping system was calculated by equation 3 based on the average price of spring maize, broomcorn millet, foxtail millet and fresh potato from 2001 to 2010.

$$GI = \sum_{t=1}^{T}\sum_{i=1}^{I}(P_i Y_{it} - C_{it}) \tag{3}$$

Where GI was the gross income, P_i was the average price of crop i from 2001 to 2010; Y_{it} was the yield of crop i in year t; C_{it} was the variable costs associated with crop i in year t. Variable costs of different crops included expense for seed, plowing, fertilizer, seeding, operating machinery.

Effects of preceding crop on potato yield were calculated using equation 4.

$$E_i = \frac{1}{N}\times\sum_{n=1}^{N}\frac{YP_{it} - Y_{pc}}{Y_{pc}} \tag{4}$$

Where N was the times of potato planted after crop i during 1965 to 2000. YP_{it} was the potato yield after crop i in year t. Y_{pc} was the actual mean potato yield in Yulin from 2008 to 2010 (5.9 t/hm^2).

2 Results

2.1 Crop Yield for Different Rotation Cropping Systems

Taking mean yield and gross income into account, MMMP and MFMP were better rotation cropping system within group A . Mean potato yield of MFMP reduced 0.52 t/hm^2 comparing with that of MMMP and increased 0.14 and 1.20 t/hm^2 comparing with that of MBMP and FMBP. Mean maize yield of MFMP was 0.45 t/hm^2 higher than that of FMBP, and was 0.10 and 0.05 t/hm^2 lower than MMMP and MBMP. Mean broomcorn millet yield of MFMP increased 0.46t/hm^2, comparing with that of FMBP. From 1965 to 2000, farmers got more gross income by MFMP rotation system, which increased 9.0% and 16.6% comparing with that by MBMP and FMBP. For MFMP rotation system, 8.4% of gross income produced by potato, which was higher than that of MMMP (6.0%) and MBMP (6.6%) rotation system and was lower than that of FMBP (9.5%).

MMP and MFP generated more yield and more gross income, comparing with other rotation systems within group B. MMP rotation system produced the highest yield of maize (1.66 t/hm^2) and potato (6.89 t/hm^2). From 1965 to 2000, MMP cropping system generated the highest gross income of 16026 Yuan/hm^2, MFP was the second with the value of 15321 Yuan/hm^2. For MMP rotation system, potato contributed 89.7% of the gross income, which was significantly higher than that for MFP (84.7%).

Within group C, MPFP rotation system got the highest yield of maize (1.56 t/hm^2) , foxtail millet (1.89 t/hm^2) and potato (6.00 t/hm^2). From 1965 to 2000, gross income of spring maize, foxtail millet and potato in MPFP cropping system was 580, 1035, 18782 Yuan/hm^2 respectively, which was significantly higher than that in MP and FP cropping system. For MPFP cropping system, potato contributed 92.1% of the gross income, which was significantly higher than that for MP (94.8%) and was significantly lower than that for FP (90.0%).

2.2 Effect of Preceding Crop on Potato Yield

Effect of preceding crop on potato yield was affected by crop species and its rotation sequence in potato based cropping system (table 3). Effect index of spring maize on potato yield was the highest in MMMP rotation system, while was the lowest in MP rotation system. MFP was the best cropping system with the effect index value of 0.054 for foxtail millet to increase potato yield, however, effect index of foxtail millet in FP indicated that foxtail millet had a negative impacts on the production of potato yield. By comparing among all the simulated crops, as a preceding crop, spring maize was better to increase potato yield, while broomcorn millet has a negative impacts on potato yield.

Table 3. Effects of preceding crops on potato yield in different rotation systems in Yulin

Preceding crops	Rotation sequence	Effect index of preceding crop
Spring maize	MMMP	0.286
	MFMP	0.198
	MMP	0.168
	MBMP	0.154
	MP	-0.003
Broomcorn millet	FMBP	-0.005
	FBP	-0.034
	MBP	-0.049
Foxtail millet	MFP	0.054
	MPFP	0.017
	FP	-0.024

2.3 Soil Water for Different Cropping Systems

2.3.1 Available Soil Water for Different Cropping Systems

Different cropping systems have different capacity to use soil water in depth and extension, therefore, the difference of available soil water between different rotation system was significant. Mean available soil water in 0-10m soil were 1016, 1018 and 1034mm for group A, group B and group C respectively. FMBP, FMP and FP was the best rotation system within group A, group B and group C respectively, with the mean available soil water of 1039, 1073 and 1072 mm respectively .

Available soil water in 0-10m soil reduced with the increasing of cultivation years of different rotation cropping systems in Yulin. Comparing with that during 1965-1976, available soil water in 0-10m soil during 1989-2000 reduced 250, 201, 161 and 197mm for MMMP, MBMP, FMBP and MFMP respectively; 193, 187 188 and 40mm for MMP, FMP, MBP and FBP respectively; 200, 47 and 171mm for MP, FP and MPFP respectively. FP was the best cropping system to use soil water sustainably, with the lowest reducing of soil water from 1965 to 2000. FBP was the second and FMBP was the third rotation system, considering the sustainable use of soil water.

2.3.2 Soil Water in Different Soil Layers for Different Cropping Systems

Steady dry soil layers built in 2.0-4.0m soil, and soil moisture changed with the change of rotation sequence and rainfall in different years. Steady dry soil layers built in 2.0-5.0 m soil for MMMP, MMP and MBMP cropping system, in 2.0-3.0 m soil for FP cropping system, in 2.0-4.0m soil for other cropping systems (table 4).

Table 4. Stable dry soil layers and required years for soil moisture grade reach to V in different potato based rotation system in Yulin

Rotation pattern	*Required years for soil moisture grade reached to* V						*Maximum restored soil depth(m)*
	0-0.5 m	0.5-1.0 m	1.0-2.0 m	2.0-3.0 m	3.0-4.0 m	4.0-5.0 m	
MMMP	—	8	5	12	15	26	2-5
MBMP	—	14	7	16	25	32	2-5
FMBP	—	18	12	22	30	—	2-4
MFMP	—	16	11	19	28	—	2-4
MMP	—	10	8	12	16	28	2-5
MFP	—	13	12	18	26	—	2-4

Table 4. *(Continued)*

MBP	—	15	10	17	24	—	2-4
FBP	—	18	13	18	24	—	2-4
MP	—	16	12	19	23	—	2-4
FP	—	24	17	27	—	—	2-3
MPFP	1	18	15	18	28	—	2-4

Crop specie was an important factor to impact the change of soil water for potato/cereal rotation system. Simulation results showed the depletion of soil water in spring maize field was maximum, in broomcorn millet field was the second, in foxtail millet field and potato field was the following. Soil water reduced in 0-1.5m soil when potato or foxtail millet was planted, and it reduced in 0-2m soil when spring maize was planted.

Rain fall was another factor to influence the changing of soil water in potato based rotation field. When annual rainfall was less than 200mm, dry soil layer developed into the deeper soil both in spring maize field and in broomcorn millet field. When annual rainfall was more than 400mm, soil water recovered in 0-1.5m soil both in foxtail millet field and in potato field. When annual rainfall was more than 500mm, soil water recovered in 0-1.0m soil both in spring maize field and in broomcorn millet field.

3 Discussion

Rotation system can help break pest cycles, reduce pesticide use, and help control weeds [8]. In addition, rotation systems that include cereals can help improve soil organic matter level, thereby improve soil structure and productivity [9]. Results of this paper showed potato yield increased with the reduction of potato frequency in rotation system. Mean potato yield of group A was the highest with the value of 6.84t/hm^2, group C was the lowest with the value of 5.38 t/hm^2. Jatoe et al. reported that rotating other crops with potato reduced farm returns above variable costs in potato based rotation system [6]. Our results indicated that farmer's gross income increased, when the frequency of potato decreased in the potato based rotation sequence. Mean gross income of group A was the highest, while group C was the lowest.

Depth and distribution character of root in soil results the different capacity of different crops to use soil water, and therefore results the different ability of different potato based rotation system to use soil water [10]. Results of simulated soil water showed, within group B, available soil water of FBP was the highest, with the value of 1073mm; MFP was the second (1010mm), and MBP was the third (1008mm), MMP was the lowest (981mm). Considering only from sustainable use of soil water, foxtail millet was the best crop to rotate with potato, and Broomcorn millet was the second. However table 3 showed, as a preceding crop, spring maize was the best crop, and foxtail millet was the second to increase potato yield. This may be a result of nutrition cycle, pest cycle and fungi development from preceding crops to potato within a rotation sequence [11].

To determine whether one rotation system was a best and sustainable cropping system, economic factor and environmental factor must be considered simultaneously. Soil water, was one of the key environmental factors to influence the sustainable development of agriculture in the arid regions of the Loess Plateau [12]. Jatoe et al. believed that rotation length, rotation sequence and crop specie together results the effect of rotation on environment and famer's income of potato based agricultural system [6]. Results of this paper showed, 1) FP and MPFP rotation system was better to produce more food for the farmer; 2) MP, FP and MPFP was better to increase farmer's gross income; 3) MP, FP, MPFP and FBP was better to utilize soil water sustainably. Considering from food production, sustainable utilization of soil water and gross income together, FP and MPFP rotation system was the better potato based rotation system for famers in the arid region of the Loess Plateau.

4 Conclusion

1) By comparing, spring maize was better to increase potato yield, however it used more water than foxtail millet and broomcorn millet; Foxtail millet was better to sustainable use of soil water, however its impacts on potato yield were depended on its sequence in potato based rotation system.
2) Spring maize/potato was not a sustainable potato based rotation system in the arid region of the Loess Plateau, foxtail millet/potato and spring maize/potato/foxtail millet/potato were two better rotation systems to use soil water sustainably and to increase farmer's gross income for a long period.

Acknowledgement. This study was sponsored by the Foundation of SWUST university for new teachers with doctor degree (No. 11zx7140). We are grateful to Wu-Qiang Li who had provided many information about potato production in the northern region of Shaanxi Province.

References

1. Wang, X.C., Li, J.: Evaluation of crop yield and soil water estimates using the EPIC model for the Loess Plateau of China. Mathematical and Computer Modeling 51, 1390–1397 (2010)
2. Statistics Bureau of Shaanxi Province: Statistical yearbook of Shaanxi Province of 2011, pp. 314–336. China Statistic Press, Beijing (2012)
3. Liu, P.S., Jia, Z.K., Li, J., Wang, J.P., Han, Q.F.: Effects of different alfalfa-crop rotation patterns on water recovery of degradation alfalfa grassland. Transactions of the CSAE 26(2), 95–102 (2010)
4. Li, J., Wang, L.X., Shao, M.A., Fan, T.L.: Simulation of autumn grain crop productive potential on Loess Plateau region of China. Journal of Northwest Sci.-Tech. University of Agriculture and Forest (Nature Science Edition) 29(5), 56–60 (2001)

5. Wang, X.L., Chen, M.C., Li, F.M., Li, Y.J.: Effects of additional water and increased soil temperature on potato yields in Plastic-covered Ridge Rainwater-harvesting system. Agricultural Research in the Arid Area 25(3), 11–16 (2007)
6. Jatoe, J.B.D., Emmanuel, K.Y., Alfons, W., Clark, J.S.: Economic and environmental impacts of introducing land use policies and rotations on Prince Edward Island potato farms. Land Use Policy 25(3), 309–319 (2008)
7. Li, J., Chen, B., Li, X.F., Zhao, Y.J., Ciren, Y.J., Hu, W., Jiang, B., Cheng, J.M., Shao, M.A.: Effects of deep soil desiccation on artificial forestlands in different vegetation zones on the Loess Plateau of China. Acta Ecologica Sinica 28(4), 1429–1445 (2008)
8. Edward, B., Robert, P.L., Stellos, T., Erich, S.M., Andrei, A., Gary, S., Andrew, L., Serena, D.G.: Compost, rapeseed rotation and bio-control agents significantly impact soil microbial communities in organic and conventional potato production systems. Applied Soil Ecology 52, 29–41 (2012)
9. Nelson, K.L., Lynch, D.H., Boiteau, G.: Assessment of changes in soil health throughout organic potato rotation sequences. Agriculture, Ecosystems and Environment 131, 220–228 (2009)
10. Xiang, H.T., Feng, N.J., Du, J.D., Ma, G.S., Liu, B., Zheng, D.F.: Effects of plant growth regulators on physiological and biochemical characteristics of potato roots. Plant Nutrition and Fertilizer Science 15(6), 1481–1485 (2009)
11. Peters, R.D., Sturz, A.V., Carter, M.R., Sanderson, J.B.: Developing disease-suppressive soils through crop rotation and tillage management practices. Soil & Tillage Research 72, 181–192 (2003)
12. Chen, L.D., Huang, Z.L., Gong, J., Fu, B.J., Huang, Y.L.: The effect of land cover/vegetation on soil water dynamic in the hilly area of the loess plateau, China. Catena 70(2), 200–208 (2007)

Using Sequential Gaussian Simulation to Assess Geochemical Anomaly Areas of Lead Element

Fengrui Chen[1], Shiqiang Chen[2], and Guangxiong Peng[3,4,*]

[1] Key Laboratory of Geospatial Technology for the Middle and Lower Yellow River Regions, Ministry of Education, Henan University, Kaifeng, China
[2] The Center for Yellow River Civilization and Sustainable Development of Henan University, Kaifeng, China
[3] School of Geosciences and Info-Physics, Key Laboratory of Metallogenic Prediction of Nonferrous Metals, Ministry of Education, Central South University, Chansha, China
[4] Qinghai Institute of Geological Survey, Xining, China
fruich@gmail.com, pgxcsu@csu.edu.cn

Abstract. Mapping geochemical anomaly is essential for prospecting. Kriging is often used to characterize the spatial variability of geochemistry. However, due to its smooth effects, it is incapable of detecting multi-location uncertainty of geochemistry estimate. Sequential gaussian simulation (SGS) can model the spatial uncertainty through generation of several equally probable stochastic realizations. In this study, SGS is used to map Pb spatial distribution and to provide a quantitative measure of its spatial uncertainty in particular. The results show that the spatial pattern of Pb is well recognized using the SGS. The different among realizations is used to provide a quantitative measure of spatial uncertainty of Pb estimate. Knowledge of spatial uncertainty is helpful to evaluate anormal areas of Pb element.

Keywords: geochemical anomaly, geostatistics, sequential Gaussian simulation.

1 Introduction

With the development of exploration geochemistry, the evaluation of regional geochemical anomaly such as regional stream and secondary halo attract more and more attention[1]. How to evaluate effectively the geochemical anomaly has been the focus for many exploration geochemists[2]. For geochemical elements carry some mineralization information. The fine spatial analysis of elemental geochemistry can reveals the tracing of mineralization directly or indirectly. The study of geochemical element distribution is one of the important ways to reveal the spatial variation, mineralization and enrichment of element[3, 4]. Geochemical anomaly analysis is the important work for prospecting, whose core content is to find, filtrate and evaluate anomaly.

---

* Corresponding author.

D. Li and Y. Chen (Eds.): CCTA 2012, Part II, IFIP AICT 393, pp. 69–76, 2013.

For better understanding of geochemical anomaly, it is very important to predict the spatial distribution of geochemistry over the study area. Geostatistics has been used for spatial variability characterization and prediction of geochemistry. The most well-known geostatistics estimation technique is ordinary kriging (OK) called also the "best linear unbiased estimator" [5, 6]. However, the smoothing effect of OK results in less variation in the estimated values than in the observed values [7-9]. The estimation at each unsampled location includes an estimation error or uncertainty. This uncertainty should be quantified and accounted for in subsequent decision-making. For geochemical anomaly analysis, assessing the reliability of delineating an area as "normal" or "anomaly" requires the knowledge of estimation uncertainty at several locations simultaneously. Geostatistics simulation has the ability to not only predict the spatial distribution of the property at hand but also to assess both local and spatial uncertainty. The main objective of this study is to assess the uncertainty of delineating geochemical anomaly. For this the feasibility of using sequential Gaussian simulation (SGS) [10-12]is examined.

2 Materials and Method

2.1 Study Area and Data Source

The study area is located in the Mohailaheng Pb-Zn deposit of Zaduo County, Qinghai Province, China. It is located in 95° 33′ - 95° 59′ E and 32° 39′ -32° 58′ N, near the North of Sanjiang metallogenic belt (Fig. 1). Numberous of Pb- Zn-Ag-Cu ore deposits with significant economic values occur in Lanping basin and Yushu and Tuotuohe areas on the eastern and northern margins of Tibetan orogenic belt. Being hosted by sediments and having no obvious affinity to igneous activities, these deposits occur in continent-continent (Indo-Asian) collisional orogen and are controlled by thrust-nappe structures, thus obviously different from the well known sediments-hosted base metal deposits in the world[13]. They were formed in Tertiary foreland basins developed from Paleo-Tethyan rifting basins and Mesozoic depression basins. During mineralization, thrust-nappe structures juxtaposed Mesozoic strata over

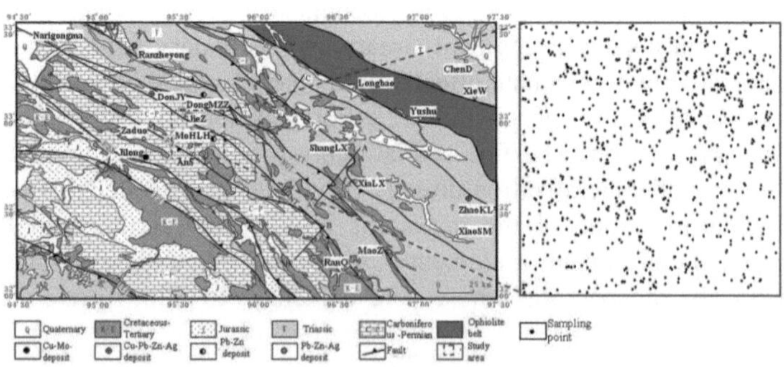

Fig. 1. Geological map of south Qinghai and the spatial distribution of sampling points

Tertiary strata through a series of parallel reverse/ thrust faults in the basins. The deposits are strictly controlled by thrust-nappe structures. The rock types of magmatic are complete with frequent magmatic activity, while the exposed area is small with scattered distribution. Sampling points get from geochemical data of stream sedimentation in Yushu region of South Qinghai in 2009. This data set includes 864 Sampling points with 13 elements of Ag, As, Au, Ba, Bi, Cd, Cu, Hg, Mn, Mo, Pb, Sb, Sn, W and Zn.

2.2 Kriging

Kriging is a linear interpolation procedure that provides a best linear unbiased estimation for quantities, which varies in space, Kriging estimates are calculated as weighted sums of the adjacent sampled concentrations[14, 15]. That is, if data appear to be highly continuous in space, the points closer to those estimated receive higher weights than those farther away.

In geostatistical analysis, the semivariogram is used to quantify the spatial dependence of the Pb concentration:

$$r(h) = \frac{1}{2N(h)} \sum_{i=1}^{N(h)} [z(x_i) - z(x_i + h)]^2 \tag{1}$$

where h is the distance between both locations, x_i and x_i+h, and N(h) is the number of pairs for x_i and x_i+h. The experimental semivariogram was then fitted with a suitable theoretical model: spherical, exponential or gaussian. The models provide information about the spatial structure as well as the input parameters for kriging interpolation. Kriging is considered as the optimal spatial interpolation for making best linear unbiased estimation of regionalized variables at unknown locations. The spatial prediction of the value of Pb at an unknown point is calculated as a weighted average:

$$z^*(x_0) = \sum_{i=1}^{n} \lambda_i z(x_i) \tag{2}$$

where $z^*(x_0)$ is the value to be estimated at the location x_0; $z(x_i)$ is the known value at the sampling site x_i and λ_i is the weight. There are n sites within the search neighborhood around x_0 used for estimation, and the magnitude of n will depend on the size of the moving search window and on user's definition. Kriging differs from other methods, such as inverse distance weighted, in that the weight function λ_i is no longer arbitrary, and is calculated from the parameters of the fitted variogram model with unbiasedness and minimized estimation variance for interpolation.

2.3 Sequential Gaussian Simulation

The basic idea of sequential simulation is that the conditional distribution of the observed variable can be used for the simulation of subsequent grid points. The SGS is based on multi-Gaussianary assumption of a random function model[16]. The data requires then a prior normal score transformation to ensure the normality of at least the univariate distribution of data. The normal score transformation[5] for a continuous variable z at location x_a, $_a$=1, ..., n, is given as:

$$y(x_a) = G^{-1}\left[F^*(z(x_a))\right] \quad a = 1, \cdots, n \tag{3}$$

where $G^{-1}(.)$ is the inverse Gaussian cumulative distribution function (CDF) of the random function Y(x), and F* is the sample CDF of z. A back transformation of the normal scores to the original space is achieved by applying the inverse of the normal score transform introduced in equation (3).

The SGS consists in defining a regularly spaced grid, covering the region of interest and establishing a random path through all grid nodes, such that each nodes is visited only once in each sequence. For every grid node the normal score conditional cumulative distribution function (CCDF) is estimated according to a simple kriging system conditioned to all Pb known data and previously simulated grid nodes within a neighborhood of the location to be simulated. In this stage the semivariogram model of Pb normal scores is used. Afterwards a simulated value is drawn from the estimated CCDF and then added to the conditioning data set to be used for simulating other grid nodes. This procedure will be continued until all the grid nodes are simulated. The simulated normal score CCDFs then need to be back transformed to obtain the CCDFs of Pb in original space. These sequential steps build up only the first realization, which is only one of equally probable models of Pb spatial distribution. To generate multiple realizations, this sequence should be repeated with different random paths passing through all nodes for each realization. Indeed the only difference among the several individual simulations is the random-number seed, which initiates the simulation process and affects the random drawing of a value from the CCDF used to generate each simulated value. Therefore, in different realizations different values will be assigned to the same location. Pixel-by-pixel averaging of generated maps of Pb spatial distribution provides the so-called E-type[17] estimate at each grid node. The corresponding conditional variance, which represents the spread of CCDF and the probability of exceeding or not exceeding a critical threshold of Pb at each grid node also can be produced through post processing.

3 Results and Discussions

3.1 Descriptive Statistics and Spatial Autocorrelation Analysis

Descriptive statistics and the histogram of Pb observations are presented in Table 1 and Fig. 2, respectively. The study area has a large range of Pb element, and the maximum and minimum Pb elements are 6.2 and 944.2 g/t. The skewness of Pb observations is 14.15, which indicates that the data is anormal distribution. Therefore, before the implementation of kriging, the Pb data should be transformed. The skewness of ln(Pb) is 0.8, and the ln(Pb) data are almost normal distribution. In Fig. 2(b), the Pb element data have been transformed by a normal score transformation. The statistics (i.e. the mean value of 0 and the variance of 1) of transformed data check the correctness of the transformation.

Table 1. Summary statistics of Pb data and its transformation

	mean	*STD*	*variance*	*maximum*	*minimum*	**skewness**
Pb	31.9	46.3	2144.4	6.2	944.2	14.15
Normal score of Pb	0	1	1	3.24	3.24	0
ln(Pb)	1.4	0.224	0.05	0.79	2.98	0.68

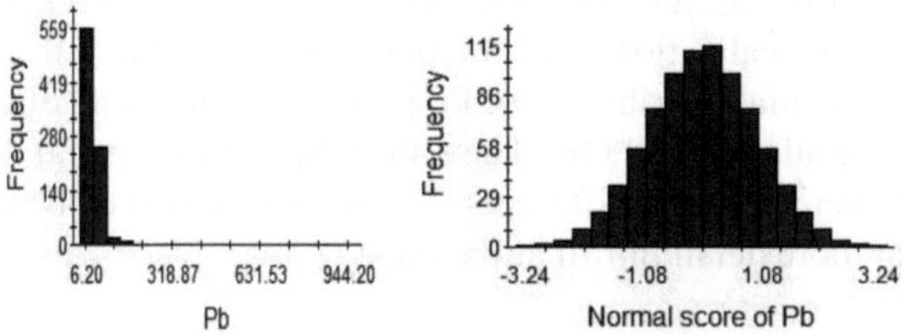

Fig. 2. Histogram of the raw and normal score transformed data for Pb

The experimental semivariogram of ln() and normal transformed data are presented in Table 2 and Fig. 3. Also plotted on each semivariogram there is the best-fitted model to the experimental points. As it is shown in Fig.3, the isotropic semivariogram of normal score of Pb follows an exponential structure with a nugget effect of 0.29, and it reaches to a sill of 0.70 at a range of 4125 m. Compared with the semivariogram model of normal score of Pb, the semivariogram of ln(Pb) follows again an exponential

Table 2. The semivariogram analysis of ln() and normal score transformed data for Pb

	model	*nugget*	*range*	*sill*	**partial sill/sill**
Normal score of Pb	Exponential	0.29	4125	0.79	0.63
ln(Pb)	Exponential	0.01	3084	0.039	0.74

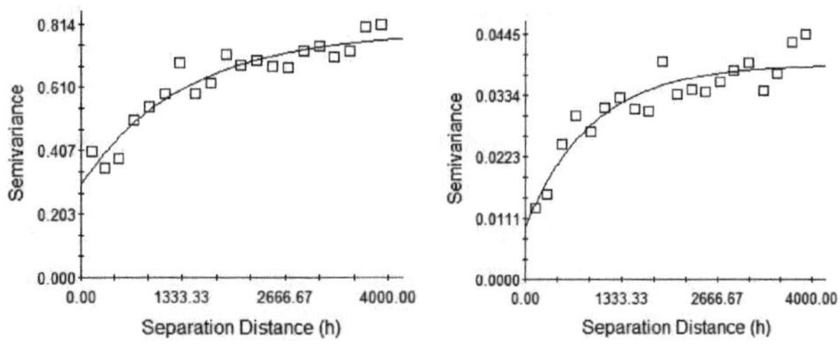

Fig. 3. Experimental semivariogram and best fitted model of the ln() and normal score transformed data for Pb

structure, and have a more or less continuity at space (3084). However, its partial sill/sill (0.74) is larger than that of former (0.63), which indicates that ln(Pb) has more spatial correlation.

3.2 Simulation Result

Three hundred realizations with equal probability of Pb spatial distribution are generated using SGS on a 30-m grid over the study area. Simulation is performed using the simple kriging estimator, and the semivariogram model of Pb normal scores. The four randomly select realizations are display in Fig.4 (a)-(d). Each realization represents a realistic spatial distribution of Pb without a smoothing effect. Fig. 4(e) and (f) is the E-type of simulation and OK. They all indicated the spatial distribution of Pb element; however, there are a lot differences among them. It is evident that simulation realizations present more detail information than OK map. The maps of E-type and OK are smoother.

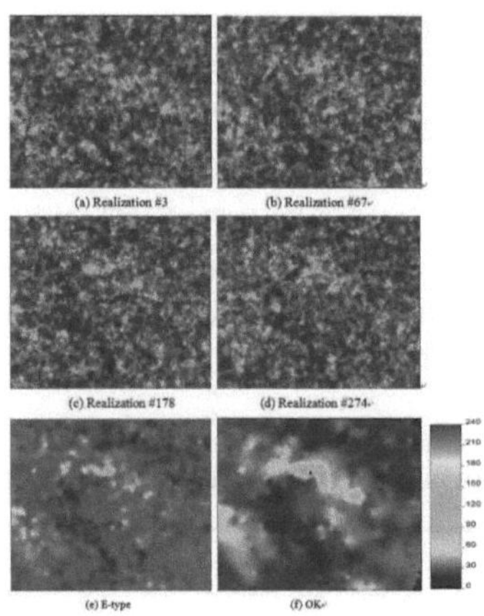

Fig. 4. The spatial distribution of Pb by sequential Gaussian simulation and OK over the study area

Table 3 represents summary statistics on Pb output result by both SGS and OK. SGS individual realizations well preserve the range of variation of the measured Pb data compared to OK map where the minimum value is overestimated while the maximum value is underestimated. This reveals the smoothing effect, a typical property of kriging. This is also evident from the kriging variance, which is much smaller than the actual variance. OK dose not reflect the true variability, and hence is not appropriate for the underlying outlook to this study. Summary statistics of E-type are similar to those for the kriging map. This is an expected result in the sense that it is well-know that when the number of realization is large enough, the mean of the realizations converges to the estimation result.

Table 3. Basic statistics of Pb output results

	mean	*minimum*	*maximum*	**variance**
Actual	31.9	6.2	944.2	2144.4
Realization#3	32.1	6.2	944.2	2205.1
Realization#67	31.6	6.2	944.2	2123.7
Realization#178	31.5	6.2	944.2	2189.4
Realization#274	32.3	6.2	944.2	2167.6
E-type	31.5	9.1	683.7	1653.4
OK	32.2	9.1	723.4	1878.3

As mentioned before, the critical value is 124.5 g/t, which is determined by mean plus two times STD for Pb data. The OK estimation and Probability map of exceeding critical value is in Fig. 5. The OK map is the unique value, therefore, it critical area for OK map is certainty. However, there must be some degree of uncertainty for estimation. For example, an estimated value in OK map is larger than critical value; if it has a smeller probability of occur (0.001), then this value can be assigned to critical area. SGS can solve this problem, it provide the uncertainty of estimated value by all simulation realizations. From Fig. 5(b), the critical area is not certainty, and it is the represent of probability. The larger the value, the higher it belong to critical area. The critical area can be determined by setting a certain threshold which can take into account the situation of economic, social and management.

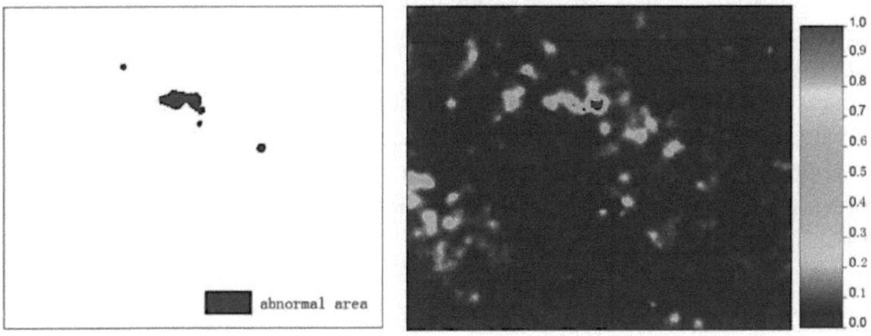

(a) Abnormal area of OK estimation (b) Probability map of SGS

Fig. 5. OK estimation and probability map of exceeding 124.5 g/t for Pb

4 Conclusion

Mapping geochemical anomaly is essential for prospecting. Traditional methods often make use of the interpolation results to attain anomalistic areas, without considering the uncertainty of estimate. In this study, SGS is used to evaluate the anomalistic areas of Pb element. The simulated maps are summarized to provide the probability map of Pb exceeding the critical value. The probability map is used to delineate the anomalistic areas. The difference among 300 realizations provides a quantitative measure of spatial uncertainty of Pb estimate. Eventually SGS is recommended as a powerful and convenient simulation-based algorithm for not only spatial prediction but also for evaluating both local and spatial uncertainty, which should be incorporated in many decision-making processes.

References

1. Ruiting, W.: Facts and problems of the evaluation of regional geochemical anomalies. Chinese Geology 23, 168–175 (2005)
2. Xueqiu, W.: Exploration geochemistry: past achievements and future challenges. Earth Science Frontiers 10, 239–248 (2003)
3. Shuyun, X., Zhengyu, B.: The method of moments and its application to the study of mineralization in Shao guan district, north Guangdong, China. Journal of Jilin University: Earth Science Edition 33, 443–448 (2003)
4. Qiuming, C.: Multifractal theory and geochemical element distribution pattern. EarthScience. Journal of China University of Geosciences 25, 311–318 (2000)
5. Goovaerts, P.: Geostatistics for natural resources evaluation. Oxford University Press, New York (1997)
6. Isaaks, E., Srivastava, R.: An introduction to applied geostatistics. Oxford University Press, New York (1989)
7. Pardo-Iguzquiza, E., Atkinson, P.: Modelling the semivariograms and cross-semivariograms required in downscaling cokriging by numerical convolution econvolution. Computers and Geosciences 33, 1273–1284 (2007)
8. Kyriakidis, P.: A Geostatistical Framework for Area-to-Point Spatial Interpolation. Geographical Analysis 36, 259–290 (2004)
9. Goovaerts, P.: Kriging and semivariogram deconvolution in the presence of irregular geographical units. Mathematical Geology 40, 101–128 (2008)
10. Castrignan, A., Buttafuoco, G.: Geostatistical stochastic simulation of soil water content in a forested area of south Italy. Biosystems Engineering 87, 257–266 (2004)
11. Dimitrakopoulos, R., Luo, X.: Generalized sequential Gaussian simulation on group size *v* and screen-effect approximations for large field simulations. Mathematical Geology 36, 567–591 (2004)
12. Delbari, M.: Using sequential Gaussian simulation to assess the field-scale spatial uncertainty of soil water content. Catena 79, 163–169 (2009)
13. Zengqian, H.: Thrust-controlled, sediments-hosted Pb-Zn-Ag-Cu deposits in eastern and northern margins of Tibetan orogenic belt. Geological features and tectonic model. Mineral Deposits 27, 123–145 (2008)
14. Webster, R., Oliver, M.: Geostatistics for environmental scientists. John Wiley & Sons Inc., New York (2007)
15. Wackernagel, H.: Multivariate geostatistics: an introduction with applications. Springer, Berlin (2003)
16. Goovaerts, P.: Geostatistics in soil science: state-of-the-art and perspectives. Geoderma 89, 1–45 (1999)
17. Deutsch, C., Journel, A.: GSLIB: Geostatistical software library and user's guide. Oxford University Press, New York (1998)

Study on Bee Product Quality Control Chain Based on Agent

Yue E[1], YePing Zhu[1], and YongSheng Cao[2]

[1] Research Institute of Agriculture Information, Chinese Academy of Agricultural Science, Key Laboratory of Agri-information Service Technology, Ministry of Agriculture, 12, ZhongGuanCunNanDaJie, Beijing, China, 100081
eyue@mail.caas.net.cn
[2] Institute of Crop Science, Chinese Academy of Agricultural Science, 12, ZhongGuanCunNanDaJie, Beijing, China, 100081

Abstract. In China, the bee farms are only organized loosely, so human factor has a great impact on bee product quality (BPQ). At the same time, the quality control methods related to information and intelligence technology did not give a useful support on BPQ. As a result, it is not easy control BPQ during production process of bee products. Therefore, how to control BPQ safety effectively always is a hot topic in BPQ control research field. The BPQ control chain is proposed in this paper. BPQ control and emergency mechanism is designed based on Agent technology, and the Agent logical architecture, the data mapping and conversion model, work data flow are detailed. In the architecture, the mapping relationship of Agent and BPQ control objects and attributes is defined. The strategies of control BPQ are also proposed.

Keywords: Bee products, Agent, Coordination, BPQ control chain.

1 Introduction

In China, Bees industry and bee products output is one of the traditional and the unique industry. At the same time, we should also be awareness of the dynamic and complex bee products industrial chain exist some adverse factors, directly or indirectly caused many problems of BPQ, and affect China's export of bee products and domestic market efficiency. For example, In 2002 the European Union and some countries to our country to stop the honey sales into the trade barriers, to our country bee products caused a loss, affect the normal development of the our country bee products industrial. These events cause supervision departments of BPQ high attention concern (Zhao Jing, 2011).

One of the unquestionable the situation is: There are many influence factors about BPQ in China, include production factors, environmental factors, human factors and so on, directly led to the management control difficulties. From the perspective of industry, bee industry is information degree lowest industry in China, popularization of computer and network is still relatively low in bee production area and production

D. Li and Y. Chen (Eds.): CCTA 2012, Part II, IFIP AICT 393, pp. 77–83, 2013.

enterprise, production process of data and files often lost or none at all. At the same time, in the BPQ control need to coordinate unit very much, operation and maintenance costs expensive. With China's bees industry quality improvement and the rapid development of modern information technology, we fully capable of using modern information technology to development of bee quality control system.

2 Bee Products Quality Control

According to the BPQ control, because of a long chain of bee products, there are many influence factors of the BPQ, so BPQ safety management and control needs further research in China.

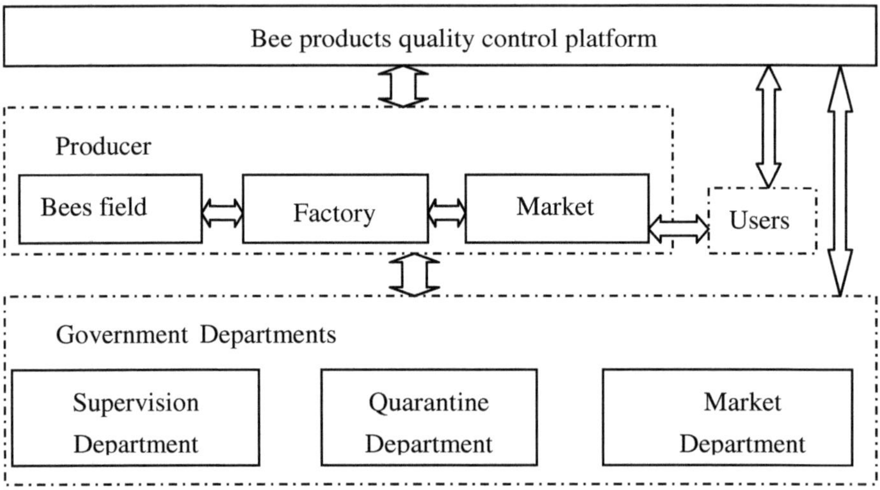

Fig. 1. BPQ control platform

From the graph, we can look at: bee industrial chain structure is complex, the main link includes: production links (bee field), purchase link (acquisition business), processing link (factory) and sales link (supermarkets, shops), and so on. Each link has its own function, between each link on each other, form complex open, distributed system. The bee products production and consumption system tends to be complicated and internationalization, any link to appear BPQ problem caused by the quality and safety of bee problems and harm consumers. Complex structure and characteristics of the system produce great resistance about BPQ control problem solving.

The influence factors on the BPQ, also have technical, economic level, also have the exogenous factors or endogenous system of joint action. These factors are in different levels influence the quality of bee. Therefore, we must first have a clear understanding of factors influencing the quality of bee products, and then we can research the quality control system of bee products.

(1) From the point of view of breeding bees,, there are many feed materials in the process of the bees feeding, such as sugar and honey, the quality of the bee medicine, bees epidemic prevention, and related the hygiene of instruments, water quality, air quality, soil quality, keepers feeding technology skills and so on, all of this influence the quality and safety of bee products and harm consumers health.

(2) From the monitoring, Bee product quality control method and risk monitoring means of the government must improve in bee product quality control chain.

(3) From bee industrial chain observation, chain node is various. It included production links (bee field), purchase link (acquisition business), processing link (factory) and sales link (supermarkets, shops). In the whole bee industry chain, have linear nodes, also have mesh nodes, the whole process complex and coordinate and management difficult.

Through the in-depth analysis, it is found that bee products traceability system could realize the problems of product recall with modern information technology, but still has the following problems remain to be solved: the quality and safety of bee products risk analysis to be further strengthen, while an bee products harm accident, we must scientific evaluation and carry out emergency measures. Therefore, in view of the current BPQ control, it could risk diagnosis and security warning before the event, harm emergency when event occurs, and assessment after the event. This paper proposes the BPQ control based on Agent and spread of related research work.

3 Bee Product Quality Control Chain Based on Agent

The BPQ control has the characteristics of complex, the quality safety emergency occurrence, development, change also has a certain dynamic, many event of evolution characteristics (Zhang Dong Ling, 2009). When a certain level or links occur BPQ events, it must affect other links different physical entity operation and production with the events associate and product a chain reaction. That is means the BPQ emergencies event are not completely isolated in the BPQ control chain, the event has direct contact and may have an impact with other links of BPQ control chain.

The BPQ control chain based on the Agent, according to the function is divided into quality monitoring layer, information analysis layer, emergency command layer, emergency directors and communication coordination of five levels.

(1)The monitoring layer responsible for monitor the specific monitoring objects, monitoring factors related information data acquisition, and evaluation according to users set parameters. In the front end of BPQ control chain, mainly complete information collection and processing, equivalent to a quality observatory of nature. Monitoring layer Agent called monitoring Agent, such as the production monitoring Agent, sales monitor Agent and so on.

(2)Analysis layer is mainly responsible for accept data from monitoring layer in the Agent, and according to the types of data, nature, carries on the analysis, the assessment and to determine the detection of links (or critical control point) presence quality safe

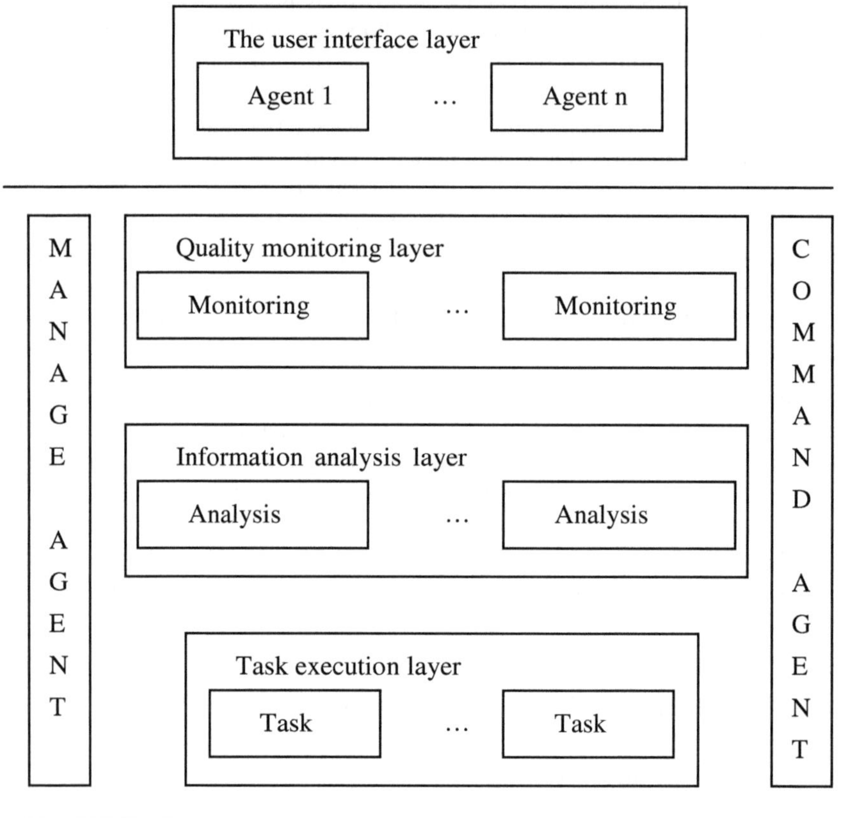

Fig. 2. BPQ control group based on Agent

hidden trouble, whether serious harm (determine harm degree level), whether start quality emergency measures, etc. Analysis layer also contains morel analysis Agent, such as abnormal data analysis Agent, harm rating Agent and so on. It plays the role of quality evaluation experts.

(3)Command layer responsible for the concrete implementation of BPQ emergency disposal measures, and control related executed Agent. Usually, a task for emergency disposition, only one emergency command Agent. Specific tasks including the emergency disposal scheme determination, task decomposition, coordination strategy formulation, implementation and the monitoring of the implementation of the Agent. Command layer is central nervous of BPQ control chain.

(4)Execution layer mainly responsible for specific emergency measures of performance, such as quality tracking, accident liability sure, release emergency report and so on. Execution layer contains different function executive Agent, according to emergency plan, independently run to finish the task of distribution, and will be submitted to the related information of management department, direct responsibility department, etc.

(5)Communication coordination layer is mainly responsible for control interaction communication between all of Agent. For each one of the specific task, it can product a communication Agent. Communication Agent is equivalent to a station, the relevant data of the emergency response, information, task, and the results will be submitted to the communication coordination Agent, it is equivalent to a resource sharing platform, will be responsible for the upload information issued.

Because other related links in the BPQ control based on Agent has similarities, therefore, mainly in bee products production process as an example, this paper describes a BPQ control process based on Agent.

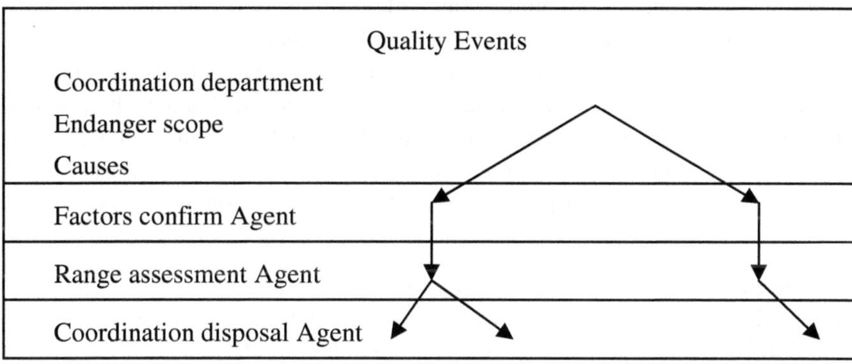

Fig. 3. BPQ Emergency process based on Agent

(1) The user input corresponding critical control points (elements) parameters, such as bee drug.

(2) And then the system automatically start the corresponding quality control Agent, forecast Agent and assessment Agent, input data parallel detection processing.

(3) If the value in safety assessment scope, on the one hand, to users test information; on the other hand, do not product the alarm behavior.

(4) The system of the related Agent to be automatic cancellation, or place to hang up state; the user testing the loop ends.

(5) If, in (2) executive process, found the quality safety problems, immediately start emergency Agent.

(6) According to the quality safety emergency Agent event situation, on the one hand, to convey quality event information users; On the one hand, to be automatic or user input of the related information of physical objects, such as the other bees fields, the business, the processing factory, sales, pharmaceutical factory and so on.

(7) Start decision-making Agent, evaluation Agent and Quality safety evaluation of other aspects.

(8) To the corresponding inspection institution, supervision department report analysis and test emergency report.

(9) To the user Agent submits information.

The BPQ control chain contains two aspects: the earlier evidence collection, after the emergency period. In the system structure based on Agent, the earlier evidence collection main application of related business process analysis and simulation based on Agent. After the emergency period main application for emergency treatment measures harm the analysis, communication and rehabilitation evaluation based on Agent.

4 Conclusion

The BPQ control chain based on the Agent is an organic whole include relevant organization, personnel, communication, information and so on. It is provide decision-making information, knowledge for the BPQ management and control. In the intelligent decision-making system, based on each link collection data related information, use of computer, Agent, network information technology help managers monitor BPQ, once found BPQ safety problems, through the coordinated decision to de whether start the emergency response mechanism and how to choose a specific emergency plan measures for BPQ control of the effective implementation.

Acknowledgments. This research was supported by National Scientific and Technical Supporting Programs Funded by Ministry of Science and Technology of China (nyhyzx07-041). National Natural Science Foundation about Agent-based quality control of agricultural products (60972154).

References

1. Badjonski, M., Ivanovic, M.: A multi-agent system for the determination of optimal hybrids in crop production. Computers and Electronics in Agriculture 25(3), 233–243 (2000)
2. Bentham, M.J.: Farm Smart 2000: A multi-agent decision support system for crop production. In: 7th International Conference on Computers in Agriculture, pp. 469–479 (1998)
3. Parrott, L., Lacroix, R., Wade, K.M.: Design considerations for the implementation of multi-agent systems in the dairy industry. Computers and Electronics in Agriculture 38(2), 79–98 (2003)
4. Ossowski, S., Klugl, F., Bazzan, A.L.C.: Agents in traffic and transportation Preface. Transportation Research Part C-Emerging Technologies 18(1), 69–72 (2010)
5. Hindriks, K.V., Jonker, C.M., Tykhonov, D.: Towards an Open Negotiation Architecture for Heterogeneous Agents. In: Klusch, M., Pechoucek, M., Polleres, A. (eds.) CIA 2008. LNCS (LNAI), vol. 5180, pp. 264–279. Springer, Heidelberg (2008)
6. Huynh, T.D., Jennings, N.R., Shadbolt, N.: Antrust and reputation model for multi-Agent systems. Autonomous Agents and Multi-Agent Systems 13(2), 119–154 (2006)

7. Huynh, T.D., Jennings, N.R., Shadbolt, N.R.: An integrated trust and reputation model for open multi-agent systems. Autonomous Agents and Multi-Agent Systems 13(2), 119–154 (2006)
8. Zhao, J.: Bees products traceability system and testing technology system combined with the construction of the model and the applied research. The fourth session of Beijing, China international food security summit BBS, pp. 11–22 (2011)
9. Zhang, D., Li, Z., Gao, Q.: Product design on the basis of fuzzy quality function deployment. Journal of System Engineering and Electronics 19(5), 1165–1171 (2008)

A Study of Agricultural Zoning of Huang-Huai-Hai Plain Based on GIS*

Manping Hou[1], Jinmin Hao[2,**], Ying Shi[3], Jun Yang[4], Qian Wen[5], Mingzhu Cha[3], and Lingkun Xiong[4]

[1] China Institute of Defense Science and Technology, Beijing 101601, China
[2] China Agricultural University, Beijing 100193, China
[3] Tianjin Polytechnic University, Tianjin 300387, China
[4] Hunan Agricultural University, Changsha 410128, China
[5] Henan Agricultural University, Zhengzhou 450002, China
jmhao@cau.edu.cn

Abstract. Many scholars in China have done a lot of research on regionalization of agricultural resources in the Huang-Huai-Hai Plain, but there is no quantitative research on agriculture of the whole plain. By using qualitative and quantitative analysis based on the basic data collected over 300 counties in the Huang-Huai-Hai Plain, considering such indicators as the natural resources of the plains and the yields of major crops and agricultural GDP, this paper designed the agricultural zoning of the plains. The Plain was classified into high-yield areas and low-producing areas. High-yield areas were divided into advantageous regions of foodstuff, advantageous regions of cotton, advantageous regions of oil, advantageous regions of foodstuff and cotton, advantageous regions of foodstuff and oil, advantageous regions of cotton and oil, and advantageous regions of foodstuff, oil and cotton. And these areas have been divided specifically. The research is of referential values for agricultural development of the Huang-Huai-Hai Plain, both practically and theoretically.

Keywords: Agricultural zoning, Producing Region Distribution, Agricultural Pattern, Huang-Huai-Hai Plain.

1 Introduction

Huang-Huai-Hai Plain is the largest plain in China. Land area accounts for about 30% of the plain area of the whole country and arable land accounts for 1/6. Its climate is advantageous for both the north and the south, and has great potential for agricultural development with higher level of modern agricultural technology. Agricultural zoning layout plays a decisive role in agricultural production in this region. Researchers abroad have done many studies in the regionalization of agricultural resources[1-7]. The author collected agricultural economic data of each county of the Plain, and used GIS

* Supported by National Scientific and Technological Project(2004BA508B01).
** Corresponding author.

D. Li and Y. Chen (Eds.): CCTA 2012, Part II, IFIP AICT 393, pp. 84–91, 2013.

platform for property analysis, and drew a more reliable quantitative foundation, and provided the basis for sustainable development of agriculture in the Plain.

2 Data Sources and Research Methods

2.1 Data Sources

The research data included more than 300 counties' crop yields and county agricultural economic data from 2000 to 2006, natural agricultural resources and distribution of property, administration zoning map in the Plain. The data comes mainly from the China Statistical Yearbook, the China Agricultural Statistical Yearbook and regional investigation team.

2.2 Data Processing and Research Methods

Soil properties, water conditions, and crop yields for cotton, foodstuff and oil and agricultural GDP are main indicators in this research. Firstly, oil distribution and its suitability, the status of water resources were analyzed, and similar property areas were classified in the same district. According to the level of natural resources, the Plain is divided into piedmont plains, Heilonggang region, Shandong west plains, Henan southeastern plain, Anhui northern plain. Secondly, using Mapgis, the author scanned related county's administrative zoning map, and digitized each county boundary, then synthesized polygons and checked the polygons and formed a complete polygon library. Data attributes that included county agricultural production and agricultural economic data were given to each polygon in the database. Considering the yield of foodstuff, oil and cotton from 2000 to 2006 in the whole country and this district, the production partition standard was determined as high yield: foodstuff yield$\geq$5250kg·hm^2, cotton yield $\geq$1275kg·hm^2, oil crops yield $\geq$3000kg·hm^2; middle and low yield: foodstuff yield <5250 kg·hm^2, cotton yield <1275 kg·hm^2, oil crops yield <3 000 kg·hm^2. Based on this data, using Mapgis Spatial analysis function, production partition advantage figure was obtained. The advantageous figure included advantageous regions of cotton, advantageous regions of foodstuff, advantageous regions of oil, advantageous regions of foodstuff and oil, advantageous regions of cotton and oil, advantageous regions of foodstuff and cotton, advantageous regions of foodstuff, oil and cotton, middle and low yielding regions of foodstuff, oil and cotton (Figure 1).

In the end, considering the two processes, the area was designated as high-yielding area, low-producing areas. High yielding regions was subdivided into seven more favorable areas and those production areas are laid and planed.

3 The Huang-Huai-Hai Plain Agricultural Zoning

3.1 High-Yield Region Agricultural Overall Zoning

Mainly high-yielding area in the Plain related to the favorable conditions of water, soil, light, heat and other natural agricultural resources. High yielding regions were divided into seven advantageous regions: advantageous regions of foodstuff, advantageous regions of cotton, advantageous regions of oil, advantageous regions of foodstuff and cotton, advantageous regions of foodstuff and oil, advantageous regions of cotton and oil, advantageous regions of foodstuff, oil and cotton. In order to stabilize the total wheat production in the areas, the result indicated that the optimum method is to moderately reduce wheat acreage of the severe water shortage areas, and vigorously develop high-quality wheat, and expand the area of high-protein, high-fat, high starch content of corn, and accelerate the development of feed industry, brewing industry, and promote the conversion of corn processing, and actively develop high-protein soybean and other oilseeds, and form a stable cotton focus area[6]. Under the premise of ensuring food security, the overall layout is that the plain should focus on diverse production and operation where agricultural products (especially special agricultural products) plays a dominant role, and that the plain can achieve a large-scale production and industrial management, by scale land management and using modern technology and the developing socio-economic conditions.

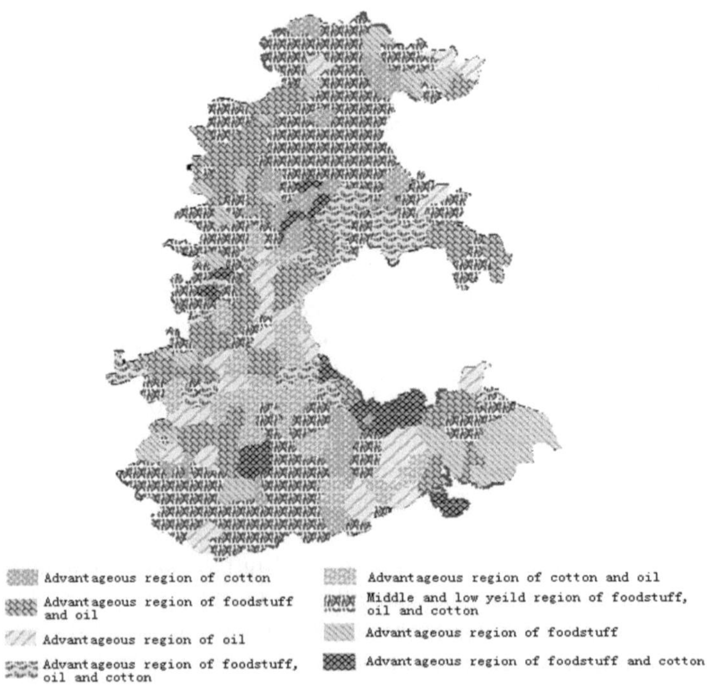

Fig. 1. Distribution Map of Foodstuff, Oil and Cotton Producing Region in Huang-Huai-Hai Plain

3.1.1 High-Yield Area Zoning

Advantageous regions of foodstuff, oil and cotton mainly include 14 counties in northern of Shandong west plains, 6 counties in Henan southeast plains, 4 counties in piedmont and the Anhui north plains. Advantageous regions of foodstuff and oil mainly include 25 counties in piedmont, 16 counties in Henan southeast plains, 10 counties in Shandong west plains, 6 counties in Anhui north plains. Advantageous regions of foodstuff and cotton were smaller in numbers, mainly including 5 counties in heilonggang region, 2 counties in Anhui plain, 2 counties in Henan southeast plains and 2 counties in Shandong west plains. Advantageous regions of cotton and oil mainly include 8 counties in Henan east plains, 7 counties in southern of Shandong west plains, 3 counties in Anhui north plains and 1 county in the southern of piedmont. Advantageous regions of foodstuff mainly include 11 counties in piedmont plains, 6 counties of Henan southeast plains, 9 counties in Anhui north plains and 1 county in Western Henan. Advantageous regions of cotton mainly include 8 countries in heilonggang region, 6 countries in Anhui north plains and 4 countries in Shandong west plains. Advantageous regions of oil mainly include 4 counties in the piedmont, 9 counties of Henan southeast plains, 5 counties in west Shandong plain and 5 counties in Anhui north plains. In addition, from the analysis of wheat, corn favoring areas distribution, wheat favoring areas are mainly in the piedmont plains, corn favoring areas are mainly in Shandong west plains. Wheat and corn favoring areas are mainly in the central region of the piedmont.

3.1.2 Agricultural Pattern of the High Yield Region

Based on water-saving cultivation, high-yielding zones had food production advantages, and optimized food variety and quality of structure and established top quality and high yield and efficient bulk agricultural commodity base of animal husbandry products. Advantageous regions of foodstuff and oil crops mainly produce special wheat, corn, oil and other pollution-free agricultural products. Advantageous regions of cotton produce high quality cotton in the main stream. Meanwhile, high-yielding zones moderately develop aquaculture industry to ensure food security objectives and quality efficiency objectives of the plains and the country are reached. In high-yielding areas of favorable conditions, production of mechanization, information technology and standardization was firstly implemented and then a model for the mechanization and modernization of agriculture was set in the country. High-yielding zones should develop processing industries to increase agricultural production and famers' income, and realize sustainable development of agriculture, the rural economy and ecological environment ultimately.

①Advantageous region of foodstuff, oil and cotton: soil and water resources in here is very advantageous. It is suitable for planting wheat, corn, cotton, oil etc. The area should be featured with a variety of crops of high quality varieties of foodstuff, oil and cotton in the future. Taking advantage of high yield of wheat and rich soil and water resources, the region should develop both summer and autumn planting, but can't expand the total acreage of wheat. In order to improve resource utilization, the region should be developed mainly in autumn. This region should form a scale and specialized cultivation of wheat, corn, cotton and oil quality product, also take advantage of companies (enterprises) with farmers and other forms, and form large-scale product processing bases, and moderately develop fodder cultivation, forest products and aquatic products, and form small-scale animal husbandry. In addition, it should increase planting of other crops with a dominant market position, and develop export-oriented agriculture. This area will become the first agricultural product advantageous district in the Plain.

From 2000 to 2006 in this area, average proportion of the acreage of foodstuff, cotton, oil was 12.50：2.45：1, and wheat and corn is 1.48:1; Average ratio of total output value of agriculture, forestry, animal husbandry, fisheries was 28.50:1:12:2.1. As for improving foodstuff, cotton and oil cultivation structure, addition of feed grown, improvement of water-saving efficiency, ratio of foodstuff, cotton, oil, and feed may reach 8:3:1.5:1. Average ratio of total output value of agriculture, forestry, animal husbandry, and fisheries may reach 25：1：11：2.5 recently. And structure of wheat and corn is basically unchanged.

②Advantageous regions of foodstuff and oil: the zones are mainly located in the piedmont, the southeast Henan plains, Shandong west plains and Anhui north plains. The area of its water resources are relatively abundant [7]. Its soil is mainly cinnamon soil, and then moisture soil. Piedmont does not only have the most advantageous soil and water conditions, but also is good farmland construction areas. It has a high degree of land use. Its arable land reserve resource is rare, and is the more favorable areas of wheat. Northern piedmont is the common advantageous areas of wheat, corn-producing. Henan southeast plains areas are the more favorable areas of the wheat, the Shandong west plains area and Anhui north plains are the more favorable areas of the corn. Piedmont in northern areas should be based on the development of wheat, corn, oil and should make a modest reduction in cotton. Southern piedmont areas and Henan southeast plain areas should be based on the development of wheat, oil and should make a modest reduction in corn and cotton. Shandong west plains and Anhui north plains should be based on the development of corn and oil crops and should make a modest reduction in wheat and cotton. At the same time, the region should intensify production in the drought and saving water special crops, and implement large-scale cultivation, and develop intercropping systems and increase the proportion of summer crops, the acreage of forage crops and other economic crops. Forest in piedmont and southeast Henan plains areas is appropriate to increase the acreage of fruit trees. The region should achieve the integrated development of food, cash crops (except cotton) and fodder crops, and use industrial production and grain and oil processing industry to promote agricultural production. Henan southeast plains, Shandong west plains, Anhui north plains should be increased mechanization acreage.

From 2000 to 2006, average proportion of the acreage of foodstuff, cotton, oil, fruit in this area is 28.2:1:6.5:3.70, and wheat and corn is 1.2:1; Average ratio of total output value of agriculture, forestry, animal husbandry, fisheries is 150:1:98:8. As for improvement of foodstuff, cotton and oil cultivation structure, addition of feed grown, improvement of water-saving efficiency, ratio of foodstuff, cotton, oil, fruit may reach 24:1:8:7 recently. The proportion of wheat and corn in the southern region is 1:1. Average ratio of total output value of agriculture, forestry, animal husbandry, fisheries may reach 100:1:70:10 recently. And structure of wheat and corn in northern areas is basically unchanged.

③Advantageous regions of cotton and oil: the area soil is mainly moisture soil, pickled moisture soil, low-lying, easy to form the floods. But this region must strengthen water conservancy construction and improve soil fertility. The plant structure is large-scale cultivation of high-quality fuel, high-quality cotton, compressed wheat and corn acreage, moderate development of fodder cultivation, increasing areas of mechanized operations. Meanwhile, this region should depend mainly on the development of cotton and oil processing industry and develop export-oriented industries.

From 2000 to 2006 in this area, average proportion of the acreage of foodstuff, cotton, oil was 4.5:1:1.1; Average ratio of total output value of agriculture, forestry, animal husbandry, fisheries is 27:1.5:12:1. The ratio of foodstuff, cotton, oil may reach 3:1:1.1 recently. Average ratio of total output value of agriculture, forestry, animal husbandry, fisheries may reach 20:2.5:14:1 which was the optimized structure recently.

④Advantageous regions of foodstuff and cotton: the soil is moisture soil with saline. It should develop water-saving foodstuff and cotton production, and modestly reduce cotton acreage and expand corn acreage. While Anhui north plain areas have obvious advantages of water resources, should increase the special corn production. This area should plant food, cash crops and fodder crops, and Anhui north plains should increase a modern planting effort.

From 2000 to 2006, in Anhui north plains, average proportion of the acreage of foodstuff, cotton, oil is 14:1:1.2, among them, wheat and corn was 3:1; Average ratio of total output value of agriculture, forestry, animal husbandry, fisheries is 15:1:7:2.5. The ratio of foodstuff, cotton, oil may reach 12:3:1. Average ratio of total output value of agriculture, forestry, animal husbandry, fisheries may reach 16:1:12:4 which is the optimized structure.

⑤Advantageous region of foodstuff: This area is only suitable for growing food crops, unsuitable for cotton and oil crops. The zones should strengthen a combination of water and soil transformation, especially transformation of the coastal saline soil in Anhui north plain. Anhui north plain areas should enhance scale and specialized planting, while other districts should develop comprehensive planting, reducing sown areas of cotton ordinary oil.

From 2000 to 2006, in Anhui north plains, average proportion of the acreage of foodstuff, cotton, oil is 15:1:2. Average ratio of total output value of agriculture, forestry, animal husbandry, fisheries is 20:1:9:4. The ratio of foodstuff, cotton, oil may reach 20:1:3 recently. Average ratio of total output value of agriculture, forestry, animal husbandry, fisheries may reach to 18:1:10:5 which is the optimized structure recently.

⑥Advantageous region of oil crops: this region is distributed in the plain areas of southeast Henan, Shandong west plain areas and Anhui north plains. The main soil is moisture soil. Areas of this type are only suitable for planting oil-bearing crops, low-producing areas for cotton and food crops. Its agricultural development should lay stress on the combination of water and soil. The goal is to form large-scale oil cultivation and oilseeds areas where oil crops are main cash crops.

From 2000 to 2006, in this area, average proportion of the acreage of foodstuff, cotton, oil, fruit tree is 21:2:6:1. Average ratio of total output value of agriculture, forestry, animal husbandry, fisheries is 25:1:10:2. The ratio of foodstuff, cotton, oil, fruit tree may reach 10:3:10:1 recently. Average ratio of total output value of agriculture, forestry, animal husbandry, fisheries may reach 16:1:12:2 which was the optimized structure recently.

⑦Advantageous region of cotton: the main soil is moisture soil. Areas of this type are only suitable for growing cotton, low-producing areas for foodstuff and oil crops. In Heilonggang areas should water-saving cropping patterns should be developed, combining water-saving with soil improvement. Anhui north plains should base on large-scale, specialized cultivation and develop cotton-based of various business models.

From 2000 to 2006, in this area, the average proportion of the acreage of foodstuff, cotton, oil, fruit, vegetables is 12:3:1.5:1.2:1. Average ratio of total output value of agriculture, forestry, animal husbandry, fisheries is 27:1:11:2. The ratio of foodstuff, cotton, oil, fruit, vegetables may reach 6:4:1:1:1 recently. Average ratio of total output value of agriculture, forestry, animal husbandry, fisheries may reach 12:1:5:1 which is the optimized structure recently.

3.2 Agricultural Zoning Layout of Middle and Low Producing Areas

Middle and low producing areas in the Plain areas is quite large, covering more than 110 counties (cities). They were mainly affected by drought, waterlog, salinity, wind and sand, lime concretion, barren and other barriers element[8]. Due to a poor efficiency of the ecological environment, the region should improve the ecological environment through ecological structural adjustment. By using salt-tolerant, bio-cover and bio-drainage to change the movement of soil water and salt, it will gain ecological and economic benefits. The region can plant forest and foodstuff, vegetable in salinity lighter areas, and plant salt-tolerant vegetation in salinity areas. By using water-saving irrigation techniques, straw, organic fertilizers to improve soil structure and increase soil fertility, this region should save and protect water resources to achieve the target of high and stable yield.

In the northern parts of the region is mainly moisture soil and bog soil, while in the south is mainly moisture soil and lime concretion black soil. Controlling saline-alkali soil in the area is a long-term arduous task. The region needs to increase agricultural inputs and strengthen water conservancy measures. In addition, improvement of saline-alkali soil is also very important. The control of sand is a focus of the area, which should be based on biological measures and engineering measures, and combine forestry fruit with flowers and grass. Then the region can establish a sand ecological and economic protection system. The southern part should focus on improving lime concretion black soil and created farmland shelterbelts and establishing three-dimensional forest shelter, fertilizing organic materials and planting green manure. The northern parts should focus on water-saving agriculture, and developing various patterns of water-saving cropping [9].

From 2000 to 2006, in this area, the average proportion of the acreage of foodstuff, cotton, oil, fruit tree, vegetables was 20:1:2:1.2:2. Average ratio of total output value of agriculture, forestry, animal husbandry, fisheries is 30:1:16:2. The ratio of foodstuff, cotton, oil, fruit tree, vegetables may reach 15:1:5:4:5 recently. Average ratio of total output value of agriculture, forestry, animal husbandry, fisheries may reach 25:1:15:2 which is the optimized structure recently.

4 Conclusion

This is a quantitative research, so the paper has got one step further than previous studies. Because of the huge workload of quantitative research, the author can only make a streamlined introduction in this paper. Accurate quantitative studies have yet to be further deepened. However, this study played a guiding role in agricultural development of the Huang-Huai-Hai Plain.

References

1. Luo, F., Lin, Y.: The application and research of GIS technology in the Fujian agricultural resources. Fujian Soil and Water Conservation 16, 5–9 (2004)
2. Tang, H.: Remote Sensing Technology in Rural Resources Planning and Prospects. China Agricultural Resources and Regional Planning 20, 23–26 (1999)
3. Bi, Y., Jiang, W., Xu, B.: Rational use of agricultural resources and ideas of protecting zones' research. China Agricultural Resources and Regional Planning 25, 58–61 (2006)
4. A-Amrieh, Sabbah, W., Isaac, J.: Impact of irrigated agricultural practices on environmental quality and human health in the West Bank. American Journal of Alternative Agriculture 14, 165–170 (1999)
5. Robertson, A.I.: The gaps between ecosystem ecology and industrial agriculture ecosystems. American Journal of Alternation Agriculture 3, 413–418 (2000)
6. Ministry of Agriculture: Regional distribution planning of competitive agricultural products. China Agriculture Press, Beijing (2003)
7. Wu, K., Huang, R.: Sustainability assessment of soil and water resources use, development potential and Countermeasures in Huang-Huai-Hai plain. Geographical Sciences 21, 391–395 (2001)
8. Sun, H., Shi, Y.: China's natural agricultural and regional development. Jiangsu Science and Technology Press, Nanjing (1994)
9. Hou, M., Hao, J., Ding, Z.: Research of low consumption resources of ecological agriculture in Huang-Huai-Hai plain. Chinese Ecological Agriculture Journal 13, 189–191 (2005)

Particle Swarm Optimization Algorithm Establish the Model of Tobacco Ingredients in Near Infrared Spectroscopy Quantitative Analysis

Bibo Ma[1,2,*] and Haiyan Ji[1,2,**]

[1] College of Information and Electrical Engineering, China Agricultural University, Beijing 100083, China
[2] Key Laboratory of Modern Precision Agriculture System Integration Research, Ministry of Education, China Agricultural University, Beijing, China 100083
Mabibo1986@163.com, instru@cau.edu.cn

Abstract. 57 tobacco samples were chosen, the near-infrared diffuse reflectance spectra at five different parts of tobacco leafs were averaged as the original spectra, the range of 4000-9000 wavenumber of spectral information was selected after wavelength selection, first-order differential was used to eliminate the impact of offset, PCA was used to reduce the dimensions and select the most representative of the principal components of the sample information. After these, use PSO and the data of chemical composition of nicotine and total nitrogen, to establish the models of near-infrared spectra of the main ingredient in tobacco quantitative analysis. In which, nicotine and total nitrogen quantitative analysis of modeling set correlation coefficients were 0.8886,0.9290; the prediction set correlation coefficients were 0.8786,0.8487. RMSEC values were 0.4737,0.2215; RMSEP values were 0.6417,0.2677. The results show that: PSO can be used to establish the model of nicotine and total nitrogen by near infrared spectroscopy quantitative analysis in tobacco.

Keywords: Tobacco, Particle Swarm Optimization (PSO), Principal Component Analysis (PCA), Near infrared spectroscopy.

Introduction

Near infrared reflectance spectroscopy does not require any chemical pretreatment for the sample. Compared with traditional chemical analysis and other spectroscopy methods, it has many advantages, such as cheap, convenient, fast and non-destructive characteristics[1].

Currently, there are some methods in the near-infrared spectroscopy quantitative analysis, such as partial least squares (PLS), support vector machines (SVM) and artificial neural network (ANN). In 2007, Yan Qing Duan, Jia Jun Wang, and others use PLS method for quantitative analysis for the chemical composition of five kinds of

---

[*] Research field: Spectral analysis.
[**] Research field: Spectral analysis and sensing technique.

D. Li and Y. Chen (Eds.): CCTA 2012, Part II, IFIP AICT 393, pp. 92–98, 2013.

tobacco sheet .The results showed that: near-infrared predictive value and the measured values have a good correlation, the model has better prediction accuracy[2] ; In 2007, Xu Liu, Hua Chen use the PCA-SVR and the ANN for a quantitative analysis of total sugar, reducing sugar, total nitrogen and nicotine content in tobacco. The results show that: the SVR model has a better prediction[3] ; In 2009, Yong Zhang, Qian Cong use the SVM for a quantitative analysis of total sugar, reducing sugar, nicotine and total nitrogen in tobacco. The results show that: the SVM model's classification accuracy rate can up to 100%, the quantitative analysis of four components of the prediction coefficient of determination (R^2), the prediction mean square error (RMSEP) and the average relative error (RME) show that: the model has a better prediction [4].

In this paper, PSO method is used in near-infrared spectroscopic quantitative analysis of total nitrogen and nicotine content in tobacco, in order to prove that the PSO method can be used in the near-infrared quantitative analysis in tobacco.

1 The Basic Principles

1.1 The Basic Principle of PSO

PSO, the English name is Particle Swarm Optimization, is proposed by Kennedy and Eberhart in the study of birds and fish group's behavior in 1995. It's a swarm intelligence algorithm [5]. In PSO, each optimization problem can be a bird in the search space (we can call it a particle). All the particles have an optimized to determine the fitness value by a function and have a speed to determine their direction and speed for flying.

Firstly, the algorithm initializes a group of random particles, and then finds the optimal solution by iteration. In each iteration, the particles update their speed and position by individual extreme and global extreme. The algorithm assumes that all particles are moving in the personal best position and global best position, and mark $x_i = (x_{i1}, x_{i2} \dots x_{in})$ to initialize the position of the particle, mark $v_i = (v_{i1}, v_{i2} \dots v_{in})$ to initialize the velocity, mark $p_i = (p_{i1}, p_{i2} \dots p_{in})$ as the best result for searching so far for the particle of i, mark $p_g = (p_{g1}, p_{g2} \dots p_{gn})$ as the best result for searching so far for the all particles. In each iteration, the particles update themselves by tracking the two optimal values, which is the value founded by particles themselves and the value founded by the entire population so far. After finding the two values, particles update their velocity and position according to the following formula[6] :

$$v_{id}(new) = w \times v_{id}(old) + c_1 \times r_1 \times (p_{id} - x_{id}) + c_2 \times r_2 \times (p_{gd} - x_{id}) \tag{1}$$

$$x_{id}(new) = x_{id}(old) + v_{id}(new) \tag{2}$$

$$w = w_{\max} - t \times (w_{\max} - w_{\min}) / \max DT \quad (3)$$

In this formula, w is constant of non-negative, we call it the inertia factor, which is used to balance the global searching and the local searching. Shi and other people for the first time introduce to update formula for the inertia weight of the speed in PSO[7]. In this paper, we propose $w_{\max}$ =0.9, $w_{\min}$ =0.4[8] ; c_1=2.8, c_2 =1.3 ; r_1 and r_2 are random number in [0, 1], $\max DT$ is the maximum number of iterations , in order to control the closing conditions of the algorithm; t is the current number of iterations. In this paper, w decrease gradually from [0.9,0.4] [9].

Firstly, calculate the new speed of the particles by the distance of the original particle velocity and position between the two optimal locations according to the formula (1). Then particles move to a new location according to the formula (2). So back and forth. If it reached the specified number of iterations, the program terminates.

1.2 The Processes of PSO

(1) Initialize the initial position and velocity of a group of particles and set the number of iterations for 5000

(2) Set the mean square error as the fitness value and get the fitness value of each particle by calculating

(3) For every particle, compare it's fitness value with the best position (pbest,) if better than pbest, use it as the current optimal particle position (pbest);

(4) For every particle, compare it's pbest with the best position gbest of all particles, if better than gbest, use it as the current optimal particle position (gbest);

(5) According to the formula (1) and (2), adjust the speed and position of the particle;

(6) If it has not reach the end of the conditions, turn to step 2;

Here, the termination condition is that the required iterations can get the number of 5000.

2 Experimental Part

2.1 The Sources of Material

In this paper, the tobacco data is K326 tobacco varieties, which is coming from the national 18 origins and is divided into three tobacco grades (B2F, X2F, C3F).

2.2 The Collection of Spectral Information

In this paper, the spectra data is collected on five different parts of the tobacco leaf, and then take the average of the five parts as the spectrum. Tobacco original near infrared spectroscopy is showed in Figure 1.

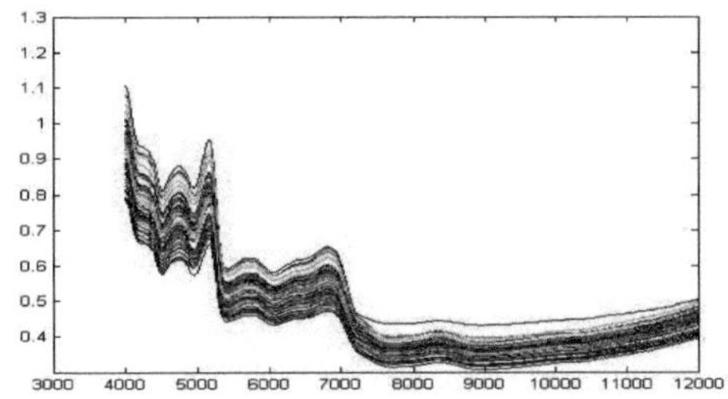

Fig. 1. Tobacco original near infrared spectroscopy

2.3 The Preprocess of Spectral

It can be seen from Figure 1, the main crest of tobacco information is concentrated in the 4000-9000 wavenumber. Therefore, in order to save modeling time, and eliminate the unwanted ingredients in the spectral information, in this paper, the wavenumber in the range of 4000-9000 cm^{-1} information are selected for the study.

The spectrum is showed in Figure 2 after wavelength selection. In this paper, use the first difference to eliminate the impact of the offset. The spectrum is showed in Figure 3 after the first difference. Then use the principal component analysis (PCA) to reduce the dimension and get the principal component of the spectrum. Use the PSO to create a system model. In the model, the main components of the spectral data is as the variable input of the model, the chemical values of nicotine and total nitrogen content is as the output of the model [10]. Use the data of mean square error between the calculated results and the real chemical value as the fitness function, in order to adjust the speed and position of the particle in the particle swarm and determine the global optimal solution. In the paper, the number of the principal components, the range of chemical value content of nicotine and the total nitrogen in modeling and in the prediction are showed in Table 1.

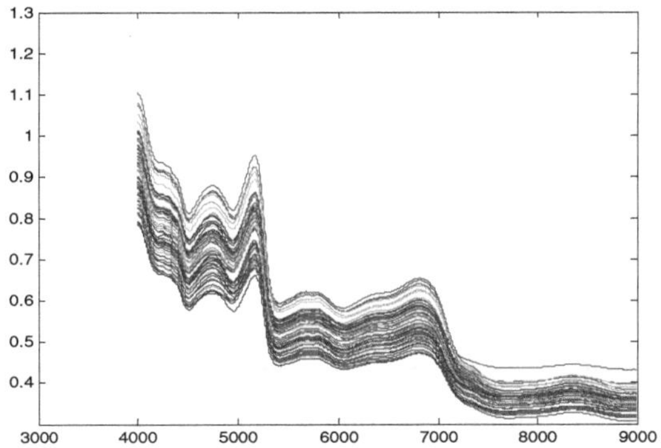

Fig. 2. Spectra after the wavelength selection

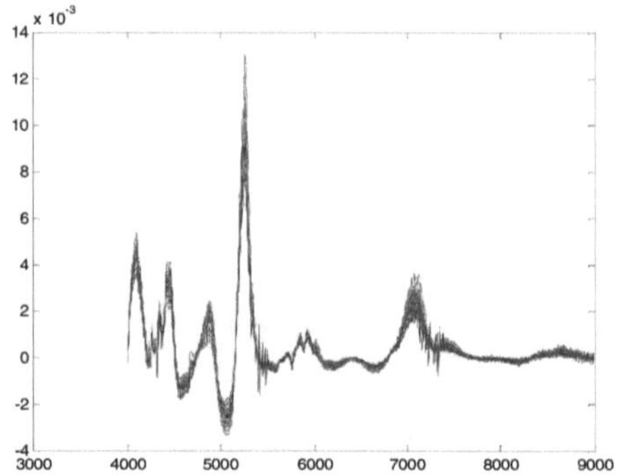

Fig. 3. Spectra after first-order Differential

Table 1. Range of chemical content and the number of principal components

Sample Name	Samples number	The range of model (%)	The range of prediction (%)	Principal components
Nicotine	57	0.88-4.68	0.94-4.67	14
total nitrogen	57	1.45-3.83	1.58-3.55	15

2.4 The Parameter Settings of the Model

In PSO, the parameter setting is very important. In order to improve the ability of modeling and prediction, in this paper, the algorithm parameter settings are showed in Table 2.

Table 2. The parameter settings of the model

The parameter settings of the model	
The location of the particles (X)	Random
The velocity of the particle (V)	Random, but between[-1,1]
Maximum number of iterations (MAXDT)	5000
Number of particles (N)	50
c1,c2	c1=2.8,c2=1.3

2.5 The Establishment and Evaluation of the Model

In this paper, in all of 57 K326 tobacco data, use 40 samples as the set of modeling, 17 samples as the set of prediction. In the model, the total nitrogen modeling set of correlation coefficient is 0.9290, the prediction set of correlation coefficient is 0.8487; nicotine modeling set of correlation coefficient is 0.8886, the correlation coefficient of prediction set is 0.8786. Other related indicators of the rest of the model are showed in Table 3. Figure 4 is PSO regression of nicotine modeling set and prediction set. Figure 5 is PSO regression of total nitrogen modeling set and prediction set.

Table 3. The indicators of evaluating the model

Sample Name	Modeling R	Modeling RMSEC	Prediction R	Prediction RMSEP
Nicotine	0.8886	0.4737	0.8786	0.6417
total nitrogen	0.9290	0.2215	0.8487	0.2677

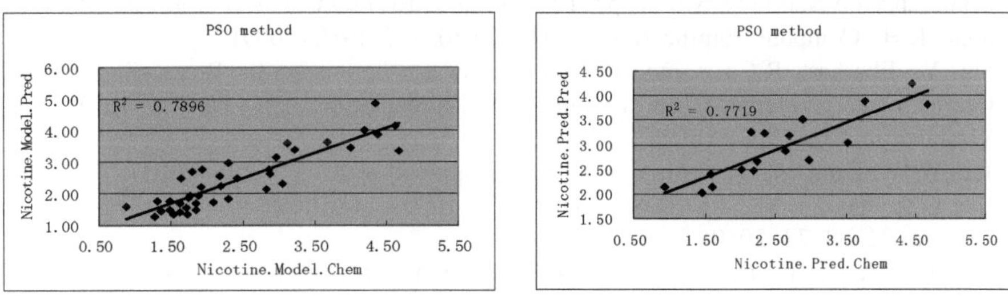

Fig. 4. PSO regression of nicotine modeling set and prediction set

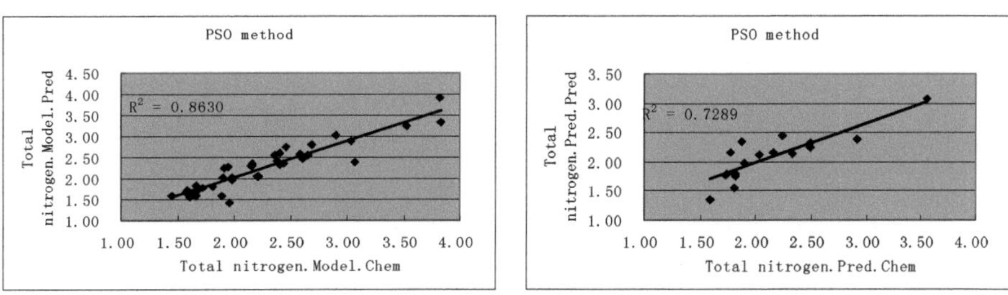

Fig. 5. PSO regression of total nitrogen modeling set and prediction set

3 Results and Discussion

57 tobacco samples were chosen, the near-infrared diffuse reflectance spectra at five different parts of tobacco leafs were averaged as the original spectra, the range of 4000-9000 wavenumber of spectral information was selected after wavelength selection, first-order differential was used to eliminate the impact of offset. PCA was used to reduce the dimensions and select the most representative of the principal components of the sample information. After these, use PSO and the data of chemical

composition of nicotine and total nitrogen, to establish the models of near-infrared spectra of the main ingredient in tobacco quantitative analysis. In which, nicotine and total nitrogen quantitative analysis of modeling set correlation coefficients were 0.8886,0.9290; the prediction set correlation coefficients were 0.8786,0.8487. RMSEC values were 0.4737,0.2215; RMSEP values were 0.6417,0.2677. The results show that: PSO can be used to establish the model of nicotine and total nitrogen by near infrared spectroscopy quantitative analysis in tobacco.

References

1. Heng, L.Y., Long, L.Z., Dong, H.H., et al.: China Light Industry Press, 1–6 (2005)
2. Yan, Q.D., Jia, J.W., et al.: Laser and Infrared 37, 1058–1061 (2007)
3. Xu, L.H., Cai, C., et al.: Spectroscopy and Spectral Analysis 27, 2460–2463 (2007)
4. Yong, Z., Qian, C., et al.: Journal of Chemical 30, 697–700 (2009)
5. Kennedy, J., Eberhart, R.C.: Particle swarm optimization. In: Proceedings of the IEEE Conference on Neural Networks, pp. 1942–1948. IEEE Press, Washington, DC (1995)
6. Shao, R.H.: Computer Engineering and Design 30, 977–1980 (2009)
7. Shi, Y., Eberhart, R.C.: A modified particle swarm optimizer. In: Proceedings of IEEE International Conference on Evolutionary Computation, pp. 69–73. IEEE Press, Washington, DC (1998)
8. Jun, W.W., Ding, W.W.: Journal of Systems Engineering 20, 194–198 (2005)
9. Jiu, F.X., Guo, L.M., Jin, R.N., Xiao, T.C.: He'nan Institute of Technology (Natural Science) 22, 68–72 (2010)
10. Xin, J.W., Yuan, L., Dan, Z., Chong, E.T.: Liaoning University of Technology 34, 18–20 (2007)

Grain Moisture Sensor Data Fusion Based on Improved Radial Basis Function Neural Network

Liu Yang, Gang Wu, Yuyao Song, and Lanlan Dong

[1] College of Engineering,
China Agricultural University,
Beijing,100083, China
zhijunr@gmail.com,{yangliu,maozhihuai}@cau.edu.cn
[2] Department of Research & Development
Beijing Beiyi Innovation Vacuum Technology CO. Ltd
Beijing, 102600, China
Donglanlan2005@126.com

Abstract. Difficulty was known to get satisfactory measurement effect on precision in capacitive grain's moisture measurement due to many influencing factors, such as temperature, species, compaction and so on. The data confusion method of Radial Basis Function (RBF) nerve network is adopted. With improved orthogonal optimal method, the RBF nerve network's weight factors can be obtained. This method can avoid artificially selected the number of hidden units, which can cause low learn precision or over learn. Tests showed that the improved RBF network algorithm reduces the network structure, greatly enhances the learning speed of calculation. By using of the improved RBF nerve network, the precision for wheat's moisture measurement has been improved.

Keywords: Radial Basis Function Never Network, moisture measurement, k-means clustering, hidden layer neuron.

1 Introduction

There are some methods for grain moisture measurement, e.g. resistance measuring, capacitance measuring, measuring with Infrared Radiation, micro wave, neutron spectrum, NMR, etc [1~3]. The advantages of low cost, high sensibility, wide dynamic range and simplicity for real time measuring make the capacitive moisture meter as one of the common method used for grain moisture measurement. The Dielectric Constant (DC) of dry grain is 2~5, while the DC for pure water is 81. As the moisture in the tested sample grows, the DC value also grows. This is the basis for Capacitive Moisture Meter (CMM) measuring [4].

Different type of grain has different DC. The CMM measuring is affected by sample's temperature, category and compactness. The conventional way of measuring the capacitive moisture only measures the capacitor value to get the moisture level, which is not accurate. Temperature compensation for the moisture measuring is actually to solve the accuracy problem by multi-sensor information fusion.

D. Li and Y. Chen (Eds.): CCTA 2012, Part II, IFIP AICT 393, pp. 99–108, 2013.

There are a number of methods for such data fusion applied in industrial technology, e.g. Multi-dimensional Regression Analysis, Bayesian Derivation, Artificial Neural Network, D-S Theory and Fuzzy Set Theory. It is as important academically as it is practically to get the best data fusion based on different models.

In recent years, the BP Neural Network has been used for linearization or temperature compensation for measured data. Same as the other forward network algorithms, BP Neural Network algorithm can be trapped by local optimum. In late 80's last century, J. Mody [5] and C. Darken [6] proposed the Radial Basis Function (RBF) Neural Network [7~9]. RBF Network has a character called optimal approximation, which will avoid local optimum trap during network establishing. This makes RBF a more effective and fast responding neural network, with a strong nonlinear processing capability. It is presented here the mathematical model for RBF Neural Network applied on grain moisture measuring.

2 Structure of RBF Neural Network for Grain Moisture Measuring

2.1 Structure of Neural Network

There are a few major factors affecting the measuring accuracy of grain moisture: the ambient temperature, the testing frequency and the compactness of the grain. This model composes by input layer, hidden layer and output layer. The input layer has 3 neurons, the hidden layer has *I* neurons, the output layer has one neuron. The neural network topology is show in Figure 1.

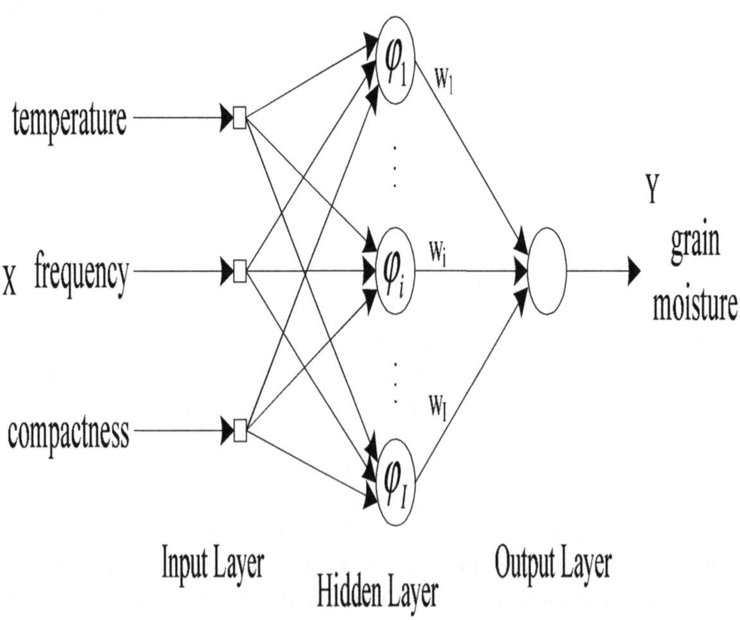

Fig. 1. RBF network structure of moisture measurement system

2.2 Mathematics Model

The input vector is $\boldsymbol{X}=[x_1,x_2,x_3]$; x_1,x_2,x_3 are temperature, frequency and compactness. The output vector is **Y** = y, y is grain moisture.

RBF network hidden layer output (the ith node):

$$u_i=\varphi(\|X-t_i\|). \tag{1}$$

In the sentence, RBF function (φ) is Gauss function, t_i is the centre of ith hidden node, $\|\cdot\|$ is Euclidean Norm.

From the mathematical model, it is known that:

$$y=\sum_i w_i u_i-\theta . \tag{2}$$

In the sentence, w_i is the ith connection weight between the ith output u_i and the system output y, θ is respectively the neurons threshold of output.

W, as a Vector with I dimensions, is given by weights matrix composed of output connection weights:

$$\boldsymbol{W}=[w_1,w_2,\cdots w_i\cdots w_I]. \tag{3}$$

To get RBF network for a measuring system, which has N groups of inputs and outputs, can be described in mathematical model as: to get the optimum weights matrix W so as to achieve minimum error between output of RBF network and the expected output of the samples. The target function is:

$$\min \quad B=\frac{1}{2}\sum\|\boldsymbol{d}_k-\boldsymbol{Y}_k\|^2=\frac{1}{2}\sum_{k=1}^{N}(d_k-y_k)^2. \tag{4}$$

$\boldsymbol{Y}_k$ is kth output vector of RBF network, $\boldsymbol{d}_k$ is kth expected output vector of training samples.

3 Training Algorithm of RBF Neural Network

There are 2 stages for this algorithm. Stage 1 is to determine the centre of the radial primary function for the hidden layer; stage 2 is the get the optimum weights matrix from nonlinear programming model. The method that the centre is chose randomly, selected by self-organization and K-means clustering algorithm are methods that determine the centre of the radial primary function. Comparing to the other methods,

K-means clustering algorithm has a advantage that it can reduce the sample space by automatically classifying the sample space to k classes, automatically getting the centre of the radial primary function for every class. The stage 2 of K-means clustering algorithm get the weight vector by Orthogonal Least Squares (OLS), but small k may lead to bad accuracy; big k will lead to bad accuracy and complex network. The improved RBF network OLS algorithm is proposed to solve this problem in the text.

3.1 Conventional RBF Network Algorithm

Three parameters are needed to be computed and learned: center vector in basis function, variance vector and weight value matrix.

1) Calculating the Centre Vector in the Basis Function: K-means clustering algorithm can calculate the centre vector in the basis function as follows:

First, initializing cluster centre. Select I different samples randomly from training samples set as the centre vector $\boldsymbol{t}_i(0),(i=1,2,\cdots,I)$ according to experience, and make the iteration step n=0.

Second, input training samples $\boldsymbol{X}_k$ randomly.

Third, find the clustering centre $i(\boldsymbol{X}_k)$ that has minimum distance with $\boldsymbol{X}_k$. In another word, making

$$i(\boldsymbol{X}_k)=\arg\min_i \left\|\boldsymbol{X}_k-\boldsymbol{t}_i(n)\right\| \qquad (i=1,2,\cdots,I) \tag{5}$$

In the sentence, $t_i(n)$ is ith centre vector of nth iteration basis function.

Fourth, adjust the centre vector of basis function according to the follow method.

$$\boldsymbol{t}_i(n+1)=\begin{cases}\boldsymbol{t}_i(n)+\eta\left[\boldsymbol{X}_k(n)-\boldsymbol{t}_i(n)\right] & i=i(\boldsymbol{X}_k) \\ \boldsymbol{t}_i(n) & \text{other}\end{cases} \tag{6}$$

In the sentence, η is a step and $0<\eta<1$.

At last, judge whether all training samples are learned and distribution don't change. If so, calculation is over, or making $n=n+1$, turning to second step. Final $\boldsymbol{t}_i(i=1,2,\cdots,I)$ is the final ith centre vector of the basis function.

2) Calculating the Variance: The approximate formula that calculate the variance is as follow when Gaussian function is selected as RBF function:

$$\sigma_1=\sigma_2=\cdots=\sigma_I=\frac{D_{\max}}{\sqrt{2I}} \tag{7}$$

In the sentence, I is the number of hidden node, and $D_{\max}$ is maximum distance between centres selected.

3) Calculating the weight vector: OLS and Givens OLS are the general methods to calculating the weight vector $\boldsymbol{W}$.

3.2 Improved RBF Network OLS Algrithm

Structural optimization in network is a difficult. The general structure has large numbers of hidden nodes and leads to over-learn. The general orthogonal method is the traditional Gram-Schmidt which has rounding error. The hidden network structure is optimized with the improved method in the article.

I eigenvectors are obtained by K-mean clustering [10]. Suppose the output testing samples column vector is:

$$\boldsymbol{d} = [d_1, d_2, \cdots, d_I]. \tag{8}$$

The hidden layer output vector is:

$$\boldsymbol{U} = \left[u_1, u_2, \cdots, u_I\right]^T. \tag{9}$$

The expected output vector can be expressed by hidden layer function output vector:

$$\boldsymbol{d} = \boldsymbol{UW} + \boldsymbol{E}. \tag{10}$$

In the sentence, $\boldsymbol{E}$ is error vector.

Make $\boldsymbol{U}$ as orthogonal triangular factorization with Gram-Schmidt orthogonal algorithm [11]:

$$\boldsymbol{U} = \boldsymbol{QR}. \tag{11}$$

The elements in upper triangular matrix $\boldsymbol{R}$ are computed in row but not in line in improved Gram-Schmidt method which lead to lower rounding error. For details, viewing $\boldsymbol{q_I}$ as the result of $\boldsymbol{u_I}$, at the same time $\boldsymbol{u_2}, \cdots, \boldsymbol{u_I}$ minus the parallel component of $\boldsymbol{u_I}$ in advanced:

$$\begin{cases} R_{11} = u_{11}, \boldsymbol{q_I} = \boldsymbol{u_I} / \|R_{11}\| \\ R_{1j} = \boldsymbol{q_I^T} \boldsymbol{u_j}, \boldsymbol{u_j^{(1)}} = \boldsymbol{u_j} - \boldsymbol{q_I} R_{1j} \qquad 2 \le j \le I \end{cases}. \tag{12}$$

After the computation, $\boldsymbol{u_2^{(1)}}, \boldsymbol{u_3^{(1)}}, \cdots, \boldsymbol{u_I^{(1)}}$ are orthogonal with $\boldsymbol{q_I}$.

Then, Ortho-normalized $\boldsymbol{u}_2^{(1)}$: $\boldsymbol{u}_3^{(1)},\cdots,\boldsymbol{u}_I^{(1)}$ minus the parallel component of $\boldsymbol{u}_2^{(1)}$:

$$\begin{cases} R_{22} = u_{22}^{(1)}, \boldsymbol{q}_2 = \boldsymbol{u}_2^{(1)} / \|R_{22}\| \\ R_{2j} = \boldsymbol{q}_2^T \boldsymbol{u}_j^{(1)}, \boldsymbol{u}_j^{(2)} = \boldsymbol{u}_j^{(1)} - \boldsymbol{q}_2 R_{2j} \quad 3 \le j \le I \end{cases} \tag{13}$$

Thus, $\boldsymbol{u}_3^{(2)},\cdots,\boldsymbol{u}_I^{(2)}$ are orthogonal with $\boldsymbol{q}_1, \boldsymbol{q}_2$. Repeating the steps, the orthogonal matrix $\boldsymbol{Q}$ and upper triangular matrix $\boldsymbol{R}$ are obtained.

The structure network is optimized with improved Gram-Schmidt method and detail solving process is designed in the article. The improved optimized network structure is obtained by sentence (10):

$$\boldsymbol{d}^T\boldsymbol{d} \approx \sum_{i=1}^{I} g_i^2 + \boldsymbol{E}^T\boldsymbol{E} . \tag{14}$$

In the sentence, $\sum_{i=1}^{I} g_i^2 = \boldsymbol{G}^T\boldsymbol{G}$

$$\boldsymbol{G} = (g_1, g_2 \ldots, g_i, \ldots, g_I) = \boldsymbol{R}\boldsymbol{W} \approx \boldsymbol{Q}^T\boldsymbol{d}$$

Suppose the compression ratio is:

$$[err]_i = g_i^2 / \boldsymbol{d}^T\boldsymbol{d} \qquad 1 \le i \le I . \tag{15}$$

Above result show that as the rounding error is lower in the improved Gram-Schmidt, the numbers of the hidden layer nodes can be computed precisely. The computation amount is decreased and over learned problem is avoided.

4 Data Analysis and Processing

Comparing experiments have been done to verify the function of the improved RBF algorithm[13~14] in this article, which is supposed to achieve higher moisture detecting accuracy.

BP network, traditional RBF network and improved RBF methods are used to compare for data fusion process.

300 groups of wheat samples are prepared for the experiments. Among which, 250 samples are used for learning algorithm, and 50 groups are used for measuring.

4.1 Compare the Measuring Results With the Method of Improved RBF to That With Method Without Data-fusion Process

Testing values and standard values of the two methods are listed in the Table 1 and 2.

Table 1. Compare the Standard Values to the Results Processed without Data Fusion

No.	*y* (%)	*d* (%)	Error (%)	No	*y* (%)	*d* (%)	Error (%)
1	6.29	6.2	1.5	9	14.02	14.5	-3.3
2	7.26	7.1	2.2	10	14.74	15.2	-3.0
3	8.15	8.3	-1.8	11	15.55	16.1	-3.4
4	8.82	9.0	-2.0	12	18.41	17.7	4.0
5	10.02	10.2	-1.8	13	19.72	19.0	3.8
6	10.78	11.0	-2.0	14	21.19	20.3	4.4
7	12.67	12.3	3.0	15	6.60	6.5	1.6
8	14.02	13.6	3.1	16	7.45	7.3	2.1

Table 2. Compare the Standard Values to the Results Processed By Improved RBF Neural Network

No.	*y* (%)	*d* (%)	Error (%)	No	*y* (%)	*d* (%)	Error (%)
1	6.29	6.2	1.45	9	14.56	14.5	0.41
2	7.12	7.1	0.28	10	15.43	15.2	1.51
3	8.32	8.3	0.24	11	16.44	16.1	2.11
4	9.03	9.0	0.33	12	17.66	17.7	-0.26
5	10.22	10.2	0.20	13	18.96	19.0	-0.21
6	10.20	11.0	1.82	14	20.50	20.3	0.99
7	12.68	12.3	3.09	15	6.61	6.5	1.69
8	13.19	13.6	-3.01	16	7.34	7.3	0.55

The Tables show that with the method without data-fusion, the error is ±5.0%. With the method of improved RBF, the error is reduced to ±1.9%.

4.2 Compare the Measuring Results with the Method of Improved RBF to That with BP Network Method

Figure 2 shows the compared curves of the measuring result using two methods. The solid line and dotted line show respectively the measuring error made by the improved RBF method, and that made by BP network method[15~16].

The maximum error and minimum error is respectively ±1.9% and ±0.2% when measuring with the improved RBF method. The maximum error and minimum error is respectively ±4.0% and ±0.3% when measuring with the BP network method.

4.3 Compare the Measuring Results with the Method of Improved RBF to That with Regular RBF Method

Computational scales for these two methods are compared with the same initial conditions. The initial conditions are: three input variables and one output variable with 300 learning samples. The conventional algorithm takes the same feature

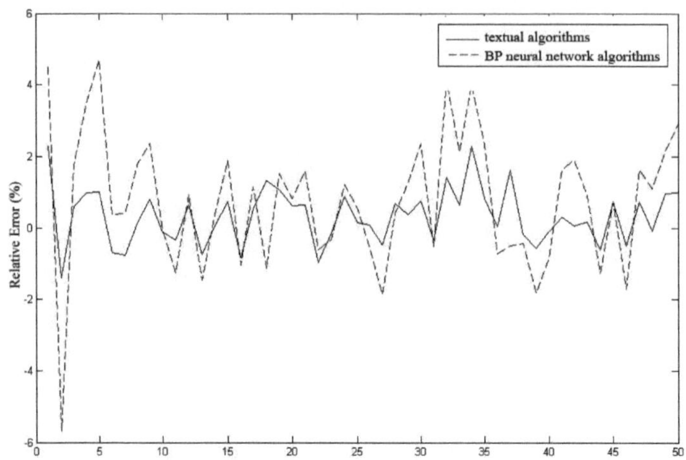

Fig. 2. Comparison of measure results between textual algorithms to BP neural network algorithms

samples as one group by clustering algorithm, reducing the hidden nodes from 250 to 164. The improved Gram-Schmidt method[17~18] is used to solve network weight and confirm the numbers of the hidden nodes automatically, reducing the hidden nodes to 56.

Figure 3 shows the compared curves of the measuring result using two methods. The solid line and dotted line show respectively the measuring error made by the improved RBF method and regular RBF method.

The maximum error range error is respectively ±1.9% and ±3.1% when measuring with the improved RBF method and regular RBF method.

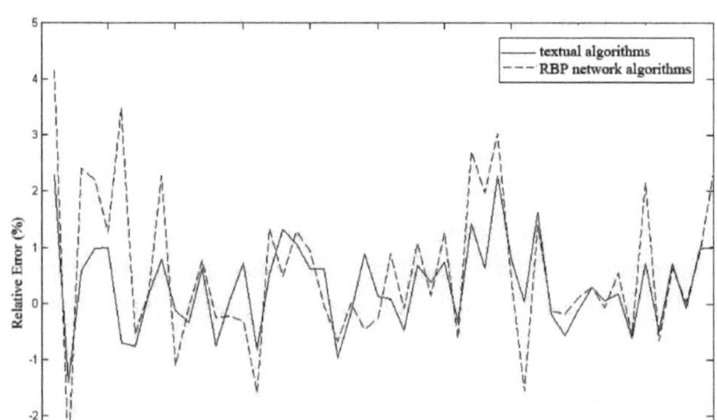

Fig. 3. Comparison of measure results between textual algorithms to RBF network algorithms

5 Conclusion

The data fusion method of RBF neural network is adopted in this study. With improved orthogonal-optimizing method, the study shows, while the RBF nerve network's weighing factors are obtained, the number of hidden units can be acquired. This method can avoid too few nerve elements that will result has been approved for its advantages over the ordinary methods with laboratory tests on grain of wheat, rice, corn, etc[19~20].

References

1. Teng, Z., et al.: Study of New Instrument for Quick Measuring Moisture Content of Cereals. Journal of HuNan University Nayural Sciences 26(3), 61–65 (1999)
2. Cheng, W., et al.: An On-line Measurement and Monitoring System of Grain Moistureduring Drying Process. Transactions of the Chinese Society of Agricultural Machinery 31(2), 54–55 (2000)
3. Yuan, Z., et al.: Fast Neutron Water-Content Mater. Nuclear Electronics & DetectionTechnology 19(1) (1999)
4. Zhai, B., Chen, Q.: Data Processing of Moisture Content Measurement Based on Data Fusion. Journal of Liaoning Insitute of Technology 26(3), 158–160 (2006)
5. Hardy, R.: Multiquadric Equations of Topography and Other Irregualr Sufaces. Journal of Geophysics Research, 1905–1915 (1987)
6. Harder, R., Desmarais, R.: Interpolation using surface splines. J. Aircraft. 9, 189–191 (1972)
7. Moody, J., Darken, C.: Fast learning in networks of locally-turned processing units. Neural Computation 1(2), 281–294 (1989)
8. Yan, P., Zhang, C.: Artificial Neural Networks and Evolutionary Computing. Tsinghua University Press, BeiJing (2002)
9. Billings, S., Zheng, G.L.: Radial basis function network configuration using genetic algorithms. Neural Networks 8(6), 877–890 (1995)
10. Liu, J.: Study on RBF Neural Network Improvement and its Application. Lanzhou University, Lanzhou (2008)
11. Huan, Y., Di, C., Zhu, S.: Matrix Theory and its Application. University of Science and Technology of China Press, HeFei (2005)
12. Zhu, C., Zhao, X.: RBF algorithm and its Application in Multi-Reach Water Quality Simulation. College of Urban Construction Hebei University of Engineering Handan (2009)
13. Wang, X., Lai, H.: An Improved RBF Algorithm for Text Classification. Communication Technology 44(12), 156–158 (2011)
14. Li, B., Lai, X.: An Improved GGAP-RBF Algorithm and Its Application to Function Approximation 20(2), 231–235 (2007)
15. Qiu, C., Zuo, X., Wang, C., Wu, J.: A BP neural network based information fusion method for urban traffic speed estimation, School of Computer. Beijing Univemity of Posts and Telecommunications 8(1), 77–83 (2010)
16. Fu, Q., Li, G., Wang, Z.: Method of Chaotic Prediction Based on Wavelet BP Network and Its Application. Journal of System s Science and Information 6(1), 55–61 (2008)

17. Zhao, T., Liu, J., Chi, X.: Gram-Schmidt Algorithm and Its Parallel Implementation. Microelectronics & Computer 9(9), 137–140 (2007)
18. Vanderstraeten, D.: An accurate parallel block Gram-Schmidt algorithm without reorthogonalization. Numer. Linear Algebra 7(4), 219–236 (2000)
19. Yang, L., Zheng, Y., Jiang, Z., Ren, Z.: Improvement of the Capacitive Grain Moisture Sensor. Beijing Jiaotong Vocational Technical University (2011)
20. Li, Z., Zhang, Y., Zhang, L.: Grain Moisture Sensors on Line. Journal of Liaoning University 33(3), 241–245 (2006)

Application of Quadratic Rotation-Orthogonal Composite Experimental Design to Assess the Relationship between Growth Environment and Ultraweak Luminescence of Mint

Yao Zhan[1], Yi Lin[1], Chunfang Wang[1], Jianping Li[1], and Yong Yu[2,*]

[1] College of Biosystems Engineering and Food Science, Zhejiang University, Hangzhou, Zhejiang, 310058

[2] College of Biosystems Engineering and Food Science, Zhejiang University, Hangzhou, Zhejiang, 310058, China

yyuzju@zju.edu.cn

Abstract. Quadratic rotation-orthogonal composite experimental design was applied in this experiment to obtain the ultraweak luminescence property of mint affected by multilateral growth environmental factors including illumination, temperature, humidity, and moisture. The influence models of four factors on luminescence were established. Analysis showed that different factors and their interactions had different effect on luminous value, among which the interaction of illumination and humidity had the most significant influence on luminescence of mint ($p \leq 0.01$). And the maximum luminous value was obtained when ten stick of lamp tube (intensity of illumination value 4229 Lx) was used, mean temperature, humidity and water were 35 °C, 85% and 90 mL/d respectively. On the base of that, linear effects and their interactions that affected ultraweak luminescence were also studied.

Keywords: mint, growth environment, ultraweak luminescence.

Introduction

It is one century since people know the ultraweak luminescence of plants. Although the mechanism has not been completely revealed, researches have already demonstrated that ultraweak luminescence is a dynamic indicator that reflects the metabolism, function mediation and information exchange inner and outer the cells in organisms [1-2]. The discovery of the intimate positive correlation between plant luminescence and vitality is one of the most basic achievements and is one of the most important detecting techniques applied in modern farming management [3].

For instance, viability and luminescence of soybeans were improved significantly after electromagnetic radiation. But the viability of Mung bean was brought down by

* Corresponding author.

D. Li and Y. Chen (Eds.): CCTA 2012, Part II, IFIP AICT 393, pp. 109–116, 2013.

using metabolic inhibitor, and the same as the luminescence [4]. Similar consequences were obtained in previous studies with rice, peanut leaf and spinach [5].

On the basis of that, many researches aiming at viability support and detection were done to study rules and factors that influence the bio-photon emission. Plant's viability and luminescence are affected by some physical (oxygen, temperature, irradiation, etc.) [6-9], chemical (oxidant, metabolic inhibitor, etc.) factors [10], hormone and environmental stress [11-13]. Benefited from the development of electronic and information technology, the alteration of plant's luminescence will be identified preferentially before the change of viability property is observed. It facilitates the application of ultraweak luminescence technology in the fields of industrial seeding nursery, intelligent cultivation and breed.

However, the study of plant's ultraweak luminescence is not complete, especially the luminescence properties affected by multilateral growth environment factors, and the effect of interaction of different factors is not clear to plant luminescence, which significantly restricts the application in the fields mentioned before.

In this article, quadratic rotation-orthogonal composite experimental design was used to study the luminescence properties of mint that affected by multilateral growth environment factors, and the relationship between luminescence properties and different factors and their interaction was explored. Correlative knowledge was prepared to facilitate the application of ultraweak luminescence testing technology in modern farming management.

1 Materials and Methods

1.1 Materials and Processing

15 bonsai of mints were supplied by Zhongtai nursery stock base of Hangzhou New Blue-Sky Landscape Group, mints grew in a greenhouse with identical environment and nutrient conditions for 45 days, and the degrees of dense foliage were kept similar. Mints were put into an artificial climate incubator (Artificial Climate Box, RXZ) after collected, kept in 30°C and 90% in humidity and exposed in light before use. The experiment was carried out on three replicates.

1.2 Range of Environment Factors

The environment factors of osmanthus tree seedings include illumination, temperature, relative humidity and moisture. These factors were controlled by RXZ except for moisture that artificially controlled by using a injector with scales. Ranges of these four factors were formulated as follows:

Illumination: according to the average illumination of the greenhouse, illuminance was regulated by ten 40 W fluorescent tubes, intensity was controlled by opening different numbers of tubes (intensity of one 40 W fluorescent tube is 422.9±10.1 Lx).

Temperature: the temperature range was automatically controlled by RXZ at 15 °C ~ 35 °C, accuracy was ±0.5 °C.

Humidity: the relative humidity was controlled by RXZ at 65%~85%, accuracy is ±1%.

Moisture: according to the moisture range (10~90 mL/d) obtained from preliminary test, moisture was controlled artificially by using a injector with scales at the same time of everyday.

1.3 Measurement of Ultraweak Luminescence Value

The optimum detect condition of ultraweak luminescence was obtained from pre-experiment. mint was cultivated in given conditions for 3 days before test, 0.2 g of mint (leaves) was detected under 35 °C to get the maximum luminescence value.

Photo-induced delayed luminescence was observed at the beginning of test, emission of mint leaves in constant time interval decreased rapidly and waved evidently. After 3 ~ 5 min, delayed luminescence disappeared gradually, emission tended to be steady. Hence, the whole data acquisition time of each experiment was 10 min, and data of last 5 min was adopted (detection cup without samples as control).

1.4 Experimental Design and Data Analysis

Quadratic rotation-orthogonal composite experimental design was used to analyze the effect of illumination, temperature, relative humidity, moisture, and interaction to the luminescence of samples. Quadratic rotation-orthogonal composite experimental design in coded and actual level of variables and the luminous value of each sample were detected (data not shown). 36 experiments were carried out with 12 replicates at the center point.

The result was analyzed to build the model of environmental factors to ultraweak luminescence by using Data Processing System (DPS), the optimum values of environment variables were obtained when luminous value was maximum, and the effects of influential factors and interactions on luminescence of mint were further studied.

2 Results and Analysis

2.1 Regression Analysis and Optimal Parameter Analysis

Polynomial regression equation with linear, quadratic and interaction terms was obtained by using DPS, analysis of variance (ANOVA) of each effect was shown in Tab.1.

Tab. 1 showed effects of single factors and interactions of four environment factors of four environment variables on the luminous value of mint. the regression equation of response function had a high correlation coefficient (r=0.993), can be used to predict luminous value of mint in trial stretch.

Table 1. Model and condensed ANOVA (analysis of variance) table in coded level of variables

Source of variations	Coefficient of polynomial	F-Value
	1.83119	
x_1	0.04471	0.03496 ^{NS}
x_2	0.12061	0.25445 ^{NS}
x_3	-0.62317	6.79296**
x_4	0.67242	7.90896**
x_1^2	0.48344	5.45088**
x_2^2	0.07280	0.12361 ^{NS}
x_3^2	0.44216	4.55968**
x_4^2	0.48463	5.47771**
$x_1 x_2$	0.14325	0.23931 ^{NS}
$x_1 x_3$	-0.84385	8.30386***
$x_1 x_4$	0.54536	3.46835*
$x_2 x_3$	0.20839	0.50640 ^{NS}
$x_2 x_4$	0.77622	7.02619**
$x_3 x_4$	-0.08052	0.07561 ^{NS}
r	0.933	—

Note: Variables: x_1—light, x_2—temperature, x_3—relative humidity, x_4—moisture.

^{NS} Non-significant at $p > 0.10$.

* Significant at $p \leq 0.10$.

** Significant at $p \leq 0.05$.

*** Significant at $p \leq 0.01$.

To determine the value of four factors including illumination, temperature, moisture and humidity required by mint, optimal calculation of luminous value was carried out based on the intimate positive correlation between luminous value and viability of plants. Results showed that the maximum luminous value of mint (18.7 photon/s) was acquired when 10 lamps were used (illumination value was 4 229 Lx), temperature, humidity and moisture were 35 °C, 65% and 90 mL/d respectively. Because of the intimate positive correlation between luminous value and viability of plants, viability of mint was higher under the optimum environmental condition than that under any other conditions.

2.2 Analysis of Effect of Each Factor and Interaction on Ultraweak Luminous Value

In order to facilitate further analysis of single factors and interaction effects on luminous value, the regression model shown in Tab.1 was corrected. First, non-significant terms in Tab.1 were rejected, the rest of terms (coded level) constitute the relational model with the luminous value of mint (Eq.1). The correlation coefficient was further elevated (r=0.953).

$$Y=1.83119-0.62317x_3+0.67242x_4+0.48344x_1^2+0.44216x_3^2+0.48463x_4^2-0.84385x_1x_3 +0.54536x_1x_4+0.77622x_2x_4 \quad (1)$$

2.2.1 Effect of Each Factor on Ultraweak Luminous Value of Mint

Fig.(1-3) showed that luminous value of mint increased with illumination and moisture, but decreased with humidity when the other factors were optimum. The maximum luminous value of mint was obtained when 10 lamps were used (mean illumination value was 4 229 Lx), temperature, humidity and moisture were 35 °C, 65% and 90mL/d respectively, and because of the positive correlation between ultraweak luminescence and viability of plants, the viability of mint also reached the maximum.

Luminous value (viability) of mint was higher while illumination increased, that is because light is energy source of photosynthesis, and is the essential growth condition for chloroplast development and chlorophyll synthesis. Light can also regulate the activity of some enzymes in carbon cycle of photosynthesis. At some extent, a linear increase was observed in photosynthetic rate with the increase of illumination. Furthermore, as the main external factor that affects transpiration, illumination facilitates stoma open and reduces stomatal resistance. In addition, illumination increased vapor pressure gradient and transpiration rate by enhancing air temperature and leaf temperature. Consequently, luminous value of mint was elevated with increased illumination, namely viability of mint was increased when illumination increased.

Luminous value (viability) of mint decreased when humidity increased, that was because relative air humidity influences transpiration rate directly. When humidity decreased, vapor pressure decreased, differential pressure inside and outside leaves increased, then transpiration rate increased finally. As the main power of water absorption and transportation, transpiration also facilitates the transmission and distribution of mineral salts and other water-soluble substances in plants. Hence, on the premise of ensuring plants' normal life, slight decrease of humidity promotes photosynthesis rate and viability of plants. In other words, luminous value increased when humidity decreased.

Luminous value (viability) of mint increased with moisture, as a great deal of water was needed in growth of mint. Appropriate irrigation increased the growth rate, enlarged the leaf area and photosynthesis area, and it also improve the activity of roots, absorption of water and mineral substances and photosynthesis rate. Therefore, proper increase of moisture elevated viability and luminescence of mint.

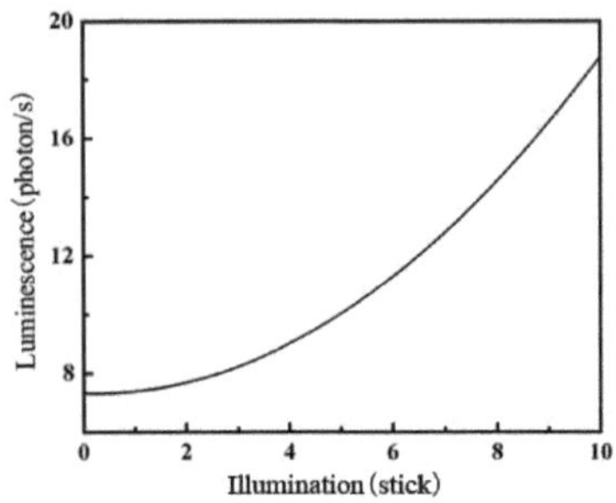

Fig. 1. Effect of illumination on the bio-illumination of mint sample

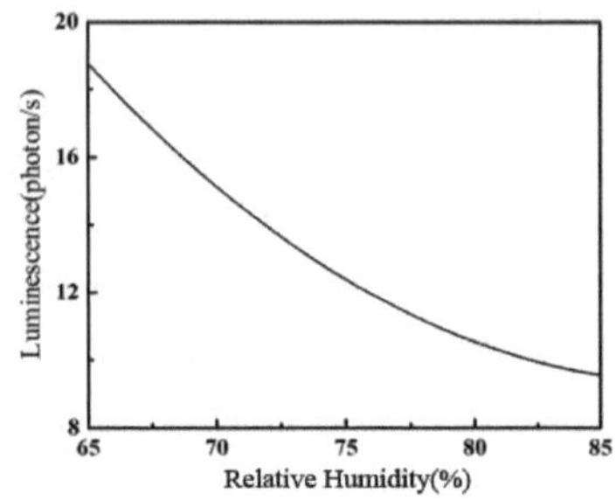

Fig. 2. Effect of relative humidity on the bio-illumination of mint sample

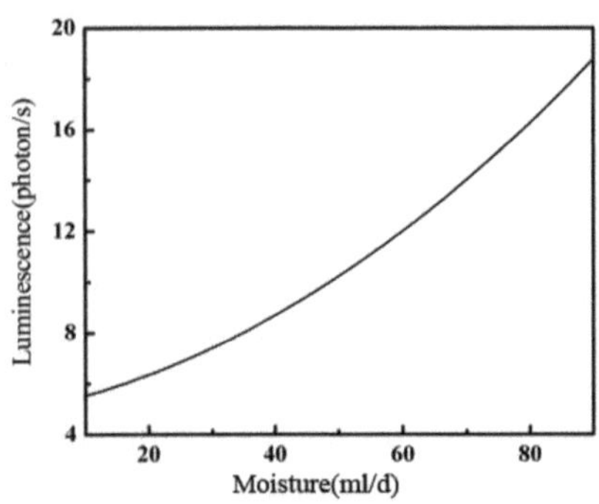

Fig. 3. Effect of moisture on the bio-illumination of mint sample

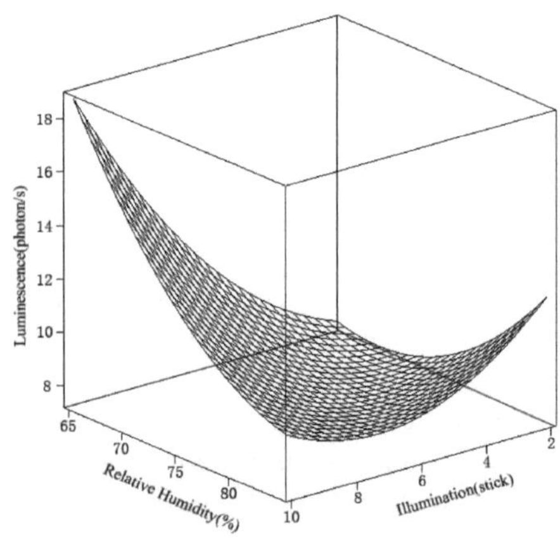

Fig. 4. Effect of relative humidity and illumination on the bio-illumination of mint sample

2.2.2 Interaction Effect of Three Factors on Ultraweak Luminous Value of Mint

Fig. 4 showed that the variation trend of luminous value changed with varied illumination. At first, luminous value of mint increased with illumination, with the increase of humidity, luminous value became initially decreased then increased. With higher illumination, similarly, the variation trend of luminescence changed with varied humidity. At first luminescence increased with humidity, with the increase of illumination, luminescence became decreased with humidity. These were explained as different effects that humidity and illumination attached on luminous value of mint, and the interaction between them had significant effect on mint's luminescence (Tab. 1).

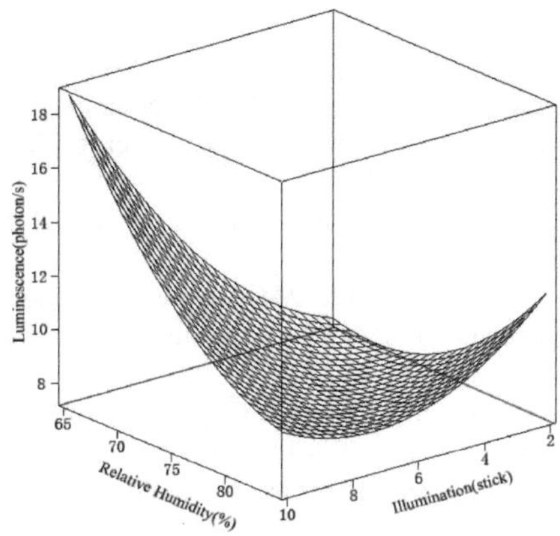

Fig. 5. Effect of moisture and illumination on the bio-illumination of mint sample

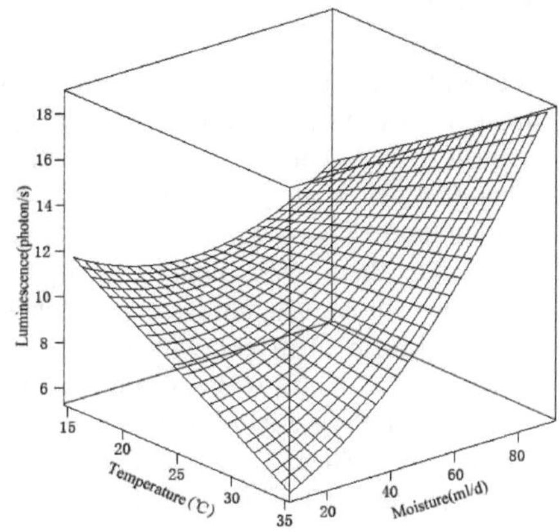

Fig. 6. Effect of temperature and moisture on the bio-illumination of mint sample

Fig. 5 showed that the luminous value changed with moisture, firstly luminous value initially decreased then increased with moisture, with the increase of temperature, luminous value became increased with moisture. Similarly, luminous value of mint decreased with the increase of temperature when moisture was in low level, but with the increase of moisture, luminous value became increased with temperature. It indicated that environmental temperature affected the luminous value, but the impact was performed by the interaction of temperature and moisture, and the interaction had significant effect on luminous value of mint (Tab. 1).

3 Conclusion

Single factors and interactions of illumination, temperature, humidity and moisture all have effect on the luminous character of mint leaves. For single factors, moisture and humidity have more significant effect on luminescence of mint than others. The interaction of illumination and humidity has the most significant effect on luminescence. Besides, the mathematical model established in this study effectively reflects the impacts of various environment factors on ultraweak luminous properties of mint.

Acknowledgements. This research was supported by a grant from the Research Fund for the Doctoral Program of Higher Education of China (RFDP) (No. 20090101120081), the major science and technology project of Science Technology Department of Zhejiang Province, China (No. 2008C02006-4), and the project of Education Department of Zhejiang Province, China (No. Y20080327).

Reference

1. Ramamoorthy, V., Viswanathan, R., Raguchander, T., Prakasam, V., Samiyappan, R.: Induction of systemic resistance by plant growth promoting rhizobacteria in crop plants against pests and diseases. Crop Protection 20(1), 1–11 (2001)
2. Zhang, X.H., Yang, H.Q.: Ultraweak bioluminescence in plant. Journal of Shandong Agricultural University 34(4), 605–608 (2003)
3. Lie, G.H., Zhao, P.F., Zhang, X.M., Lu, H.B., Jiang, T., Li, W.D.: Laser microbeam technique, biological ultraweak photon emission and biological vitomystery. Journal of Qinghai Mormal University (Natural Science Edition) 1, 78–81 (2003)
4. Zhang, Z.L.: Ultra-weak Chemiluminescence Analytical Technology Principle and Application. Progress in Biochemistry and Biophysics 27(1), 102–104 (2000)
5. Zhang, J.P., Zhang, X.Z., Gong, Z.H.: Application of ultraweak bioluminescence in vegetable research. Chinese Agricultural Science Bulletin 22(1), 220–222 (2006)
6. Cheng, H.P., Wang, J.H., Chi, H.C., Zhu, M.Y.: Study on ultraweak luminescence of pisum sativum seeds at the stage of germination. Journal of Zhejiang University (Science Edition) 28(6), 682–685 (2001)
7. Tan, S.C., Xing, D., Tang, Y.H., Li, D.H.: Spectral studies of ultra-weak biophoton emission from Plant's leabes. Acta Photonica Sinica 29(11), 961–965 (2000)
8. Ma, F., Feng, M., Gao, Y., Feng, X.M., Wang, Z.G., Luo, S.S., Ge, C.L.: Vegetable seed radiosensitivity and kinetic analysis of super-weak bioluminescence. Journal of Nuclear Agricultural Sciences 17(5), 382–387 (2003)
9. Gao, Y., Feng, M., Wang, Z.G., Lu, H.Y., Ma, F., Shi, Z.L., Ge, C.L., Luo, S.S.: A preliminary study of irradiation disturbance on the rapid bacteria assay with ATP bioluminescence technique for agricultural products. Journal of Nuclear Agricultural Sciences 21(1), 48–51 (2007)
10. Chen, W.L., Xing, D., Wang, J., He, Y.: Rapid determination of rice seed vigour by spontaneous chemiluminescence and singlet oxygen generation during early imbibition. Luminescence 18(1), 19–24 (2003)
11. Strand, J.F.: Some agrometeorological aspects of pest and disease management for the 21st century. Agricultural and Forest Meteorology 103, 73–82 (2000)
12. Yazgan, M.S., Tanik, A.A.: New approach for calculating the relative risk level of pesticides. Environment International 31, 687–692 (2005)
13. Li, Z.H., Li, P.P.: The study of a system for the diagnosis and prevention of pests, diseases and nutritional deficiency of greenhouse tomato. Journal of Agricultural Mechanization Research 2, 187–191 (2003)

Application of Quadratic Rotation-Orthogonal Composite Experimental Design to Assess the Relationship between Growth Environment and Ultraweak Luminescence of Osmanthus Tree Seedings

Yong Yu[1], Yao Zhan[1], Yi Lin[1], Chunfang Wang[1], and Jianping Li[2,*]

[1] College of Biosystems Engineering and Food Science, Zhejiang University, Hangzhou, Zhejiang, 310058, China
[2] College of Biosystems Engineering and Food Science, Zhejiang University, Hangzhou, Zhejiang, 310058, China
jpli@zju.edu.cn

Abstract. Quadratic rotation-orthogonal composite experimental design was applied in this experiment to analyze the ultraweak luminescence property of osmanthus tree seedings affected by multilateral growth environmental factors. The influence models of four factors on luminescence were established including illumination, temperature, humidity, and moisture. The ANOVA (analysis of variance) analysis showed that different factors and their interaction had different effect on luminescence property, among which the linear effects of four variables, the quadratic effects of illumination, relative humidity and moisture, and the interaction effect of illumination × moisture and temperature × relative humidity all have significant effect on the luminous value of osmanthus ($p \le 0.01$). The maximum luminous value was obtained when ten stick of lamp tube (intensity of illumination value 4229 Lx) was used, mean temperature, humidity and water were 35 °C, 85% and 90 mL/d respectively. On the base of that, linear effects and interactions that affected ultraweak luminescence were further studied.

Keywords: Osmanthus tree seedings, Growth environment, Ultraweak luminescence.

Introduction

Ultraweak luminescence is a kind of low level biological photon emission which was first discovered by a former Soviet Union scientist G. Gurwitsh in his famous onion experiment in 1923. Although the mechanism is not revealed completely up to now, it has been demonstrated that ultraweak luminescence is a dynamic indicator that reflects the metabolism, function mediation and information exchange inner and outer the cells

---

* Corresponding author.

D. Li and Y. Chen (Eds.): CCTA 2012, Part II, IFIP AICT 393, pp. 117–125, 2013.

in organisms [1-2]. The discovery of the intimate positive correlation between plant luminescence and vitality is one of the most basic achievements and is one of the most important detecting techniques applied in modern farming management [3].

Researches showed that viability and luminescence of soybeans were improved significantly after electromagnetic radiation. The viability of Mung bean was brought down by using metabolic inhibitor, and the same as the luminescence [4]. Similar consequences were obtained in previous studies with rice, peanut leaf and spinach [5].

On the basis of that, many researches aiming at viability support and detection were done to study rules and factors that influence the bio-photon emission. Plant's viability and luminescence are affected by some physical (oxygen, temperature, irradiation, etc.) [6-9], chemical (oxidant, metabolic inhibitor, etc.) factors [10], hormone and environmental stress [11-13]. Benefited from the development of electronic and information technology, the alteration of plant's luminescence can be identified ahead of the change of viability. It effectively facilitates the application of ultraweak luminescence technology in the fields of industrial seeding nursery, intelligent cultivation and breed.

However, the study of plant's ultraweak luminescence is not complete, especially the luminescence properties affected by multilateral growth environment factors, and the effect of interaction of different factors is not clear to plant luminescence, which significantly restricts the application in the fields mentioned before.

In this article, quadratic rotation-orthogonal composite experimental design was used to study the luminescence properties of osmanthus tree seedings that affected by multilateral growth environment factors, and the relationship between luminescence properties and different factors and their interaction was explored. Correlative knowledge was prepared to promote the application of ultraweak luminescence testing technology in modern farming management.

1 Materials and Methods

1.1 Materials and Processing

45 bonsai of osmanthus tree seedings were supplied by Zhongtai nursery stock base of Hangzhou New Blue-Sky Landscape Group, seedings grew in a greenhouse with identical environment and nutrient conditions for 30 days, and the degrees of dense foliage were kept similar. Seedings were put into an artificial climate incubator (Artificial Climate Box, RXZ) after collected, kept in 30℃ and 90% in humidity and exposed in light before use. The experiment was carried out on three replicates.

1.2 Range of Environment Factors

The environment factors of osmanthus tree seedings include illumination, temperature, relative humidity and moisture. These factors were controlled by RXZ except for moisture that artificially controlled by using a injector with scales. Ranges of these four factors were formulated as follows:

Illumination: according to the average illumination of the greenhouse, illuminance was regulated by ten 40 W fluorescent tubes, intensity was controlled by opening different numbers of tubes (intensity of one 40 W fluorescent tube is 422.9±10.1 Lx).

Temperature: the temperature range was automatically controlled by RXZ at 15 °C ~ 35 °C, accuracy was ±0.5 °C.

Humidity: the relative humidity was controlled by RXZ at 65%~85%, accuracy is ±1%.

Moisture: according to the moisture range (10~90 mL/d) obtained from preliminary test, moisture was controlled artificially by using a injector with scales at the same time of everyday.

1.3 Measurement of Ultraweak Luminescence Value

The optimum detect condition of luminescence was obtained from pre-experiment. Osmanthus seedings were cultivated in given conditions for 3 days before test, 2 g of seedings (leaves) were detected under 35 °C to get the maximum luminescence value.

Photo-induced delayed luminescence was observed at the beginning of test, emission of seeding leaves in constant time interval decreased rapidly and waved evidently. After 3 ~ 5 min, delayed luminescence disappeared gradually, emission tended to be steady. Hence, the whole data acquisition time of each experiment was 10 min, and data of last 5 min was adopted (detection cup without samples as control).

1.4 Experimental Design and Data Analysis

Quadratic rotation-orthogonal composite experimental design was used to analyze the effect of illumination, temperature, relative humidity, moisture, and interaction to the luminescence of samples. Quadratic rotation-orthogonal composite experimental design in coded and actual level of variables and the luminous value of each sample were detected (data not shown). 36 experiments were carried out with 12 replicates at the center point.

The result was analyzed to build the model of environmental factors to the luminescence by using Data Processing System (DPS), the optimum values of environment variables were obtained when luminous value was maximum, and the effect of influential factors and interaction on luminescence was further studied.

2 Results and Analysis

2.1 Regression Analysis and Optimal Parameter Analysis

Polynomial regression equation with linear, quadratic and interaction terms was obtained by using DPS, analysis of variance (ANOVA) of each effect was shown in Tab.1.

Tab.1 showed that linear effects of 4 environment factors significantly ($P \leq 0.01$) affected the luminous value of leaves, according to P-value, the impact of each linear effect followed the order: moisture > humidity > temperature > illumination and the impact of nonlinear effect followed the order: moisture > humidity > illumination > temperature. Hence, moisture, humidity, illumination, temperature were all significant factors that influenced the growth of seeding leaves, and should be paid attention in modern farming management.

Table 1. Model and condensed ANOVA (analysis of variance) table in coded level of variables

Source of variations	Coefficient of polynomial	F-Value
Constant	4.3258	—
x_1	1.4379	21.02781***
x_2	1.6296	27.00721***
x_3	1.5571	24.65757***
x_4	3.2204	105.47528***
x_1^2	1.0128	13.90988***
x_2^2	0.7103	6.84170**
x_3^2	0.9153	11.36067***
x_4^2	1.3853	26.02321***
$x_1 x_2$	0.7056	3.37585*
$x_1 x_3$	0.5469	2.02774 NS
$x_1 x_4$	1.3894	13.08803***
$x_2 x_3$	-1.1431	8.85978***
$x_2 x_4$	1.0469	7.43062**
$x_3 x_4$	0.9381	5.96701**
r	0.976	—

Note: Variables: x_1—light, x_2—temperature, x_3—relative humidity, x_4—moisture. Non-significant at $p > 0.10$; * Significant at $p \leq 0.10$; ** Significant at $p \leq 0.05$; *** Significant at $p \leq 0.01$.

Besides, interactions of four environment factors had different influence on the luminous value of seeding leaves. It was revealed that moisture was a crucial factor which affected the growth of seedings, its importance was reflected not only in it as a single factor, but also in the strong interaction between it and the other three factors. Nevertheless, illumination had no strong interaction with temperature or humidity, therefore, regulation of moisture and illumination should be attached great importance in modern farming management of osmanthus tree seedings.

Tab.1 also showed that regression equation of response function had a high correlation coefficient(r=0.993), can be used to forecast luminous value of seedings in trial stretch.

To determine the value of four factors including illumination, temperature, moisture and humidity required by osmanthus tree seedings [3], optimal calculation of luminous value was carried out based on the intimate positive correlation between luminous value and viability of plants. Results showed that the maximum luminous value (48.11 photon/s) was acquired when 10 lamps were used (illumination value was 4 229 Lx), temperature, humidity and moisture were 35 °C, 65% and 90 mL/d respectively. Experiments showed that luminous value was 45.76 photon/s under this optimum condition, slightly lower than the predicted value, but higher than that under other experimental conditions.

2.2 Analysis of Effect of Each Factor and Interaction on Ultraweak Luminous Value

In order to facilitate further analysis of single factors and interaction effects on luminous value, the regression model shown in Tab.1 was corrected. First, non-significant terms were rejected, the rest of terms (coded level) constitute the relational model with the luminous value of osmanthus (Eq.1).

$$Y=4.3258+1.4379x1+1.6296x2+1.5571x3+3.2204x4+1.0128x12+0.7103x22+0.9153x32 \\ +1.3853x42+0.7056x1x2+1.3894x1x4-1.1431x2x3+1.0469x2x4+0.9381x3x4 \quad (1)$$

The correlation coefficient was further elevated (r=0.983).

2.2.1 Effect of Each Factor on Ultraweak Luminous Value of Osmanthus Tree Seedings

Fig.1 and Fig.2 showed that luminous value of seedings first decrease then increase with four single factors when the other factors were zero. The maximum luminous value was obtained when factors reached maximum respectively, and because of the positive correlation between ultraweak luminescence and viability of plants.

Luminous value (viability) was larger while illumination increased, that is because light is energy source of photosynthesis, and is the essential growth condition for chloroplast development and chlorophyll synthesis. Light can also regulate the activity of some enzymes in carbon cycle of photosynthesis. Furthermore, as the main external factor that affects transpiration, illumination facilitates stoma open and reduces stomatal resistance. In addition, illumination increased vapor pressure gradient and transpiration rate by enhancing air temperature and leaf temperature. To osmanthus seedings, illumination of 6 ~ 8 h is required. Under this illumination condition, luminous value of seedings was elevated with increased illumination. Meanwhile, luminous value increased with increasing of temperature, humidity and moisture.

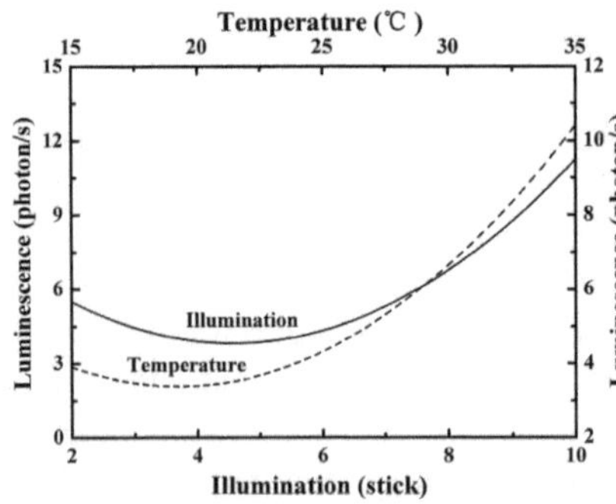

Fig. 1. Effect of illumination and temperature on the bio-illumination of osmanthus tree seedings sample

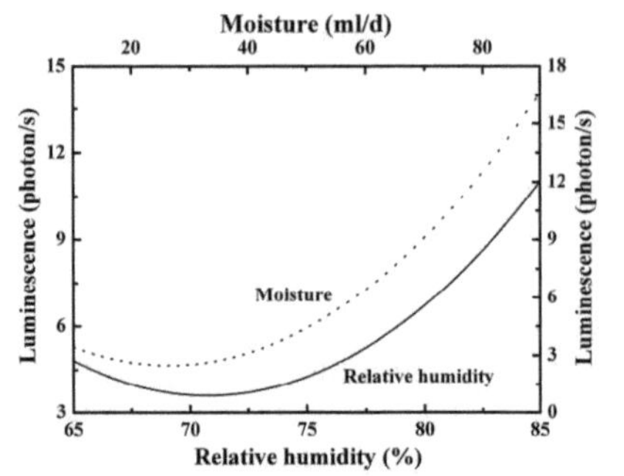

Fig. 2. Effect of relative humidity and moisture on the bio-illumination of osmanthus tree seedings sample

2.2.2 Interaction Effect of Four Factors on Ultraweak Luminous Value of Osmanthus Tree Seedings

Fig.3(a) showed that, with the increase of environmental humidity, the variation trend of luminous value remained unchanged with varied illumination, luminous value first decreased then increased with the increase of illumination. Nevertheless, the increasing rate of luminous value was enhanced significantly when illumination increased at a high humidity. Likewise, with the increase of illumination, luminescence was firstly decreased then increased when humidity increased. When illumination was high, the increasing rate of luminescence was elevated significantly with the increase of humidity. This fully demonstrated the interaction between humidity and illumination.

Compared with Fig.3(a), the other interactions had similar effects on luminescence of seedings. When illumination, temperature, moisture and humidity were in high level, the variation rate of luminous value increased with variation of factors which significantly correlated with luminescence. Luminous value that higher than maximum luminescence was obtained, and the same as the viability of osmanthus tree seedings.

Different from effects of other interactions, interaction of temperature and humidity was unique to luminescence of osmanthus. Almost linear correlation was observed between temperature and luminous value when osmanthus tree seedings were in low humidity, with the increase of humidity, luminous value became initially decreased then increased. Furthermore, the maximum luminous value appeared at the maximum temperature in the condition of low humidity, in contrast the maximum luminescence showed at the minimum temperature when it was in high humidity. Meanwhile, the minimum luminescence in high humidity was much higher than that was in low humidity. The same trend was observed in humidity.

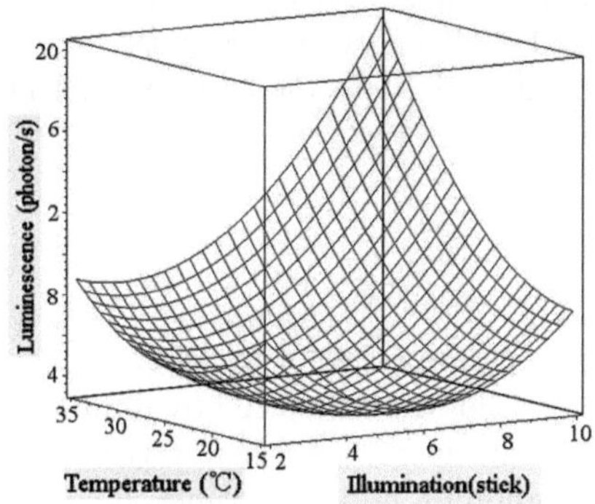

(a) Temperature and illumination

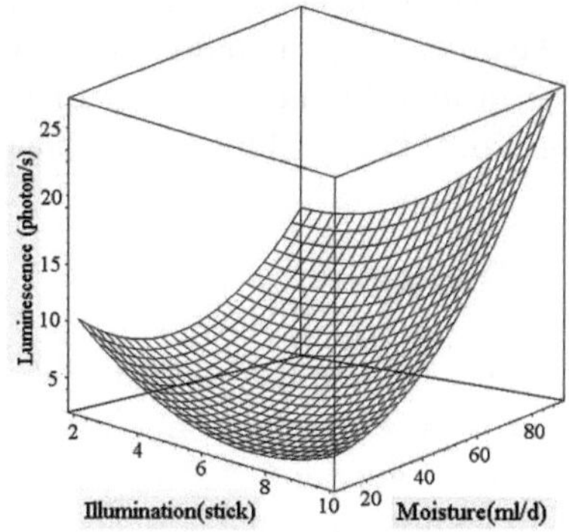

(b) Illumination and moisture

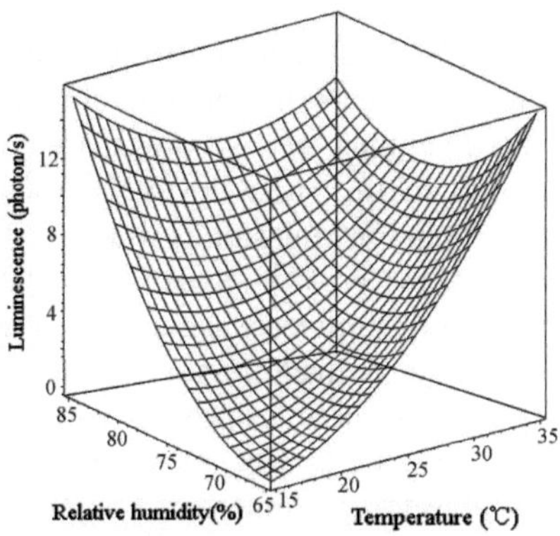

(c) Relative humidity and temperature

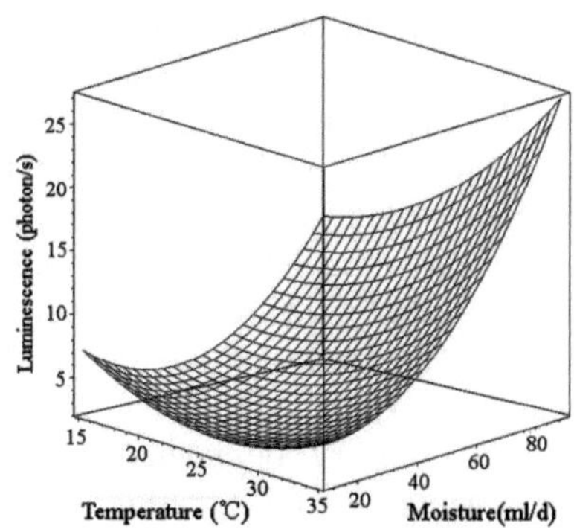

(d) Temperature and moisture

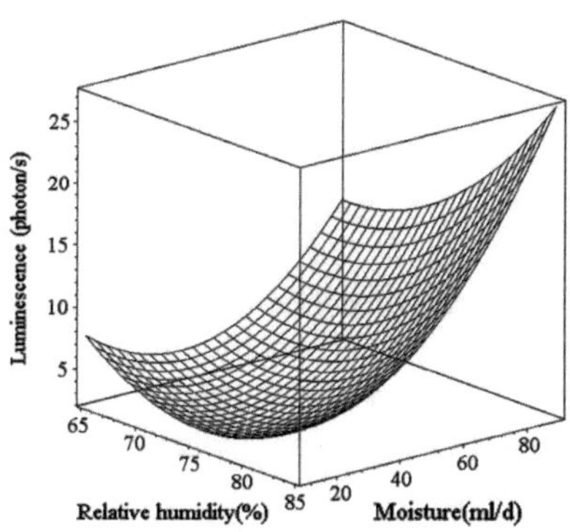

(e) Relative humidity and moisture

Fig. 3. Effect of different factors on the bio-illumination of osmanthus tree seedings sample

So it should be noticed in modern farming management of osmanthus tree seedings that the influence of humidity on viability reduced while in high temperature, and when it was in high humidity, effect of temperature decreased in the same way. Hence, energy could be saved by controlling one of the two factors.

3 Discussion

Fig.1-Fig.3 showed that no matter single factor response curve or two-factor response surface, there was a minimum luminous value of seedings influenced by illumination, temperature, humidity and moisture. Luminescence increased when variables were higher or lower than that corresponding to the minimum luminous value. It could not explained that growth of osmanthus tree seedings was promoted when four variables were higher or lower than certain critical values.

It can be explained by Liu [14] who demonstrated the theory of coherent states and its application in physical stimulation of acupoint [15-16]. According to the theory, acupoint is commonly existed in organisms, and it affects bio-photon emission. Generally, most normal cells in organisms like osmanthus tree seedings are in a sub-change state, and photons can be easily stored in this state. In other words, osmanthus seedings are in the most steady state when photon emission is lowest.

When environmental variables are higher or lower, the growth status is destroyed and then pathological phenomena such as the increase of luminescence are observed. For instance, oversized and undersized cells both indicate pathological changes in human body. However, this pathological phenomenon is sometimes profitable when it comes to sensory demanded plants and animal products. For example, when environmental variables are higher than the most stable level, osmanthus tree may be taller, having larger leaf area and more buds, though it is a pathological phenomenon strictly. These can be explained by the physiological resistance of bio-systems when the environment condition deviates the most stable state. For modern farming management, it is to take advantage of this bio-resistance to get products that meet these sensory needs.

4 Conclusion

Single factors and interactions of illumination, temperature, humidity and moisture have effect on the luminous character of osmanthus tree seeding leaves. This is explained by the variation of acupoint state when the environment factors change. For single factors, moisture and humidity are more significant than illumination and temperature. For interactions, the interactions of illumination and moisture, temperature and humidity have the most significant influence on bio-emission, and the interaction of temperature and humidity has particular effect comparing with others. In addition, the mathematical model established in this study effectively reflects the impacts of various environmental factors on ultraweak luminous properties of osmanthus tree seedlings.

Acknowledgements. This research was supported by a grant from the Research Fund for the Doctoral Program of Higher Education of China (RFDP) (No. 20090101120081), the major science and technology project of Science Technology Department of Zhejiang Province, China (No. 2008C02006-4), and the project of Education Department of Zhejiang Province, China (No. Y20080327).

References

1. Ramamoorthy, V., Viswanathan, R., Raguchander, T., Prakasam, V., Samiyappan, R.: Induction of systemic resistance by plant growth promoting rhizobacteria in crop plants against pests and diseases. Crop Protection 20(1), 1–11 (2001)
2. Zhang, X.H., Yang, H.Q.: Ultraweak bioluminescence in plant. Journal of Shandong Agricultural University 34(4), 605–608 (2003)
3. Lie, G.H., Zhao, P.F., Zhang, X.M., Lu, H.B., Jiang, T., Li, W.D.: Laser microbeam technique, biological ultraweak photon emission and biological vitomystery. Journal of Qinghai Mormal University (Natural Science Edition) 1, 78–81 (2003)
4. Zhang, Z.L.: Ultra-weak Chemiluminescence Analytical Technology Principle and Application. Progress in Biochemistry and Biophysics 27(1), 102–104 (2000)
5. Zhang, J.P., Zhang, X.Z., Gong, Z.H.: Application of ultraweak bioluminescence in vegetable research. Chinese Agricultural Science Bulletin 22(1), 220–222 (2006)
6. Cheng, H.P., Wang, J.H., Chi, H.C., Zhu, M.Y.: Study on ultraweak luminescence of pisum sativum seeds at the stage of germination. Journal of Zhejiang University (Science Editton) 28(6), 682–685 (2001)
7. Tan, S.C., Xing, D., Tang, Y.H., Li, D.H.: Spectral studies of ultra-weak biophoton emission from Plant's leabes. Acta Photonica Sinica 29(11), 961–965 (2000)
8. Lin, G.Y., Huang, Z.F., Zhang, C.H., Zheng, C.S.: Changes in ultraweak luminescence intensity, respiration rate and physiological metabolism of chrysamthemum during floral differentiation. Acta Horticulturae Sinica 35(12), 1819–1824 (2008)
9. Guo, Y., Yang, H.Q.: Effects of high temperature on bioluminescence and energy metabolism of malus hupehensis (Pamp) rehd.var.pingyeensis jiang. Acta Horticulturae Sinica 35(1), 99–102 (2008)
10. Chen, W.L., Xing, D., Wang, J., He, Y.: Rapid determination of rice seed vigour by spontaneous chemiluminescence and singlet oxygen generation during early imbibition. Luminescence 18(1), 19–24 (2003)
11. Strand, J.F.: Some agrometeorological aspects of pest and disease management for the 21st century. Agricultural and Forest Meteorology 103, 73–82 (2000)
12. Yazgan, M.S., Tanik, A.A.: New approach for calculating the relative risk level of pesticides. Environment International 31, 687–692 (2005)
13. Li, Z.H., Li, P.P.: The study of a system for the diagnosis and prevention of pests, diseases and nutritional deficiency of greenhouse tomato. Journal of Agricultural Mechanization Research 2, 187–191 (2003)
14. Liu, C.Y., Liu, R., Zhu, L., Yuan, J.Q., Hu, M., Liu, S.H.: Homeostatic photobiomodulation. Front Optoelectron China 2(1), 1–8 (2009)
15. Gu, Q.: Biophotonics, pp. 169–193. Science Press, Beijing (2007)
16. Liu, C.Y., Liu, S.H., Hu, X.L.: Time theory on meridian: I. Acupoint research. Journal of South China University (Natural Science) (1), 40–45 (1998)

Construction of Monitoring System of Dangerous and Harmful Species of Import Taiwan Fruits and Vegetables Based on GIS

Hong Chen[1], Qiyong Weng[2], Meixiang Chi[2], Rongzhou Qiu[2], and Jian Zhao[1,*]

[1] Institute of Digital Agriculture Research, Fujian Academy of Agricultural Sciences, 247 wusiRd., Fuzhou, Fujian, P.R. China, 350003
zhaojian@faas.cn

[2] Institute of Plant Protection Research, Fujian Academy of Agriculture Sciences, 247 wusiRd., Fuzhou, Fujian, P.R. China, 350003

Abstract. In order to improve the efficiency of dangerous and harmful species monitoring of import Taiwan fruits and vegetables and achieve informatization, networking and visualization of monitoring, this paper uses geography information system technology to develop monitoring system of import Taiwan fruits and vegetables dangerous and harmful species on the basis of the analysis of monitoring business needs. The system realizes some useful functions, such as showing monitoring data in map, map operation, monitoring data report and monitoring data management. This paper supply informatization supporting platform to dangerous and harmful species monitoring of import Taiwan fruits and vegetables and provide information support for scientific decision-making of alien invasive biological prevention and control department.

Keywords: GIS, import Taiwan fruits and vegetables, dangerous and harmful species, monitoring system.

1 Introduction

In recent years, with the cross-strait relations gradually easing, between two sides of the agricultural cooperation and trade also have entered a development of the fast lane. From the CPC Central Committee and the State Council departments at all levels have issued a series of preferential policies to promote cross-strait agricultural exchanges and cooperation and to encourage Taiwan farmers to the mainland venture capital. Due to the expansion of cross-strait agricultural trade require fast inspection of Taiwan's agricultural products and implement a policy named "green clearance" to fruit and vegetable of Taiwan on entry inspection and quarantine. The fruit and vegetable of Taiwan planting area expands unceasingly in the mainland, constantly increasing the risk of pests and diseases of the Taiwan fruits and vegetables at the introduction. The high-risk diseases and pests within Taiwan fruits and vegetables have found on the mainland, which create potential big losses for fruit and vegetable

* Corresponding author.

D. Li and Y. Chen (Eds.): CCTA 2012, Part II, IFIP AICT 393, pp. 126–134, 2013.

planting in the mainland[1]. In response to this form, to carry out the entry of dangerous and harmful species monitoring of Taiwan fruits and vegetables and to keep abreast of the occurrence and distribution of the epidemic have great significance to formulate scientific prevention and control measures.

Recently, quarantine departments have been monitoring and researching dangerous and harmful species found in fruits and vegetables entering Taiwan and have achieved certain results, in coastal provinces importing a large number of Taiwan fruits and vegetables[2],[3],[4],[5],[6]. However, most of the monitoring also used handwritten notes, which are not conducive to data preservation and could easily damage and loss records. This will impact analysis and research of monitoring data in the future. In addition, most of the monitoring data are detailed position information, reflecting the significant spatial distribution characteristics. The monitoring of the spatial distribution and the route of transmission of invasive alien species is one of the core content of the work of prevention and control. How about making the monitoring data of the spatial distribution characteristics to get a better display, data management and application issues need to be addressed. In view of the above consideration, this paper proposes building a monitoring system of import Taiwan fruits and vegetables dangerous pests based on GIS. With GIS spatial data processing and display capabilities, combined with the techniques of database technology, network technology and rich client technology, develop data reporting and management platform for dangerous and harmful species monitoring, which provides decision support for quarantine and pest prevention and control.

2 System Design

2.1 System Construction Goal

Dangerous and harmful species monitoring monitors and detects the occurrence and development of a certain period of the disease, pests or weeds within a certain range[7]. However, a complete dangerous and harmful species monitoring system should include a standardized monitoring processes and standardized data management system. According to the monitoring requirements, the goal of constructing the monitoring system of dangerous and harmful species of import Taiwan fruits and vegetables includes below elements.

1. Establish monitoring data standard of dangerous and harmful species of import Taiwan fruits and vegetables and establish a unified database.
2. Apply GIS technology to realize dangerous and harmful species monitoring data online publishing, data display and spatial queries, providing decision-making of prevention and control with spatial distribution data.
3. Realize multi-source collaborative dangerous and harmful species monitoring data reporting, auditing, and management, supplying unified data management and sharing platform to monitoring.

2.2 System Framework

The development of the monitoring system of dangerous and harmful species of import Taiwan fruits and vegetables uses the browser/server framework of multi-tier architecture. Improving system maintainability, security, and extensibility, system reasonably splits functionality and business logic in different layers, as fig.1. show The framework of system business functions is divided into four layers except system users. User performance layer is responsible for system interaction with the user, receiving user input or data reported and returning the results of the system running. Data service layer sets up map data and business data transmission channel between the user and the system. Business logic layer achieves all business processing functions of the system. Data storage layer is used to store various types of system information. At the same time, monitoring data standards and data management mechanisms ensure the efficient operation of the system and standardized management.

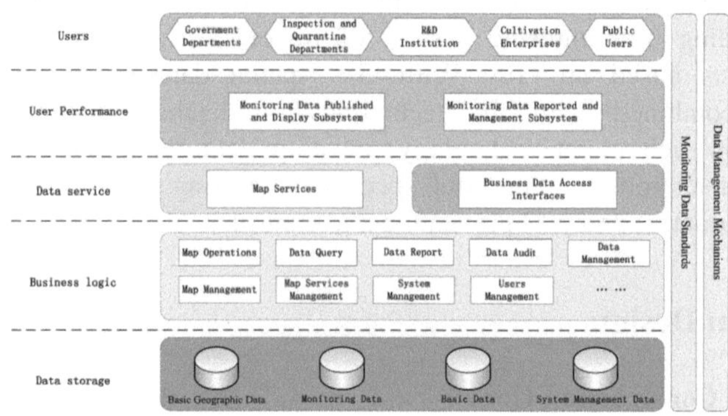

Fig. 1. The system framework

2.3 System Business Process

In the business process, the system operation and management staff can be divided into: fields investigator, data manager, data auditor and administrator. All categories of staff have different permissions, as fig.2 show. The field investigator is responsible for monitoring the occurrence and spread of dangerous and harmful species and submitting monitoring data to the system through the network. Monitoring positions and monitoring objects depend on the basic data included in the system database. In practice, if the monitoring positions and the monitored objects change, investigator data must require the administrator to apply for basic data changes. The data manager is responsible for basic data management and change. The administrator is responsible for managing system users, user permissions and the map services. The data auditor is responsible for auditing and publication of the data. The monitoring data and the basic data changes passed audit can be entered into the database and data that has not been approved will be sent back to the data submitter to modify. Only approved data will be released.

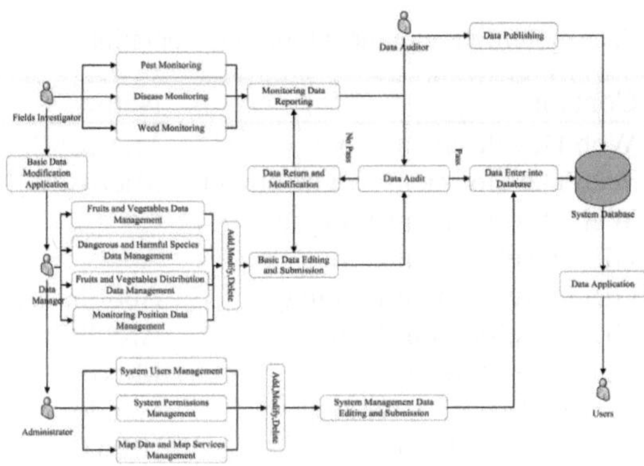

Fig. 2. The system business process

2.4 Data Standards and Database Design

In order to ensure the standardization of the system data resource construction, this paper developed pest monitoring data collection standards and basic data collection standards based on the monitoring practice. Among them, pest monitoring data collection standards include: pest survey standard, disease survey standard, weed survey standard and specimen labeling standards. The basic data collection standards, including: data standards of fruit and vegetable, fruit and vegetable cultivation data standards, data standards of monitoring positon and species data standards. In addition, data forms were specially designed for each type of data content.

The system database design is based on data standards, to meet the needs of business operation of the system. The system databases are made up of business databases and spatial databases. The business databases manage dangerous and harmful species monitoring and systems running data, including five aspects: investigation of dangerous and harmful species data(pests, diseases, weeds, specimen labels), fruit and vegetable varieties of resources data, fruit and vegetable planting data, monitoring positions data(plantation, processing enterprises, distribution center, the port of entry), and system management data(users information, system permissions).The spatial databases store basic geographic data for the map display, including national administrative divisions data (province, municipality and county), national administrative center data(province, city, county, township and village), railways, roads and rivers, etc.

3 System Realization

3.1 Development Tools and Runtime Environment

After fully considering the construction goal and the function design, system of the development tools and runtime environment in the following table:

Table 1. The system development tools and operation environment

Classification	Content	Name
Development Tools	Web Development Tool	Microsoft VisualStudio 2008
	Web Development Environment	Microsoft .NET
	Web Development Language	C#
	GIS Development Tool	Adobe Flex 4.0
	GIS Development Language	Actionscript
	GIS Development Component	ArcGIS API for Flex 2.5
Operation Environment	Database Platform	Microsoft SQL Server 2005
	Server Operation System	Windows Server 2003 SP2
	Web Service Platform	IIS 6.0
	GIS Service Platform	ArcGIS Server 9.3
	Client Plug-in	Adobe Flash Player

3.2 Key Technology Research

3.2.1 RIA Technology

The WebGIS rely principally on the map interface for data display and interaction. In the traditional Web application, a number of complex applications often require multiple data response and transaction processing between the client and server, which cause the page repeatedly refresh and affect the system user experience effect[8]. At the same time, traditional Web applications due to the scalability, interaction, semantic, page links and other problems, resulting in inefficient system development, difficult maintenance, that limits the system application[9]. To solve these problems, the Rich Internet Application(RIA) technologies used in the development of the system to compensate for the lack of traditional Web applications. The RIA effectively integrates desktop applications powerful user interaction with the Web application cost-effective deployment and excellent cross-platform features, representing the new trend of the development of network application[10]. System WebGIS interfaces used Adobe's RIA framework - the Flex, with ESRI's ArcGIS API for Flex development components, achieving rich spatial data display and user experience.

3.2.2 GIS Service Technology

GIS service is a set of functional entities of geographic information software in the network environment, exposing package functionality through the interface[11]. It combines the Web Service technology and GIS technology with the characteristics of cross platform, telescopic, fully functional, system integration and low cost, and high degree of standardization, that provides an effective solutions to the GIS system of internal coupling high ,lack of interoperability, difficulty integrate with other systems and other issues[12]. This paper used ArcGIS Server software releases REST standard data interface as GIS services. Independent service for different functional applications is deployed to achieve the efficient sharing of multi-source geographic data, providing basic geographic information support to unified system data display.

3.2.3 Tile Map Technology

If the client browses the map using real-time rendering mode, each response from the server to the client's request must be each real-time rendering maps and results are returned to the client display. Considering the application environments of the system which are mostly in the relatively backward areas of rural that have terrible network infrastructure conditions,real-time map rendering is not only inefficient but also affect the user operation. Therefore, the deployment of the system used a tile map technology to cache map in the server. When the client initiates a request can be directly read static images in the cache, which can effectively improve the mapping and transmission efficiency and provide fast map display effect[13].

3.3 System Function Realization

The monitoring system of dangerous and harmful species of import Taiwan fruits and vegetables is divided into: monitoring data published and display subsystem and monitoring data reported and management subsystem, as fig.3 show:

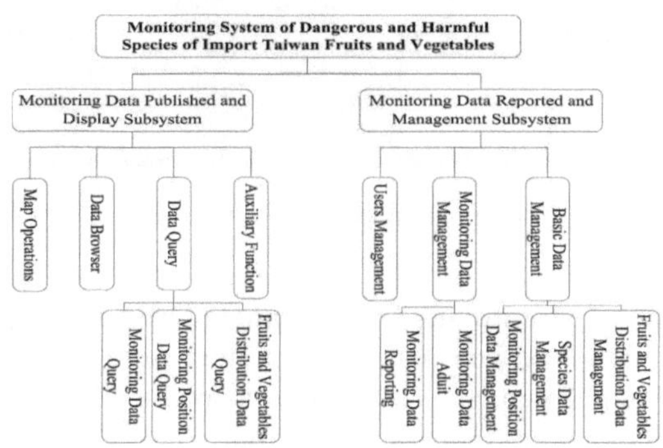

Fig. 3. The system function structure

3.3.1 Construction of Monitoring Data Published and Display Subsystem

The monitoring data published and display subsystem using GIS technology is responsible for displaying data to public users and realizes data browse and query based on map interface, as fig.4 show. As the use of RIA technology makes interface with drag controls, smooth scrolling animation, effect enhances the display of maps and get rid of the traditional Web page monotonous user operation. This subsystem has main functions below:

1. Map operations: the realization of the basic map interface operations. For example: zoom in, zoom out, pan, full extent, overview map, layer management and so on.

2. Data browser: do not enter any query to display data to help the user quickly view the data on the map. When you open a type of data query window, the subsystem will automatically display in the map for all of this type data spatial distribution points. Click on one point will pop-up detail information box of the data.

3. Data query: the subsystem data query function has three types, including monitoring data query, monitoring position data query and fruit and vegetable distribution data query. Different types have different terms in the query. Subsystem uses separate query window for each query type and the query results also uses a different marker style to show the distinction to increase the contrast effect of the spatial distribution of different data types. In addition, most of the type of query can be subdivided into subclass, in the same query window using different page tab to arrange query condition. When you close the query window, the subsystem will clear the results of the map and free up system resources.

4. Auxiliary function: provide mapping, positioning bookmark, print and other complex map operations. These features improve the ability to interact with the map interface and a supplement to map operations.

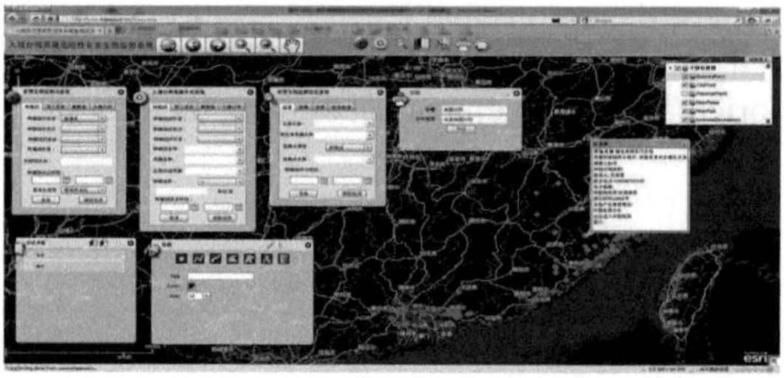

Fig. 4. The monitoring data release and display subsystem interface

3.3.2 Construction of Monitoring Data Reported and Management Subsystem

The monitoring data reported and management subsystem is responsible for system management and data maintenance, which don't open to the public users. The user groups of the subsystem are limited to the four kinds of system managers. The subsystem through the user account and access control to restrict the operation of the different users, preventing data breaches caused by the different users cross operation. This subsystem has main functions below:

1. Users management: supply the management of the system user data and permissions configuration. Due to the particularity of the system data reporting and management of business, users management operations only by the system administrator.

2. Monitoring data management: this function includes data reporting and audit. Data reporting page provides data furnished, the data submitted and data status view. When submitting large amounts of data, the system can save the data reported in batches until all data furnished. Monitoring data audit page lists the reported monitoring data in list form.The data auditor can select a data record to view detailed information about it. If the data is qualified by the audit, it will be entered into the system database. If the data is not approved, it will be displayed in the data status view page in order to further modify by fields investigator.

3. Basic data management: provide the management operations of the system basic data(monitoring positon data, species data and fruit and vegetable distribution data) such as add, modify, and delete. Three kinds of data are located in a different page tab, arranged in a list and switched through the navigation bar.

3.4 Construction of Basic Data

The basic data accumulation is a necessary precondition for carrying out applications. This paper has collected some basic data about Fujian and Guangdong which have imported a large number of Taiwan fruit and vegetable. At present, the system has collected 24 records of fruit and vegetable data; 66 records of dangerous and harmful species data; 132 records of various types of monitoring position data; 230 records of fruit and vegetable distribution data.

4 Conclusion and Discussion

This paper introduced GIS technology into dangerous and harmful species prevention and control work, analyzed the needs of the business of import Taiwan fruits and vegetables dangerous and harmful species monitoring data management and proposed construction scheme of monitoring system. The system effectively integrates monitor data reporting, management and presentation of business processes based on collaborative division of multi-role labor. In the system construction of the integrated use of RIA, GIS services, tile maps and other information technology, make the interactive experience of the system map interface has been enhanced significantly. The system could provide the prevention and control of decision-making with data in support of import Taiwan fruits and vegetables dangerous and harmful species occurrence and spread, provide information channels for reporting of monitoring data and supply information platform to inter-regional or cross-sectoral monitoring data sharing.

The dangerous and harmful species prevention and control as a systemic project, monitoring is a key component. However, the lack of the functions of epidemic prediction and analysis in support of the IT technology will restrict the application of the system. How is monitoring data effectively applied? How to play GIS technology advantages in spatial analysis? How to combine the monitoring data and prediction models to provide better information services for the prevention and control of decision-making? All of these issues will be addressed in the further focus.

References

[1] Network collaboration platform of Dangerous and Harmful Species of Import Taiwan Fruits and Vegetables Prevention and control technology research and demonstration, `http://www.mtpest.cn/news/NewsContent.aspx?ID=78`

[2] Huimin, L., Yan, C., Qinghai, F.: A Preliminary Report on the Monitoring of Bactrocera in Fuzhou. Entomological Journal of East China 15(4), 272–275 (2006)

[3] Furong, L., Jianguo, S., Yuan, W., et al.: Detection and Pathotype Identification of Pepper Mild Mottle Virus in Sweet Pepper Seeds from Taiwan. Chinese Journal of Tropical Crops 32(6), 1128–1135 (2011)
[4] Furong, L., Shiming, L., Zhipeng, F., et al.: Detection and Identification of Acidovorax Citrulli from Imported Watermelon Seeds. Plant Protection 36(6), 135–138 (2010)
[5] Furong, L., Mujin, G., Shiming, L., et al.: Detection and Identification of the Tomato Mosaic Virus in Imported Tomato Seeds. Plant Protection 37(5), 124–128 (2011)
[6] Pengying, H., Linlin, L., Zhengqing, Y., et al.: Taiwan's Wax Apples Intercepted Anselmella miltoni Girault. Plant Quarantine 22(3), 178–179 (2008)
[7] Wancai, L., Yuying, J., Yuejing, Z., et al.: Thought on Promote Agricultural Pest Monitoring and Warning Development. China Plant Protection (8), 28–31 (2009)
[8] Junwei, W., Ke, C., Zhifang, G., et al.: Primary Designing and Implementation of Monitoring Information System for Epidemic Situation of Quarantine Pest Based on ArcGIS and .NET. Plant Quarantine 24(4), 1–5 (2010)
[9] Yonggang, L.: Discussion of WebGIS Reseach Based on Flex_RIA and REST. Geomatics & Spatial Information Technology 34(6), 87–88 (2011)
[10] Liang, X., Liang, Z., Bingfang, W., et al.: Research on Hydro-environment Analysis System of Three Gorges Reservoir Based on RIA. Computer Engineering 37(23), 244–247 (2011)
[11] Dayu, C., Faliang, W., Zhiwu, Z., et al.: Research and Realization of Composition of Geographic Information System Services Based on BPEL. Science of Surveying and Mapping 35(6), 259–261 (2010)
[12] Xiaojun, Y., Weirui, W., Jianping, L., et al.: Design and Implementation of the Spatial Decision Support System for Agriculture. Transactions of the Chinese Society of Agricultural Engineering (9), 257–262 (2010)
[13] Menglong, H.: Study on the Application of Tile Map Technology in Desktop GIS. Geospatial Information (4), 149–151 (2011)

Design and Experiment of NIR Wheat Quality Quick Detection System*

Lingling Liu**, Bo Zhao***, Yinqiao Zhang, and Xiaochao Zhang

National Key Laboratory of Soil-Plant-Machine System,
Chinese Academy of Agricultural Mechanization Sciences, Beijing, China, 10083
whyisp@sina.com, zhaoboshi@126.com

Abstract. In this paper, NIR wheat quality quick detection system (NIR-WQDS) was developed on the base of grating technology with a scanning range of 900-1700nm.46 wheat samples were analyzed to compare performance of NIR-WQDS and MPA FT-NIR spectroscopy (MPA). Experiment results of NIR-WQDS show that the coefficient of determination R^2 is 94.44%, root mean square error of cross validation RMSECV is 0.3460, ratio of performance to standard deviate RPD is 4.240, and root mean square error of prediction RMSEP is 0.2430, respectively. The model of MPA spectroscopy gave an R^2 of 95.95%, a RMSECV of 0.2950, a RPD of 4.970, and an RMSEP of 0.2270 respectively. It is found that spectra of NIR-WQDS have a roughly same changing trend. NIR-WQDS can detect wheat quality with good accuracy, repeatability and stability, which is similar to MPA. The results show that NIR-WQDS can work steadily with a good performance.

Keywords: NIR spectroscopy, Wheat quality, Quick detection system, Quantitative analysis, Predictive performance.

1 Introduction

Near infrared spectroscopy (NIRS) technique is characterized by simple operation, short period and environmental friendliness. Compared with Kjeldahl nitrogen method, NIRS technique can detect protein of wheat quickly and nondestructively, which is meaningful for wheat quality inspection, processing and breeding. NIRS technique is constituted by NIR instrument, chemometrics software and calibration model [1]· As NIRS instrument is becoming miniature, professional, steady and excellent, domestic and foreign famous companies have developed kinds of the special purpose instruments focused on inspecting grain, feed, soil and oil [2]. Domestic researchers have already gained some achievements in wheat spectroscopy research, but our research is still far

* This work was supported by national science and technology support program for the 12th five-year plan (2011BAD20B06), and national basic research program of China (2010CB735707).

** Major in automatic measuring and control technology.

*** Corresponding author.

D. Li and Y. Chen (Eds.): CCTA 2012, Part II, IFIP AICT 393, pp. 135–141, 2013.

lower than international level [3]. In order to analysis wheat quality professionally and quickly, NIR-WQDS was developed on grating technology and it works in the 900-1700nm region. In this paper 46 wheat samples were detected by MPA spectroscopy and NIR-WQDS to test the performance of NIR-WQDS.

2 Development of System

NIR-WQDS includes optical system with fixed grating, CCD drive and signal acquisition system, CCD temperature control system, light source control system, sample rotator control system, and computer software system. NIR absorption spectra are recorded in the 900-1700nm (11111.1—5887.8cm^{-1}) range. CCD is controlled to work at 0°C and its temperature controlling precision is ±0.1°C. The light intensity accuracy can be controlled within 0.02% and the SNR is 4000:1. The system structure of NIR-WQDS is shown in figure 1.

Following are working principles of NIR-WQDS spectroscopy:

(1) Light source control system is based on current negative feedback. It controls light intensity of source and provides stable near infrared light to the whole instrument.

(2) After light concentrated and collimated, it penetrates wheat sample directly in a sample rotator. Then, diffuse transition light is obtained.

(3) In optical system, diffuse transition light is concentrated and collimated. Then, light is projected onto fixed optical grating and a continuum spectrum is produced.

(4) Cooled CCD detector is driven to sample the spectrum signals and converts it to electric signals. The working temperature of CCD detector is cooled and balanced by CCD temperature control system.

(5) Signal sampling system performs fast, real time data gathering and communicates with upper computer. Then software system in the upper computer processes signals acquired.

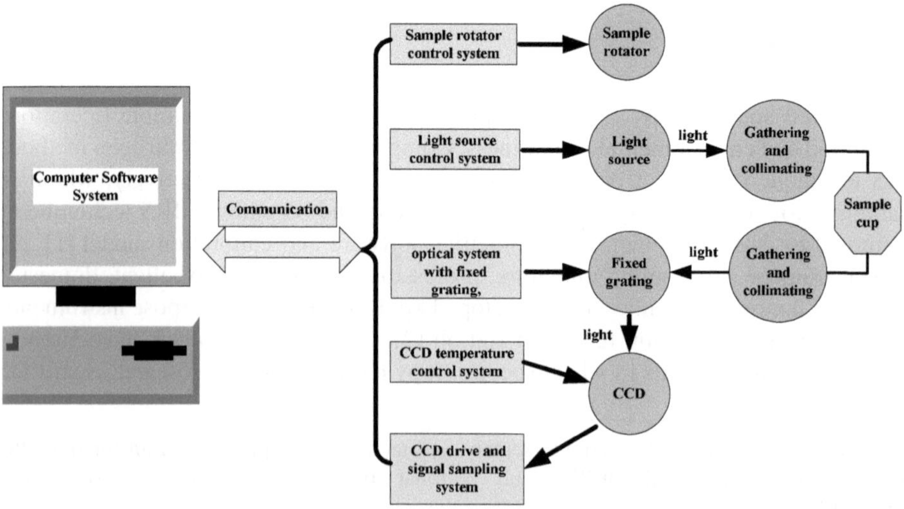

Fig. 1. Structure of NIR-WQDS

The main characteristics of NIR-WQDS are as follows:

(1) To collect more hydrogen information in wheat groups, spectrum wavelength ranges from 900nm to 1700nm.

(2) To eliminate influence caused by temperature shift and ensure high measurement precision, CCD temperature control system was designed to ensure CCD's working temperature low and steady.

(3) Light intensity control system was designed to ensure stable light with accuracy 0.02%, which reduces error from light intensity shift.

(4) Wheat sample is heterogeneous and inhomogeneous; therefore sample rotator was designed to eliminate error by increasing number of scans [1].

3 Materials and Methods

3.1 Experiment Materials

46 wheat kernel samples were brought from Heilongjiang academy of agricultural sciences. After milling, the samples were sealed in plastic bags. Protein standard values of samples were measured by national standard method, and it varies representatively and widely from 9.830% to 16.94% with a 14.18% average.

MPA was selected as the target analyzer, which is a fourier transform infrared spectrometer produced by Bruker Optics Company. It has a scanning range of 12800-3600 cm^{-1}. The NIR spectra of 46 wheat powder samples were collected using quartz sample cup on sample rotator in this experiment. The main parameters and setting of two spectrometers are shown in table 1.

Table 1. Main parameters and setting of two instruments

Spectrometer	Spectra Range (cm^{-1})	Sample Cup	Detector	Modeling software
NIR-WQDS	11111.1—5887.8	quartz	InGaAs	OPUS6.5
MPA	12800- 3600	quartz	PbS	OPUS6.5

3.2 Experiment Methods

The spectra were acquired with MPA and NIR-WQDS at room temperature (20~25°C) respectively. In order to decreasing operation differences, both instruments were started at the same time for about half an hour. After completing all measurements, the obtained data was processed into absorbance value by using MATLAB 7.0 software. Wheat NIR spectral data was finally analyzed using OPUS 6.5 software. In OPUS6.5, partial least squares (PLS) regression was used to build calibration models.

4 Results and Analysis

4.1 Wheat NIR Spectra

46 spectra collected by both instruments were saved as OPUS and showed in OPUS6.5 software, see figure 2. Because of different component contents, there is no absolutely general overlapping among all spectra. All spectra have a good identity without any anomaly. Ranging from 11111.1 cm^{-1} to 5887.8cm^{-1}, each of NIR spectra has a roughly same changing trend and intendancy with a little difference in peak shape and peak position. In both pictures, wheat protein absorption band with peaks appears at about 10990cm^{-1}, 9804cm^{-1}, 6623cm^{-1} and 5051cm^{-1}. It shows that NIR-WQDS has a stable performance and a good reproductively of sampling wheat NIR spectra.

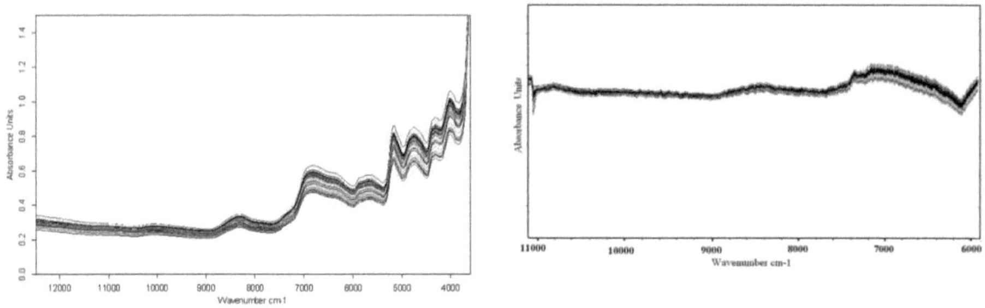

(a) (b)

Fig. 2. 46 wheat sample spectra: (a) MPA (b) NIR wheat quality quick detection system

Comparing with the MPA scans, NIR-WQDS scans are thorny and not smooth with spike-waves at local wavelength and extreme ends, which shows obvious noise and low SNR. To improve the quality of spectra, we can increase scans and improve precision of light control system, and decrease affection caused by CCD dark current.

4.2 Peprocessing and Modeling Methods

In this paper, PLS were selected for building equation between spectra and protein standard values. R^2, RMSECV and RDP were computed to select the best model. Higher R^2 represents that the predictive ability of model is better. RMSECV (cross validation) and RMSEP (external validation) indicate the rate of deviation between predictive value and true value. When they get smaller, models get better. RPD can state the stability of model. A model has a good stability when its RPD>3.

In order to eliminate differences, OPUS 6.5 software was used to process spectra obtained by two instruments and to build PLS calibration model. More excellent evaluating indicators and wider modeling wavelength ranges are used to select a better calibration model [4]. Table.2 shows preprocessing algorithm, wavelength ranges and rank when modeling.

Table 2. Preprocessing methods, spectral range and rank of the calibration models

Instrument	Preprocessing	Regions(cm^{-1})	Rank
NIR-WQDS	First derivative	7500.8-5887.8	5
MPA	First derivative +Vector normalization(SNV)	12489.5-5446.3	6

4.3 Results and Analysis

The relationship between predictive values and standard values in both models is shown in Fig. 3. We can acquire that both models have an excellent goodness of fit, but the linear correlation of MPA is more significant than that of NIR-WDQS.

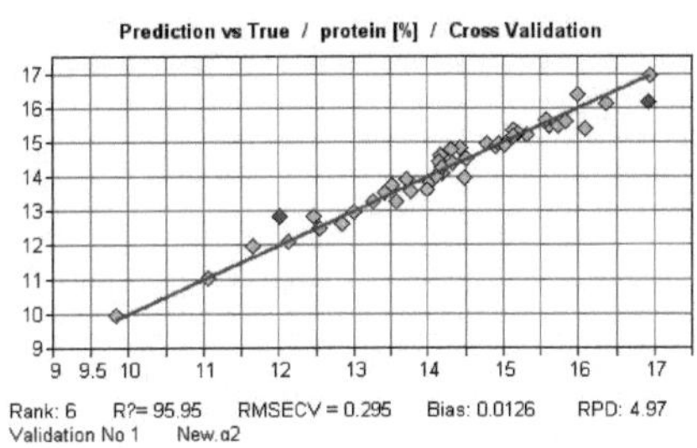

(a)

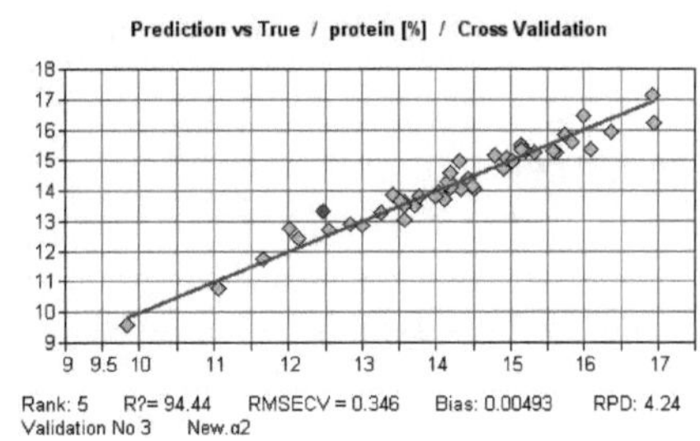

(b)

Fig. 3. The relationship between prediction value and standard value: (a) MPA (b) NIR-WQDS

Evaluating parameters of two models have been shown in Table.3. According to models acquired from NIR-WQDS and MPA, R^2 are 94.44% and 95.95%, RMSECV are 0.3460, 0.2950, RMSEP are 0.2430, 0.2270, and RPD are 4.240 and 4.460, respectively. According to experiment results, forecast performance of NIR-WQDS is good but less excellent than MPA. In the NIR-WQDS, there is a significant linear correlation between standard values and prediction values and the model gets a good predictability and stability [3]. It proves that NIR-WQDS spectroscopy has a good performance with high stability, accuracy and repeatability and it is practical to inspect wheat quality.

Table 3. Optimization results of both models

Instrument	R^2	RMSECV	RMESP	RPD
NIR-WQDS	94.44%	0.3460	0.2430	4.240
MPA	95.95%	0.2950	0.2270	4.970

5 Conclusion

NIR wheat quality quick detection system was designed based on grating technology in a spectra range of 900-1700nm. The spectrometer consists of optical system with fixed grating, CCD drive and signal acquisition system, CCD temperature control system, light source control system, sample rotator control system, and computer software system. In this paper, the instrument is proven to be accurate, stable and reliable by analyzing wheat protein content in 46 wheat samples.

The spectra acquired by NIR-WQDS are thorny with spike-waves at extreme ends, which may be caused by noise from slightest vibration of sample rotator, or precision of temperature control system. To improve performance of NIR-WQD, suggestions of the next step are put forward as follows:

(1) Continuing to reduce working temperature of CCD detectors to low dark current.

(2) Adding more processing algorithms and modeling algorithms to perfecting supporting computer software system.

References

1. Yan, Y.-L.: Foundation and application of near-infrared spectroscopy analysis. China Light Industry Press, Beijing (2005) (in Chinese)
2. Chu, X.-L., Wang, Y.-B., Lu, W.-Z.: Present situation and prospect of near infrared spectrometer in China and abroad. Analytical Instrumentation 4, 1–4 (2007) (in Chinese)
3. Zhang, X.-C., Wu, J.-Z., Xu, Y.: Near infrared spectral analysis technology and its application in modern agriculture. Publishing House of Electronics Industry, Beijing (2012) (in Chinese)
4. Li, J., Zhang, X.-C., Yuan, Y.-W., Zhang, J.-N.: Analysis of soil nutrient content based on near infrared reflectance spectroscopy in Beijing region. Transactions of the CSAE 28(2), 176–179 (2012) (in Chinese)
5. Xiang, X.-Y., Wen, Z.-Y., Long, Z.-C., Hong, M.-J., Liang, Y.-Q., et al.: Research on the near-infrared spectrometer system. Spectroscopy and Spectral Analysis 28(8), 2286–2290 (2009) (in Chinese)
6. Yin, H.-M., Wu, W.-F., Fu, Y., Zhang, Y.-Q., Liu, L.-T.: Designed and test of near infrared detector for grain component based on virtual instrument. Transactions of the Chinese Society for Agricultural Machinery 41(5), 115–119 (2010) (in Chinese)
7. Sun, X.-R., Liu, C.-L., Wu, J.-Z., Suo, S.-Z., Wu, S.-N.: Research of rapid and undamaged assessment of flour quality using near infrared spectroscopy. Journal of Beijing Technology and Business University 29(4), 68–70 (2011) (in Chinese)
8. Zhang, Y.-Q., Zhao, B., Wang, H., Mao, W.-H., Zhang, X.-C.: Technology of near infrared spectroscopy analysis system design based on fixed grating and charge-coupled device. Transactions of the Chinese Society of Agricultural Engineering 25(9), 176–181 (2009) (in Chinese)
9. Jin, H.-L., Xu, C.-H., Xu, Z.-L.: Determination of Protein Content in Wheat Grain by Near Infrared Spectroscopy. Journal of Henan University of Technology (Natural Science Edition) 31(6), 21–24 (2010) (in Chinese)

10. Balabin, R.M., Safieva, R.Z., Lornakina, E.I.: Wavelet neural network (WNN) approach for calibration model building based on gasoline near infrared (NIR) spectra. Chemometrics and Intelligent Laboratory Systems 93(1), 58–62 (2008)
11. Sundaram, J., Kandala, C.V., Govindarajan, K.N., Subbiah, J.: Sensing of moisture content of in-Shell Peanuts by NIR Reflectance Spectroscopy. Journal of Sensor Technology 2, 1–7 (2012)

Research on Prediction Model and Characteristic Parameters on Dry Matter Accumulation in Wheat Based on Normalized Method and Grey System

Juan Liu[1], Xiaoli Zhao[2], Shuping Xiong[1], Xinming Ma[1,2,*], Yanfeng Wang[1], and Jing Wang[1]

[1] Key Laboratory of Physiology, Ecology and Genetic Improvement of Food Crops in Henan Province, College of Agronomy, Henan Agriculture University, Zhengzhou Henan, China
[2] College of Information and Management Science, Henan Agriculture University, Zhengzhou Henan, China
Xinmingma@126.com, Liujuanviolet@163.com

Abstract. To investigate a model to simulate wheat dry matter accumulation, three wheat cultivars with different tillering abilities were grown at three densities each in a field experiment. Five simulation models with high correlation coefficients for relative dry matter accumulation were established by the method of normalized. Among these models, the Richards equation was the best in fitting and forecasting, i.e., y=1.1435/(1+e$^{0.2776\text{-}4.6558x}$)$^{1/0.1130}$, r=0.9927. Correlation coefficient of grey comprehensive relationship degree between actual dry matter accumulation in occurrence time of maximum rate of dry matter accumulation and dry matter weight was highest, so a higher actual dry matter weight in occurrence time of maximum rate of dry matter accumulation played an important role in stabilizing and improving dry matter weight of wheat.

Keywords: Wheat, Dry matter accumulation, Simulation model, Characteristic parameters.

1 Introduction

Wheat is one of the most important grain crops in China; over 95% of its yield is based on photosynthesis. Dry matter accumulation is the final product of wheat photosynthesis, and closely related to the yield formation [1-2]. Regulation and control of many cultivation factors affect the wheat dry matter accumulation, among them density is the important factor which affect the wheat growth and development and yield [3]. So it has great significance to clear the high yield or even super high yield in wheat by investigating the relationship between density and dry matter accumulation. Dry matter accumulation appears S-shape during the whole life of wheat, and has obvious differences in different densities [4]. There are great differences among those tillering abilities of wheat in different cultivars, so it is

[*] Corresponding author.

D. Li and Y. Chen (Eds.): CCTA 2012, Part II, IFIP AICT 393, pp. 142–149, 2013.

necessary to study the evolution law of dry matter accumulation process with different densities and tillering abilities of wheat cultivars.

Crop simulation technology can quantitatively describe the process of crop growth [5]. At present, there has been created some simulation models on photosynthetic production and dry matter accumulation of wheat [6-7], but those models have stronger mechanism, and more parameters which were more difficulty to get, so their practicability was obvious limited. Empirical models mostly used Richards or Logistic equation to simulate the process of dry matter accumulation in wheat [8-9], However, some other researchers hold that the time scale, which measured on the process of crop growth and development, should be selected reasonably[10]. For example, ecological variables such as accumulated temperature, are more representative than time variables that of days, and they can also reflect the status of crop growth better. In order to investigate the relationship between density, tillering ability and dry matter accumulation of wheat, provide the theoretical basis for prediction on wheat growth and culture of high yield, making research based on field experiment, and taking the accumulated temperature as time scale, a prediction model, which is suitable for dynamic dry matter accumulation with different densities and tillering abilities of wheat, was tried to create with the method of normalization[11].

2 Materials and Methods

2.1 Experimental Design

Field experiments were conducted during 2009~2011 at the Experimental Station of the Henan Agricultural University, Zhengzhou, PR China. The cultivars studied were medium tillering ability and high earing rate wheat (YM49-198), stronger tillering ability and low earing rate wheat (LKAZ8) and stronger tillering ability and high earing rate wheat (YZ4110). Seeds were sown on 14th October 2009 and 15th October 2010, which produced with densities of 750, 000, 2, 250, 000, and 3, 750, 000 plants hm^{-2} in randomized block design. The field was also managed as same as high yield wheat field.

2.2 Measure Items and Methods

Choosing 200 strains wheat seedling in uniform growth state at three leaf period of wheat, marking those wheat seedling with label, and sampling every 10 days from three leaf period to wintering stage, every 30 days from wintering stage to returning green stage, and every 10 days from returning green stage to maturity, dry matter of 10~20 samples were measured each time. Weighing and calculating the dry matter accumulation of aerial part in unit land area (*DMA* for short).

2.3 Meteorological Data

The meteorological data was come from Meteorological Bureau of Henan, mainly including maximum temperature, minimum temperature daily during the growth period of wheat and so on. Calculation method of daily mean temperature is using an equation T_i=(T_{imax}+$T_{(i+1)\min}$)/2, where T_i is mean temperature of the *i* day, $T_{i\max}$ is the maximum temperature of the *i* day and $T_{(i+1)\min}$ is the minimum temperature of the *i*+1 day. Calculation method of accumulated temperature is $AT_i = \sum T_i$, where AT_i is the accumulated temperature of the *i* day, while if $T_i \leq 0$°C or $T_i \geq 35$°C, then T_i=0°C [12].

2.4 Normalized Treatment of Data

Define the maximum dry matter accumulation ($DMA_{\max}$) the maximum accumulated temperature ($AT_{\max}$) to 1 (using the data of 2009~2010), the relative dry matter accumulation ($RDMA_i$) and relative accumulated temperature (RAT_i) were obtained with the method of normalization [11], using the equations as follows: $RDMA_i=DMA_i/DMA_{\max}$, $RAT_i=AT_i/AT_{\max}$. The DMA_i is the dry matter accumulation in different growth stages, $RDMA_i$ is the relative dry matter accumulation in different growth stages, RAT_i is the relative accumulated temperature in different growth stages. The value of $RDMA_i$ and RAT_i were 0~1. The experiment data have been fitted by soft of Curve Expert 1.38, statistical analyzed by SPASS 13.0, and mapped by Sigmaplot in this paper.

2.5 Model Validation

Using the experiment data of 2010~2011 as validation data, the accuracy of simulation model of RDMA was tested [13].

3 Results and Analysis

3.1 Established and Screened of RDMA Models

Treating *AT* and *DMA* that from seeding stage to maturity period of wheat by the method of normalization, and fitting the relationship between *RAT* and *RDMA* by soft of Curve Expert 1.38, five mathematical equations in Table 1 which have higher correlation coefficient were got. The limit value of the 5 fitting equations show that simulation equations of (1), (2), (3) could not explain the *DMA* process, but (4) and (5) could reflect the law of *DMA* dynamic changes. Equation (5) is a special case of (4), and it could be transformed from (4) when *d*=1. So equation (4), the Richards equation was chosen to simulate the relationship of *RDMA* and *RAT* in this paper. The Richards curve equation was fitted as follows:

$$y = 1.1435/(1+e^{0.2776-4.6558\cdot x})^{1/0.1130} \quad . \tag{6}$$

Table 1. Relative dry matter accumulation models of wheat

Simulated equations	Coefficients				Standard deviation	Correlation coefficient	y value
	a	b	c	d	SD	r	$x\to\infty$
$y=a+bx+cx^2+dx^3$. (1)	0.130	-1.573	5.863	-3.412	0.044	0.993**	∞
$y=a+b\cos(cx+d)$. (2)	0.512	0.497	3.645	2.622	0.044	0.993**	∞
$y=(ab+cx^d)/(b+x^d)$. (3)	0.009	0.179	1.206	3.573	0.045	0.993**	1.3
$y=a/(1+e^{b-cx})^{(1/d)}$.(4)	1.144	0.278	4.656	0.113	0.045	0.993**	1
$y=a/(1+be^{-cx})$.(5)	1.047	89.652	7.698		0.048	0.992**	1

[1] In these equations x and y denoted *RAT* and *RDMA*, respectively. ** denoted P<0.01. a denoted relative maximum increment; b denoted initial value of parameter; c denoted growth rate parameters; d denoted shape parameter; The same below.

3.2 Validation of RDMA Models in Wheat

The *RDMA* models of wheat were tested using the data of 2010~2011 as validation data, Fig. 1 shows that the simulated values exhibited significant positive correlation with the measured values, correlation coefficients were all more than 0.97. The accuracies of simulation were all over 0.91, and increased with the increase of density. Through the deduced of Richards equation, characteristic parameters with biological significance would be obtained, and they could be used to analyze the dynamic characteristics of *DMA* in wheat quantitatively.

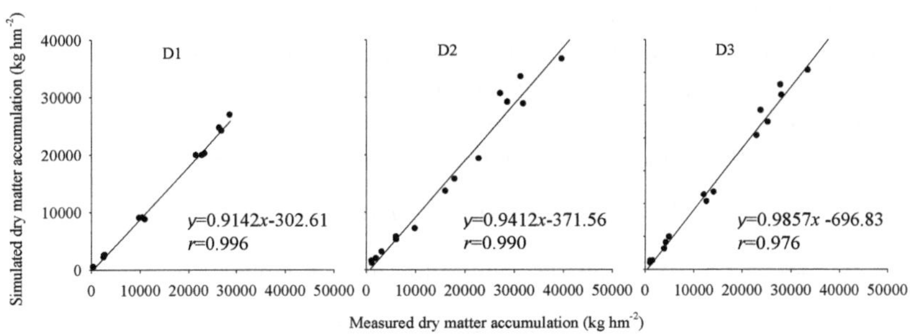

Fig. 1. Relationship between the simulated and measured dry matter accumulation of wheat

3.3 Characteristic Parameters Analysis of Wheat

After derivation calculus to the equation $y=a/(1+e^{(b-cx)})^{1/d}$, variation rate of *RDMA* equation was got as follows:

$$\frac{dy}{dx}=\frac{a\cdot c\cdot e^{(b-c\cdot x)}}{d\cdot(1+e^{(b-c\cdot x)^{(d+1)/d}})} \tag{7}$$

Mean relative growth rate of *DMA* and the maximum relative accumulation rate of dry matter during the whole growth were obtained by integral to equation (7).

Meanwhile, the characteristic parameters, such as *RAT* and *RDMA*, at the fastest stage were also derived [14-15].

$$G = \frac{1}{a}\int_0^a \frac{dy}{dx}dx = \frac{a \times c}{2(d+2)} \quad . \tag{8}$$

$$ARAT = (b - \ln d)/c \quad . \tag{9}$$

$$AR_{\max} = \frac{a \times c}{(1+d)^{(d+1)/d}} \quad . \tag{10}$$

$$ARDMA = \frac{a}{(1+d)^{1/d}} \quad . \tag{11}$$

Where G is the mean relative growth rate of *DMA*, *ARAT* is *RAT* in occurrence time of maximum rate of *DMA*, AR_{max} is the maximum relative accumulating rate of dry matter accumulation, *ARDMA* is *RDMA* in occurrence time of maximum rate of dry matter accumulation. The values of these characteristic parameters were obtained by using the value of a, b, c, d in equation (4), G=1.26, *ARAT*=0.53, AR_{max}=1.86, *ARDMA*=0.44. So *RAT* is 0.53 when AR_{max} appears, and at this time *DMA* was almost half of total *DMA*. The actual accumulated temperature in occurrence time of maximum rate of dry matter accumulation (*ARAT'*), actual dry matter accumulation in occurrence time of maximum rate of dry matter accumulation (*ARDMA'*), mean actual growth rate of dry matter accumulation (G'), and maximum actual accumulating rate of dry matter accumulation (AR_{max}') could be derived by using the value of DMA_{max} and AT_{max} when AR_{max} appears.

In order to determine the relative importance of characteristic parameters to dry matter weight, using the grey system comprehensive relationship degree to analyze the relationship between dry matter weight and characteristic parameters. Consider the dry matter weight as system characteristic behavior sequence:

$$X_i = (x_i(1), x_i(2), ..., x_i(n)) \,. \tag{12}$$

Consider characteristic parameters as related factors sequence:

$$X_j = (x_j(1), x_j(2), ..., x_j(n)) \,. \tag{13}$$

Calculated grey comprehensive relationship degree between each of them. Grey comprehensive relationship degree not only embodied the similar degree between the two sequences, but also reflected the approaching of the change rate with characteristic behavior sequence and related factors sequence relative to the starting point respectively. Meanwhile, it was a quantitative index to characterize more comprehensive if the relationship of sequences are close, and the calculation formula is

$$\rho_{0i} = \theta\varepsilon_{0i} + (1-\theta)\gamma_{0i}, \theta \in [0,1] \,. \tag{14}$$

Where ε_{0i} is absolute correlation degree, γ_{0i} is relative correlation degree, and $\theta = 0.5$. From Table 2, the results of correlation degree analysis show that the correlation coefficients between dry matter weight and characteristic parameters of wheat *ARAT'*, AR_{max}', *ARDMA'*, *G'*were all above 0.6. The correlation coefficients of *ARDMA'*and dry matter weight was 0.861, and then decreased in order of: AR_{max}', *G'*and *ARAT'*. Therefore, the effect of *ARDMA'*on dry matter production was most important, higher *ARDMA'*plays a very important role to improve dry matter weight of wheat.

Table 2. Analysis on dry matter weight and characteristic parameters of wheat based on grey comprehensive correlative degree

Item	Correlation coefficients			
	ARAT'	AR_{max}'	*ARDMA'*	*G'*
Dry matter weight	0.647	0.724	0.861	0.718

4 Conclusion and Discussion

4.1 Established Model and Its Characteristics of Wheat DMA

Wu et al [16] established *DMA* mechanism models, which had stronger interpretability and the parameters had physical meaning. However there were too many parameters in the model. Empirical models with less parameters, but they were only used to describe. Crop growth model should be established by combined the method of mechanism and empirical [17]. Analyzing the characteristics of dynamic *DMA* in wheat, mainly using time variable as the scale of growth and development in the past [18]. Therefore, using the normalized method and making accumulated temperature as time scale. Richards equation was established in this paper as $y = 1.1435/(1+e^{0.2776-4.6558 \cdot x})^{1/0.1130}$ (r=0.9927), it improved generality of the model, overcome the defects of model parameters changes that cultivation measures and varieties caused. The Richards equation has less parameters and good biological significance. The production process of wheat could be predicted by *DMA* which be simulated through the model and accumulated temperature in any growth period.

4.2 Research on Characteristic Parameters of Dry Matter Accumulation in Wheat

By deduced of simulation equation, characteristic parameters with biological significance would be obtained. Dynamic characteristics of crop dry matter accumulation have been analyzed by many scholars [19-20]. As the results show, *RAT* at the maximum rate of dry matter accumulation was 0.53, meanwhile, *DMA* was almost half of total *DMA*. This proved that strengthening field management in early stage of wheat growth, cultivating sound seedlings and constructing proper population structure were important to high yield of wheat. Correlation coefficient of grey comprehensive relationship degree between AR_{max}', *ARAT'*, *G'*and final dry matter weight were more than 0.6. The highest correlation coefficient was between *ARDMA'*and dry matter weight which was 0.861 proved that *ARDMA'* effects the dry matter weight most. A higher *ARDMA′* played an important role in stabilizing and improving dry matter weight of wheat.

Acknowledgments. This research was supported by grants from the Special Research Foundation of Agriculture Industry (201103001), National Natural Science Fund (30771266) and the System of Industrial Technologies for Modern Agriculture of Henan Province (S2010-01-G04).

References

1. Yao, S.-M., Kang, Y.-H., Liu, H.-J.: Studies on dry matter accumulation, partitioning and translocation in winter wheat under sprinkler and surface irrigation conditions. Agricultural Research in the Arid Areas 26(6), 51–56 (2008) (in Chinese)
2. Dordas, C.A., Sioulas, C.: Dry matter and nitrogen accumulation, partitioning, and retranslocation in safflower (Carthamus tinctorius L.) as affected by nitrogen fertilization. Field. Crop. Res. 110(1), 35–43 (2009)
3. Hiltbrunner, J., Streit, B., Liedgens, M.: Are seeding densities an opportunity to increase grain yield of winter wheat in a living mulch of white clover? Field. Crop. Res. 102(3), 163–171 (2007)
4. Su, Z.-G., Yang, W.-B., Tian, J.-C., Song, G.-Z.: The relationships between soluble sugar contents of plant and kernel and dry matter accumulation in different high yielding wheat genotypes. Chinese Agricultural Science Bulletin 23(9), 307–311 (2007) (in Chinese)
5. Cao, H.-X., Zhao, S.-L., Ge, D.-K., Liu, Y.-X., Liu, Y., Sun, J.-Y., Yue, Y.-B., Zhang, Z.-Y., Chen, Y.-L.: Discussion on development of crop models. Scientia Agricultura Sinica 44(17), 3520–3528 (2011) (in Chinese)
6. Liu, T.-M., Cao, W.-X., Luo, W.-H., Wang, S.-H., Yin, J.: A simulation model of photosynthetic production and dry matter accumulation in wheat. Journal of Triticeae Crops 21(3), 26–31 (2001) (in Chinese)
7. Zheng, Y.-F., Yan, J.-Y., Wan, C.-J., Ge, H.-M., Yang, Z.-M.: Numerical study of wheat photosynthetic production. Journal of Nanjing Institute of Meteorology 18(4), 566–571 (1995) (in Chinese)
8. Goudriaan, J., Van Laar, H.H.: Modelling Potential Crop Growth Processes. Textbook with Exercises. Kluwer Academic Publishers, Dordrecht (1994)
9. Cao, W., Moss, D.N.: Modelling phasic development in wheat: a conceptual integration of physiological components. J. Agr. Sci. 129(2), 163–172 (1997)
10. Lü, X.: Studies on Effects of Ecological Factors on Growth of Maize and Establishment of Climate Ecology Model and Appraisement System. Shandong Agricultural University, Taian (2006) (in Chinese)
11. Zhang, B., Zhao, M., Dong, Z.-Q., Li, J.-G., Chen, C.-Y., Sun, R.: Establishment and test of LAI dynamic simulation model for high yield population. Acta Agronomica Sinica 33(4), 612–619 (2007) (in Chinese)
12. Xiao, J., Li, N., Jiang, H.-F.: Calculation and stability of accumulated temperatures in the growing season of winter wheat. Journal of Meteorological Research and Application 3(12), 64–67 (2010) (in Chinese)
13. Fu, X.-L., Zhao, M., Zhou, B.-Y., Cui, G.-M., Ding, Z.-S.: Optimal model for dynamiccharacteristics of grain weight commonly used in wheat and maize. Acta Agronomica Sinica 35(2), 309–316 (2009) (in Chinese)

14. Li, X.-L., Zhao, M., Li, C.-F., Ge, J.-Z., Hou, H.-P., Li, Q., Hou, L.-B.: Effect of sowing-date and planting density on dry matter accumulation dynamic and establishment of its simulated model in maize. Acta Agronomica Sinica 36(12), 2143–2153 (2010) (in Chinese)
15. Li, Y.-D., Tang, L., Chen, Q.-C., Zhang, Y.-P., Cao, W.-X., Zhu, Y.: Dry matter accumulation in rice aboveground part: quantitative simulation. Chinese Journal of Applied Ecology 21(6), 1504–1510 (2010) (in Chinese)
16. Wu, R.-J., Zheng, Y.-F.: A dynamical model of photosynthetic production and dry matter accumulation in wheat under increment UV-B radiation. Journal of Agro-Environment Science 27(6), 2325–2331 (2008) (in Chinese)
17. Cao, H.-X., Zhao, S.-L., Ge, D.-K., Liu, Y.-X., Liu, Y., Sun, J.-Y., Yue, Y.-B., Zhang, Z.-Y., Chen, Y.-L.: Discussion on development of crop models. Scientia Agricultura Sinica 44(17), 3520–3528 (2011) (in Chinese)
18. Hou, Y.-H., Chen, C.-Y., Guo, Z.-Q., Hou, L.-B., Dong, Z.-Q., Zhao, M.: Establishment of dry matter accumulation dynamic simulation model and analysis of growth characteristic for high-yielding population of spring maize. Journal of Maize Sciences 16(6), 90–95 (2008) (in Chinese)
19. Huang, Z.-X., Wang, Y.-J., Wang, K.-J., Li, D.-H., Zhao, M., Liu, J.-G., Dong, S.-T., Wang, H.-J., Wang, H.-J., Yang, J.-S.: Photosynthetic characteristics during grain filling stage of summer maize hybrids with high yield potential of 15000 kg ha^{-1}. Scientia Agricultura Sinica 40(9), 1898–1906 (2007) (in Chinese)
20. Li, G.-Q., Tang, L., Zhang, W.-Y., Cao, W.-X., Zhu, Y.: Dynamic analysis on response of dry matter accumulation and partitioning to nitrogen fertilizer in wheat cultivars with different plant types. Acta Agronomica Sinica 35(12), 2258–2265 (2009) (in Chinese)

Spatial Optimization and Mode Analysis of Primary Industry Structure in Yellow River Delta

Ping Yang[1] and Yujian Yang[2,*]

[1] The institute of sustainable development of Shandong Agricultural Science Academy
[2] S & T Information Engineering Technology Center of Shandong Academy of Agricultural Science, Information Center of Agronomy College of Shandong University
S&T Information Engineering Research Center, Number 202 Gongye North Road, Licheng District of Jinan, 250100, Shandong Province, P.R. China
yyjtshkh@gmail.com

Abstract. With the support of the total output of primary industry, secondary industry and tertiary industry and industry value of cultivating industry, forestry industry, animal husbandry industry and fishery industry of 19 country-level units in Yellow River Delta in 2007, the paper calculated the variation index of economic development based on country level, the MDS analysis was explored for primary industry, secondary industry and tertiary industry and compared with spatial difference characteristics of 19 country-level units in Yellow River Delta. The study carried out the analysis of combination analysis of primary, secondary and tertiary industry, and optimized the cultivating industry, forestry, animal husbandry and fishery industry. Autocorrelation characteristics was clarified by Moran`s I calculation, the new development mode of agricultural industrial structure was systemically concluded that the different mode presented in the different country units. The results showed that the industry evolution stage of primary, secondary and tertiary industry structure of 19 country units is in "secondary, primary and tertiary industries " stage and " secondary, primary and tertiary industries " stage according to industry structure stages theory, the obvious characteristics is that the cultivating industry value has spatial variability based on classical statistics combinig with universal Kriging. The emphasis of industry adjustment should be the focus on the secondary industry driving transformation into the primary, secondary and tertiary industries jointly driving economic growth, straight-pushing mode and tilting mode were put forward to new ideas about promoting industrial optimization and coordination development in Yellow River Delta. The study drew a conclusion that the measures were took to optimize industrial structure of cultivating industry, forestry industry, animal husbandry and fishery industry by combing straight-pushing mode with tilting mode, during the coordination development, focused on association effects of land and marine industry, extended agriculture industrialization chain and formed the diversification mode.

Keywords: Industry structure, Yellow-river Delta, Spatial analysis, Universal Kriging, New mode.

---

* Corresponding author.

D. Li and Y. Chen (Eds.): CCTA 2012, Part II, IFIP AICT 393, pp. 150–160, 2013.

1 Introduction

Industry structure was influenced by historical factors, natural factors and social& economic factors, and its spatial heterogeneity is very close to the unbalance economic development. Differences of regional industrial structure which reflected the coordination of regional economic development is the result of long-term evolution a variety of factors, optimizing and upgrading of industry structure is helpful for the target of the regional economic development[1]. Accelerating transformation and adjustment of industry structure is important opportunity promoting development of rural areas and agriculture in Yellow River Delta, and adjustment of agriculture structure is also essential of rural areas and agriculture economic development.

The spatio-temporal model is important method for economic and social life study of the clear alternative to traditional analysis model. The spatial variability characteristics of industries value of primary, secondary and tertiary industry, and the cultivating industry, forestry industry, animal husbandry industry and fishery industry was explored and analyzed in Yellow River Delta in 2007, spatial difference of industry structure was developed to classify into different combination types, the continuity surface was used to establish model and prediction of the industries value. Further, the study provided the foundation for grasping the spatial difference of industry structure. On this basis, new mode was put forward to adjustment and upgrading of industry structure in Yellow River Delta.

2 Material and Methodology

2.1 Overview in Yellow River Delta

The yellow river delta was taken shape on the basis of the yellow river alluvial plain and the adjacent sea regions in the north parts of Shandong province, extension the round regions, including Dongying city, binzhou city, Hanting district, Shouguang city, Changyi city in Weifang city, Leling and Qingyun in Dezhou city, Gaoqing in Zibo city and Laizhou inYantai city. The total area is 26500Km2, the 1/6 of the total area in Shandong province. Speeding up the high-efficient and high-ecological development in yellow river delta, is not only promoting the regional economic development in Shandong, but also accelerating open-up of Binhai new district in Tianjin, at the same time, Establishment of ecological zone accelerating the development of Bohai economic zone and cultivating new economic growth pole in the lower reaches of the Yellow River. Some countries in Yellow River Delta are a crossing and overlay regions located in "high-efficient ecological zone in Yellow river delta and the blue economic zone of peninsula". It is crucial issue for adjustment and upgrading of industry structure during the constructing high-efficient ecological zone in Yellow river delta, not only beneficial from of the industry significance from macro perspective, but also engaging in high-efficiency, high-ecological and high-industrialization[2].

2.2 Universal Kriging

The Ji wheat 22 yield obtained from the actual yield of the observation area, with the Universal Kriging is a high precision prediction method one of Kriging methods, theoretical explanations of the method and technical aspects referred to the relevant literatures [3]. On the whole, this method not only removed the trend effect, but also has obvious characteristics of the residual prediction, the mathematical expectation of the regional variables was regarded as function of spatial position, effectively dealing with non-stationary, or drift characteristics of the regional variable. During the universal Kriging analysis. The general steps, firstly, analysis of data trends that exist in original data to the fit-goodness model. Secondly, the residual data (original data minus the trend data) was used to develop Kriging analysis. Finally, the comprehensive analysis of trend surface analysis and residual analysis of Kriging using the second-order trend surface equation, the final result is obtained by the separation of trend effects.

3 Results and Analysis

3.1 Characteristics Analysis of Industry Structure in Yellow River Delta

Variation index effectively reflected the difference of country-level economy in Yellow River Delta, as the following description: $S_t = [\sum (Y_{ti} - Y_t)^2 / N]^{1/2} / Y_t$

Where, N stands for country-level units number in Yellow River Delta, Y_{ti} represents per GDP in the ith country in tth , the larger variation index indicated the more difference of economic development in the regions[4,5].

As is illustrated in Figure 1, the variation index results of 19 country-level units in Yellow river delta referred to the following sequences: Kenli > Guangrao>Yangxin> Zouping> Dongying> Leling>Huimin>Hekou>Gaoqing>Qingyun> Laizhou> Zhanhua>Hanting> Bincheng>Shouguang> Changyi>Wudi> Boxing> Lijin,according to the sequence, the difference of economic development level presented smaller and smaller trend, the weak heterogeneity of economic development and the strong similarity, the difference of variation index reflected the spatial pattern of economic development and industry structure, the results indicated that there is balanced development characteristics, and is not in polarization situation of industry structure pattern in Kenli country, Guangrao country and Yangxin country. The relatively large heterogeneity difference of the regional economic developmen in Wudi Country, Boxing Country and Lijin Country. For the status, transformation and upgrading of industry structure should be utilized by the different policy and measures.

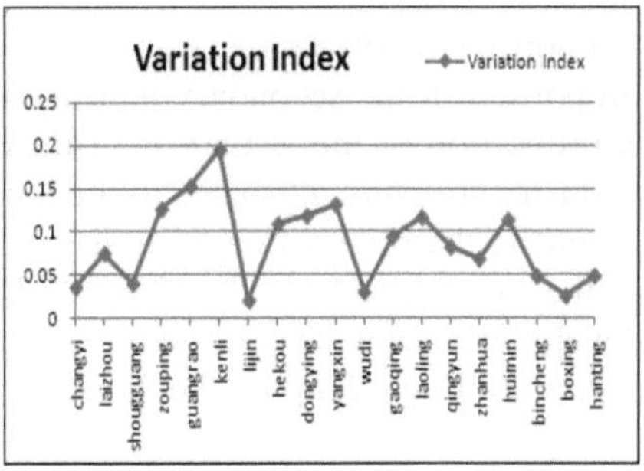

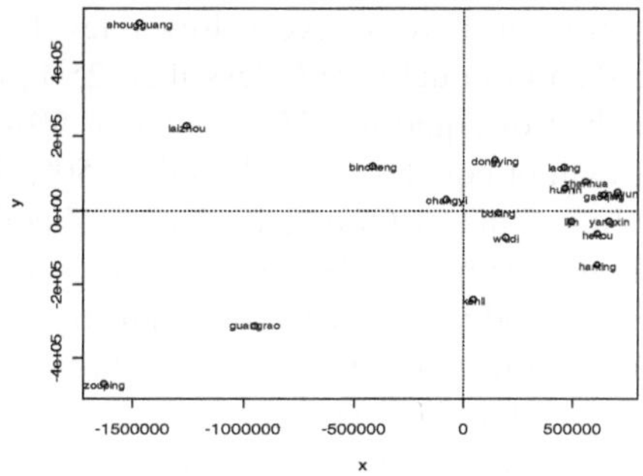

Fig. 1. The curve linear of 19 country units (left)

Fig. 2. The location plot based on multidimensional scaling analysis (right)

Multidimensional scaling method (multidimensional scaling, MDS) is a generic term that includes many different specific types, which can be classified according to the form of two-dimensional figure, its results was generally validated by the stress, the stress is less than 0.2, which showed the reasonable MDS results, the stress is more than 0.05, or equal to 0.05 and the stress is less than 0.1 or equal to 0.1, the results showed that MDS results are considered of a good sequence, the stress is less than 0.05, with good representation.By the R system based on open-source software, the MDS value of primary industry, secondary industry, tertiary industry was calculated in Yellow River Delta, referred to figure 2. MDS can explain the coordination degree of industry structure and the similarity relationship of industry values[5,6].Generally, the total industry values and the coordination degree has the similarity characteristics located in the same quadrant or the near interval distance of country-level units, the results of MDS showed that there is similar to industry values mode about Shouguang and zouping, Guangrao and Laizhou,Wudi and Dongying,Laoling and Hanting and zhanhua, Boxing and wudi, Huimin gaoqing qingyun yangxin liji and hekou, Changyi and kenli has the special mode characteristics. In summary, the relationship between coordination goodness of industry structure and the total value of the primary, secondary industry and tertiary industry is very complex, the results also clearly showed that the goodness of industry structure is probably poor in the developed country units, but in the smaller total values country units, there is coordination goodness of industry structure. So industry restructuring should be in accordance with industry structure development and evolution in essence, and be treated differently.

In the study, the proportion of agricultural industrial structure and primary industry, secondary industry, tertiary industry was the following description in Yellow River Delta. In the definition the proportion of 0-25% is lower level structure, at the scope from 25% to 50% is defined as the lower level structure, the proportion scope of 50% -75% is defined as the high level, but the proportion of 75%-100% is regarded as the higher structure [7]. According to the above classification, the types combination of industry structure is classified into Figure 3, as follows:

Low level +lower level+ lower level, the proportion of the primary industry is more than or equal to 0%, less than 25%; the proportion of the secondary industry is more than or equal to 25%, less than 50%; the proportion of the tertiary industry is more than or equal to 25%, less than 50%.The industry structure attributes to the type in Leling country, Qingyun country, Zhanhua country, the obvious characteristics is the lower value of the primary industry and the considerable proportion of the secondary industry and tertiary industry. The development emphasis in the next step is integration with city economy, increasing of the total economy of the secondary industry and tertiary industry.

Low level +higher level+ low level. The proportion of the primary industry is more than or equal to 0%, less than 25%; the proportion of the secondary industry is more than or equal to 25%, less than 75%; the proportion of the tertiary industry is more than or equal to 0%, less than 25%. The industry structure belongs to this type in Changyi country, Guangrao country, Lijin country, Hekou district, Yangxin country, Wudi country, Huimin country. The characteristics is the higher proportion of secondary industry, the lower proportion of primary industry and the tertiary industry, the development focus is on increasing proportion of tertiary industry based on industrial foundation.

Low level +higher level+ lower level. The proportion of the primary industry is more than or equal to 0%, less than 25%; the proportion of the secondary industry is more than or equal to 50%, less than 75%; the proportion of the tertiary industry is more than or equal to 25%, less than 50%. The industry structure belongs to this type in Laizhou city, Shouguang country, Dongying district, Gaoqing country, Bincheng district, Boxing, Hanting country. The characteristics is the higher proportion of the secondary industry and the tertiary industry, the lower proportion of the primary industry. The emphasis is increasing development of tertiary industry, strengthening the second industry and decreasing environmental damage.

Low level +high level+ lower level. The proportion of the primary industry is more than or equal to 0%, less than 25%; the proportion of the secondary industry is more than or equal to 75%, less than 100%; the proportion of the tertiary industry is more than or equal to 0%, less than 25%. The industry structure belongs to this type in Zouping country and Kenli country. The characteristics is the lower proportion of the primary industry and the tertiary industry, and the higher proportion of the secondary industry, the focus of development is upgrading the second industrial, accelerating the tertiary industry development.

During the evolution process of industry structure, adjustment and upgrading of it in developed regional showed the obvious evolution stages, " primary, secondary and tertiary industry "→" secondary, primary and tertiary industry "→" secondary, tertiary and primary industry "→ " tertiary, secondary and primary industry ", the industry structure of "tertiary, secondary and primary industry" was regarded as the goodness level of industry structure. By calculation, there is " secondary, primary and tertiary industry " industry development stage in Lijin country, the industry evolution stage is " secondary, tertiary and primary industry " in Changyi city, Laizhou city, Shouguang country, Zouping city, Guangrao country, Kenli country, Hekou district, Dongying district, Yangxin country, Wudi country, Gaoqing country, Leling city, Qingyun country, Zhanhua country, Huimin country, Bincheng district, Boxing country, Hanting district. So the achievement of the superior level of industry structure was considered of the following country-level units: Laizhou city, Shouguang city, Dongying district, Gaoqing country, Leling city, Qingyun country, Zhanhua country, Bincheng district and Boxing country. In a word, three industry restructuring should depend on "high-efficient, ecological and economic zone plan in Yellow River Delta " and "the memorandum about joint promoting eco-agriculture demonstration zones in Yellow River Delta", the industry structure was carried out to mainly rely on the driving of secondary industry transformation into the collaboration driving of the primary industry, second industry and the tertiary industry, based on structural optimization and upgrading the joint development of the region and highlighting intensive utilization of resources, eco-efficient industrial system was preliminary constructed.

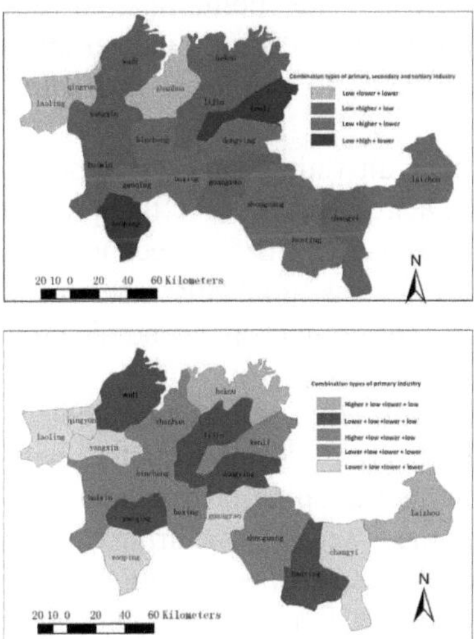

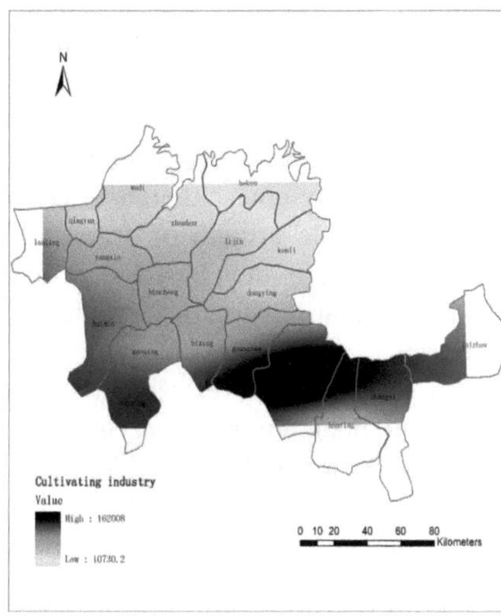

Fig. 3. Industry structure types of 19 country units in Yellow River Delta in 2007 (left)

Fig. 4. The spatial prediction map of cultivating industry value of 19 country units in Yellow River Delta in 2007 (right)

3.2 Optimization and Variability of Primary Industry Structure

3.2.1 Optimization Analysis of Primary Industry Structure

"Agriculture, forestry, animal husbandry and fishery" industry is one of the 16 major industries reflecting the economic function [8]. The proportion of cultivating industry, forestry industry, animal husbandry and fishery industry value in Yellow River Delta showed that cultivating industry is the superior industry in Changyi city, Shouguang city, Zouping country, Guangrao country, Lijin country, Yangxin country, Gaoqing country, Leling, Qingyun country, Huimin, Bincheng district, Boxing country. Fishery industry is relatively prominent industry in Laizhou city, Kenli country and Hekou district. Industry structure types of cultivating industry, forestry industry, animal husbandry industry and fishery industry in Yellow River Delta is illustrated in Figure 3, as follows:

Higher level+ low level +lower level+ low level. Cultivating industry is prominent industry in Changyi, Zouping, Guangrao, Yangxin, Gaoqing, Leling, Qingyun, these country-level units are not only grain and cotton production base, but also the relatively good animal husbandry. Adjustment and upgrading of industry structure centered “Number 1” file in 2011, expanded the effective irrigation area, promoted high- efficient and high-quality of cultivating industry, moreover, the industry chain of vegetable, winter jujube was developed and extended, animal husbandry base was planned to construct, animal epidemic prevention system was built and the dairy chain was extended to construct the relatively reasonable economic structure.,

Lower level + low level +lower level+ low level. The industry structure belongs to the type in Dongying, Lijin, Wudi country, Hanting. For the four countries, Industrialization development should be added to the contents, which included "high technology, low resource consumption, little environmental pollution", so the new path of agriculture industrialization of information leading industrialization, industrialization promoting information was constructed to accelerate upgrading and adjustment of agricultural industry structure.

Higher level+low level+lower level+low level. Cultivating industry is prominent industry in Shouguang, Zhanhua, Huimin, Bincheng, Boxing. At the premise of ensuring the stable growth of grain, high quality and high efficient economic crops should be positive developed and fodder crops s was gradually constructed to transform cultivating structure into ternary structure of grain crop, economic crop and fodder crop.

Lower level +low level +lower level+ lower level. High salinization content and fragile ecological environment is gradually improved in Kenli Country, with the comprehensive technology development of saline and alkaline land reclamation and water-saving demonstration development. In the next step, restructuring of industry structure should emphasize the coordinated development of agricultural industries centered industry effects of 1 adding t 1 is more than 2. For the new planning of the ecological function in the country, economic development and industry adjustment are the important thing, improving wetland ecological areas and cultivation of fishery industry, construction of aquaculture bases, the introduction of high-yield, high-quality, high stress resistance breeding new varieties of sea water are the other emphasis in Yellow River Delta.

Lower level+ low level +lower level+ lower level. the kind of industry structure represented the good fishery foundation in Laizhou city and Hekou district, the aquaculture base, which is ecological health of the breeding base should be constructed in the country-level units, and the upgrading and restructuring of marine industries and traditional industries promoted the regional economic development.

There is obvious agglomeration characteristics of industries value of primary, secondary and tertiary industry, and the cultivating industry, forestry, animal husbandry and fishery industry, which suggested spatial association characteristics of industry values existed in 19 country-level units. Moran’s I was used to analyze the cluster pattern of industries value. Theory and technology referred to the correlation literatures (Anselin,2004 ; YujianYang, 2010), Moran's I value is equal to 0.1034, 0.0484 and -0.0761 of primary industry value, secondary industry value and tertiary industry value, respectively, and Moran's I value is equal to -0.0083, 0.0989, 0.2090, 0.2202 and -0.0761 of primary industry value, secondary industry value and tertiary industry value, respectively. The biggest Moran's I value indicated the highest spatial autocorrelation characteristics, so the study results showed that animal husbandry value and the primary industry have the higher spatial autocorrelation, other industries has the lower autocorrelation characteristics, The autocorrelation characteristics of industries value was caused by the complicated factors, such as natural factors, social & economic factors and historical factors[9,10].

3.2.2 Spatial Variability of Cultivating Industry Value

The classical statistics analysis was developed for the value of cultivating industry, forestry industry, animal husbandry industry and fishery industry of 19 country units in Yellow River Delta, the range reflected the regional difference of industry value, fishery industry value > forestry industry value > cultivating industry value > animal husbandry industry value, and the variation coefficient is 0.657,1.028,0.807,0.537, respectively, for cultivating industry value, fishery industry value, forestry value and animal husbandry value, the results showed that there existed in the big difference among industry values. The normality suggested that distribution of the above industry value is higher than normal distribution, the large standard deviation and coefficient of variation showed that relatively large differences existed in agriculture industrialization. The large skewness coefficient indicates the mean value of four industry value is the right side of the peak value, showing a positive (right) skewness, kurtosis coefficients is more than 0 showed that the distribution of four industry value is higher than normality, kurtosis coefficient of fishery industry and fishery industry value is more than 3 showed the fitting curve is steep.

The analysis of the Pearson`s correlation coefficient among cultivating industry value, forestry industry value, animal husbandry value, and fishery industry value showed that the value is equal to 0.886 between cultivating industry value and animal husbandry value in confidence interval of 99%, the value is equal to 0.532 between forestry industry value and fishery industry value in confidence interval of 95% and the value is equal to 0.463 between animal husbandry value and fishery industry value in confidence interval of 95%. The striking result is that the better correlation characteristics among cultivating industry value, forestry industry value, animal husbandry value, and fishery industry value.

The spatial variability of industries value is not able to be clarified by the classical statistics, but using geostatistics combining with GIS. The continued surface of industry value using Universal Kriging referred to Figure 4. The results reflected the total distribution of cultivating industry value, based on the centroid coordinates of 19 country-level units, not only considering of sample size, shape, relationships and spatial location distribution, but also the non-stationary characteristics of industry value of samplings, centroid coordinates is calculated by the average weighted method and the formular is the following description:

$$X_G = \frac{\sum_{i=1}^{n} W_i X_i}{\sum_{i=1}^{n} W_i}, \quad Y_G = \frac{\sum_{i=1}^{n} W_i Y_i}{\sum_{i=1}^{n} W_i}$$

Where X_G, Y_G represents the centroid coordinates; X_i, Y_i stands for the discrete coordinates of ith objective, W_i represents the weight of ith objective.

3.3 Mode Analysis of Primary Industry Structure

There is a lack of proper balance and the low ratio among forestry industry, animal husbandry industry and fishery industry in Yellow River Delta and are still serious structural problems during the economy development. There are a lack of competitive industries and products, the low-quality varieties and the lag status of agricultural products processing and circulation, and so on. To solve these issues, we should improve the organization degree of agricultural industry, and build a new mode for regional industrial development [11].The new mode of the regional development combing straight-pushing mode with tilting mode was applied in the economy development in Yellow River Delta. Straight-pushing development mode emphasized the synchronous development of industries, or balanced development in the balanced regions of industry proportion, for example, Kenli country. Tilting development mode emphasized the development of the leading industry, which is the ideal regional model of industrial development comparative with straight-pushing development mode, the mode is suitable to Changyi city, Zouping country, Guangrao Country, Yangxin Country, Gaoqing Country, Laoling city and Qingyun country, dominant industry of cultivating industry. So tilting mode was also applied in the superior level of fishery industry in Laizhou, Kenli and Hekou[12].

We will continue to follow a new path of building industrial zone involved in marine industries and land industries, moreover, and we should accelerate the industry clustering concentrating on upgrading the industrial structure. The non-linear characteristics of 1 adding 1 is more than 2 from the association effects of industry, so we should emphasize the achievement development of industry clustering regions in Changyi, Shouguang, Laizhou, Wudi, Zhanhua and Bincheng district. With the support of economic technical development area, high-efficient and high-ecological agricultural demonstration zone, modern industrial clusters which has distinctive, specialized characteristics and complete service platform was constructed on basis of industry chain as a link, and industry demonstration area as the main object[13]. Agricultural industrial chain is loose, fragile in Yellow River Delta, and the regional chain is identical, short and narrow, the lack of diversity. Considering of it, the agricultural industrialization mode should combine with the service mode of agricultural information, such as "professional market (company) and farmers," "farmers and cooperative economic organization", "leading enterprise and farmers". The variety and quality of the agricultural products, the regional pattern of agriculture was carried out to adjust in all aspects, achieve the strategic adjustment and upgrading of agricultural structure, finally, the country industry model was formed, industry association characteristics of modern agriculture demonstration site of China of "Shouguang city mode" and industrial clusters of "Zouping city mode"[14,15].

4 Conclusions

In a word, the low level of industrial structure, the bigger proportion of traditional agriculture, the low-tech industry and the behind service industry were reflected in Yellow River Delta. There is the difference of country economy in 19 country-level units from variation index and multidimensional scaling index analysis. As is known to us, industrial structure types reflected the degree of industry optimization and provided the foundation of industry structure adjustment. By analyzing the combination types of industry structure of primary, secondary and tertiary industry, and the cultivating industry, forestry, animal husbandry and fishery industry of agricultural industrial structure, we drew a conclusion that the type of primary, secondary and tertiary industry structure of 19 country units located in Yellow River Delta is in the industrial structure "second, first and third" stage and "second, third and first" stage according to industry structure stages theory, which explained the necessity of upgrading and adjustment of industry structure. The results also showed that the value of cultivating industry has spatial variability based on classical statistics combining with universal Kriging, and the continued surface from point scale to the regional scale, the results also indicated that the optimization of industry structure in Yellow River Delta should take some important measures, such as the optimization of the agricultural industry chain and the boosting effects of the rural areas and agriculture information, the study also put forward to new ideas which straight-pushing model is combing with favorable model of country economy about promoting industrial optimization and coordination development in Yellow River Delta. Correspondingly, we should accelerate efforts to upgrade the industrial structure and make independent innovations, and the emphasis of industry adjustment should be the focus on the secondary industry driving development transformation into the primary, secondary and tertiary industry jointly driving development, achievement of the transition from secondary industry serving as the major driving farce to primary, secondary and tertiary industries jointly driving economic growth, and accelerating the development of modern industrial system, integrating IT application with industrialization.

References

1. Cui, G., Wei, Q., Chen, Z.: Regional analysis and planning. High Education Press, Beijing (1999) (in Chinese)
2. Development and reform committee of Shandong province, the high-efficient and high-ecological economic zone in Yellow River Delta (2008)
3. Using ArcGIS Geostatistical Analyst. GIS by ESRI (2008)

4. Wu, D.: The advanced tutorial of regional analysis and planning. High Education Press, Beijing (1999) (in Chinese)
5. Wang, B.: Multivariate statistical analysis and modeling of R language. Jinan University Press, Guangzhou (2010)
6. Li, W.: Application of multivariate statistical analysis. Beijing University Press, Beijing (2008)
7. Lijun: Geographical information and the multi-objective planning in the regional scale. Electronics Industry Press, Beijing (2006)
8. Wang, J., Lian, Y., Liuxin: Spatial ananlysis tutorial. Science Press, Beijing (2010)
9. Anselin, L., Florax, R.J.G.M., Rey, S.J.: Advances in spatial econometrics: methodology, tools and applications. Springer, Berlin (2004)
10. Yang, Y., Zhu, J., Wan, S., Zhang, X.: The spatial and temporal prognosis of oilseed yield in Shandong Province Springer. Computer and Computing Technologies in Agriculture, Daoliang li et al. pp. 146–157
11. `http://www.moa.gov.cn/zwllm/zwdt/201008/t20100803_1611790.html`
12. Wang, C.: Collaboration development model of Country economic industry. China Economic Press, Beijing (2006)
13. Zhang, M.: Growth mechanism of agricultural industrialization chain. Science Press, Beijing (2010)
14. He, G., Wang, X.: Economic and industrial pattern of marine industry. Ocean Press, Beijing (2010)
15. Li, P., Wang, X.: Harmonious development —path of China count-level development. The People Press, Beijing (2010)

The Implementation of Satellite Data (Land SAF) in the INCA Surface Temperature for Austria

Jing Liu[1], Jingyu Bai[1], Yong Wang[2], and Alexander Kann[2]

[1] Public Weather Service Center, Chinese Meteorology Administration, Beijing, China
shunying0309@163.com, baijy@weather.com.cn
[2] Central Institute for Meteorology and Geodynamics, Vienna, Austria
yong.wang62@gmail.com, alexander.kann@zamg.ac.at

Abstract. The Ground Temperature module (TG) of the current INCA only used the observations from stations in Austria. In this paper, the combinations of stations and satellite observations from 4th to 26th December 2011 are used in the TG module, and its results are analyzed. The retrievals from satellite can be used as the complement of observations under specific surface conditions, and which kinds of conditions are feasible to retrieve LST products from satellite have to be explored. There are lower errors for satellite in the areas of less than 500m, and higher errors in the areas from 500m to 1000m. The satellite retrievals are lower than the values from stations, maybe it exist systematic errors in the algorithm of satellite. INCA is a relatively young system still under development. The further improvements and refinements will be added in a step by-step process in the future.

Keywords: INCA, satellite data, surface temperature.

1 Introduction of INCA

The Integrated Nowcasting System for the Central European Area (INCA-CE) project aims at reducing adverse effects of weather-related natural disasters (e.g. windstorms, flooding, icing, drought) by establishing a state-of-the-art, high resolution, real-time analysis and forecast system on atmospheric, hydrological, and surface conditions. The project will implement a transnational information system as well as applications for different social and economic sectors to reduce risks of major economic damage.

The INCA is being developed primarily as a means of providing improved numerical forecast products in the nowcasting range (up to +4 h) and very short range (up to about +12 h) even though it adds value to NWP forecasts up to +48 h through the effects of downscaling and bias correction [1-4]. The INCA system adds value to the classical NWP forecast by providing (a) high-resolution analyses (1km), (b) nowcasts, and (c) improved forecasts both within and beyond the nowcasting range. The first version of the INCA system became operational in 2005. In the years 2006-2008, several major extensions and modifications were made to the system [5-7], such as the additional field precipitation type [5], the use of a surface layer index in the temperature and humidity analysis, or the elevation dependence of precipitation. Based on station

D. Li and Y. Chen (Eds.): CCTA 2012, Part II, IFIP AICT 393, pp. 161–167, 2013.

observations and remote sensing data it now provides three-dimensional hourly analyses and nowcasts of temperature, humidity, and wind, two-dimensional hourly analyses and nowcasts of global radiation, and two-dimensional 15-min analyses and nowcasts of precipitation, precipitation type, and cloudiness.

INCA is a relatively young system still under development. The further improvements and refinements will be added in a step by-step process in the future [8]. The Ground Temperature module (TG) of the current INCA only used the observations from stations in Austria. In order to improve the 3D temperature analysis in complex terrain, especially during inversion situations, using the observations from satellite is the best available way to solve this difficult problem. In this paper, the combinations of stations and satellite observations are used in the TG module, and its results are analyzed.

2 Materials and Methods

2.1 NWP Model Output

For the three-dimensional INCA analyses of temperature, humidity, and wind, NWP forecast fields provide the first guess on which corrections based on observations are superimposed. The NWP fields are 1-hourly, at a resolution of 9.6 km, with 45 levels in the vertical (Wang et al. 2006). Four ALADIN forecast runs per day are performed (00Z, 06Z, 12Z, 18Z). The 00Z and 12Z runs are integrated to +72 h, the 06Z and 18Z runs to +60 h. Post-processed fields are available roughly 4 h after analysis time. ALADIN forecast fields used in INCA are geopotential, temperature, relative humidity, u-, v-, w- wind components (3-d fields), 2m temperature and relative humidity, u-, v- 10m-wind components, precipitation, total cloudiness, low cloudiness, and surface temperature (2-d fields). The three-dimensional fields are provided on pressure surfaces with a vertical resolution of 50 hPa up to 600 hPa, and aresolution of 100 hPa above.

2.2 Surface Station Observations

The single most important data source for the INCA system are surface stations. About 275 automated stations observations are used. In the vertical, the network of stations spans most of the topographic range in Austria, with highest stations Brunnenkogel (3440 m), and Sonnblick (3105 m). Although the distribution of stations is biased towards valley locations there is a sufficient number of mountain stations to construct three-dimensional correction fields to the NWP model output, based on observations. The station density versus elevation roughly corresponds to the area height distribution of the topography up to about 1500 m. At higher elevations, the station density is lower than it ideally should be according to the area-height distribution.

2.3 Satellite Data

The ECUMETSAT Polar System (EPS) is Europe's first polar orbiting operational meteorological satellite, the first of which was successfully launched on October 19,

2006. The retrieval of LST is in the thermal infrared window (MetOp/AVHRR-3 channels 4(10.3-11.3μm) and 5(10.5-11.5μm)). The version2.0 for AVHRR LST products is valid on 10/Jan/2008, and its characteristics are described in Table 1. Further details may be found in the website (http://landsaf.meteo.pt). Theoretically, LST values can be determined twice per day from MetOp. However, fewer observations are generally available due to cloud cover contamination.

Table 1. Summary of Product Characteristics – AVHRR LST

coverage	Resolution		Range	Factors	Appended data
	Temporal	Spatial			
Europe & High latitudes	1/2 day	1*1km at nadir	-80-70℃	100	Quality control information

The LST data in this study are from 4th to 26th December 2011. In order to matching the observations from stations, the time window of satellite data is ±30 minutes at exact hour. Running the ground temperature module (TG) needs the 2 meters temperature (T2m), which is not retrieved from AVHRR. The available solution is obtaining T2m fields from INCA analysis fields interpolated by the output of ALADIN model to match the satellite LST values, and then run the INCA TG module to get the TG analysis field.

3 Results

The different input data from 4th to 26th December 2011 are used to run INCA TG module, obtaining the analysis fields of ground temperature, which are based on only model data, model and stations data, model and satellite, total data, respectively.

The better cases are shown in Fig 1and Fig 2, and the worse cases are shown in Fig 3. Most of the ground temperature values based on the satellite data are lower than those values on the stations data. The order of magnitude for good cases is much smaller than that for bad cases. Comparing the values of satellite with stations, most of values for good cases are within 5°C (Fig 1 and Fig 2), and many values for bad cases are beyond 10°C. At 20:00 on 13th December, the mean deviation is -1.76°C, and the mean absolute difference is 2.54°C. About 89% of the values are less than 5°C, and about 65% are less than 2°C. The best case is at 19:00 on 23th December, the mean deviation is -1.23°C, and the mean absolute difference is 1.96°C; Only two values exceed -5°C, and the others are within 5°C (about 94% of the total). The worst case is at 20:00 on 21th December, the mean deviation is -13.70°C; all values of satellite are lower than those of stations; more than half of the differences are larger than 10°C.

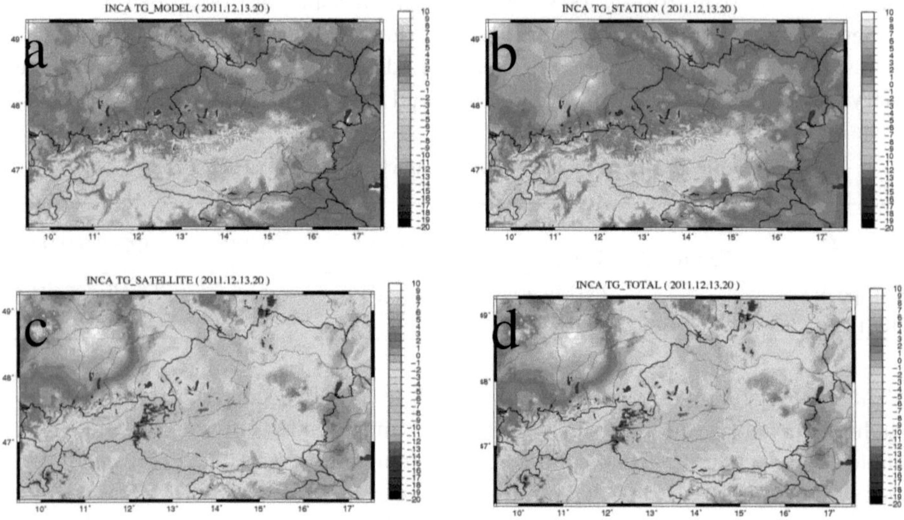

Fig. 1. Different Analysis fields of INCA ground temperature, a: only model data, b: model and stations data, c: model and satellite data, c: model, stations and satellite data

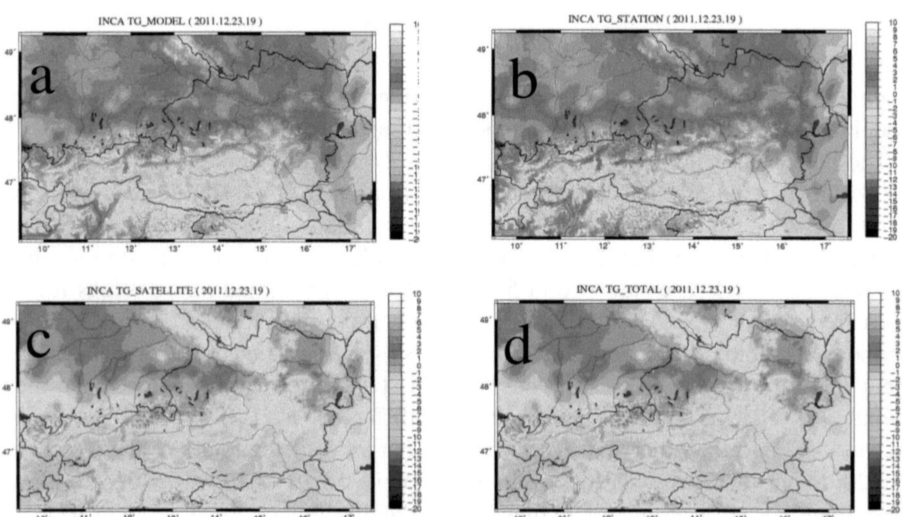

Fig. 2. The same as above

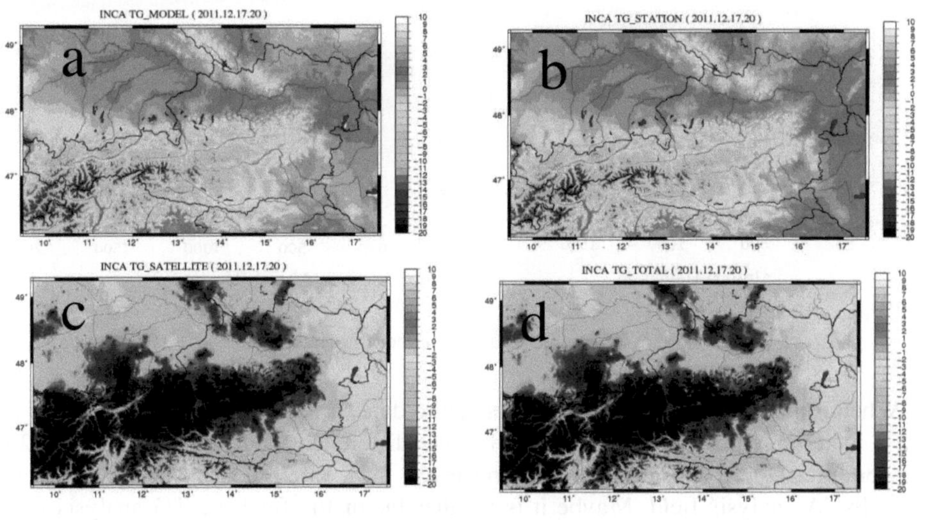

Fig. 3. The same as above (a bad case)

4 Discuss

Based on the analysis of above cases, LST qualities are sometimes good and sometimes bad. The satellite retrieval results may be affected by many factors, such as altitude, aerosols, weather conditions, surface types and so on. Here, the relations between TG differences and altitude height are discussed(Fig 4 and Fig 5).

The lower deviations of TG appear in the areas of less than 500m. With the altitude increasing from 500m to 1000m, the differences of TG have a strong fluctuation, emerging the extreme values for most of the cases. That is to say, the quality of LST retrieval is not good in the case of higher altitude. The reason maybe is the surface types or vegetations are difficult to predicate. For low-altitude areas, it is possible to use the LST data as the complement of observations; for high altitude, it needs to do many works when using the LST data.

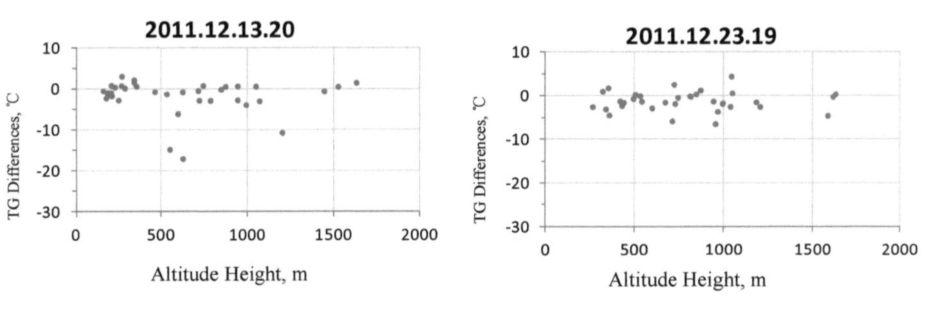

Fig. 4. Two good cases for the differences of ground temperature (satellite minus stations) varying with the height of altitude

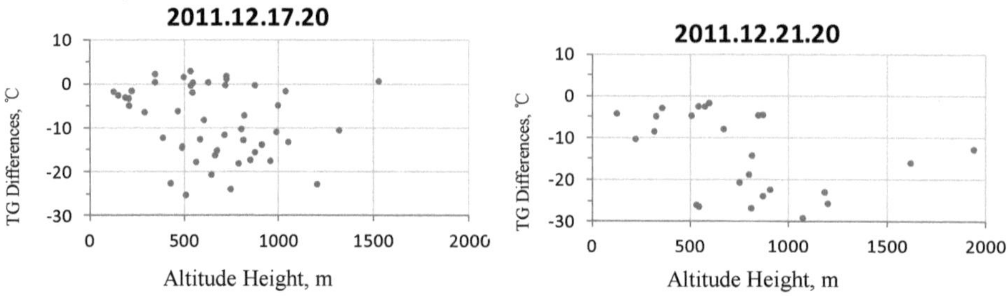

Fig. 5. The same as above, except for bad cases

When running TG module based on satellite data, it needs the input parameters of the 2 meters temperature (T2m) to derive the differences field of ground and T2m temperature. Due to no T2M product retrieved from satellite, we obtain T2M data from INCA analysis field. Maybe it is another factor to affect the TG analysis field derived by model and satellite data.

5 Conclusions

1) The LST differences of satellite and stations for good cases are lower than those for bad cases. For two good cases of this study, the mean error is about -1.5°C; about 89% of the values are less than 5°C, and about 65% are less than 2°C. That is to say, the retrievals from satellite can be used as the complement of observations under specific conditions. We have to explore which kinds of conditions are feasible to retrieve LST products from satellite.

2) The qualities of satellite values are related with altitude. There are lower errors for satellite in the areas of less than 500m, and higher errors in the areas from 500m to 1000m. For low-altitude areas, it is possible to use the LST data as the complement of observations; for high altitude, it needs to do many works when using the LST data.

3) Whether good or bad cases, the satellite retrievals are lower than the values from stations. Maybe it exist systematic errors in the algorithm of satellite. We can analyze the factors affecting the algorithm of satellite, and find out the systematic errors.

4) More than one year of satellite LST should be analyzed and verified. Many factors such as surface types, weather conditions, aerosols, and so on, can affect the LST results retrieved from satellite. According to a long series data, the systematic errors in the data can be obtained for different regions with seasons.

Acknowledgment. The Land SAF data were obtained from http://landsaf.meteo.pt. This work was supported by the integrated Nowcasting System for the Central European Area (INCA-CE) project and the infrastructure project of China Meteorological Administration (141161000016). I'd like to thank Benedikt Bica and Xin Yan for help me dealing with the problems of the INCA-TG module, and thank Alex and andrew for helping me prepare satellite data.

References

1. Golding, B.W., Nimrod: A system for generating automated very short range forecasts. Meteorol. Appl. 5, 1–16 (1998)
2. Haiden, T., Pistotnik, G.: Parameterization of elevation effects in short-duration precipitation analysis. In: Preprints, 13th Conference on Mountain Meteorology. Amer. Meteor. Soc., Whistler 4p. (2008)
3. Haiden, T., Kann, A., Pistotnik, G., et al.: Integrated nowcasting through comprehensive Analysis (INCA) System description (2009)
4. Sharples, J.J., Hutchinson, M.F., Jellett, D.R.: On the horizontal scale of elevation dependence of Australian monthly precipitation. J. Appl. Meteor. 44, 1850–1865 (2005)
5. Skoda, G., Filipovic, N.: Multi-data precipitation analysis. Austrian Academy of Sciences, Final report, Vienna, 17 p. (2007)
6. Steinacker, R., Ratheiser, M., Bica, B., Chimani, B., Dorninger, M., Gepp, W., Lotteraner, C., Schneider, S., Tschannett, S.: Downscaling meteorological information over complex terrain with the fingerprint technique by using a priori knowledge. Mon. Wea. Rev. 134, 2758–2771 (2006)
7. Wang, Y., Haiden, T., Kann, A.: The operational limited area modelling system at ZAMG: ALADIN-AUSTRIA, Österr. Beiträge zu Meteorologie und Geophysik, Heft 37, 33p. (2006)
8. Whiteman, C.D., Haiden, T., Pospichal, B., Eisenbach, S., Steinacker, R.: Minimum temperatures, diurnal temperature ranges and temperature inversions in limestone sinkholes of different sizes and shapes. J. Appl. Meteor. 43, 1224–1236 (2004)

Rice Blast Area Monitoring Based on HJ-CCD Imagery

Litao Wang, Jidong Xiong, and Yagang Du

Institute of Research on Science and Technology
Information Academy of Land Reclamation Sciences, Harbin, China
{Wanglitao2003,xjd}@163.com, duyagang@yahoo.com.cn

Abstract. It is difficult to determine the health of rice for simple vegetation index (NDVI) thresholding method which is widely used through remote sensing technology in crop disaster monitoring. The study selected binary logistic regression which were respectively established vegetation index getting from measured spectral and the relationship model between health status. The results show that the triangular vegetation index TVI model is with better reliability. When remote sensing monitoring rice blast was taken into account, geographical range was widely involved and rice-growing conditions were existing obvious differences in the local area, using a 3 × 3 pixel neighborhood consistency assumption to eliminate differences in the local environment. Applying China's own property "environmental disaster satellite" CCD sensor data into the model and the stress range of extracting rice blast was basically consistent with Plant Protection Institute of Heilongjiang Academy of Agricultural Reclamation Sciences as well as the ground measured results, among which TVI model results accuracy reached 76.47%, which can meet remote sensing monitoring requirements. of the blast area.

Keywords: HJ-CCD, regression analysis, rice, rice blast, vegetation index.

1 Introduction

Rice is one of the world's major food crops, and Chinese rice sown area accounts for 1 / 4 of the whole food crops, the second production accounts for more than half. However, rice blast is the main factor affecting rice production as well as one of the major diseases[1], and in the general disease epidemics year cut 10% to 20%, 40% in severe damage to 50%, or even failure. Therefore, it is important for food security to rapidly and timely mastery the occurrence of rice blast conditions. Traditional rice blast monitoring is mainly based on the field survey and sampling which requires considerable human and material resources, and has no real-time monitoring capabilities. However, the remote sensing technology with rapid, real-time, large area, without undermining and other characteristics is showing the traditional blast monitoring methods can not compare with it.

Domestic and foreign scholars have done a lot of method studys about remote sense monitoring rice blast, which are mainly to analyze the characteristics waveband of spectral reflectance or the relationship between vegetation index getting from

D. Li and Y. Chen (Eds.): CCTA 2012, Part II, IFIP AICT 393, pp. 168–176, 2013.

characteristics band combination and health status. Qin et al[2] used multi-spectral remote sensing technology to perform remote sensing monitoring of rice heath blight; Qi Long et al from Jilin University studied on visual-based and multi-spectral classification technology of the rice blast resistance detection. In addition, many scholars have explored the relationship between the chlorophyll-related band (or band combination) and the blast. Although remote sensing for rice blast has done many studies, but few about the real business applications. Constraints is that, firstly, data acquisition costs is too much; Secondly, for a wide range of crop growing areas, the growth of their crops is affecting by the very complex environmental factors[3-4], resulting in difficult promotion and business-oriented approach.

"Environmental Mitigation satellite" is small satellite constellation system with Chinese independent intellectual property rights. The wide-band satellite equipped with a CCD sensor whose spatial resolution is 30 m, which is ideal data source of the remote operational monitoring blast. Using wide-band CCD sensor can detect the abnormal information of the blast, however, to determine the exact level of blast resistance, and the type and stage of stress and other issues, but also with the analysis of hyperspectral data[5] to determine the red edge position drift. The purpose of this study is to use the measured spectral data to model and analyze the relationship between the coerced and the health of rice, and based on analyzing the differences about vegetation index, to preliminarily explore the large-scale monitoring of ground blast occurred area method based on HJ-CCD data complex area environment.

2 Study Area and Data

2.1 Study Area

This choice of study area is located in Heilongjiang Reclamation Qianjin farm which is at latitude 47°34', longitude 132°17' in the southeasten Sanjiang Plain; and the farm whose cultivated area is 830 thousand mu is based on growing rice, and a large state-owned mechanized farms. In recent years, large-scale cultivation rice variety Kongyu 131 accompanies weak resistance, high frenquency blast occurred within the region, and then which causes serious rice production. Especially in 2010, it is found that rice blast large-scale outbreak in the agricultural census. In addition, Plant Protection Institute of Heilongjiang Land Reclamation Sciences randomly surveyed 300 district fields in blast hardest-hit areas in 2010, and found that 125 fields have infected with rice blast, rhiziomania fields rate achieving on average 41.7% as a result of a direct threat to rice production safety.

2.2 Data Acquisition and Processing

2.2.1 Field Spectroscopy

In 2010, Information Institute of Heilongjiang Academy of Land Reclamation Sciences conducted a field spectroscopy in study regional, and a total of 58 sites were measured data, and observation stations were covering the entire farm, and focused on coercion by the rice blast area. All stations, 26 sites rice was threatened by the blast stress, and 32 sites rice was not endangered. For the site with rice blast coercion, taking four group crop canopy spectral data; taking respectively two groups of the rice with the blast and the adjacent normal rice data.

Ground field spectroscopy instruments used ASD FieldSpec and Spectroradiometer for ProFR2500, and the spectral range is 350 ~ 2500 nm, and sampling interval in the 350 ~ 1000 nm is 1.4 nm, 1000 ~ 2500 nm for 2 nm. When observing, the probe was vertically downward from the ground level 1.3 m, view angle for 25 °. In the working process, the mean value choice was repeated 30 times within the field of view limits, and before each spectral measurement, a standard whiteboard was used to correct[6] .

Studies adopted the method to wide-band CCD image achieving the rice blast area monitoring, and based on the band width of satellite images, all the measured spectral reflectance of rice canopy was calculated mean value: 520 ~ 600 nm (green band), 630 ~ 690 nm (red band), 760 ~ 900 nm (near infrared), so as to match the ground truth spectral data and remote sensing image data.

2.2.2 Remote Sensing Image Data

Study mainly selected two almost no cloud layer images which was acquried by A star CCD2 sensor (referred to as HJ-1A-CCD2) of small satellite constellation system for the environment disaster reduction, and used GPS ground control points to perform the geometric correction, and controlled positioning precision within half a pixel.

3 Models and Methods

3.1 The Mechanism of Remote Sensing Monitoring for Blast Area

This paper discusses vegetation index monitoring methods to blast stress area. In a wide range of rice growing areas, different regions will make the rice growth conditions difference because of the affection of the growth conditions (varieties, planting time, fertilization status, etc.), and from the spectral point of view, if not discuss it separately, it is difficult to distinguish rice blast paddy from normal rice. On the other hand, QianJin farm grows a broad range of rice, most farmland for large-scale bulk distribution. Based on this, we propose a algorithm assumption based on 3 × 3 neighborhood consistency: assuming, within 3 × 3 pixel area, rice varieties, planting time, and the same growing environment in the vicinity range will largely reduce vegetation index difference resulting in the growing conditions through comparison among adjacent inter-pixel.

3.2 Logistic Regression Model Building

The study introduced three vegetation indices: Normalized Difference Vegetation Index, NDVI, Renormalised Difference Vegetation Index, RDVI and Triangular Vegetation Inde, TVI to establish a logistic regression model for the measured spectrum (Table 1). NDVI is the most widely used vegetation index in monitoring green vegetation respect. However, in the process of field observation, it is found that rice leaves were relatively sparse in more serious areas of rice blast, so it could not be ignored in the contribution of the spectrum to soil, therefore, the study introduced Lu and Bu Ruiang's RDVI, because it respectively took the dense and sparse vegetation sensitive advantages of NDVI and DVI (difference vegetation index). TVI, by Broge and Leblanc[7] proposed, could well detect the crops spectral reflectance rate minor or moderate increase caused by the disease stress in the red band, and decreases in the green band, as well as sharp fall in the infrared and so on feature information.

Table 1. Applied vegetation indexes

Abbreviations	Name	Formula	References
NDVI	Normalized Difference Vegetation Index	NDVI=(NIR-R)/(NIR+R)	Rouse Et(1973)
RDVI	Renormalised Difference Vegetation Index	RDVI= (NDVI·DVI)1/2, (DVI= NIR-R)	Roujean Et(1995)
TVI	Triangular Vegetation Index	TVI=0.5×(120(NIR − G) − 200(R − G))	N.H.BrogeEt(2000)

Notes:NIR—Near infrared band;R——Red light band;G—Green light band.

In the study, the health of rice was devided into the normal growth and rice stess coercion. Adopting the method to evaluation for stress rice and nomoral rice vegetation index in order to calculate the stress level of rice blast, and get a relative amount, but also to some extent reduce the environmental factors effect because of different sites. Spectral data for each measured site respectively included two sets of normal rice spectrum and two group of stress rice spectrum, and then to calculate ratio between vegetation index getting from canopy spectra of stress rice and the normal cross rice vegetation index. Obviously, as for as normal rice was concerned, this ratio was equal to 1, and the ratio of stress rice was between 0 and 1.

Rice health condition from normal state to change by the stress was non-linear process, and coerced probability was not sensitive in the vicinity of p = 0 and p = 1, but rice blast stress probability meeted logical curve. In order to get the distribution area of rice blast stress, choosing to use binary logistic regression method to take the ratio to stress rice and neighborhood normal rice vegetation index through the above-mentioned calculating method as independent variables, and to establish binary logistic regression model of rice blast monitoring. The basic logistic regression expression as follows[8]

$$Y=\text{logit}(p)=\ln\{p/(1-p)\}. \quad (1)$$

$$Y=a+bX+\varepsilon. \quad (2)$$

In it: X - covariable, dimensionless, is the ratio of normal growth and under stress rice vegetation index; a, b - coefficient, dimensionless, as sample logistic regression coefficient through these vegetation index ratio; p- the probability of rice blast under stress, dimensionless; ε is infinitely small quantities, dimensionless. Probability p will be represented by Y, the result is

$$P=ey/(1+ey)=ea+bx+g/(1+ ea+bx+g). \quad (3)$$

when the ratio of vegetation index is 1, p is 0, otherwise p will be between 0 and 1, which means no stress. When probability p was more than 0.5, the logic is true, and rice is under stress. logit (p) is the odds ratio of the dependent variable Y or natural logarithm of the log likelihood ratio, and is called log odds ratio, log likelihood ratio or logit. After model established, introducing goodness of fit and forecasting accuracy to determine the adequacy of the model[9]. Goodness of fit was used to evaluate the consistency between model predictions and the corresponding observations. Used covariates was taken as the ratio vegetation index, which belongsed to a continuous variable. Because the number of covariant type was relatively larger, which resulted in a number of covariant type with small sample observations, so the goodness of fit test method proposed by Hosmer and Lemeshow (1989) was selected. The basic idea of this method was to group data based on the predicted probability values, if the significance level P value was less than 0.05, then the model goodness of fit was higher. Predictive accuracy used pseudo determination coefficient of Cox and Snell (1989), Nagelkerke, the greater the coefficient was, the higher degree of curve fitting was, ie the model was more credible.

3.3 The Model Application of Satellite Remote Sensing Images

When model was established, ground truth spectral data used the ratio of vegetation index, so before the logistic regression model was applied to remote sensing data analysis, you needed to choose pixel vegetation index to operate ratio calculation. As the the range of rice blast stress only accounted for a small region part, so, after calculation unit transformed to the neighboring 3 × 3 pixels, it is believed that there was at least one pixel within the nine pixels existing in a healthy condition. The selection way to rice healthy growth pixel is to compare vegetation index of 9 pixel, and then to select one of the largest pixel as rice health growing areas, on this basis, to perform ratio calculating for all the vegetation index in each calculation unit and healthy growth pixel vegetation index so as to access the ratio vegetation index of each pixel in the calculation unit. Obviously, the pixel ratio vegetation index of the health rice was 1, while the other pixel values were between 0 and 1. After using 3 × 3 window to obtain ratio vegetation index for the whole scene images, using the established logistic regression model to calculate the probability of each pixel rice blast stress, and as a criterion, when the probability p was more than 0.5, rice was under stress, otherwise the rice of normal growth, then to count the rice number of blast stress pixels in order to get rice affected area.

4 Results and Validation

4.1 Extraction of Rice Vegetation Range

In this study, 2010 satellite imagery was selected, and supervised classification method to extract the scope of rice cultivation in the study area was adopted. In the supervised classification process, the transformation dispersion values between the rice samples and samples of other features reached 2000, indicating the separation degree between rice and other surface classes was very good[10], on the above base, we can look forward to getting better classification results using this sample.

Figure1 shows the classification results, where the green area stands for the rice-growing range. Mergering non-crop types in the map with only considering classification accuracy of rice and non-rice, and to select 51 ground reference validation samples to perform accuracy verification. Table 2 shows the classification accuracy evaluation form[11], and the overall classification accuracy reaches 94.12%, and kappa coefficient is 0.8 or more. Rice cultivating area statistics in classification results is 74.6 acres, which shows a difference of 3.18% from recorded 72.3 hectares plan cultivation area of Qianjin farm in 2010, indicating that classification accuracy can meet the requirements.

Fig. 1. Classification results of HJ-1A- CCD2 image, 2010

Table 2. Precision evaluation of the supervised classification

Category	Refer Data	Number Of Random Sample	Correct Number of Samples	Producer accuracy	User Accuracy	Kappa Coefficient
Other	36	35	34	94.44%	97.14%	0.903
Rice	15	16	14	93.33%	87.50%	0.823
The Overall Accuracy			94.12%			

4.2 Ratio Vegetation Index Based on Logistic Regression Model

The use of coercion by the blast and the normal growth of neighborhood rice NDVI, RDVI and TVI ratio vegetation index establish the binary logistic regression model. 78 groups samples obtained in the 26 measured sites, applicability of the model results shown in Table 3. It can be seen, although pseudo determination coefficient of NVDI modle is high. Cox & Snell R2 and Nagelkerke R2 reached 0.665 and 0.887, but the Hosmer-Lemeshow goodness of fit P is higher, the model was abandoned. RDVI and TVI established logistic regression model, although the pseudo coefficient determination is not high, but statistically very significant, P is 0, the equation of goodness of fit is high. Therefore, RDVI and TVI model is used.

Table 3. Goodness of fit and reliability of the logistical models

Vegetation Index	Cox & SnellR2	Nagelkerke R2	Hosmer-Lemeshow P Value
NDVI	0.665	0.887	0.473
RDVI	0.554	0.739	0
TVI	0.480	0.640	0

4.3 Blast Range CCD Image Extraction and Validation Results

The use of TVI and RDVI extract information about the exception of blast from the satellite images to obtain the probability map of occurrence blast, probability greater than 0.5, considered by the blast hazards, The resulting stress range distribution of rice by the blast is shown in Figure 2.Figure 2a is the result of extraction by the RDVI model, Figure 2b is the result of extraction by the TVI model. By the statistical analysis, blast area of the rice-growing area extracted by the RDVI model and TVI model were 15.6% and 18.7%.

In 2010, Heilongjiang Academy of Agricultural Reclamation Plant Protection had a blast of field survey on the QianJin Farm, the results showed that incidence of rice blast is relatively high. The information about the exception of blast extracted from the satellite images was consistent basically with the census results of Plant Protection Station.

To further analyze the accuracy of the logistic regression model, the rice –growing state measured with rice range stressed of blast by matching the spatial location, is results from the model in line with the actual observation?The results shown in Table 4.Measured at 58 sites, There are 26 sites at which the rice is stressed by the blast, 32 sites at which the rice is healthy. When the blast information is extracted through the RDVI model, obtained14 sites stressed and 22healthy sites, the overall accuracy of Extraction is 62.07%. When the blast information is extracted through the TVI model, obtained 20 sites stressed and 24 healthy sites, the overall accuracy of Extraction is 75.86%.Both RDVI and TVI model can obtain good results, In contrast, TVI model better meet the monitoring requirements of rice blast.

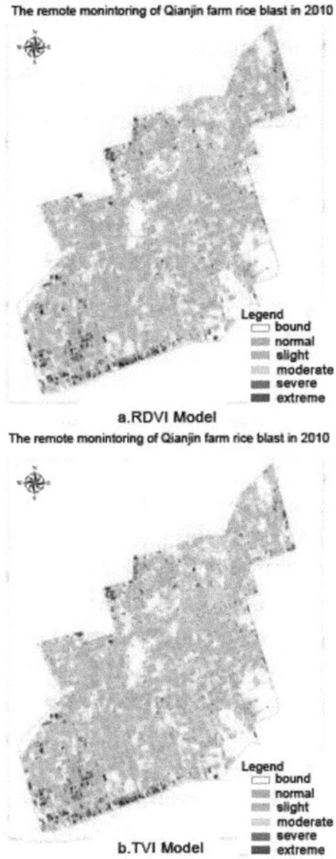

Fig. 2. Distribution map of ShengLi farm stressed rice

Table 4. Precision of the extracted rice blast area by logistical model

	sites stressed	healthy sites	Accuracy %
RDVI	14	7	61.76
TVI	20	6	76.47
number of sites	26	8	——

5 Conclusion

For the measured spectra of field, due to the combined effects of various environmental factors, the rice of different regions showed different spectral characteristics. In the present study, for each site affected by rice blast stress gets the 4 set of spectral data, 2 groups spectral data affected by the stress rice and the 2 set normal spectral data. During data analysis, the vegetation indices stressed and adjacent normal rice seek ratio, putting forward a kind binary logistic regression method to the vegetation index ratio as indpendent variables, model to monitor the health status of rice. The results show that the goodness of fit of binary logistic regression model was created by TVI ratio is better. When rice blast information monitoring of remote sensing images with this model, we proposed a hypothetical algorithm based on the consistency of the 3 × 3 neighborhood. For the elimination of local environmental differences which can be ignored, Better to avoid the impact of a variety of environmental differences which exist between the different regions in the whole scene images.

This study, preliminary using China's HJ-1A- CCD2 data to the remote sensing monitoring of the rice blast。 HJ-CCD data with extensive coverage, short revisit cycle, the higher resolution et, offers the possibility of periodic monitoring of crop growth status. This study Proposed ideas, can help to use the data to operational monitoring of health status in rice.

References

1. Ou, S.H.: Rice Diseases. Kew Surrey, UK (1985)
2. Qin, Z.-H.: Detection of rice sheath blight for in-season disease management u-sing multispectral remote sensing. International Journal of Applied Earth Observation and Geoinformation 7(2), 115–128 (2005)
3. Xing, S., Zhang, G.: Application status quo and prospect of agriculture remote sensing in China. Transactions of the CSAE 19(6), 174–178 (2003)
4. Ma, Y.P., Wang, S.L., Zhang, L., et al.: Monitoring winter wheat growth in North China by combining a crop model and remote sensing data. International Journal of Applied Earth Observation and Geoinformation 10(4), 426–437 (2008)
5. Lelong, C., Pinet, P., Poilve, H.: Hyperspectral imaging and stress mapping in agriculture: A case study on wheat in Beauce. Remote Sensing of Environment 66(2), 179–191 (1998)
6. Li, F., Gnyp, M.L., Jia, L.L., et al.: Estimating N status of winter wheat using a handheld spectrometer in the North China Plain. Field Crops Research 106(1), 77–85 (2008)
7. Broge, N.H., Leblanc, E.: Comparing prediction power and stability of broadband and hyperspectral vegetation indices for estimation of green leaf area index and canopy chlorophyll density. Remote Sensing of Environment 76(2), 156–172 (2000)
8. Huang, B., Zhang, L., Wu, B.: Spatiotemporal analysis of rural-urban land conversion. International Journal of Geographical Information Science 23(3), 379–398 (2009)
9. Wang, J., Guo, Z.: Logistic Regression model: Methods and Applications. Higher Education Press, Beijing (2001)
10. Jensen, J.R.: Introductory Digital Image Processing: A Remote Sensing Perspective. Prentice Hall, Upper Saddle River (1996)
11. Fitzpatrick-Lins, K.: Comparion of sampling procedures and data analysis for a land-use and land cover map. Photogrammetric Engineering and Remote Sensing 47(3), 343–351 (1981)

Urban Wetland Change Detection Using Time-Series Remote Sensing Data

Lin Liu[1,2], Yapeng Zhou[3], Li Wang[4,*], Jianchun Hou[5], and Mingquan Wu[4]

[1] School of Land Science and Technology, China University of Geosciences, Beijing 100083, China
[2] Shijiazhuang Engineering Technology School, Shijiazhuang 050061, China
[3] College of Natural Resources & Environment Sciences, Agricultural University of Hebei, Baoding 071001, China
[4] State Key Laboratory of Remote Sensing Science, Jointly Sponsored by the Institute of Remote Sensing Applications of Chinese Academy of Sciences and Beijing Normal University, Beijing 100101, China
[5] China National Petroleum Corporation, 100007 Beijing, China
Liulin7801@126.com, Wangli@irsa.ac.cn

Abstract. Wetland, as one of the weakest ecosystems in the world, is deteriorating rapidly in many regions. Longfeng wetland is the biggest urban wetland in China, located in the inner of Daqing city, Heilongjiang Province. It is divided into two areas by Wolong road, the east is provincial natural reserve and the west is unprotected. This paper describes a remote sensing change detection approach used to assess Longfeng wetland change. Four images were used, i.e., Landsat MSS in August 1979, Landsat TM in July 1990 and in August 2001, LISS-III in July 2006. Linear spectral mixture model (LSMM) method was used to classify the wetland and the results were satisfying. The results showed that the land cover types of Longfeng wetland had been greatly affected by human civilities (such as city extension and fishery), which was more deeply in unprotected area than in natural reserve.

Keywords: Change detection, Remote sensing, Time-series images, Urban wetland.

1 Introduction

Wetlands account for only three to six percent of the earth's land surface, but they provide service of great value to human beings, for example, they can control flood, protect coastal zones, replay water supplies, improve water quality, produce agricultural product, and provide plant and wildlife habitat [1,2,3]. Despite these proven advantages, wetlands have been one of the weakest ecosystems in the world. In many regions wetlands are deteriorating rapidly [3,4].

The Longfeng wetland is located in the inner of Daqing city, Heilongjiang Province of China (figure 1). It is the biggest urban wetland in China. It is divided into two areas

* Corresponding author.

D. Li and Y. Chen (Eds.): CCTA 2012, Part II, IFIP AICT 393, pp. 177–186, 2013.

by Wolong road, the east is provincial natural reserve and the west is unprotected. Daqing city is the largest petrochemical industry base in China. Over the last two decades (1985–2005), Daqing has undergone a high-speed economic development and environmental deterioration as well.

The objective of the present study was to monitor the changes of Longfeng wetland using multi-temporal and multi-source remote sensing data, to understand its change patterns, and to analyze the relationships between natural and human factors and the deterioration of Longfeng wetland. To achieve this goal, the following 3 objectives were presented:

1) Choose appropriate methods to classify these four multi-temporal images, and evaluate the classification accuracy;

2) Monitor Longfeng wetland dynamics using four classified remote sensing images;

3) Study the urban wetland dynamics; analyze the relationship between natural and human factors and these changes.

2 Study Site

The printing area is 122 mm × 193 mm. The text should be justified to occupy the full line width, so that the right margin is not ragged, with words hyphenated as appropriate. Please fill pages so that the length of the text is no less than 180 mm, if possible.

Covering an area of 8,000 ha, the Longfeng wetland is located in the middle of Songnen Plain in Heilongjiang Province in the northeast of China, between 46°35′—46°41′ N latitudes and 125°20′—125°26′ E longitudes(figure 1). The study area has typical characteristics of continental monsoon climate.

Longfeng wetland is encompassed by two districts of Daqing city, i.e., Longfeng and Wolitun, traversed by Wolong road and Binzhou railway, approximately 8km from the centre of Daqing city. It is very helpful to improve and maintain ecological system and environment of Daqing city. However, the fast development of Daqing city leads to the rapid reduction and deterioration of the Longfeng wetland.

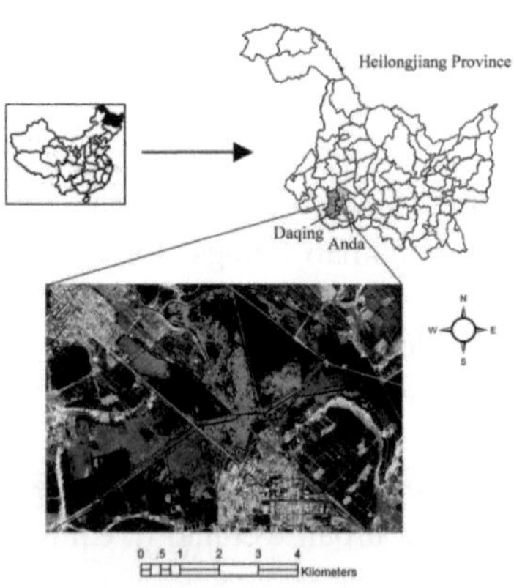

Fig. 1. Location of Longfeng wetland in Heilongjiang province, China

3 Methodology

In our papers, one hybrid classification approach was used. Firstly, based on the linear spectral mixture model (LSMM), the imagery was transformed into corresponding end-member bands (including water body, bare land, and green vegetation). In the result of linear spectral unmixing, three classes, i.e. water body, bare land and green vegetation, can be easily extracted, because the pixel values of these three classes are highest in corresponding band. To make change detection, only three classes are not far enough. According to the characteristics of Longfeng wetland, the imagery would be classified into 6 classes, and the ranges of corresponding pixel value of the 6 classes were shown in table 3. Through comparing the value of 3 end-member bands of each pixel, the 6 classes were very easily classified.

3.1 LSMM Method

Linear spectral mixture model is based on the idea that a continuous scene can be modeled as the sum of the radiometric interactions of individual cover types weighted by their relative proportions [5]. Mathematically, the linear spectral mixture analysis framework can be expressed as follows:

$$R_i = \sum_{j=1}^{n} f_j R_j + \varepsilon_i \tag{1}$$

$$0 \le f_j \le 1 \quad \text{and} \quad 0 \le \sum_{j=1}^{n} f_j \le 1 \tag{2}$$

Where R_i is the composite reflectance of the mixed spectrum in band i, f is the weight of end-member(i) in the mixture, R_j is the reflectance of that end-member in band j, j is the number of end-members, ε_i is the error in the sensor band $_i$. Eq.(3) constraints the weight and the sum of the weight is allowed to be between 0 to 1.

According to these formulas the value of a pixel must equal the sum of the values of the end-members times their abundance. The fit of the model can be assessed by the root mean square error (RMSE):

$$RMSE = \sqrt{\frac{\sum_{i=1}^{n} \varepsilon_i^2}{n}} \tag{3}$$

Where i is the number of bands used in the spectral unmixing. Higher values of RMSE indicate regions that could contain lacking end-members.

3.2 End-Members Selection

The key to a successful mixture model is the selection of appropriate end-members [6,7]. These end-members principally being pure reflectance spectra that are derived by a specific target material with no mixing with any other materials are usually selected either from spectral libraries built from field surveys [8] or from the image data.

In Longfeng wetland, bare land (BL), water body (WB) and green vegetation (GV) were chosen to represent the spectral variability. In order to successfully complete the spectral unmixing, it was very essential to accurately estimate the spectral end-members for each component of 3 end-members. These end-members were determined from the two-dimensional scatter plot of RED and NIR bands. In order to get the purest pixels, the MNF transformation was employed to the four images, and then PPI was computed. Those pixels with PPI great than 10 were extracted and projected to the two-dimension of RED and NIR bands. The PPI images and two-dimensional scatter plot were shown in figure 2. Figure 2 (b) showed that rather symmetrical patterns in pixel distributions allowed the appropriate identification of three dominant spectral end-members, bare land (BL), green vegetation (GV) and water body (WB), at the extremes of image feature space of red (MSS B2, LISS-III

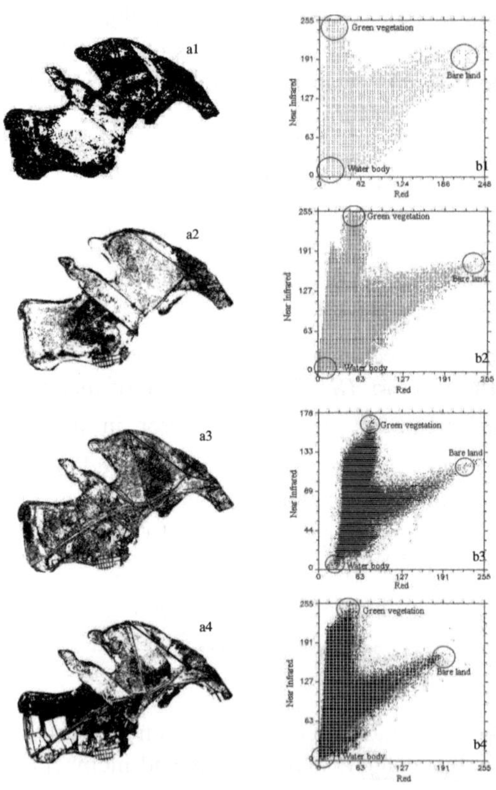

Fig. 2. PPI images (a) and the corresponding two-dimensional scatter plots of red and NIR bands(b). a1, a2, a3 and a4 denote 1979 MSS, 1990 TM, 2001 TM and 2006 LISS-III image. In b1, red band is MSS B2 and NIR is MSS B4; In b2 and b3, red band is TM B3 and NIR is TM B4; In b4, red band is LISS-III B2 and NIR is LISS-III B3. End-members derive from b1, b2, b3 and b4.

B2, TM B3) and NIR (MSS B4, LISS-III B3, TM B4) bands. An average of pixels of these vertices were calculated and used in LSMM. The BL end-member was identified at the top right vertex while GV end-member on the left vertex of the scatter plots. The WB end-member was selected from the pixels at the origin of the scatter plots.

End-member spectra were shown in figure 3, and used in the LSMM. After using image normalization and the same end-member selection method, the end-member spectra of 4 dates were very similar. Therefore, using these end-member spectra to analyze 4 scenes assured that the 4 LSMM results had more comparability.

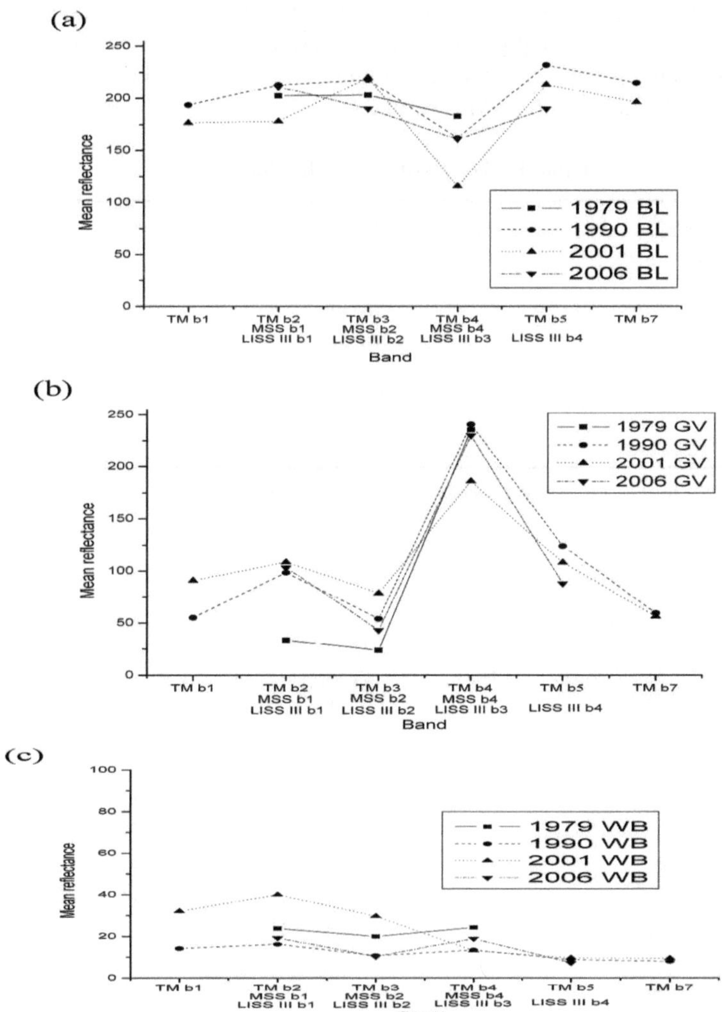

Fig. 3. End-member spectra used in LSMM. These end-member spectra were respectively derived from 4 temporal images. (a) is the spectra of bare land (BL); (b) is the spectra of green vegetation (GV); (c) is the spectra of water body (GV).

3.3 LSMM Result

The primary results of LSMM were separate images for each of the end-members (BL, GV and WB) containing an estimate of the percentage of that end-member in each pixel. When the model was run, each pixel value corresponded to the fraction of BL, GV and WB in that pixel and the range of fraction values was between 0 and 1. Figure 4 illustrated the fraction images of the 4 dates.

After achieving LSMM results of the 4 date's images, the six classes were identified in LSMM results. Some samples of 6 classes were extracted from the 2006 image, based on the investigated ground truth information and GPS location. Then the characteristics of 6 classes were obtained from 2006 image. According to these characteristics, other dates' samples of 6 classes were chosen, and the spectral curve of 6 classes at each temporal were extracted. Figure 5 is the spectral curve of 6 classes at the end-member band in 2006 LSMM result.

Table 1. The classification algorithm

Classes	Algorithm*	Step
Bare land and construction (BLC)	BL > 1	1
Dense reed (DR)	GV > 1	1
Open water (include fishpond) (OW)	WB > 1	1
Sparse reed (SR)	BL < WB & BL < GV	2
Dry grassland (DG)	WB <BL & WB < GV	2
Muddy area (MA)	GV < BL & GV < WB	2

* 'BL', 'GV' and 'WB' represent the pixel value normalized in corresponding band.

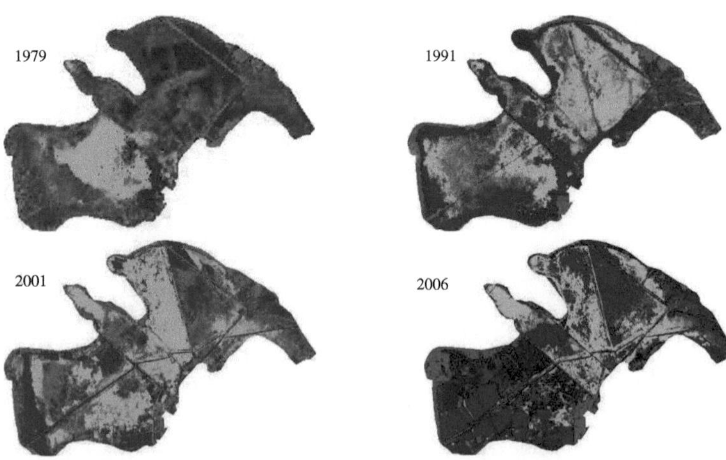

Fig. 4. Fraction images for three end-members, bare land (BL) in red, green vegetation (GV) in green and water body (WB) in blue, obtained by LSMM

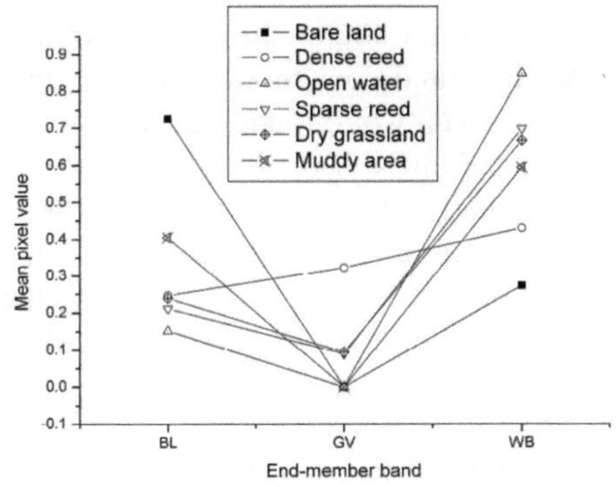

Fig. 5. Mean curves of 6 classes at the end-member band in 2006 LSMM result

4 Results and Discussion

4.1 Land Cover Change Result

We had only survey data in 2006, and land types in wetland changed greatly, so only the error in 2006 was analyzed. Because of the consistent classification method, classification accuracy of other years cannot be low. Error matrices were used to assess classification accuracy, shown in table 5 for 2006. The overall accuracy for 2006 was 90.49%, with Kappa statistics of 88.57%. User and producer accuracies of individual classes were consistently high, ranging from 75% to 100%.(Table 4)

Table 2. The error matrices of classification for 2006

Class	BLC	DR	OW	MA	SR	DG	Total	User Acc.
BLC	**171**	0	0	0	0	0	171	100
DR	0	**218**	0	0	23	34	275	79.27
OW	0	0	**181**	0	0	0	181	100
MA	1	0	0	**198**	26	0	225	88
SR	0	1	30	0	**153**	0	184	83.15
DG	0	0	0	0	0	**173**	173	100
Total	172	219	211	198	202	207	1209	
Prod. Acc.	99.42	99.54	85.78	100	75.74	83.57		
Overall Accuracy = 90.49%								
Kappa Coefficient = 88.57%								

Classification maps were generated for all four years (figure 4) and the individual class area and change statistics for the four years were summarized in table 6.

From 1979 to 2006, bare land and construction land increased approximately 500 ha, and open water increased 850 ha while muddy area decreased 1650 ha. Relatively, bare land and construction land increased 185.20% from 1979 to 2006, with the greatest increase occurring from 1990 to 2001;open water increased 494.41%, while muddy area decreased 57.36%. Compared to the changes with the former mentioned three land uses, dense reed, sparse reed and dry grassland changed slightly between 1979 and 2006. In 1979, the proportion of these three classes was, respectively, 15.37%, 24.18% and 19.19% and changed to 18.87%, 27.31% and 16.77% in 2006. The variations were only 3.5%, 3.13% and 2.42%.

To further evaluate the results of land cover conversions, matrices of land cover changes from 1979 to 1990, 1990 to 2001, and 2001 to 2006 were created (Table 5). In the table, unchanged percentages were located along the major diagonal of the matrix. These results indicated that the muddy area had the most variation with the least unchanged percentage in three periods, respectively, 12, 0 and 14. From figure 6, we also found this great change. For instance, highlighted area a (a1 and a2) in figure 6, indicated the change process MA→BLC; b (b1 and b2) showed the process MA→DR or SR→OW; c illustrated MA→DG, and d was MA→BLC→MA.

Table 5 showed the BLC was stable with almost highest unchanged percentage in three periods. This stabilization owed to the construction in BLC, which supported by highlighted a (figure 6). In highlighted area a, approved by field survey, a1 was the urban expansion of Wolitun, and a2 was a fishing workshop. Highlighted area showed if the MA were changed to BLC then the changed area were not transition but only expanded. Nevertheless, we also found the BLC conversion to other classes. In the study area, bare land was formed as a result of salinization, and its variation had close relation with rainfall. A example is highlighted area d (figure 6), showing the process MA→BLC→MA.

Table 3. Transition between the major land use types (area in %) in the Longfeng wetland

Land use type	BLC	DR	OW	MA	SR	DG
1979-1990						
BLC	**38**	2	24	16	8	8
DR	5	**21**	15	7	44	7
OW	0	35	**43**	1	20	1
MA	7	11	27	**12**	33	10
SR	1	26	21	2	**44**	5
DG	3	28	22	6	35	**6**
1990-2001						
BLC	**82**	0	1	0	4	12
DR	1	**38**	2	7	51	2
OW	10	17	**12**	7	48	6

Table 3. *(Continued)*

MA	40	2	7	**0**	19	31
SR	3	28	3	10	**54**	2
DG	17	7	1	0	25	**48**
2001-2006						
BLC	**47**	2	10	14	9	15
DR	1	**47**	7	13	24	8
OW	9	5	**55**	13	16	2
MA	2	8	23	**14**	50	3
SR	4	16	11	17	**37**	15
DG	6	3	5	12	6	**67**

5 Conclusion

Longfeng wetland located in Daqing city is one of the biggest wetlands in China. Because of the fast development of Daqing city, this wetland had been destroyed and occupied to some extent. This paper analyzed quantitatively the change pattern of Longfeng wetland using multi-temporal remote sensing data. This research aimed to explore the long-term dynamics in land cover and the change driving forces in Longfeng wetland, and revealed that there were substantial land use changes in the area from 1979 to 2006. General patterns and trends of land use change in the area were evaluated by: (1) classifying images used LSMM method and get satisfying result; (2) comparing the classified results of 4 dates; (3) analyzing the major changes in the whole area and these changes in relation to natural and human factors. In addition, the different change patterns between protected area and unprotected area were compared, which indicated the influence of human activities on this area and the protection function of natural reserve.

The drawback of this research is that there are not so many classes, owing to the limits of remote sensing image. Especially, caused by spectral similarity, it is very difficult to separate bare land into construction and water into rearing pond, which affects the correlation analysis between changes and human activity. However, in this research, according to the field survey result and the special imagery characteristics (there are many roads between ponds or constructions), the influence is minimized.

This study showed that the ecosystem of Longfeng wetland had been threatened by the rapid development of Daqing city. In unprotected area, many areas had been filled with rearing pond and urban expansion. However, in natural reserve, the wetland had been protected very well, the areas of dense reeds had consistently increased, and the urban expansion did not occupy the areas of natural reserve (figure 8). Therefore, establishment of natural reserve is a good method to prevent the degradation and improve the ecosystem of wetland.

Acknowledgments. We would like to appreciate the constructive suggestions by both reviewers and editor, which made the article more consistent. This work was funded by an NSFC grant (41001209) and China's Special Funds for Major State Basic Research Project (2010CB950603).

References

1. Acreman, M.C., Hollis, G.E.: Water management and wetlands in Sub-Saharan African. IUCN, Gland (1981)
2. Kashaigili, J.J., Mbilinyi, B.P., Mccartney, M., Mwanuzi, F.L.: Dynamics of Usangu plains wetlands: Use of remote sensing and GIS as management decision tools. Physics and Chemistry of the Earth 31, 15–16 (2006)
3. Sugumaran, R., Harken, J., Gerjevic, J.: I Using Remote Sensing Data to Study Wetland Dynamics in Iowa. In: Iowa's 6th Biennial GIS Conference, Ames, IA (2003)
4. http://www.wetlands.org/
5. Graetz, R.D.: Remote Sensing of Terrestrial Ecosystem Structure: An Ecologist 's Pragmatic View. Remote Sensing of Biosphere Functioning. Ecological Studies 77, 5–30 (1990)
6. Elmore, A.J., Mustaro, J.F., Manning, S.J., Lobell, D.B.: Quantifying Vegetation change in semiarid environments: Precision and accuracy of Spectral Mixture analysis and the Normalized Difference Vegetation index. Remote Sensing of Environment 73(1), 87–102 (2000)
7. Shanmugama, P., Ahna, Y.-H., Sanjeevib, S.: A comparison of the classification of wetland characteristics by linear spectral mixture modelling and traditional hard classifiers on multispectral remotely sensed imagery in southern India. Ecological Modelling 194(4), 379–394 (2006)
8. Asner, G.P., Heidebrecht, K.B.: Spectral unmixing of vegetation, soil and dry carbon cover in arid regions: comparing multispectral and hyperspectral observations. International Journal of Remote Sensing 23(19), 3939–3958 (2002)

Data Quality Evaluation of ZY-1 02C Satellite

Mingquan Wu[2], Jie Wang[1,*], Ni Yao[3], Zhongwei Hou[4], and Changyao Wang[2]

[1] Institute of Remote Sensing and Earth Sciences, Hangzhou Normal University, Hangzhou 310036, China
wangjie1022@163.com
[2] The State Key Laboratory of Remote Sensing Science, Institute of Remote Sensing Applications, Chinese Academy of Sciences, Beijing 100101, China
liexuewuwei@163.com
[3] Beijing University of Agriculture, Beijing 102206, China
[4] Liaoning Engineering Survey and Design Institute, Jinzhou Liaoning 121000, China

Abstract. ZY-1 02C satellite is a new Chinese civil remote sensing satellite launched on December 22, 2011. It is the highest resolution civil remote sensing satellite in China which carried panchromatic/multispectral sensor and high resolution sensor. Compared with image of SPOT5 satellite, the radiation quality of ZY-1 02C sensors data was evaluated. The result showed: the SNR of ZY-1 02C is higher than SPOT5 panchromatic band; the power spectrum of ZY-1 02C is close to SPOT5 panchromatic band; however, the mean value and variance of ZY-1 02C data are smaller than SPOT5 panchromatic band, while the skewness and kurtosis of ZY-1 02C data are higher than SPOT5 panchromatic band; the definition of ZY-1 02C data is much smaller than SPOT5 panchromatic band and the entropy is a little smaller than SPOT5 panchromatic band.

Keywords: ZY-1 02C satellite, Data quality evolution, Signal to Noise Ratio, Entropy, Definition.

1 Introduction

Chinese resource satellite (ZY-1 02C) is a new civil remote sensing satellite which was launched on December 22, 2011. It weighs about 2100 kg, 3 years design life, and equipped with panchromatic/multispectral sensor (P/MS) and High-resolution sensor (HR). Its mainly usage is to acquire high-resolution panchromatic and multispectral data for land resources survey and monitoring, disaster prevention and mitigation, forestry, water conservancy, ecological environment, and the state of major projects monitoring.

Data quality evolution is not only an important technology to inspect the design of satellite sensors, but also an important theoretical reference for remote sensing satellite sensor development and band selection[1]. In the last decades, several approaches have been proposed in this study area. Yang YY et.al. [2] evaluated extended service image of CBERS-01, Wang J et.al. [3] evaluated image of CBERS-02, Zhou YB et.al. [4] and Zhang JL et.al. [5] evaluated image of CBERS-02B, Liu R et.al. [6] evaluated image of HJ-1 satellite, Zhang JL et.al. [7] evaluated image of Hyperion sensor onboard on EO-1satellite, Wang QJ et.al. [8] evaluated image of LISS3 sensor onboard on IRS-P6

* Corresponding author.

D. Li and Y. Chen (Eds.): CCTA 2012, Part II, IFIP AICT 393, pp. 187–195, 2013.

satellite, G H Thomson[9] evaluated image of Arkon-2 satellite. Most of those methods are based on image engineering. We would like to draw your attention to the fact that it is not possible to modify a paper in any way, once it has been published. This applies to both the printed book and the online version of the publication. Every detail, including the order of the names of the authors, should be checked before the paper is sent to the Volume Editors.

2 Study Area

The test area covers a surface of about 104 km^2 and covered by images of SPOT5 and ZY-01 02C satellite. It is located in the north-east of Beijing, China (40.05°-40.58° N, 116.51°-117.21°E) (fig.1). Most of the area is flat and devoted to agriculture, with a prevalence of winter wheat, corn and urban areas.

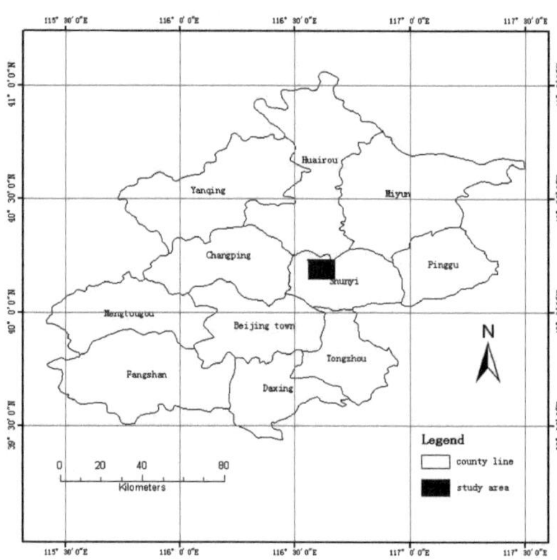

Fig. 1. Location of the study area

3 Data Introduction

Three ZY-1 02C images and one SPOT5 image were used in this study (Table 1). All of those data were acquired in clear shy conditions and covered the same area.

Table 1. Images used in this study

Satellite	ZY-1 02C				SPOT5
Sensor	02C-MS	02C-PAN	02C-HR	PAN	Sensor
Path/row	1/107	1/107	395/107	281/269	Path/row
Acquisition time	2012/02/01	2012/02/01	2012/02/13	2007/04/24	Acquisition time
Spatial resolution	10m	5 m	2.36 m	2.5 m	Spatial resolution
Sensor	02C-MS	02C-PAN	02C-HR	PAN	Sensor

3.1 ZY-1 02C Satellite

ZY-1 02C is a high spatial resolution civil remote sensing satellite. It carried one panchromatic/multispectral sensor (P/MS) and two High-resolution cameras (HR) (Table 2).

Table 2. Basic parameters of sensors carried on ZY-1 02C and SPOT5 satellite

Satellite	ZY-1 02C			SPOT5	
Sensor	P/MS		HR	PAN	MS
	PAN	MS			
Bandwidth(μm)	0.51-0.85	0.52-0.59 0.63-0.69 0.77-0.89	0.50-0.80	0.49-0.69	0.43-0.47 0.49-0.61 0.61-0.68 0.78-0.89
Spatial resolution	5 m	10m	2.36 m	2.5 m or 5m	10m
Swath Width	60km		54km	60km	
Revisit Time(day)	3-5		3-5	2-3	

3.2 SPOT5 Satellite

SPOT5 satellite is the fifth satellite of France SPOT series. It carried two High Resolution Geometric (HRG) sensors, one High-Resolution Stereoscopic (HRS) imaging instrument, and one Vegetation sensor.

Please note that, if your email address is given in your paper, it will also be included in the meta data of the online version.

4 Approach

4.1 Qualitative Evaluation

Image quality of satellite data can be evaluated qualitatively by visual interpretation approach or histogram comparison method. Visual interpretation method judged the image quality is good or bad directly by naked eyes, while histogram comparison method determined the image quality by histogram comparison. Images with good quality have a smoother and broader histogram.

4.2 Quantitative Evaluation

Radiation quality of satellite data can be evaluated quantitatively by parameters such as radiation precision, Signal to Noise Ratio, sharpness, information content, and power spectrum.

Radiation Precision. Radiation precision is an indicator that reflects the radiation situation of images. It can be measured by parameters like mean, variance, skewness and kurtosis. The calculation formulas are as follows:

$$Mean = \overline{x} = \frac{1}{N}\sum_{j=0}^{N-1} x_j \ . \tag{1}$$

$$Variance = \frac{1}{N-1}\sum_{j=0}^{N-1}\left(x_j - \overline{x}\right)^2 \tag{2}$$

$$Skewness = \frac{1}{N}\sum_{j=0}^{N-1}\left(\frac{x_j - \overline{x}}{\sqrt{Variance}}\right)^3 \tag{3}$$

$$Kurtosis = \frac{1}{N}\sum_{j=0}^{N-1}\left(\frac{x_j - \overline{x}}{\sqrt{Variance}}\right)^4 - 3 \tag{4}$$

Where N is the number of pixels, j is the index of pixels, and x_j is the value of pixel j.

Signal to Noise Ratio. Signal to Noise Ratio is the ratio of signal and noise in image. There are many different methods to calculate the image SNR, and different methods have different result. Some research showed that the image SNR and land cover closely related [10]. Variance method can determine the SNR level of sensors, but the method is influenced by surface target. However, using this method between different sensors in the same area can be used to compare the SNR level between different remote sensing sensors. The calculation formula of SNR by variance method is followed:

$$SNR = Mean / Variance \ . \tag{5}$$

Sharpness. The sharpness reflects the sensitivity of the image edge change. The more distinctive of color contrast of the image junction, the clearer of the image. A improved point sharpness algorithm was used to calculate the image sharpness1. The calculation formula is followed:

$$p = \frac{\sum_{i=1}^{m\times n}\sum_{\alpha=1}^{8}\left|df / dx\right|}{m \times n} \ . \tag{6}$$

Where *p* is the sharpness, *m* and *n* is the line and row number of pixels, *df* is the amplitude of intensity change, *dx* is the distance increment between pixels.

Information Content. Information content is a indictor that reflect the rich degree of information contained in images. It always measured by information entropy. The calculation formula of entropy is followed:

$$H = -\sum_{i=1}^{n} P_i \log P_i \quad . \tag{7}$$

Where *H* is the information entropy, *i* is the value of pixels, P_i is the probability of value *i*. Due to the cross correlation between adjacent bands, the redundant information must be removed when calculating total entropy of multiband images [11].

Power Spectrum. Image power spectrum reflects the image texture features. When the image quality depredated, the loss of high frequency components will lead to blur of the image space texture and the image power spectrum of each component will fall. Thus, the comparative evaluation of image quality can be evaluated by comparing the image power spectrum12. Image power spectrum can calculate by Fourier transform. The calculation formula is followed:

$$sum = \sum_{u=0}^{m-1} \sum_{v=0}^{n-1} \left| \hat{F}(u,v) \right|^2 \Delta u \Delta v \quad . \tag{8}$$

Where sum is image power spectrum, *u* and *v* is the index of pixels, $\hat{F}(u,v)$ is the value of pixel (u, v) after Fourier transform.

5 Result and Analysis

5.1 Result of Qualitative Evaluation

Through visual interpretation approach, we can found out that both SPOT5 image and ZY-1 02C images have a good visual effect (Fig.2). Land cover types like farm land, water and cities were very clear. However, those land cover types have a more strong contrast in SPOT5 image.

The histogram of SPOT5 PAN band image, and images of ZY-1 02C were calculated by ENVI 4.3 (fig.3). By comparing histogram of those images, we found out that the quality of SPOT5 PAN band image is better than ZY-1 02C images. The histogram of SPOT5 PAN band image is more broad and more smooth, while the histogram of ZY-1 02C images were narrow.

5.2 Result of Quantitative Evaluation

Parameters of radiation precision, Signal to Noise Ratio, sharpness, information content, and power spectrum were calculated to evaluate the radiation quality of ZY-1 02C data and SPOT5 PAN band image (table 3).

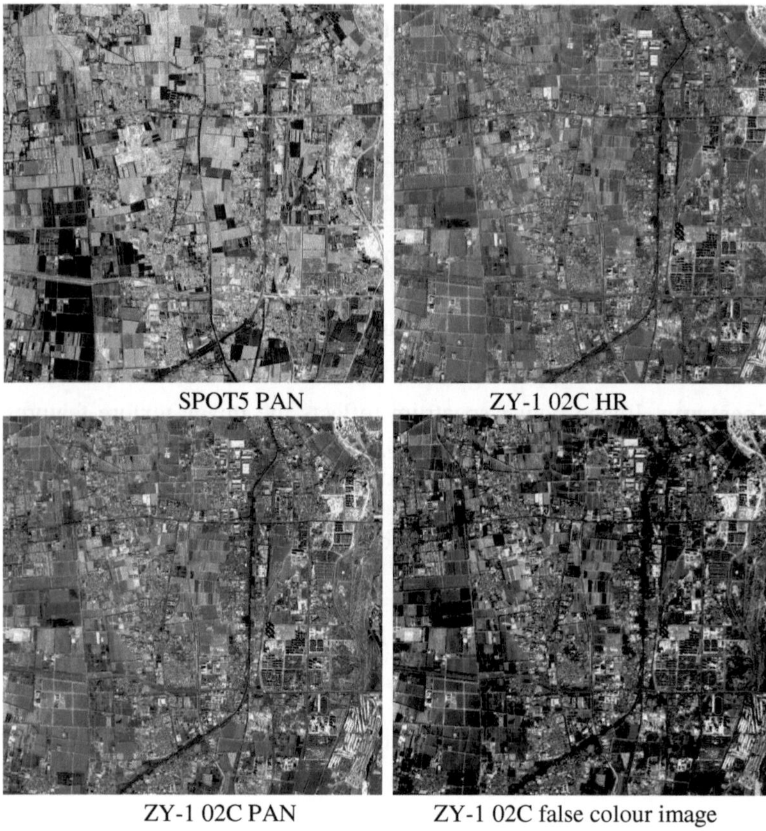

Fig. 2. Images of ZY-102C and SPOT5

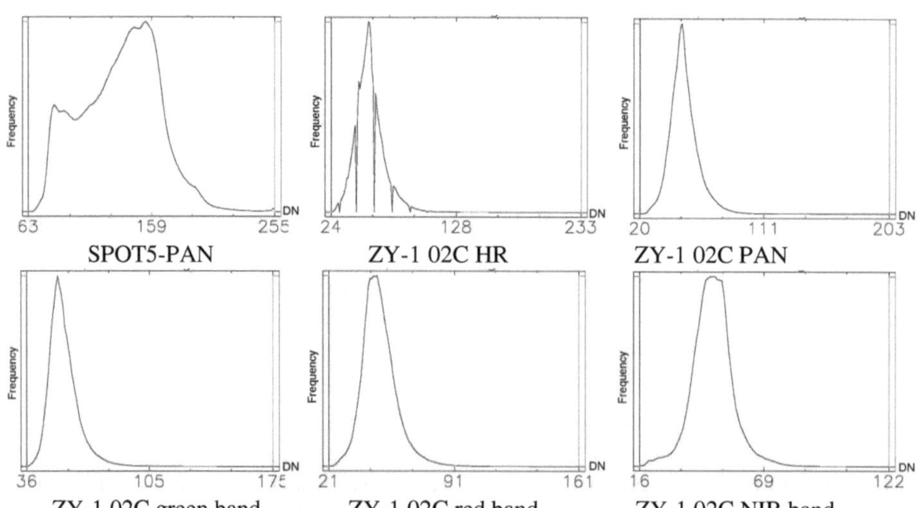

Fig. 3. Histogram comparison of different sensors

Table 3. Quantitative evaluation results of radiation quality

Parameters	SPOT	HR	Pan	MUX		
				Green	Red	NIR
Mean	135.917	57.246	52.990	56.905	49.918	47.813
Variance	882.993	134.722	101.315	66.393	85.967	64.759
Skewness	-0.087	0.984	1.023	1.698	1.058	0.286
Kurtosis	-0.284	3.449	3.742	7.058	3.435	1.866
SNR	4.574	4.932	5.264	6.984	5.384	5.942
Sharpness	33.115	16.339	19.899	17.586	20.800	17.812
Power spectrum	115.096	107.559	100.855	95.394	94.314	93.914
Entropy	6.816	5.388	5.280	4.596	4.970	5.003
Total entropy	6.816	5.388	5.280	6.649		

By comparing parameters of mean, variance, skewness and kurtosis, we found that the mean and variance of SPOT5 PAN band image is higher than any ZY-1 02C images, while the skewness and kurtosis are smaller. This showed that the value distribution of SPOT5 PAN band image is more smooth and broad, which is coincide with the result of histogram comparison.

The SNR of ZY-1 02C images are higher than SPOT5 PAN band image. The green band image of ZY-1 02C MS sensor has the highest SNR. It is 53% higher than SPOT5 PAN band image, while the red and near-infrared band is 26% and 30% higher than SPOT5 PAN band image. The SNR of pan band of ZY-1 02C is 25% higher than SPOT5 PAN band image. The HR band of ZY-1 02C has a very close SNR with SPOT5 PAN band, which is only 8% higher than SPOT5 PAN band.

The sharpness of SPOT5 PAN band is more higher than any ZY-1 02C images. The HR band of ZY-1 02C has the highest sharpness in ZY-1 02C images. However, its sharpness is only 51% of the sharpness of SPOT5 PAN band.

Images of those two satellites have a close power spectrum. The power spectrum of SPOT5 PAN band is a little higher than images of ZY-1 02C.

The SPOT5 PAN band has higher entropy than any ZY-1 02C images. The entropy of SPOT5 PAN band is about 21%、23%、33%、27% and 27% higher than the entropy of HR band, pan band, green, red and near-infrared band. Even the total entropy of MS images is also a little lower than the entropy of SPOT5 PAN band.

6 Conclusions and Discussion

6.1 Conclusions

Comparing with SPOT5 PAN band, the quality of ZY-1 02C images is similar to SPOT5 PAN band. Each data has distinct advantages and disadvantages.

The SNR of ZY-1 02C images are higher than SPOT5 PAN band. And the Power spectrum of ZY-1 02C images is close to SPOT5 PAN band. However, due to the histogram of ZY-1 02C images is too narrow and abrupt, the ZY-1 02C images cannot distinguish the differences of the surface reflectance very well. The sharpness of ZY-1 02C images is lower than SPOT5 PAN band. Moreover, the entropy of ZY-1 02C images are lower than SPOT5 PAN band.

The ZY-1 02C has not blue band images. This will limit the application of ZY-1 02C data in water monitoring. However, ZY-1 02C have two Panchromatic bands. So the panchromatic data acquisition ability of ZY-1 02C satellite is stronger than SPOT5 satellite.

6.2 Discussion

Due to the lack of higher-resolution topographic maps, the positioning accuracy of ZY-1 02C satellite was not evaluated in this paper. However, the positioning accuracy is a very important index of satellite image quality. So, we will carry out work in this area when conditions are ripe.

Due to data constraints, only SPOT5 PAN band was used as a comparative data. And comparison of multi-spectral bands between SPOT5 and ZY-1 02C satellite was not carried out. The difference of spatial resolution may influence the result of some comparison between SPOT5 PAN band and multispectral bands of ZY-1 02C satellite. Moreover, the acquire date of SPOT5 PAN band is five years earlier than ZY-1 02C images. The change of land cover was not considered in the evolution which has a certain extent effect to evolution result.

Acknowledgments. This work was supported by the opening foundation of Institute of Remote Sensing and Earth Sciences, Hangzhou Normal University, and the Major State Basic Research Development Program of China (2010CB950603).

References

1. Liu, R., Sun, J., Wang, J., et al.: Data quality evaluation of chinese HJ CCD sensor. Advances in Earth Science 26, 971–979 (2011)
2. Yang, Y., Wang, H., Wang, J., Xia, D.: Reseach on the quality of exceed images provided by CBERS-1 Satellite. Spacecraft Recovery and Remote Sensing 12, 34–39 (2003)
3. Wang, J., Yang, Y., Wang, H., Xu, Y., Xia, D.: The quality evaluation of the Image Date of CBERS-1No. 2 Satellite. Spacecraft Recovery and Remote Sensing 25, 34–38 (2004)
4. Zhou, Y., Tian, Q., Zhang, X.: CBERS-02B CCD image data quality evaluation and application potential for vegetation classification. Applications of Remote Sensing 6, 47–52 (2008)
5. Zhang, J., Zhao, Y., Zhang, J., et al.: The image quality evaluation of CBERS-02B and the application of the satellite of the interpretation of ore-forming factors of uranium deposits. Remote Sensing for Land and Resources 79, 69–73 (2009)
6. Yang, Z., Gu, S., Qiu, H., et al.: CBERS-1's CCD image quality evaluating and cross calibrating study. Journal of Remote Sensing 78, 113–120 (2004)
7. Zhou, Y., Tian, Q.: Image quality evaluation of EO-1 Hyperion sensor. Geo-Information Science 10, 678–683 (2008)
8. Wang, Q., Tian, Q.: Quality evaluation of LISS3 image from IRS-P6 satellite. Geography and Geo-Information Science 23, 11–14 (2007)

9. Thomson, G.H.: Evaluation of Russian Arkon-2 Earth observation satellite. The Imaging Science Journal 53, 163–173 (2005)
10. Liu, J., Zheng, L., Tong, Q.: Estimation of signal to noise ratio of remote sensing images. Journal of Basic Science and Engineering 7, 360–365 (1999)
11. Lin, Z., Zhang, Y.: Measurement of information and uncertainty of remote sensing and GIS data. Geomatics and Information Science of Wuhan University 31, 569–572 (2006)
12. Balboa, R., Grzywacz, N.: Power spectra and distribution of contrasts of natural images from different habitats. Vision Research 43, 2527–2537 (2003)

Intellectualized Identifying and Precision Control System for Horticultural Crop Diseases Based on Small Unmanned Aerial Vehicle*

Hongxin Cao[1,**], Yuwang Yang[2], Zhiyuan Pei[3], Wenyu Zhang[1], Daokuo Ge[1], Yiran Sha[1,4], Weixin Zhang[1], Kunya Fu[1], Yan Liu[1], Yuli Chen[1], Hongjun Dai[1], and Hainan Zhang[2]

[1] Institute of Agricultural Economics and Information Research/Engineering Research Center for Digital Agriculture, Jiangsu Academy of Agricultural Sciences, Nanjing 210014, Jiangsu, P.R. China
caohongxin@hotmail.com
[2] School of Computer Sciences, Nanjing University of Sciences and Technology, Nanjing 210014, Jiangsu Province, P.R. China
[3] Resource Monitoring Station, Chinese Academy of Agricultural Engineering, Beijing 100026, P.R. China
[4] College of Applied Meteorology, Nanjing University of Information Science and Technology, Nanjing 210044, Jiangsu Province, P.R. China
caohongxin@hotmail.com

Abstract. To explore small unmanned aerial vehicle (UAV) remote sensing identifying technology for horticultural crop diseases, and to combine it with small UAV spraying, the peach leaf blade was taken as material, the peach shot-hole disease was taken as object for spectral capture, and the intellectualized identifying and precision control system for horticultural crop diseases were developed, mainly including the identifying system for horticultural crop diseases, spraying pesticide system carried by small UAV, and the ground mobile control station system. Some sub-systems were initially tested in laboratory and outside. This study can provide an important reference for unmanned precise control of horticultural crop diseases.

Keywords: Spectrum, identification, horticultural crop diseases, precision control, unmanned aerial vehicle (UAV).

1 Introduction

The control for diseases and insect pests is the key to horticultural crop production with high yielding, good quality, better benefit, ecology, and safety, and the traditional

---

* **Foundation Information:** Jiangsu Province Agricultural Scientific Technology Innovation Fund [CX(11)2042], and Jiangsu Province Scientific Technology Supporting Program [BE2012386-1].

** Corresponding author.

D. Li and Y. Chen (Eds.): CCTA 2012, Part II, IFIP AICT 393, pp. 196–202, 2013.

control for diseases and insect pests depends on manned or machined application of pesticide, which has some obvious disadvantages such as bigger work intensity, high cost, and environmental problems, etc. [1-4].

In recent years, rapid progress on the studies of unmanned aerial vehicle (UAV) spraying facility has been made in Japan and USA, etc. The first UAV in the world was created by Yamaha Company in Japan in 1990, which was mainly used in bestrewing pesticide [5]. Since 90's in 20 century, the remote control helicopter has been used in prevention and cure for diseases and insect pests of field crops, fruit tree, and vegetables, proper fertilization and quality management of field crops, and estimation of crop growth status (rice) [6, 7]. Because the UAV has some advantages such as small in scale, agility in grasp, and good effect in spray, etc., it has been into one of development trend that the UAV is used in agricultural production. The studies of aviation spraying of UAV were also done in USA [8]. In china, the application of UAV spraying is not adaptable with needs of modern agricultural development, and has a big gap comparing with that of overseas [8].

The objectives of this study is to explore the UAV remote sensing identifying technology for horticultural crop diseases, to combine it with UAV spraying, and to create intellectualized identifying and precision control system for horticultural crop diseases based on small UAV.

2 Materials and Methods

2.1 Materials

The experiment for spectral capture was carried out in peach garden of headquarters in Jiangsu Academy of Agricultural Sciences from May to Nov. in 2012, the peach leaf blade was taken as material, and the peach shot-hole disease was taken as object for spectral capture.

2.2 Methods

Canopy spectral measurement: The spectral reflectance of the disease leaf and the happy leaf in the same peach, and the disease peach and the happy peach in the same peach garden and time were measured using MSR16R Multispectral Radiometry (16 wave bands, spectral ranges from 460 nm to 1700 nm, and manufactured by Cropscan, Inc. in USA) in the fine day's 10:00 a.m.-2:00 p.m. with above 45°solar elevation angle, the radiometry probe surface was paralleled to peach canopy or target leaf with 0.2 m upright distance, and the spectral reflectance adjusting of standard whiteboard was done around each measurement.

Expert identification of disease leaf: The severity of disease is composed of the size and number of disease spot in a leaf or a peach, in that the size and number were divided into 3 levels according to expert identification.

Analysis and using of data: First, the ranges of sensitive waveband for disease parameters were found based on correlation between disease parameters and single waveband spectral reflectance of disease leaf and peach, and the vegetation indices (VI) composed of sensitive waveband were computed based on existing research; Second, the optimum VI of the most sensitivity to disease parameters were screened out through analyzing correlation between disease parameters and VI, and the corresponding regression equations were established, in that correlation analysis, and VI computing were done using OFFICE.2007 and SPSS v 16.0.

2.3 Concept Design

First, the standard spectral reflectance of disease leaf in peach garden was captured using spectral radiometry, and the spectral reflectance of peach garden canopy was measured using spectral probe carried by small UAV. Second, the information captured was wirelessly sent to ground, analysed, and identified. Third, the intellectualized identifying and precision control system for horticultural crop diseases based on identifying-pesticide spraying integration was set up through combining the spectrum-based identifying for crop diseases and the UAV-based pesticide spraying system (Fig. 1).

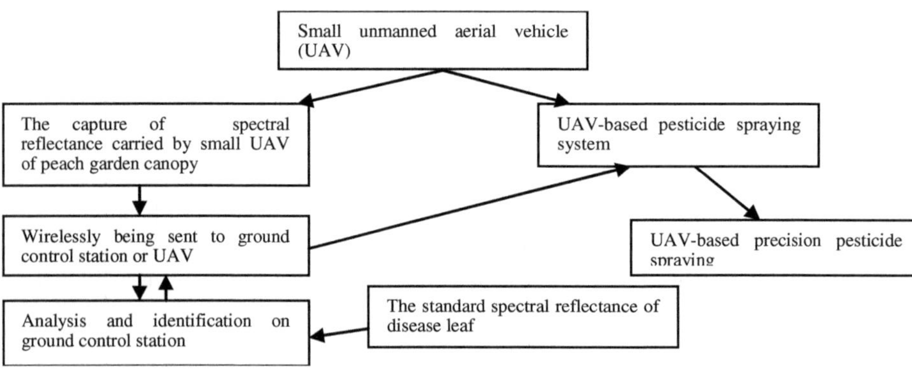

Fig. 1. Concept design of the intellectualized identifying and precision control system for horticultural crop diseases based on identifying- pesticide spraying integration

3 Results

The intellectualized identifying and precision control system for horticultural crop diseases based on small UAV mainly included identifying system for horticultural crop diseases, spraying pesticide system carried by small UAV, and the ground mobile control station system.

3.1 Identifying System for Horticultural Crop Diseases

The identifying system included the part carried by small UAV, and the ground mobile control station.

The Spectral Measurement Device Carried by Small UAV for Horticultural Crop Diseases. The focal plane probe with 200 mm large caliber, inclined 45°lens structure, large light area, and stable turning was used in spectral probe, and the head of spectral measurement probe was composed of the imaging optics system, the bearing flat, and spectral probe. The scanning spectral ranges from visible to infra-red (430-1030 nm, 60 wavebands with 10 nm resolution) were selected, its verification examples were shown in Fig.2. The linear detector-scanning program was applied in modularizing spectral meter system, and the spectral resolution, space resolution, sensitivity, and volume and weight were considered in the development of spectral measurement section. The GPS system was applied in UAV positioning, and the dynamic precision of position system can go to the meter level. The peg-top stable flat was introduced into UAV in order to insure pose stability of the spectral measurement device.

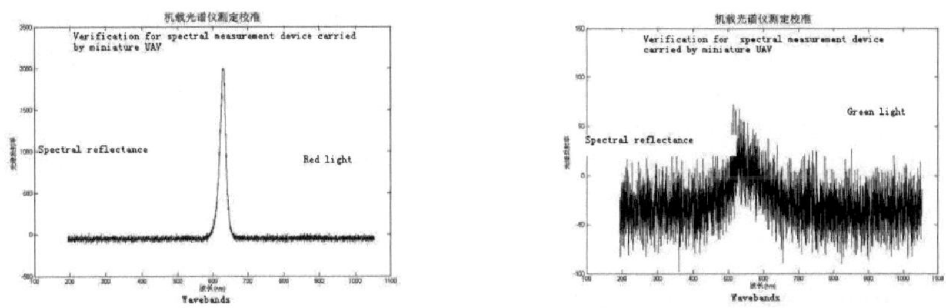

Fig. 2. Verification for the spectral measurement device carried by small UAV

Communication between the Small UAV and the Ground Mobile Control Station. Due to the large number of spectral data needed to be wirelessly transferred between the small UAV and the ground mobile control station, the bandwidth for communication should be larger. At the same time, the various statements from the ground mobile control station needed to be incepted by UAV, but because the formats of statements were to a certainty, its length was finitude, and the demand to communication bandwidth was not high. The independent uplink and downlink channel were adopted by the system, and the asymmetric bandwidths were used in transfer to avoid interfere between data and statements transfer each other. Under the normal condition, this can provide less than 3.5 Mbps of the uplink speed, and 54 Mbps of the downlink speed. The communication frequency with 2.4GHz was selected, and the wireless transceiver system developed by ourselves was used in communication device.

The Real-Time Transaction for Spectral Data and Identifying of Peach Diseases. There is a powerful computer system in the ground mobile control station, which mainly was used in pre-processing for the spectral data, such as GPS data processing and composite, calibration processing, bad line repair, pixel registration, histogram calculation, contrast stretching, regional amplification, spectral curve display, data capture, and data format conversion, etc. The spectral data from the UAV was real-timely processed, the comparison of the final spectral data and the standard spectral reflectance of disease leaf were done after a series of processing steps, and the identifying information of peach diseases were obtained.

3.2 Spraying Pesticide System Carried by Small UAV

The impact factors of UAV spraying effect included spraying device, target plant, atomization degree for droplet, droplet flow transport properties, and droplet penetrability, etc.

The Development of UAV Spraying Device. It included self-regulation for amount of spray, droplet diameter, and nozzle structure, etc. so as to fulfill low volume, ultra-low volume, and electrostatic spraying, and to help to realize double times atomization when the high speed airflow were passed through spraying nozzle.

Droplet Atomization. It can be divided into two modes, centrifugal type and hydraulic atomization. The centrifugal atomization nozzle was selected in UAV spraying device, and driven by generator in UAV. The size of droplet can be regulated using rotate speed and the dial structure of nozzle.

In addition, the atomization droplet was effected by airflow speed, state, and outside atmosphere, etc. in the flight UAV. The droplet atomization was studied in order to obtain the optimum particle size and movement state.

Sedimentation Process. In the droplet transport and sedimentation processes, due to space of flow was very complex, and it let droplets to collide and polymerize each other, which resulted in movement of droplets with huge randomicity. Therefore, the mathematic models of space of flow were developed through experiments in order to capture the integrative flow state of droplets in the space of flow.

Effects of Meteorological Conditions on Spraying. The optimum time of spraying was one of important factors for raising droplet target rate, and under the different regions, and terrain conditions, the effects of temperature and wind speed, swirl from rub of airflow and crop, etc. on distribution of droplet on crop were studied.

Effects of Droplet Size and Number on Spraying. The droplet coverage was related to many factors, especially rotor vortex, and flight velocity in UAV. The droplet size depends on the target state sprayed, the optimization of droplet size can help to raise target coverage, and use pesticide availably.

Establishment of UAV Operating Instructions for Spraying Technology. The UAV operating instructions for spraying technology were established according to the relevant FAO's rule to ensure security for flight and operators, spraying device stability, and operating precision, and reduce the negative effects on environment, nearby crop, and human and livestock.

3.3 The Ground Mobile Control Station System

The ground mobile control station system included the ground mobile control center carried by vehicle, the ground remote control and measurement system for UAV, the ground control station system for spraying pesticide, and the ground spectral signal incept system, etc., in that the ground remote control and measurement system for UAV were composed of industrial control computer, remote control transmitter, telemetry receiver, remoter, and power box, etc.

3.4 Test of the System

Test of the system mainly included the identifying precision of peach diseases, and the spraying pesticide effect, and the spraying pesticide effect was checked through comparison the spraying pesticide position carried by UAV and the observed disease position in peach garden. Some sub-systems were initially tested in laboratory and outside.

4 Discussion and Conclusions

4.1 Discussion

Identifying System for Horticultural Crop Diseases. The previous studies showed that there were some shape and physiological changes in leaf such as becoming yellow, wilt, and withered, etc. when plant were stressed by diseases, which was visible changes easily to be identified. In addition, red light, infra-red, "blue shift", and "red shift", etc. all need to be considered in this study [9].

Size and Volume of Device Carried by Small UAV. In order to reduce loading of the small UAV, the spectral data captured by small UAV were transferred to the ground control station, and real-timely processed in the ground control station. The optic structure of spectral system carried by the UAV were elaborateelaborately designed, the whole spectral system weight was less than 1 kg, and its volume was less than 0.001 m^3 (0.10m×0.10m×0.10m).

Spraying Technology in Lower Level. The spraying technology in lower level is the key to the intellectualized identifying and precision control system for horticultural crop diseases based on small UAV. Therefore, the centrifugal atomization nozzle was used in small UAV spraying device to adapt to spraying pesticide in lower level, and raise control efficiency.

The System Need to be Test Under Different Conditions. The study is still in the initial stage, the various sub-systems were not integrated into the intellectualized identifying and precision control system for horticultural crop diseases, and tested under different conditions.

4.2 Conclusions

The intellectualized identifying and precision control system for horticultural crop diseases were developed, mainly including the identifying system for horticultural crop diseases, spraying pesticide system carried by small UAV, and the ground mobile control station system. Some sub-systems, e.g. the spectral measurement device carried by small UAV, were initially tested in laboratory and outside.

References

1. Zhang, L.: The Pollution of Agricultural Environment with Chemical Pesticide and its Control. J. Nanjing Agric. Technol. Coll. 17(4), 36–38. 43 (2001)
2. State Environmental Protection Administration: China's Pesticide Pollution and Its Being Problems and Suggestions. Environ. Prot. 6, 23–24 (2001)
3. Zhu, M.Y., Zhang, J.X., Lu, Z.H., et al.: Controlling the Environmental Pollution of Chemical Pesticide, Carrying out Consecutions Protection. Jiangxi Plant Prot. 26(3), 29–31 (2003)
4. Tu, Y.Q.: Thinking of Problems of Agriculture Chemicals and Circumstance. Reclaiming and Rice Cultiv. 1, 58–60 (2003)
5. Zhang, Y.: Application of remote control helicopter in Japan agriculture. World Agric. (4), 49–50 (1997)
6. Sun, X.Y.: Multi-purposes unmanned helicopter in agriculture. Shangdong Agric. Mech. (9), 27 (1999)
7. Hunt, E.R.: Evaluation of digital photography from model aircraft for remote sensing of crop biomass and nitrogen status. Precis. Agric. 6, 359–378 (2005)
8. Gong, Y., Fu, X.M.: Aviation spraying pesticide technology in modern agriculture. Agric. Equip. Technol. 34(6), 26–29 (2008)
9. Huang, M.Y., Wang, J.H., Huang, W.J., et al.: Hyperspectral character of stripe rust on winter wheat and monitoring by remote sensing. Trans. Chin. Soc. Agric. Eng. 19(6), 154–155 (2003)

Creating Topologically Consistent 3D City Models of LOD+ with Extrusion

Yunfei Shi[1] and Biao He[2]

[1] College of Resources & Environment, Linyi University, 276000 Linyi, China
shiyunfei_001@163.com
[2] Shenzhen Municipal Committee of Urban Planning, Land Resource and Real Estate, 518040 Shenzhen, China
1688698@qq.com

Abstract. There are many methods to create 3D city model, in which the simplest one is to extrude building footprints to obtain "block-shaped" polyhedra representing buildings. While the resulting 3D city models are not topologically consistent, if the 2D topological relationships between the footprints are not taken into account. As a result, the created models only meet the purpose of visualization. In this paper, we focus on creating topologically consistent 3D city models from the footprints of heterogeneous structure buildings. The generated models are more detailed than LOD1 of CityGML and are called LOD1+, which not only construct topology between buildings but also generate internal structure topology of them. We have implemented the procedure, tested it with real-world datasets, and validated it.

Keywords: 3D city model, extrusion, topological consistent, heterogeneous structure building.

1 Introduction

3D city models have been increasingly applied as data source or visualization method in a growing number of fields such as urban planning, architecture, archaeological reconstruction, tourism, civil engineering, mobile telecommunication, energy supply, navigation, etc[1].Different applications require various detail and accuracy of 3D model. For this purpose, 3D city models were divided into five levels of details (LODs) by City Geography Markup Language (CityGML), from LOD0 where only the terrain is stored to LOD4 where buildings have detailed roofs structures, windows, rooms and even pieces of furniture [2].

There are many ways to create 3D city model, in which the simplest one is arguably with extrusion. That is, given a set of footprints representing the buildings or rooms (polygons in the plane), assign them a height and then push them upwards to create polyhedra[3]. Extrusion is indeed easy if the only thing you want to do with your 3D city model is to look at it! [2]. Many commercial software can extrude polygons to obtain buildings, but the resulting models cannot be stored in a topological data structure [3]. The reason is that the created models are not topologically consistent. In a

D. Li and Y. Chen (Eds.): CCTA 2012, Part II, IFIP AICT 393, pp. 203–210, 2013.

nutshell, that means that extruding results contains duplicate points, overlapping faces, faces intersecting where there are no points, etc[2]. Fig.1 illustrates a process where two building footprints are being extruded. Extrusion without considering other footprints only guarantee that each resulting polyhedron is valid, but not ensure that a group of footprints will generate a topologically consistent 3D city models.

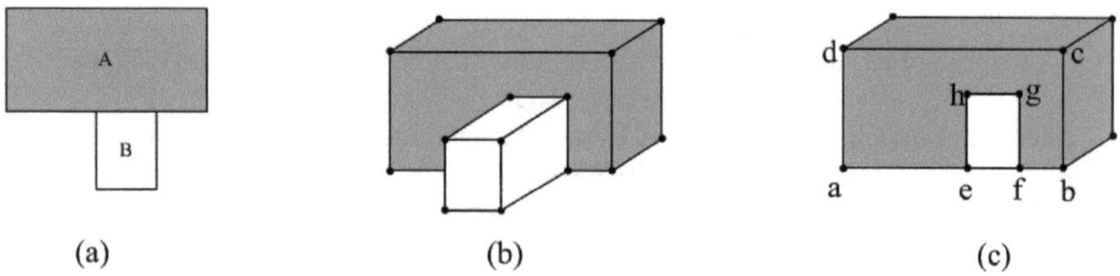

Fig. 1. (a)Two polygons in the plane.(b)Two polyhedra obtained by extrusion of the two polygons.(c) To be topologically consistent the front polygon of the polyhedron obtained by the extrusion of A should be modeled with two polygons: efgh and aehgfbcd.

Many studies [2] [3] have been developed on topologically consistent 3D city model by extrusion. Models created by existing studies belong to LOD1 in CityGML. In this level, whole building is represented as a "block-shaped" model with flat roof structures and no internal floors and rooms. However, a building often includes many floors and rooms in the real world. Existing studies only consider the topological construction between buildings, but do not take into account that of internal structure (floors and rooms) of buildings. In fact, topology of internal structure of buildings is very important for many applications, such as indoor navigation [4], safe escaping and so on. In addition, because a building contains multiple floors and the different floor corresponds to different structure (i.e. heterogeneous structure), that make the extruding process need to consider simultaneously vertical and transverse topology. Therefore, topological construction of internal structure of buildings is more complicated than that between buildings.

In this paper, we focus on creating more detailed topologically consistent 3D city models than LOD1 and are called LOD1+, which not only construct topology between buildings but also generate internal structure topology of them. Our main contribution is to create topologically consistent 3D City models of LOD1+ level from heterogeneous structure buildings. The remainder of this paper is organized as follows. Section 2 provides the basis of research. Section 3 gives our extruding approach for heterogeneous structure buildings. Section 4 gives ours experiments. Finally, section 5 offers conclusions of our work.

2 The Basis for Research

Topology is one of the mechanisms to describe relationships between spatial objects. Thus, it is the basis for many spatial operations. Let S be a set of spatial objects in $\mathbb{R}^3$, the three dimensional Euclidean space. Spatial objects in S are formed by four geometric primitives: (i) ρ^0 is a point(0 dimensionality); (ii) ρ^1 is a straight line segment (1-dimensionality), which is referred to simply as a line in the following; (iii) ρ^2 is a polygon (2- dimensionality); (iv) ρ^3 is a polyhedron (3- dimensionality).

Topological consistency: Let S be a set of spatial objects in $\mathbb{R}^3$, we say that S is topologically consistent if the following rules are valid [3]:

(1) every line segment ρ^1 in S is formed by two points ρ^0 also in S;

(2) the intersection of two lines ρ^1_1 and ρ^1_2, denoted $\rho^1_1 \cap \rho^1_2$, is either empty or is a point in S;

(3) the intersection of two polygons ρ^2_1 and ρ^2_2, denoted $\rho^2_1 \cap \rho^2_2$, is either empty or a finite set of primitives in S (points or lines).

(4) the intersection of two polyhedrons ρ^3_1 and ρ^3_2, denoted $\rho^3_1 \cap \rho^3_2$, is either empty or a finite set of primitives in S (points or lines or polygons).

The related storage structures of topological data include half-edge, winged-edge [5], Quad-edge [6] and so on. From the view of manifold, they belong to manifold topology. In addition, many researchers also propose non-manifold data structures, such as Face-edge [7], Radial edge, X-maps and so on. The paper uses half-edge as data structure to storage the data with extrusion.

3 Our Extruding Approach

This section presents our algorithm to create topologically consistent 3D city model. The algorithm is based on the idea that the floor plans taken as the input are a set of topologically consistent polygons (in 2D). We use existing GIS tools (e.g. arcgis) to create topologically consistent 2D dataset as input and output topologically consistent 3D models of buildings and a set of faces embedded in 3D. In created models, a building (polyhedra) is enclosed by its bounding faces and simply has a reference to these faces, and faces (or edges or nodes) shared by adjacent polyhedra (or faces or edges) are not duplicated.

3.1 Generation of ROOF-PLAN and FLOOR-PLAN

According to the above analysis, the generation of floor faces or roof faces is influenced by the structure of adjacent floors. Therefore, extruding process needs to consider synthetically the relationship between the extruded floor and its adjacent floors. For this reason, we divide floor plans into three types: STRUCTURE-PLAN, ROOF-PLAN and FLOOR-PLAN.

STRUCTURE-PLAN is used to create vertical walls (polygons) and determines the unit structure of the whole floor. If the extruded floor is the *nth*, STRUCTURE-PLAN will be the original floor plan.

ROOF-PLAN is used to create roof face of the extruded floor. If the extruded floor is *n*, ROOF-PLAN will be obtained by overlapping the *nth* and the *(n+1)th*

STRUCTURE-PLAN (As shown in Fig.2).If building only has ***n*** floors, ROOF-PLAN is equivalent to STRUCTURE-PLAN.

FLOOR-PLAN is used to create floor face of the extruded floor. If the extruded floor is ***n***, ROOF-PLAN will be obtained by overlapping the *nth* and the *(n-1)th* STRUCTURE-PLAN (As shown in Fig.2). If *n* is equal to 1, FLOOR-PLAN is equivalent to STRUCTURE-PLAN.

From these three definitions, we can find that ROOF-PLAN of the *nth* floor is equivalent to FLOOR-PLAN of the *(n+1)th* floor and FLOOR-PLAN of the *nth* floor is equivalent to of the *(n-1)th* ROOF-PLAN floor. For a building with n floors, it needs ***n*** STRUCTURE-PLAN and *n+1* ROOF-PLAN or FLOOR-PLAN.

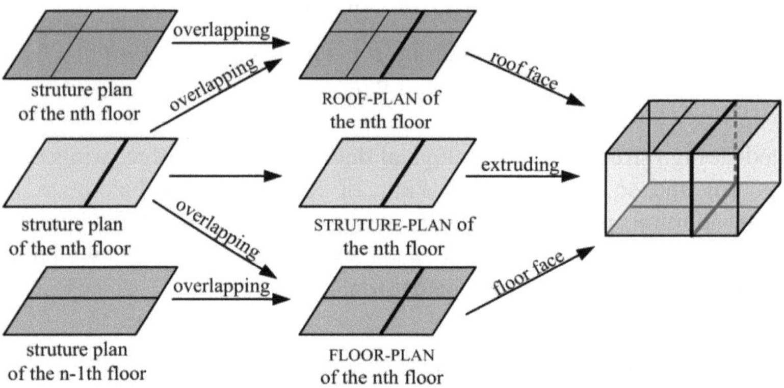

Fig. 2. Generation of ROOF-PLAN and FLOOR-PLAN

The overlap operation can use existing tools (e.g. arcgis) to implement. In order to ensure overlapped plans are topologically consistent, clean operation is used to reconstruct topology. In this way, we create topologically consistent ROOF-PLAN and FLOOR-PLAN for every floor of a building.

3.2 Topological Construction of Vertical Direction

After the generation of ROOF-PLAN and FLOOR-PLAN, the next step is to create vertical walls and body data. ROOF-PLAN and FLOOR-PLAN only determine the structure of roof face and floor face of extruded floor, but the whole floor structure is determined by STRUCTURE-PLAN. It's noteworthy that STRUCTURE-PLAN is only used as reference to create vertical walls but itself is not the topological primitive of created model, which is different from the usage of ROOF-PLAN and FLOOR-PLAN.

Let's use a simple sample to illustrate. As shown in Fig.3, vertical shaded face (wall) is created by extruding the line segment *ab*. But *ab* is not a part of the created face. The edge set of the created face is composed by the following line segments: $\{V_{a''}V'', V''V_{b''}, V_{b''}V_{b'}, V_{b'}V',, V'V_{a'}, V_{a'}V_{a''}\}$. It can be seen that each vertex in STRUCTURE-PLAN creates a vertical edge, and each edge and each polygon in STRUCTURE-PLAN creates a vertical face and a polyhedron separately.

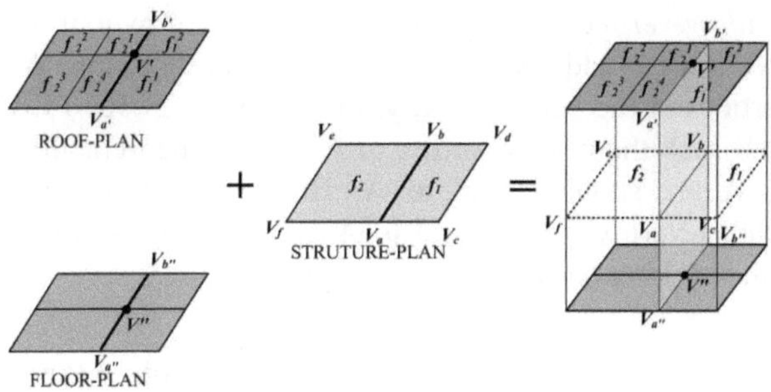

Fig. 3. Generating process of vertical wall

Generation of Vertical Elements. *Creating vertical edges.* Taking out a vertex from the vertices set of STRUCTURE-PLAN, it seeks the vertex corresponding to counterpart in ROOF-PLAN and FLOOR-PLAN respectively and forms a vertical edge. For instance, vertex V_a corresponds to vertex $V_{a'}$ in FLOOR-PLAN and $V_{a''}$ in FLOOR-PLAN respectively (Fig.3).Connecting $V_{a'}$ and $V_{a''}$, a vertical edge is created. In order to facilitate following work, the direction of edge is from the vertex in FLOOR-PLAN to the vertex in ROOF-PLAN. Iterating the process, we can obtain the vertical edge set.

*Creating vertical faces (walls).*For each edge of STRUCTURE-PLAN, corresponding edge set is sought in ROOF-PLAN and FLOOR-PLAN. An edge in STRUCTURE-PLAN may correspond to multiple edges in ROOF-PLAN and FLOOR-PLAN. For example, the corresponding edges of edge V_aV_b are $V_{a'}V'$ and $V'V_{b'}$ in ROOF-PLAN (fig.3).The found corresponding edges (crosswise edges) and the created vertical edges generate vertical faces in common. As shown in fig.3, the vertical face created by edge V_aV_b is $V_{a''}V''V_{b''}V_{b''}V'$ $V_{a'}V_{a''}$, which is formed by edge set {$V_{a''}V''$, $V''V_{b''}$, $V_{b''}V_{b'}$, $V_{b'}V'$,, $V'V_{a'}$, $V_{a'}V_{a''}$}. Here, we need to sort the edge set and make them arrangement in clockwise or counterclockwise order.

Creating polyhedra. For each face (polygon) in STRUCTURE-PLAN, corresponding face set is sought in ROOF-PLAN and FLOOR-PLAN. Similarly, a face in STRUCTURE-PLAN may correspond to multiple faces in ROOF-PLAN and FLOOR-PLAN. The corresponding faces of face f_2 are f_2^1, f_2^2, f_2^3 and f_2^4. The found corresponding face (horizontal face) set and the created walls compose polyhedra in common.

Algorithm of Seeking Corresponding Relationships. The above three steps all have a core issue: seeking corresponding relationship, namely, finding vertices, edges and faces in STRUCTURE-PLAN corresponding counterparts in ROOF-PLAN and FLOOR-PLAN separately. The seeking process of three corresponding relationship is following.

Seeking corresponding vertices. Due to the model obtained by extrusion, the extruded vertex should have same x and y coordinates in vertical direction. When we seek the corresponding vertices of the vertex A (x, y, z) in STRUCTURE-PLAN, z coordinate can be ignored and seek the nearest vertex to A (x, y). Theoretically, the coordinates obtained nearest vertex should equal to A. However, owing to storage error of computer, the nearest vertex is not equal to A and has a little offset. Therefore, the nearest vertex is the seeking corresponding object exactly.

Seeking corresponding edges. Based on the method of seeking corresponding vertices, we can obtain the corresponding vertices in ROOF-PLAN and FLOOR-PLAN of vertices of each edge in STRUCTURE-PLAN. As shown in fig.3, corresponding vertices in ROOF-PLAN of vertices of edge V_aV_b are $V_{a'}$ and $V_{b'}$. After confirming corresponding vertices, it needs to check whether $V_{a'}$ and $V_{b'}$ can make up an edge. If they can, the edge $V_{a'}V_{b'}$ is corresponding edge. Otherwise, we need to ignore z coordinate and find the vertex which distancing $V_{a'}V_{b'}$ less than given threshold in ROOF-PLAN. If the vertex is found, we think that the vertex is on the line. For example, Vertex **V'** is on the $V_{a'}V_{b'}$ in fig.3. By the way, corresponding edge set can be found.

Seeking corresponding faces. With the method of seeking corresponding edges, we can acquire the corresponding edges (in ROOF-PLAN or FLOOR-PLAN) of all edges of each face in STRUCTURE-PLAN. The obtained corresponding edges draw up a range in ROOF-PLAN and FLOOR-PLAN. If the corresponding edges just surround a face, the surrounded face is corresponding face. Otherwise, we need to judge all faces of ROOF-PLAN (or FLOOR-PLAN) and pick out those faces which drop in the tagged range of the corresponding edges. These faces compose the corresponding face set.

With above algorithm, topological consistent 3D models of every floor are created. The next step is to build the topology between floors. Because the extrusion of every floor is independent, the common elements (nodes, edges and faces in ROOF-PLAN or FLOOR-PLAN) of adjacent floors are reconstructed twice. Moreover, the duplicated common elements are same in structure. Thus, the topological construction of adjacent floors can be implemented by deleting the duplicated nodes, edges and faces and adding references to the reserved common topological primitives for adjacent polyhedra in different floors. By this way, the created 3D buildings models include a set of complete topological relationships, in which common faces are shared by adjacent polyhedrons, and common edges are shared by adjacent faces and the common nodes are shared by adjacent edges.

4 Experiments

With the proposed algorithm, we establish an experiment system which takes Visual C++ as development tools, and use Open Scence Graph (OSG) as graphic interface and choose a small residential area as test data. The data include floor plans and elevation tables of every building. Floor plans provide the structure information of every floor and elevation tables provide the height information of every floor. Fig.4 gives a process of a sample building, which only has two floors. (a) is the structure plan of the first floor; (b) is the structure of the second floor; (c) is the result of extrusion.

(a) (b) (c)

Fig. 4. (a) Structure plan of first floor; (b) Structure plan of second floor;(c)Model by extrusion

Fig.5 is the modeling process of complicated building with two floors. (a) and (b) are the structure plans of the first floor and the second floor separately. (c) is the extruding result.

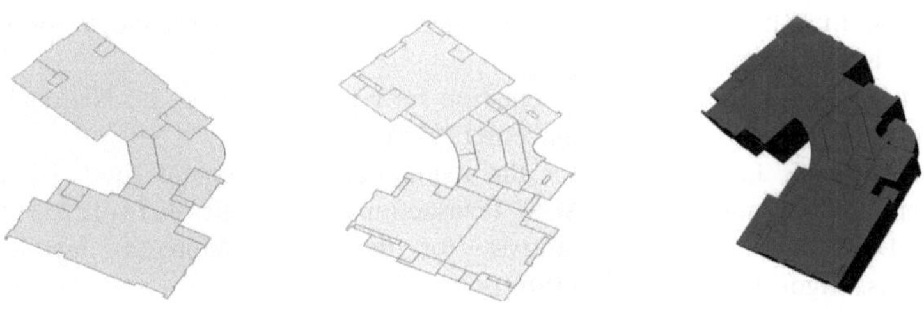

(a) (b) (c)

Fig. 5. (a)Structure plan of first floor; (b)Structure plan of second floor;(c) Model by extrusion

Fig.6(a) is a sample of whole building with extrusion; (b) is the model of a small residential area with extrusion.

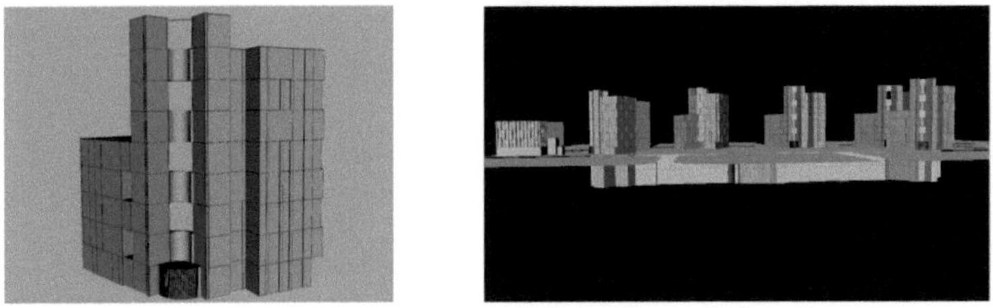

(a) (b)

Fig. 6. (a) Building with extrusion; (b) Small residential area with extrusion

5 Conclusion

The paper analyzes the application demands of 3D city model and proposes a method for constructing topological consistent 3D city model. With the method, we can quickly generate 3D data of large area buildings, avoid the re-survey of buildings and save the cost of data collection. The shortage of the method is that it can't handle the building with slope walls, which is our further goal.

References

1. Tack, F., Buyuksalih, G., Goossens, R.: 3D building reconstruction based on given ground plan information and surface models extracted from spaceborne imagery. ISPRS Journal of Photogrammetry and Remote Sensing 67, 52–64 (2011)
2. Ledoux, H., Meijers, M.: Extruding building footprints to create topologically consistent 3D city models. In: Krek, A., Rumor, M., Zlatanova, S., Fendel, E. (eds.) Urban and Regional Data Management, UDMS Annual 2009, pp. 39–48 (2009)
3. Ledoux, H., Meijers, M.: Topologically consistent 3D city models obtained by extrusion. International Journal of Geographical Information Science 25, 557–574 (2011)
4. Lee, J., Zlatanova, S.: A 3D data model and topological analyses for emergency response in urban areas. In: Zlatanova, L. (ed.) Geospatial Information Technology for Emergency Response. Taylor & Francis (2008)
5. Baumgart, B.G.: A polyhedron representation for computer vision. In: National Computer Conference, pp. 589–596. AFIPS, Anaheim (1975)
6. Guibas, L., Stolfi, J.: Primitives for the manipulation of general subdivisions and the computation of Voronoi diagrams. ACM Transactions on Graphics 4, 74–123 (1985)
7. Dobkin, D.P., Laszlo, M.J.: Primitives for the manipulation of three-dimensional subdivisions. Algorithmica 4, 3–32 (1989)

The Study of Soil Fertility Spatial Variation Feature Based on GIS and Data Mining*

Chunan Li, Guifen Chen**, Guangwei Zeng, and Jiao Ye

College of Information and Technology, Jilin Agricultural University,
Changchun, 130118, China
anchunli123@sina.com,
guifchen@163.com

Abstract. Studying on the spatial distribution and variation characteristics of soil fertility by Application of geographical information systems (GIS) and data mining(DM), with the quantitative analysis of spatial variability of soil fertility in complexity, and provide the effective way for simulations which can be closer to the farmland soil fertility variability. At the same time, interpolation method based on the spatial variability of soil can provide different accuracy data for soil database, researching of space fuzzy clustering based on precise fertilization regional fertility space can provide a decision-making basis for spatial variability of soil fertility evaluation, a precise prediction and evaluation of precision agriculture are the indispensable basic information and theoretical basis of the development and implementation of precision agriculture.In this paper, the authors take the national 863 project which in the Yushu City of Jilin province as the survey region. And then get the best prediction method and discuss the different effect of the soil nutrient under spatial interpolation prediction model on the analysis of the soil nutrient space mutation characteristics of the foundation by the GIS technology and the geography statistical method. Finally, we analysis the soil fertility status by using the weighted space fuzzy clustering as well as the soil nutrient of space mutation distribution, so as to provide basis for the study on area space mutation characteristics of cultivated land soil fertility which has an important theoretical and practical effect on the precise fertilization of crop and soil fertility evaluation.

Keywords: evaluation of soil fertility, GIS, Data Mining, fuzzy clustering.

1 Introduction

With the development of precision agriculture, soil testing and fertilizer recommendation is often used methods recently. But a long time, people use the mean

* Foundation project: Supported by National "863" High-tech Project (2006AA10A309), National Spark Project (2008GA661003) and Changchun Technology Project (09KT14), The research and application of facilities for the safety of vegetables production technology based on Internet of Things(2011-Z20) .

** Corresponding author.

D. Li and Y. Chen (Eds.): CCTA 2012, Part II, IFIP AICT 393, pp. 211–220, 2013.

of the determination of soil fertility to evaluate the soil fertility status of the entire farmland and use single fertilization method to achieve uniform fertilization. But in this way, we cannot reach the required fertilization purposes, and can't effectively improve the utilization rate of fertilizer. At this stage, precision fertilization usually use GPS to make land of an area divided into equally spaced rectangular or square, and then soil testing and sampling in the middle of the rectangle or square of the district. Sample laboratory analysis; re-use GIS technology processing into fertilizer formula grid diagram. After sample analyzed, we use GIS technology to process the fertilization formula grid map. According to the content of specific fertility in the field for each grid operating unit, variable rate fertilizer applicator adjust the corresponding amount of fertilization to achieve low cost, high-income and green environmental protection. But the method according to mesh fertilizing in lack of areas of variable rate fertilizer applicator has been limited. Therefore, according to the actual situation of the current grid fertilization and uniform fertilization, this study attempts to use GIS spatial analysis techniques and weighted spatial fuzzy clustering and decision tree method, which use fuzzy clustering analysis the soil fertility of agricultural cornfield and propose precision fertilization principles and methods, in order to provide a theoretical basis for the lack of variable rate fertilizer applicator region corn-precision fertilizer. Those provide a theoretical basis for precise fertilization of the lack of variable rate fertilizer applicator of the region to promote corn[1].

The spatial variability of soil properties is the variation of soil chemistry, physics, and biological properties in the spatial location. Even in the same soil type, soil characteristics have huge difference. The spatial variability of soil includes two aspects: one is the vertical profile of the soil spatial variability; another is the spatial variation of soil plane. Many studies have found that the nutrient with great coefficient variation showing a significant negative bias peak distribution and small coefficient of variation of the nutrient state of normal distribution. The research of spatial variability of soil property is to determine the value of spatial variability of soil property, improve the accuracy of soil mapping, and provide important scientific basis for precision agriculture [2].

2 Materials and Methods

2.1 The Overview of Research Area

The study area located in the 13th Village, GongPeng Town, Yushu City of Jilin province which is the important commodity grain area of Jilin province, and in which the soil is chernozem and the major crop is corn, soy and so on. This study area is one of the important commodity grain bases in Jilin Province. The terrain is southeast high, northwest low, southeast to northwest tilt. It is at an altitude of157-220m, longitude 126° 01′ 44 ″ -127° 05′ 09 ″, latitude 44° 30′ 57 ″ -45° 15′ 02 ″, belongs to the northeast monsoon in sub-humid temperate climate, average annual precipitation is 500-700mm, rain heat over the same period, longer daylight, conducive to the growth of crops[3].

2.2 Sampling Methods

For the implementation of precise fertilization, the first use DGPS to make land divided into grid cell with 40m × 40m, sampling methods see figure 1. We sample in his grid cell, and the sampling depth is 0~20cm. The specific approach: We collect the samples where at the center of the mesh point and the samples that around the center point about 20m away, then mix them for one sample. Using the four division method to process and separate, we collect amount of soil, and put them into the bag with a number mark. Finally, we analyze the samples in laboratory. We use the standard testing method to chemically analysis the nutrient content of pH, available nitrogen, available phosphorus and potassium in the soil samples.

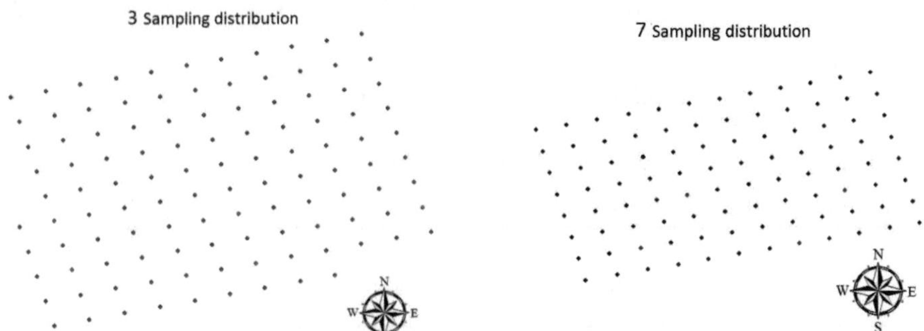

Fig. 1. Sampling distribution

2.3 Fertility Rating

According to the soil test results on the 13th Village, Gongpeng Town, Yushu City of Jilin academy of agricultural sciences agriculture information center and using the land statistical interpolation method, we analyzes the soil nutrient (PH, alkali solution nitrogen, phosphorus, all available p, total potassium, rapidly-available potassium) of space mutation characteristics. And then use fuzzy clustering analysis method as well as the decision tree to deal with the problem.

We collect the samples where at the center of the mesh point and the samples that around the center point about 20m away, then mix them for one sample. Using the four division method to process and separate, we collect amount of soil, and put them into the bag with a number mark. Finally, we analyze the samples in laboratory.

3 Results and Analysis

3.1 Analysis of Data Statistics

Analysis of research data statistics using GIS software, to get the data in Table 1 below. Table 1 lists the conventional statistical characteristics values of the study area soil nutrient: minimum, maximum, mean, standard deviation, coefficient of variation and skewed coefficient. Table 2 lists the second national soil survey grading standards of soil level.

Table 1. Soil nutrient statistic characteristic values

Project	*Min*	*Max*	*Mean*	*SD*	*CV%*	*SK*
pH	5.50	6.67	6.03	0.26	4.27	0.49
Alkaline hydrolysis N (mg/kg)	68.67	168.55	121.15	13.00	10.73	0.32
Total phosphorus (g/kg)	0.02	1.38	0.45	0.25	56.87	0.82
Available phosphorus (mg/kg)	6.58	98.21	14.86	11.01	74.14	5.16
Total K (g/kg)	6.04	27.55	18.89	5.24	27.74	-0.35
Potassium (mg/kg)	50.00	180.00	84.68	21.69	25.61	1.22

Table 2. Soil nutrient grading standards

Project	One	Two	Three	Four	Five	Six
pH	≤4.5	4.5-5.5	5.5-6.5	6.5-7.5	7.5-8.5	8.5-9.0
Alkalin hydrolysis N (mg/kg)	>150	120-150	90-120	60-90	30-60	≤30
Total phosphorus (g/kg)	>1	0.8-1	0.6-0.8	0.4-0.6	0.2-0.4	≤0.2
Available phosphorus (mg/kg)	>40	20-40	10-20	5-10	3-5	≤3
Total K (g/kg)	>25	20-25	15-20	10-15	5-10	≤5
Potassium (mg/kg)	>200	150-200	100-150	50-100	30-50	≤30

Can be seen from Table 1 and Table 2, the study area, soil pH in the range between 5.5-6.67, soil pH is a three level; soil available nitrogen content between 68.67mg/kg-168.55mg/kg, mean for 121.15mg/kg, in view from the mean, indicating that the soil nitrogen content of the study area in a two level; total phosphorus content of the soil between 0.02g/kg-1.38g/kg, mean for 0.45g/kg. In view from the mean, indicating that the soil phosphorus content in the fourth level; soil available phosphorus content in 6.58mg/kg-98.21mg/kg, the mean for 14.86mg/kg. In view of the mean, indicating that the soil phosphorus content in the three level; total potassium content of the soil between 6.04g/kg-27.55g/kg, averaging 18.89g/kg. In view from the mean, indicating that the study area soil potassium at three level; available potassium 50mg/kg-180mg/kg between, mean 84.68mg/kg, from the mean point of view, indicating that the study of soil available K in the four level. In accordance with the size of the coefficient of variation describes the degree of variation, coefficient of variation less than 10% belong to the weak variability of the coefficient of variation; greater than or equal to 10% and less than or equal to 100% of moderate variability; coefficient of variation greater than 100% of the strong variability. Regional soil nutrient coefficient of variation between 4.27% -74.14%, in addition to PH other than the other belong to the moderate variability can be seen from Table 1.

The coefficient of variation of soil available phosphorus is max, 74.14%, far greater than the other elements; pH of the coefficient of variation is minimum, 4.27%. The order of the size of the coefficient of variation of nutrient available: P> total P> K> available K> alkaline hydrolysis N> pH.

The distribution of soil nutrients type relates to the correct evaluation of the soil nutrients. The skewed reflect the frequency distribution of the deflection, SK = 0 when the distribution is symmetry, positively skewed SK> 0, negative bias, SK <0. It can be seen from Table 1, total potassium in the skewed coefficient is negative, the other are positive. For the Kriging method, if data distribution is different, require the use of the method is also different. If the data obeys the lognormal distribution, the selection of the lognormal Kriging interpolation method; If it doesn't obey simple distribution, selection of disjunctive Kriging method; When the data in the presence of dominant trends, selected Fankelige method; When you need to know is whether the attribute value exceeds a certain threshold, the choice of indicator Kriging method; When the same thing with two kinds of attribute relationship, and an attribute is not easy to obtain, choose collaborative Kriging way; If the attribute value expectations for a known constant, with simple Kriging; if the attribute value expectation is unknown, the selection of the ordinary Kriging.

3.2 Spatial Variability of Soil Fertility

One starting point of this study is that obviously difference between soil fertility of different sampling points while use the software ArcGIS, and used to guide the production is valuable, and have a clearly concise indication of existing field's fertility[4]. According to the obtained data, using GIS spatial analysis and applying statistics features of ARCMAP 10. With LPI (Local Polynomial Interpolation) Interpolation analysis fertility data of discrete sampled points to get soil available nitrogen, phosphorus and potassium, pH spatial distribution map. Take phosphorus and potassium spatial distribution map of No.7 and No.3 for comparison.(In Fig.2, Fig.3)

From the figure, there is an obvious difference of the distribution of soil fertility in the third and seventh field. The potassium and nitrogen in the seventh field was obvious marginalization and unevenly distributed. There is a little peak among the third field and with block feature. Block edges and peaks are mainly caused by the current production mode. Spatial variability of these maps are clearly show the distribution of fertility and easy to apply to the actual production.

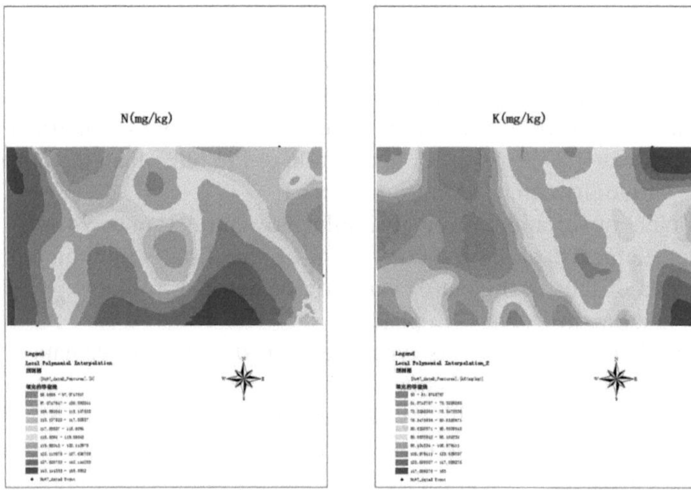

Fig. 2. Spatial distribution of soil fertility in No.7 field

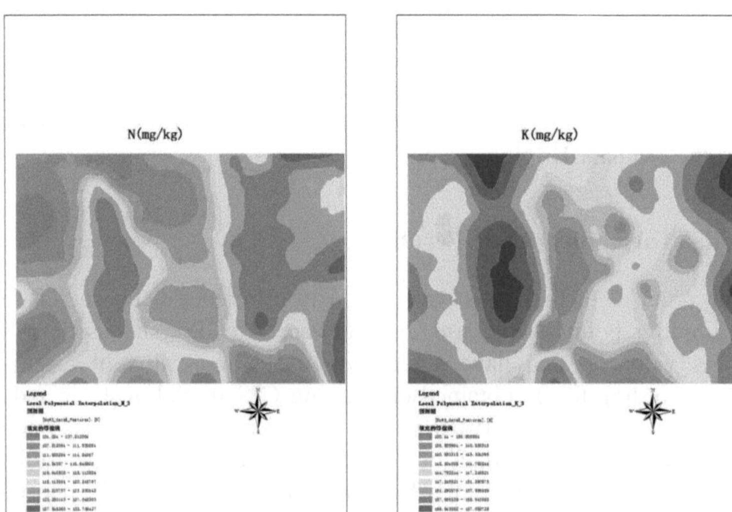

Fig. 3. Spatial distribution of soil fertility in No.3 field

3.3 Fuzzy Clustering Analysis

In fuzzy clustering (also referred to as soft clustering), data elements can belong to more than one cluster, and associated with each element is a set of membership levels. These indicate the strength of the association between that data element and a particular cluster.

Equipped with N Disaggregated indicators, M Category object $X_1, X_2, \cdots, X_m$,each category X_i have N objects $(X_{i1}, X_{i2}, \cdots, X_{im})$. There should be five steps to Fuzzy clustering the M object.

(1) Selection of the appropriate mathematical formulas to original data $\{X_{ij}\}(i=1,2,\cdots,N, j=1,2,\cdots,N)$ for standardization

(2) The selection of the appropriate establishment of mathematical formulas of the M category object based on fuzzy similarity relation matrix $R=(r_{ij})_{m\times m}$.

(3) Using this methods to established relationship matrix R generally satisfies reflexivity and symmetry, have no passing, not the needed fuzzy equivalence relations, requires the following modifications: squares R makes to $R\circ R=R^2$ and then $R^2\circ R^2=R^4$ and so on to make $R^8, R^{16}, \cdots$ until get the equation $R^{2k}=R^k$ .The R^k is a fuzzy equivalence relation, sign for $R^*=(r^*ij)_{m\times m}$ 。

(4) Gradually reduced from 1 to intercept values λ （$\lambda > \min(r_{ij})$）,to make matrix R^* , generate a dynamic cluster map from fine to coarse.

(5) Select the appropriate threshold values λ_0 , to interception matrix R^* or generate dynamic classification map.

Then multiply this matrix by weights vector $A=(a_0, a_1, \cdots, a_m)$ then make weight fuzzy matrix[5].

$$Y=\begin{bmatrix} x_{11} & x_{12} & \cdots & x_{1m} \\ x_{21} & x_{22} & \cdots & x_{2m} \\ \vdots & \vdots & \cdots & \vdots \\ x_{n1} & x_{n2} & \cdots & x_{nm} \end{bmatrix}\begin{bmatrix} a_0 & 0 & \cdots & 0 \\ 0 & a_1 & \cdots & 0 \\ \vdots & \vdots & \cdots & \vdots \\ 0 & 0 & \cdots & a_m \end{bmatrix}$$

$$=\begin{bmatrix} y_{11} & y_{12} & \cdots & y_{1m} \\ y_{21} & y_{22} & \cdots & y_{2m} \\ \vdots & \vdots & \cdots & \vdots \\ y_{n1} & y_{n2} & \cdots & y_{nm} \end{bmatrix}$$

In this paper, we use the weighted clustering algorithm to analysis the data of soil organic matter, available N, P and K nutrition which continuous variable fertility for five years (2008 year's data); Using unweighted algorithm to analysis the data of 2008 year. Clustering results as shown in the table. λ represents the degree of similarity between the data in the table, the class number represent the number of categories when the similarity is λ, , the F value is the value of F distribution that using statistics to calculate. The larger the F value, indicating the distance between classes is larger, and then the classification is better.

Table 3. Clustering result table

Weight in 2008			Unweight in 2008		
λ	Class number	F	λ	Class number	F
1.000	113	0.339	1.000	117	0.327
0.999	100	0.652	0.999	96	1.065
0.998	86	1.305	0.998	76	1.685
0.997	71	2.304	0.997	65	2.459
0.996	48	3.445	0.996	55	2.836
0.995	40	4.594	0.995	47	2.307
0.994	32	4.875	0.994	38	1.845
0.993	28	4.898	0.993	33	2.957
0.992	23	1.371	0.992	30	2.142
0.991	21	1.469	0.991	23	1.971
0.990	17	1.543	0.990	19	1.915
0.989	15	1.084	0.988	17	1.830
0.988	13	1.069	0.986	16	1.889
0.987	9	1.121	0.985	14	2.063
0.982	7	0.935	0.983	13	2.210
0.981	6	0.920	0.981	12	2.081
0.978	4	0.983	0.978	11	2.092

Comparing the weighted clustering and unweighted clustering results in 2008 year shows that: when λ take 0.993, F can has the largest value, and the classification has the best results.(Table 3) At this point, the value of weighted F is 4.898, the value of unweighted F is 2.957, indicating that the gap between classes are larger when have weighted, so when it has the weighted the classification is better. In addition, the unweighted clustering can assigned the different soil nutrient to the same class level by offsetting the differences between the soil nutrient, but when it has the weighted value it can assign to different classes, and then it can has a better reflect of the real situation of soil nutrients, provide a more reliable and more accurate basis for soil fertility classification.

4 Conclusions

In this paper, we use the Yushu City demonstration field data,ArcGIS geostatistical modules and weighted fuzzy clustering algorithm, to research the distribution, and variability of the 576 soil samples' nutrients (including the pH, nitrogen, total phosphorus, available phosphorus, potassium, potassium) in 3 and 7 field in-depth and meticulous. And the result was compared with the second national soil survey. Using ordinary Kriging, inverse distance weighted, local polynomial and global polynomial interpolation method to predict the spatial variation of soil, we compared the accuracy of the four forecasting methods and draw the following main conclusions:

(1) The demonstration field soils is acidic, its pH value is mainly distributed between 5.5-6.5; available nitrogen content in a two level (121.15mg); content of total P(0.45g) and K (84.68mg)are low, in a four level; content of total potassium (18.89g) and available phosphorus (14.86mg) in a three level; from the coefficient of variation (CV%), the coefficient of variation of soil nutrients in demonstration field in the 4.27% -74.14%, in addition to the PH the other belong to the moderate variation. The coefficient of variation of soil available phosphorus is 74.14 percent, far greater than other elements, the spatial distribution is uneven, appear high or low state; PH variation coefficient is the smallest, only 4.27%. The CV% order of nutrient is available P> total P> K> available K> alkaline hydrolysis N> pH. There is a big difference of the spatial variability of different plots of soil nutrients in the soil. The mobility of available nutrient directly affects their spatial distribution in the soil. Clear the different variation of available nutrient in the soil can be variable fertilization on soil to achieve the purpose of both saving fertilizer and to meet the demand for soil nutrients.

(2) The test results show that the weighted fuzzy clustering algorithm is an effective method of soil fertility evaluation; It can provide more reliable and more effective science basis for variable rate fertilization and soil fertility evaluation through the application of weighted fuzzy clustering algorithm, and thus in the same block, by variables fertilization to significantly reduce the regional differences of soil nutrients, and to increase production.

(3) In the ordinary Kriging, inverse distance weighted, global polynomial and local polynomial prediction method, the accuracy is the highest by using ordinary Kriging Interpolation method to predict the spatial distribution of pH and potassium. But using the local polynomial method to create the spatial variability of N, total P, available P and K can have the highest prediction accuracy. Generally speaking, the Kriging method and local polynomial interpolation method have a high precision to predict soil nutrient spatial distribution, while the global polynomial interpolation in predicting the spatial variation of soil has the least accurate, but has the minimum value.

(4) Fertility evaluation work is an important work of precision agriculture, usually, the fertility evaluation, involved less space factors; it has most experts in the field to participate. They use the experimental data to analyze, it is subjectivity, and also more difficult to express the nonlinear relationship between soil properties and environmental variables and uncertainty information. In this study, we use GIS

technology and statistical techniques, clustering algorithms to evaluate of soil fertility which provides a new way and method for spatial data mining and knowledge discovery [6].

References

1. Cao, L.: Soil Nutrient Evaluation Based on GIS and Fuzzy Clustering Analysis. Journal of Anhui Agri. Sci. 38(16), 8595–8596, 8661 (2010)
2. Fan, M.: Spatial Variability of Soil Nutrients and Comparison of Predictive Methods Using GIS, Southwest University
3. Chen, G., Wang, Y., Wang, G.: Research and application of the maize precise fertilization system. Journal of Jilin Agricultural University 28(5), 586–590 (2006)
4. Zhao, Y., Chen, G., Wang, Y.: A Study on Spatial Variability of Soil Nutrients Based on GIS. Acta Agricultural Boreali-occidentalis Sinica 14(6), 195–198 (2005)
5. Chen, G., Cao, L., Wang, G.: Application of Weighted Spatially Fuzzy Dynamic Clustering Algorithm in Evaluation of Soil Fertility. Scientia Agricultura Sinica 42(10), 3559–3563 (2009)
6. Chen, G.: Research and Application of Spatial Data Mining Technology for Precision Agriculture, Jilin University

Study on Agricultural Park Planning Methods Based on Omni-Directional Information Processing Technology*

Xueyuan Chen[1,**], Yongchang Wu[1,***], and Bingwen Zhao[2]

[1] Institute of Agricultural Resources and Regional Planning,
Chinese Academy of Agricultural Sciences, Beijing 100081, P.R. China
chenxueyuan666@sohu.com,
wuyc@mail.caas.net.cn
[2] Dalian V.R Global Vision Co., Ltd, Dalian 116025, P.R. China

Abstract. With rapid development of computer science, the means of the agricultural park planning are renewed continuously. Based on the long-term practice in the agricultural park planning, we put forward innovatively the integrated information processing technology in three dimensions in the paper, that is, information, time and spatial dimensions, On the basis of the coupling relation of multi-dimensional space, time and information, data information and time & spatial elements in agricultural park planning can be integrated and utilized to the maximum by 3D VR calculation as a means of basic technology application; so that a park planning system which is integrating plane, space, time and information resources is formed. By this way, the method is available for the multi-level and fully informational agricultural park planning, and it is an approach to explore modern agricultural park planning in thinking and technical means.

Keywords: Omni-directional information processing technology, Agricultural park, Planning methods, VR technology.

1 Introduction

The region, a highly abstract and broad concept, is a regional space divided for the human's production and life [1]. There are different choices for an area type as the regional unit of regional planning for different research purposes and decision-making backgrounds [2][3]. Aiming at different planning areal units, some disciplines emerge, such as urban planology, which is involved in the comprehensive arrangement for future urban development, the rational urban layout and overall urban arrangement of

* Foundation Project. National Key Technology R&D Program in the 12th Five year Plan of China" study and demonstration on urban carbon sinks protection and improvement of key technical integration" (2011BAJ07B05).

** Maily engaged in Land-Use study.

*** Corresponding author: Institue of Agricultural Resources and Regional Planning, Chinese Academy of Agricultural Sciences, Doctor/Professor, Maily engaged in agriculture park planning and Agriculture resources management research.

D. Li and Y. Chen (Eds.): CCTA 2012, Part II, IFIP AICT 393, pp. 221–228, 2013.

engineering construction [4]. In other words, the Urban Planology is a discipline dealing with urban and rural built environment by their land utilization and urban material spatial planning [5].

The idea of the balance urban and rural development is put forth in the Report of the Sixteenth CPC National Congress. So, many areas or parks are built successively nationwide such as agricultural sci-tech park, agricultural high and new tech area, agricultural sci-tech demonstration area, combined agricultural test area, high-tech agricultural park, agricultural demonstration area with high production and efficiency, and modern agricultural experiment park[6][7][8]. To sum up, all belong to modern agricultural area construction substantially, and are the important ways for development of modern agriculture [9][10][11]. Currently, there is no uniform evaluation standard and construction principle for planning and construction of numerous agricultural areas. Through a long-term practical study on planning and design of agricultural areas, we put forward the agricultural regional planology different from the urban planology and urban & rural planning in research object and target. The agricultural regional planology deals with the agricultural park planning, construction and development with adjustment of agricultural structure and increase of farmers' income. This regional planology offers not only a method to realize modern agricultural regional planning by multilevel and full information, but also sustainable utility of agricultural resources and means of multi-dimensional technology. This will be a better approach for modern agricultural park planning in thinking and technology.

2 Design Method of Park Planning System

The traditional agricultural park planning is expressed in the form of the planning text and plane plan. With great increase of the planning and information contents, the single expression cannot well present the planning idea and contents, and introduction to new technical means becomes a new orientation of the park planning research. The routine park planning is available from time, space and information, but with integration of social economy and combination of space and time, we set forth the concept of the omni-directional processing technology to fit for new planning form, expression and contents. For the omni-directional processing technology, the coupling relation of the multi-dimensional space with time and information should be considered fully besides dependence of objects, scale effect, anisotropy of space and information layering linear in the multi-dimensionalal space. By calculating 3D VR technology as a basic application means, the data information and time & spatial elements can be integrated and utilized in the park planning to the maximum so as to form a park planning system with plane, space, time and information resources.

The core of the omni-directional processing technology is the integration and application of information, time and spatial dimensions and resources, and the key to realize this processing is dependent on a tool system—the whole-park planning & omni-directional VR processing system. By the special VR processing engine developed specially aiming at the park planning, the multi-dimensional space is built in the planning target area, and the information and time can be overlaid and integrated with the multi-dimensional space. With application of functional components, such omni-directional planning results as a scale result presented in the form of 3D VR, planning scheme expressed by time and space, and automatically-generated maps and income from investment can be available. The omni-directional planning results include three major parts as shown in Fig. 1.

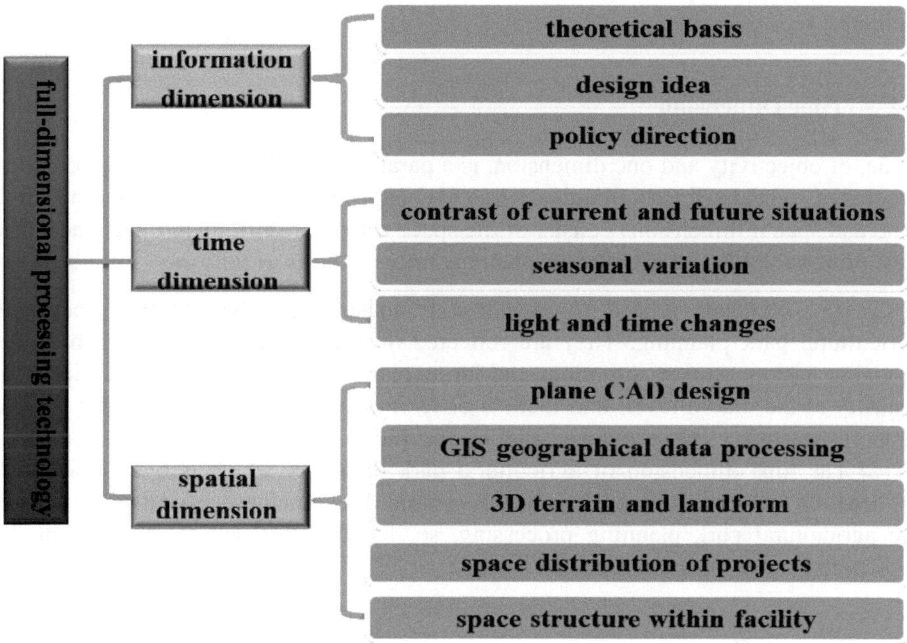

Fig. 1. Dimensional structure for omni-directional processing technology

2.1 Information Dimension

Information is a description of state and characteristics of objective things, and information itself is abstract and invisible with independency of carriers. Its transmission and effect of action are not limited to time and space. It is an independent dimension which can be beyond time and spatial dimensions defined in the physical conception.

With a long-term practical research on the agricultural park planning, it is believed that the information dimension involved in agricultural park planning refers to parts as follows:

(1) All kinds of agricultural information and data supporting the planning design in the process of agricultural park planning, including requirements and target of the region construction, land utilization, water conservancy and transport, agricultural resources, climate resources, social economy, etc.;

(2) Design idea, design thought, national and local industrial policy, industrial standard, and calculating standard for income from the region investment formed during agricultural park planning design;

Above parts belong to key ones that the information dimension covers in the omni-directional conception, and all are important ideas and data foundations forming the agricultural park planning design results based on the omni-directional processing technology.

2.2 Time Dimension

Time, of objectivity and one dimension, is a parameter used to describe the occurring process of an event. It is an imaginary line of change and development of a thing. Both time and spatial dimensions consist of the space-time carrier of matter existence. The time dimension in agricultural park planning refers to parts as follows:

(1) The elements related to agricultural industry described in the process of agricultural park planning. They are reflected in development and change by time, including history, current situations and future planning of target regional agricultural industry at the macro-level; and include difference of characters in different time in different seasons, climates even a day for a specific planning project at the micro-level;

(2) The time dimension of agricultural park planning, which is overlapped with information and spatial dimensions based on omni-directional processing technology in the agricultural park planning processing, so that the designed results present the switching process of design scheme by time.

2.3 Spatial Dimension

Space is a form of matter existence relative to time. The space geographically includes the visual contents from the one-dimensional line, two-dimensional plane to three-dimensional body. Any thing has its distributional pattern, form and layout in space with interaction and mutual effect. It is found that the traditional multi-dimension from one dimension to three dimensions can't express and cover the planning information and idea completely. So, the spatial dimension aiming at agricultural park planning refers to the following parts:

(1) Such information as 3D terrain scenes, water infrastructure, electrical infrastructure, agricultural infrastructure, civil buildings, forests and fruit trees in the planning target region contained in agricultural park planning field;

(2) The visual expression for omni-directional agricultural park planning in the planning design. The visual expression is formed by structuring above multi-dimensional spatial information of the planning target region, with an overall application of CAD tool, GIS information, 3D modeling tool, database building and integrated information system.

3 Applications of the Omni-Directional Processing Technology in the Park Planning—An Example of Zhengding of Hebei Province

3.1 Background of Zhengding Modern Agricultural Park Planning for Hebei Province

The main target for the Zhengding agricultural park planning construction is to recover and develop the geological environment of the Cihe Ancient Road Sandpit, where there are the typical criss-cross sandpits and ravines. In this case, the planning key and difficulty are how to carry out the planning of all kinds of projects in such a complicated terrain condition and how to design spatial layout and 3D levels rationally, so that the limited land and basic resources can be used sufficiently.

Supposing only the traditional planar technique is taken as the planning design method in the case, it is difficult to well express such time factors as seasonal variation and light change which impact the planning contents. For above said reasons, the omni-directional idea and method in this case can well be applied in the planning design. Meanwhile, by sufficiently making use of information, time and spatial elements obtained in study, the planning of the case can be well completed with the VR processing software which is developed independently by us.

3.2 Technical Path of Omni-Directional Processing Technology Application

In the park planning survey, it is very important to collect digital elevation terrain information, on-site live-action photos, positioning information, information of historical changes, and seasonal and climatic variation information about the planning area besides data acquisition and field survey and research. In implementation of the planning design, all kinds of information and data collected in the survey are put into the omni-directional VR processing system software which is developed independently by us, and the multi-dimensionalal spatial environment is built taking advantage of automatic processing unit analysis and processing digital elevation terrain data provided by the system. By this way, the 3D park planning design is realized. Major technical path is seen in Fig. 2.

3.3 Technical Essentials for Omni-Directional Processing Technology Application

In the realization of the case, major technical points are listed as follows:

(1) Work out the terrain maps of the planning area by the CAD application; realize the exhibition of 2D layout on the basis of the layout of agricultural regional planning;

(2) Collect types, distribution and scales of land use by GIS application and social economy data, to realize combination of spatial data and attribute data;

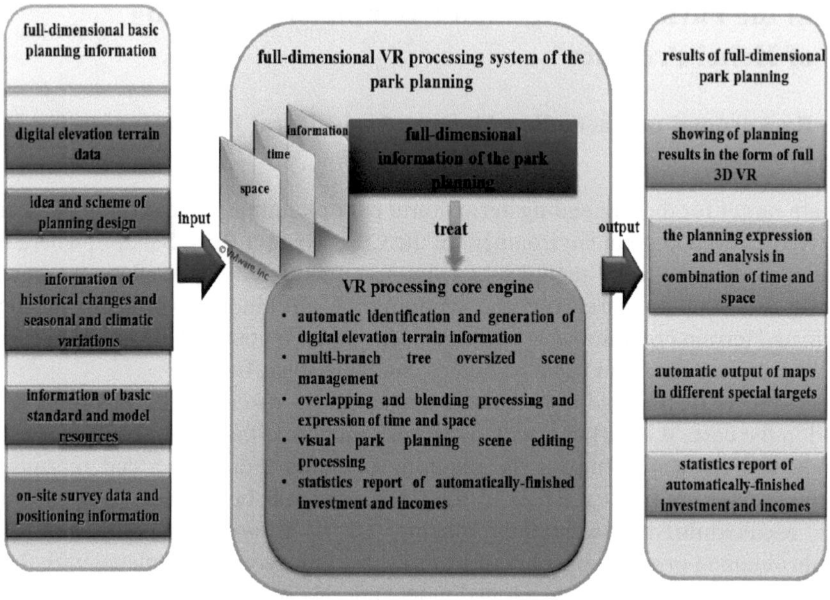

Fig. 2. Applied technology path of omni-directional processing technology for Zhengding Agricultural park planning in Hebei province

(3) Analyze digital elevation data and recover the terrains in the planning area through digital terrain processing units of the omni-directional system, so as to show the planning projects in valley economy and water-area ecological agriculture with complex spatial relation and layered distribution;

(4) Deal with a dynamic description of geographic advantages of the target area by a dynamic loading process of the supersized area scene of the system. Describe the abstract design conception directly about 3D agriculture, four-in-one ecological agriculture and channel scene transform in the planning target by making use of the system support to the 3D space expression;

(5) Realize the straight comparison of primitive state to the planning state in the target area, as well as direct exhibition of related influence on the planning projects in different seasons by overlapping process of space and time;

(6) Realize to complete the planning design based on omni-directional VR system software, as well as automatic uniform summarization output of the analytical data of planning investments and incomes.

All kinds of survey data, planning ideas and layout are integrated rationally with processing of above major key technical nodes. The major key technical nodes and system structure are shown in Fig. 3.

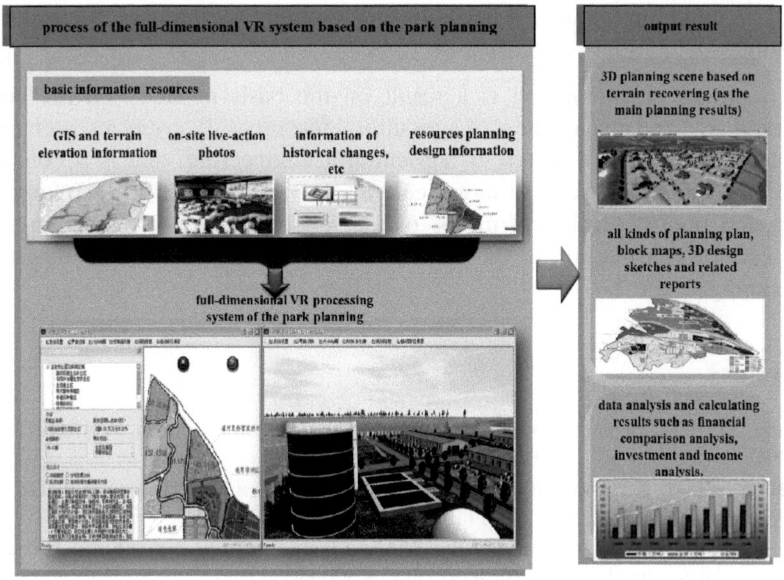

Fig. 3. Technical nodes and system structure of the omni-directional processing technology in Hebei Zhengding agricultural park planning

4 Discussion and Conclusions

The agricultural park planning based on omni-directional information processing technology should be further improved and developed in the following aspects through application and study on the combination of the traditional method and omni-directional processing technology of the agricultural park planning:

(1) The system graphic analyzing unit and module should be further improved. As the spatial and attribute data are put into the omni-directional VR processing system only by external software, for example, CAD and GIS's processing and information extraction, so this not only increases working load of data processing but also confines matching of file layout. Thus, improving the analyzing units and modules will reduce the working load of the data pre-processing and promote work efficiency;

(2) The social economy data processing and analytical system should be integrated. The social economy data in the planning area are the premise of the whole planning. The distribution and data analysis of the social economy in the planning area are the effective basis of the planning. Thus, the integration of the processing and analytical system of social economy data is an important supplement and perfection to the planning system;

(3) The coupling of units and modules in the system should be strengthened further. The omni-directional VR planning system is a new one which is a result of accumulation and research on the basis of the long-term agricultural park planning. The proposed analytical units and modules should be further adjusted and improved, and

specially in coupling relation, it needs further studying to improve efficiency of data processing;

The agricultural park planning is a result on the basis of study on the layout of agricultural area and management of agricultural resources. It covers research results in theory as well as design results of agricultural engineering, and is a concentrated reflection of the small area and large-scale agriculture. Thus, it is very important to develop and improve technical means of the modern agricultural planning, and of importance for construction of the technical and standard system of the agricultural park planning.

References

1. Wang, H.: Theory and method of regional planning. Management World (2), 149–152 (1994)
2. Yan, Z., Zhou, K.: Thoughts on the regional planning of the cross-administrative areas. Economic Geography 15(14), 1–6 (1995)
3. Li, X., Wang, H.: The Analysis of the Trinity Planning of Eco-agricultural Tourism Park. Chinese Agricultural Science Bulletin 27(25), 300–306 (2011)
4. http://baike.baidu.com/view/1952434.html
5. http://baike.baidu.com/view/5422265.html
6. Guan, L.: Design and research of modern agricultural park planning—An example of Yangling modern agricultural demonstration area (Master dissertation). North West Agriculture and Forestry University, Shaanxi (2010)
7. Liu, S.: Study on the Plan and Design of Modern Agriculture Demonstration Garden. Journal of Anhui Agri. Sci. 39(35), 21875–21876 (2011)
8. Liang, F., Liu, L., Qu, Y.: Methodology of agricultural landscape classification with rural scale and its application in planning of new rural construction. Transactions of the CSAE 11(27), 330–335 (2011)
9. Zhao, H., Zhang, F.: Analysis of agricultural landscape and the zoning layout of agricultural function in Beijing. Transactions of the CSAE 24(supp.1), 78–84 (2008)
10. Li, X., Zhu, H.: Agro landscape changes and driving factors in coastal area of southeast Fujian: A case study of Dananban farm. Resources Science 22(1), 35–39 (2000)
11. Bian, Z., Wang, Q., Yu, M.: Empirical research on theory. Transactions of the CSAE 24(2), 95–100 (2008)

Design and Realization of Agricultural Information Intelligent Processing and Application Platform

Dan Wang[1,2]

[1] Institute of Agricultural Information, Chinese Academy of Agricultural Sciences, Beijing 100081, China
[2] Key Laboratory of Digital Agricultural Early-warning Technology, Ministry of Agriculture, P.R. China
wd@mail.caas.net.cn

Abstract. The work of this paper aims to provide a practical software product, the intelligently gathering, processing and utilizing for the leaders of agricultural departments, research institutions, library and information systems and information consultancy departments, etc. The software product is an information processing platform integrated intelligent gathering, automatic indexing, and search and utilization of agriculture information. This software is tested to be accurate and can substitute people to process information to some extent.

Keywords: agricultural information, automatic collection, automatic indexing, information search, information management.

1 Introduction

After 40 years in the development of the Internet and after 20 years in the development of the World Wide Web, we have entered into the era of digital information. There is a mass of real-time and valuable information and electronic documents scattered in various departments on the Internet. Currently the effective use of the digital information is affected by the three restricting factors. Firstly, how to crawl the desired information from the mass information on Internet in real time and collect electronic documents scattered in various departments to the local database. Secondly, how to classify and deal with the digital information from the local database, make it the standard information resources so that the information retrieval and utilization. Thirdly, how to retrieve the information they need from the local database easily, accurately and quickly, to achieve the purpose of efficient information exchanging, using and spreading.

To solve the above technical problems, we have developed the platform of the Agricultural Information Intelligence Collection, Processing and Utilization, also called information processing platform (IPP). The IPP is a general platform which provides information intelligence collection, information ordering classification, subject indexing and information retrieval for the leading departments, agricultural

D. Li and Y. Chen (Eds.): CCTA 2012, Part II, IFIP AICT 393, pp. 229–237, 2013.

information and advisory organizations, the construction unit of agricultural information resources.

Agricultural information institute of the Chinese Academy of Agricultural Sciences (hereinafter referred to as AII) have made a lot of basic research work in the information organization and management of agricultural domain from 1990s, which have prepared "Agricultural Science Thesauri", "Chinese Library Classification - Agricultural Professional Classification" and "Multifunctional Agricultural Vocabulary", and developed "the Chinese information automatic indexing system". The AII has officially studied and developed IPP in order to enhance the original scientific and technological achievements with the development of the information age.

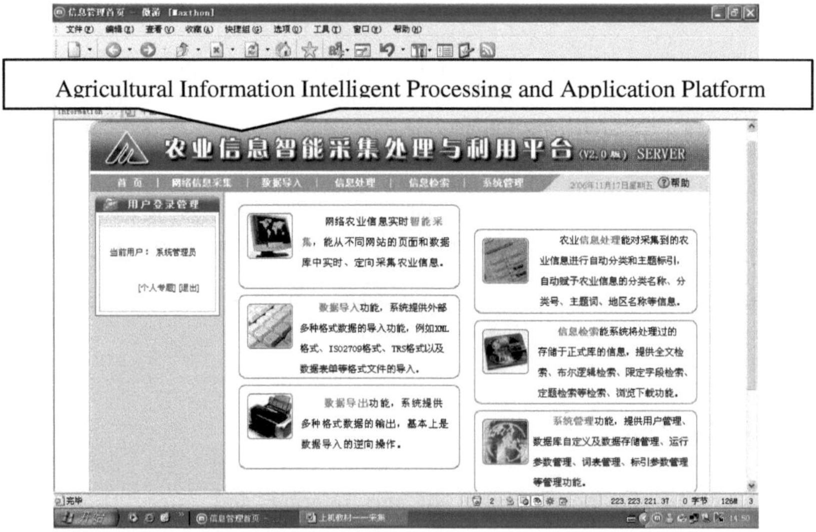

Fig. 1. Agricultural Information Intelligent Processing and Application Platform

2 Research Objectives and Content

R & D goal of the IPP is to develop agricultural information processing platform based on B/S structure for agricultural information intelligence collection, automatic classification, subject indexing and the use of agricultural information retrieval. The functions of the IPP can be classified the following parts: 1) the subsystem of integrated classification vocabulary of agricultural information management; 2) the subsystem of automatically grabbing information on the Internet and smart obtaining electronic document; 3) the subsystem of the automatic classification and subject indexing of agriculture information; 4) the subsystem of database definition and data storage; 5) the subsystem of retrieval information; 6) the subsystem of data import and export.

3 Architecture and Functional Design of the IPP

3.1 System Architecture

Information processing of the IPP is firstly a collection of agricultural information that is stored in the temporary database. Secondly, the information is automatic processed and stored in the formal database. Thirdly, information is retrieved and used. Based on the flow of information, the agricultural information Integrated classification vocabulary management subsystem, the database management subsystem, and system operating parameters participate in together to achieve the goal of the IPP.

3.2 System Functions

The main functions of the IPP can be concluded in the six following aspects:

3.2.1 Intelligent Real-Time Acquisition of the Agriculture Network Information

The IPP may intelligently collect personalized information real-time from the agricultural website. The IPP can collect requirement information on the Internet and store in the local database by way of user-set information acquisition parameters of generic and professional templates.

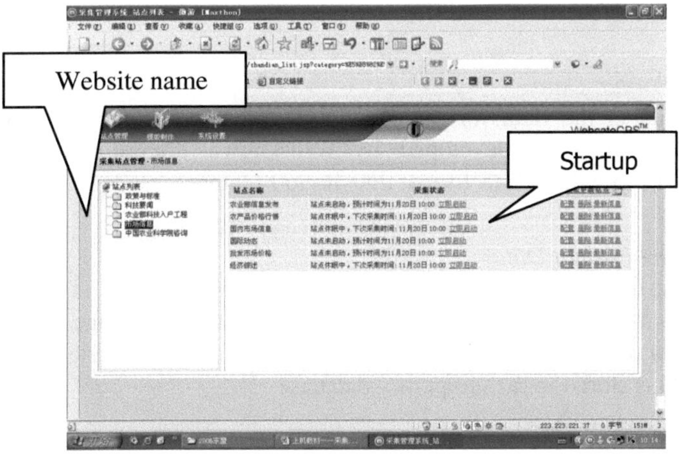

Fig. 2. The IPP may intelligently collect personalized information real-time

3.2.2 Information Import Function

Data import function of the IPP is to provide the collection function of other electronic documents and the collection function of network information. The main features are:

Fig. 3. A screen shot of information import function

3.2.2.1 XML Format Data

XML format data is read and analyzed by IPP after the data is imported into the related temporary database of the IPP.

3.2.2.2 TRS Format Data

TRS format data is the data format of import and export supported by full-text retrieval system which is developed by well-known domestic TRS company. Therefore, the IPP provides TRS format data import feature.

3.2.2.3 ISO2709 Format Data

ISO2709 format is the international standard of data format which is widely used in library and information sector for bibliographic information exchange. Firstly the IPP can read the data of the ISO format. Secondly the data can be converted to XML data. Thirdly the data is stored in the temporary database.

3.2.2.4 Electronic Text Format Data

The IPP can invoke directly WORD processing software that read DOC, PDF and other document, and edit it. Then the data is imported into a temporary database.

3.2.2.5 Information Based on Form

The IPP can automatically generate information form of the database, the user directly login and input the data which is stored in a temporary database.

The data collected from the website and import data from other electronic documents are stored in a temporary database. After information is automatic classified and subject indexed by the IPP, they are stored in a formal database of the IPP.

3.2.3 Information Processing Function

The IPP provides the function of information processing. The first step is that the data of the IPP is stored in the temporary database. Then, information is batch indexed on temporary database by the IPP selectively automatically, or which is entire database indexed or unattended indexed or artificial auxiliary indexed. At last, index result of the data is edited by the IPP. Specific indexing features:

① The IPP automatically produces classification codes (CLC, category code), the category names, geographical names, geographical codes and information keywords to pretreatment information.

② The IPP supports a variety of indexing methods, including the indexing of single articles, batch indexing, automatic indexing of the entire database as well as unattended indexing.

③ The IPP supports the functions of modify and editing for automatic indexing results.

④ Artificial auxiliary indexing functions are supported. When users index a message online, they may browse the information classified directory tree at any time and click required indexing category name. The category names and corresponding classification coding are automatic inputted the database by the IPP.

⑤ Support the function of the word frequency statistics analysis and the function of updating vocabulary of data. When users may modify correction of the results of indexing keywords, if vocabulary words are not yet logged, the IPP will automatically added, and then the IPP will compute vocabulary frequency. When vocabulary frequency of the word exceeds the prescribed threshold, the word is updated automatically on the vocabulary of data.

⑥ Flexible configuration in word field range for indexing, such as the title field, summary field, the body field.

⑦ Flexible configuration for the position weight value of the keyword, such as keyword from the title field, from summary field, from the body field, from the first or the tail paragraph of the text, and from the first few words in the paragraph.

⑧ The flexible configuration of indexing depth, that is, a literature can be marked with the number of keywords. According to the results of statistical analysis of artificial indexing, indexing depth is generally appropriate in 7-8 words.

3.2.4 Information Retrieval Functions

The IPP can retrieval information which is stored in the formal database, the specific features can be summed up:

① The user can choose a database or multiple databases at the same time for retrieval. the IPP supports for multi-database searching.

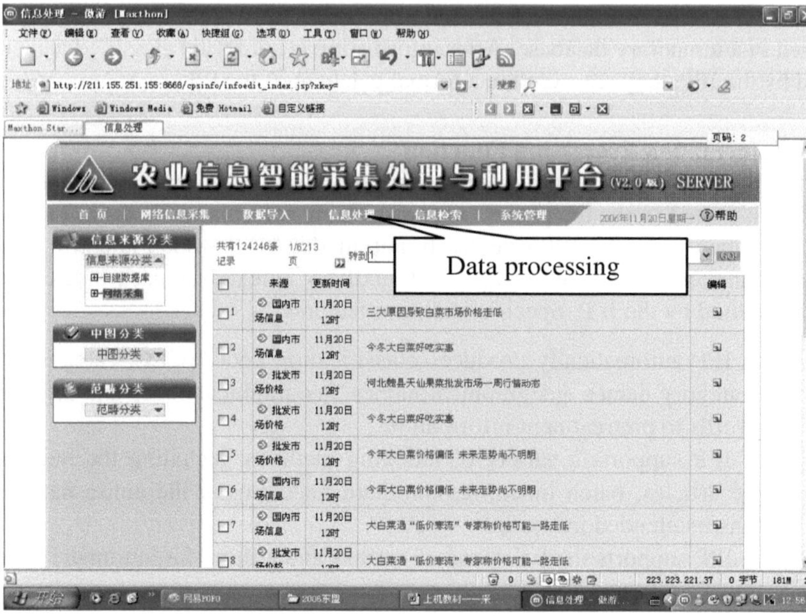

Fig. 4. A screen shot of information processing function

② Support keyword search, the second search and text search in the limited field.

③ Support boolean logic expression to search, according to user's searching requests, composition of search expression with "and", "or", "non", as well as the search expression with operator priority retrieval.

④ Support the search function of the SDI. The IPP save the user's search expression, in order to retrieve the information repeatedly to form a thematic database.

3.2.5 System Management Function

System management function contains the configuration and management operating parameters of Intelligence collection, processing and utilization from the agriculture network information, including management of users, databases of custom parameter configuration management, information collection indexing parameters of configuration management, information taxonomies, thesauri, regional tables and information areas of the table maintenance, and run log management.

User management: It Supports the user's function of increasing, deleting and modifying; It sets up and manages the user name, password, roles and permissions.

Custom database: Users can flexibly define databases structure in different type databases, describe the properties of the database to determine the taxonomic relationship between the different databases and create a user database.

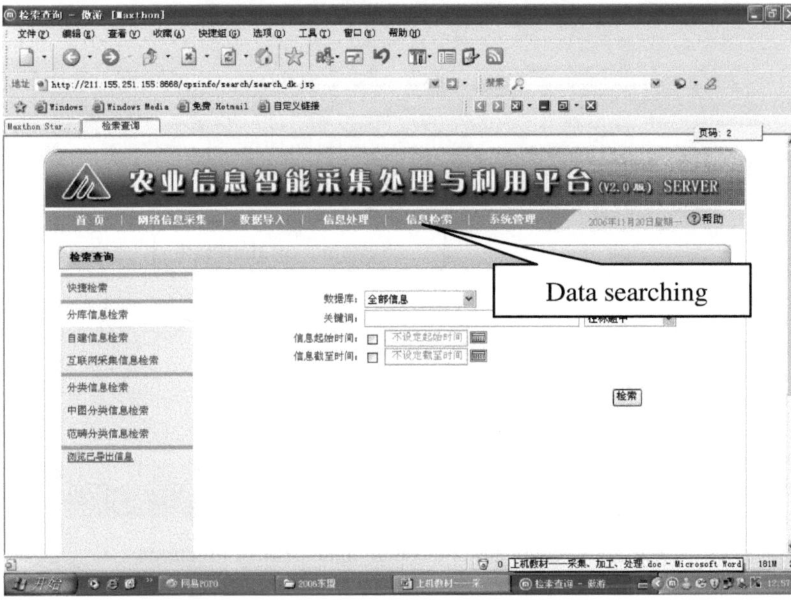

Fig. 5. A screen shot of information retrieval function

Vocabulary management: It contains maintenance and management of thesaurus of information, the scope of vocabulary, regional tables. Users can add, delete, change keywords according to the above vocabulary.

The word frequency statistics and vocabulary Automatic Updates: The IPP computes word frequency for indexing word. If indexing words does not appear in any vocabulary database or word frequency in excess of the threshold, the system will automatically input into vocabulary database.

3.2.6 System Management Functions

The IPP provides the exchange with the outside world database interface that can output the information in XML format document, TRS format documents, ISO2709 format document output.

4 Operating Environment

The platform is a software products of cross-platform operating system with JAVA language development based on B/S architecture, and operating environment is:

- ➢ hardware configuration: CPU Pentium 4 or above, CPU Clock Speed more than 1G
- ➢ Computer memory: more than 2G
- ➢ Hard disk: more than 20G
- ➢ Video card: resolution ratio support of 1028 * 768 pixels, True color refresh frequency 85HZ, Video card memory more than 32M

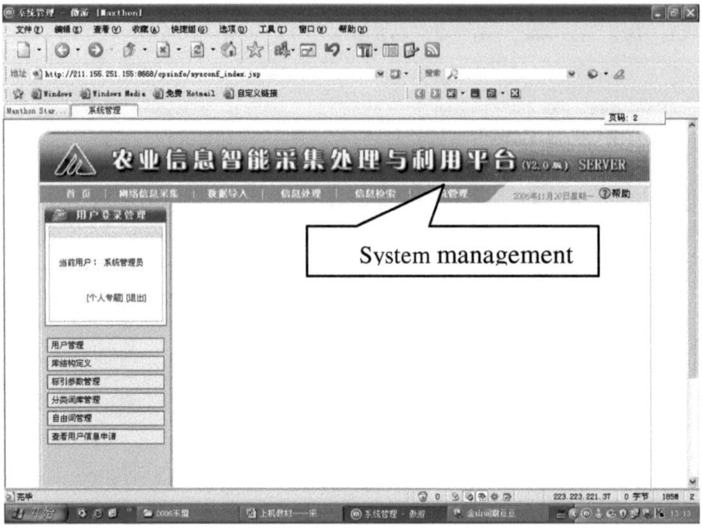

Fig. 6. A screen shot of system management functions

- Indicator: Standard display, more than 15-inch
- Operating system: Windows2000, Windows XP, Redhat, or Linux9.0
- Database: ORACLE, MYSQL, or SQLSERVER
- Development software: JAVA 1.6
- Application : IE6.0 above

5 Conclusion

The user uses login name and password to enter the information processing platform, and click on the system menu to finish related operations. It automatically crawled and annotated 10000 pieces of documents from the websites and the documents provided by Chinese Academy of Agricultural Information Institute, China Agricultural University and other units, agricultural engineering, plant protection, soil nutrients, food economy, vegetable gardening, animal husbandry and veterinary literature information. We compare the annotated result with the results of manual annotation, which proved that the method is very effective.

References

1. Feng, X.-X., Zhu, X.-F.: The Design and Implementation of Information Resources Extractor of Agricultural Websites Based on Spring. Informatization Research 37, 19–22 (2011) (06)
2. Dong, S., Yu, J., Li, H., Sun, D.: Development and Construction of Beijing Suburban Natural Resources and Socio-economic Information Management Platform. Chinese Agricultural Science Bulletin 167–171 (2011) (20)

3. Zhang, W.-H., et al.: Design of Agricultural Information Retrieval Platform Based on Chinese Word Segmentation. Journal of Anhui Agricultural Sciences 39(20), 12586–12587 (2011)
4. Zhang, Y.: On the Exploitation of Online Teaching Resources Abroad. New Century Library, 78–79 (2012) (05)
5. Zheng, Z.-H., Xu, Z.-J., Wen, H., Zheng, Z.-G.: Improved Search Engine and Its Data Structure Design. Information Science 30(2), 200–205 (2012) (2)
6. Kong, W.-T., Yan, H.-Y.: The Design of Lucene-based Automatic Answering System. Computer Development & Applications 25(4), 32–37 (2012) (4)

Study on the Spatial – Temporal Variability of Soil Nutrients during Winter Wheat Growth Season

Bei Cui[1,2], Wude Yang[1,*], Meichen Feng[1], Wenjiang Huang[2], and Xiaoyu Song[2,*]

[1] College of Agronomy, Shanxi Agricultural University, Taigu 030801, China
[2] Beijing Research Center for Information Technology in Agriculture, Beijing 100097, China
{cuibei1214,fmc101,yellowstar0618}@163.com,
sxauywd@126.com, songxy@nercita.org.cn

Abstract. The object of the study is to analyze the spatial-temporal variability of soil nutrients in the winter wheat growing season, and try to reveal the dynamic change of soil nutrients in winter wheat field. Measurements of soil nitrate nitrogen (NN), total nitrogen (TN) and available potassium (AK) in wheat field were taken at 132 sample points in April and June, 2006. Classical statistical method and geostatistics method were used in this study. The classical statistical analysis results indicate the average value of NN content decreased obviously from April to June, and the value of AK decreased slightly. But there were no change for the mean of TN content from April to June. All soil nutrient both in April and June were medium-degree variations. From the spatial distribution maps of soil variables, soil NN changed obviously on distribution between April and June. However, soil TN and AK had the similar spatial distribution that high in the northeast part and low in the southwest of the study field. In summary, winter wheat growth had great effect on soil NN content, whereas little effect on TN and AK. The differences of NN, TN and AK between June and April were negative in most areas and the absorptive capacity of winter wheat was different in different places.

Keywords: spatial - temporal variability, soil nutrients, geostatistics.

1 Introduction

Many researchers pointed that the variability of soil nutrients were affected by many factors, such as, land use types[1], soil types[2], planting types[3-8], topography[9-10], crop rotation[11-12] and so on. There were also some studies on the variability of soil nutrients variables in winter wheat field. Zhang et al[13] researched the spatial variability of soil properties including TN, available N, organic matter, available P, AK and so on in winter wheat field in March 2002. Xie et al[14] analyzed the spatial-temporal variations of purple soil ammonia and NN under application of inorganic and organic fertilizers in winter wheat field. Chen et al[15] studied the law of spatial-temporal change of soil organic matter, total N, available N, available P, and available K in farmland for last 20 years(1981～2000). However, little study studied the spatial and temporal changes of soil nutrients during winter wheat growth season.

* Corresponding authors.

D. Li and Y. Chen (Eds.): CCTA 2012, Part II, IFIP AICT 393, pp. 238–247, 2013.

This paper studied the spatial-temporal variability characteristics of soil NN, TN and AK during wheat growth period. The study tried to reveal the dynamic change rule of soil nutrients in winter wheat field, in order to prove theoretical basis for winter wheat precise management.

2 Materials and Methods

2.1 Study Area

The study were conducted at National Experiment Station for Precision Agriculture, Changping District, Beijing(116°26'10"E–116°27'05"E, 40°10'31"N–40°11'18"N). The experimental plot had an area of 5.4 ha (Figure 1), and soil type was alluvial soil. Winter wheat (Triticum aestivum L.) cultivar ('Jingdong-8') was selected in this study.

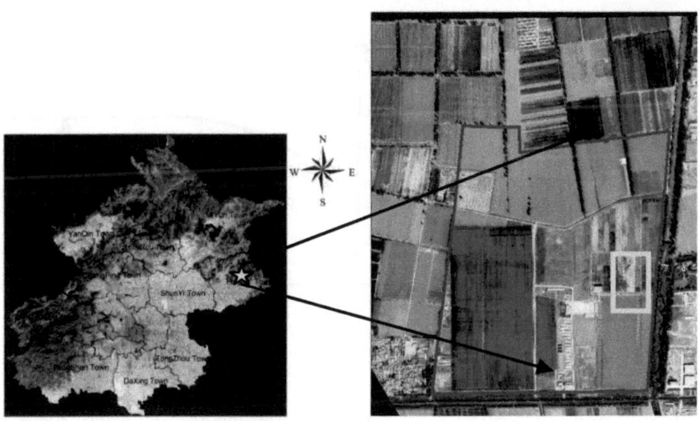

Boundary of National Experiment Station for Precision Agriculture

Study area

Fig. 1. Location of study area

2.2 Soil Sampling and Laboratory Analysis

Soil samples from 0- to 30-cm depth were collected within 51 locations on 30×30 m grid and 81 locations on 10×10 m grid (Figure 2). The first set of soil samples was taken on April 4, 2006 just before fertilizer application at the wheat elongation stage. The second set of soil samples was taken on June 16, 2006 just before the wheat harvest. The location of each soil sample was recorded by a GPS receiver with a station-based differential signal (DGPS, Trimble 5700 RTK), of which the measuring accuracy was controlled within 20cm. The soil samples were taken manually using a stainless steel probe. For each plot, a composite soil sample comprising five random samples from within a 5 m radius and a depth of 0～30 cm was taken, then placed into its designated plastic bags, mixed, and transport back to the laboratory. Soil samples were air dried, crushed, sieved, and analyzed for soil NN, TN and AK using the agro chemistry conventional analysis method[16].

2.3 Analysis Methods

In this study, classical statistical methods and geostatistics methods were used to analyze spatial - temporal variability of soil nutrients. Through the classical statistical methods, the minimum, maximum, mean, skewness, kurtosis, standard deviation and coefficient of variation were got. The CV of variables can be used to roughly estimate the variation degree: if the value of CV was<10%, the variable was considered weak variability ; if the value was >10% and <100%, the variable was considered the medium-degree variability; if the value was >100%, the variable was considered the strong variability[17].

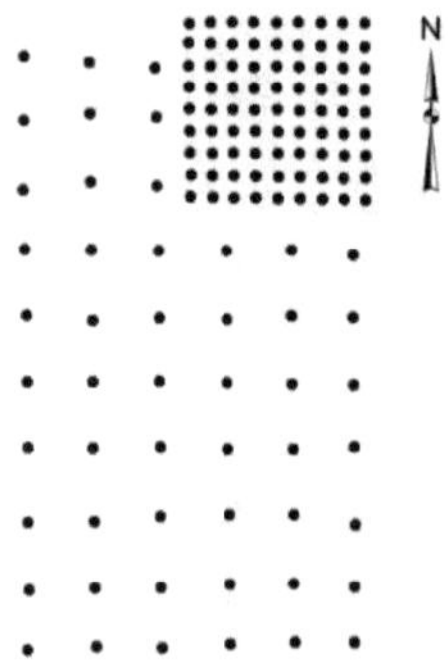

Fig. 2. The distribution of soil sampling points

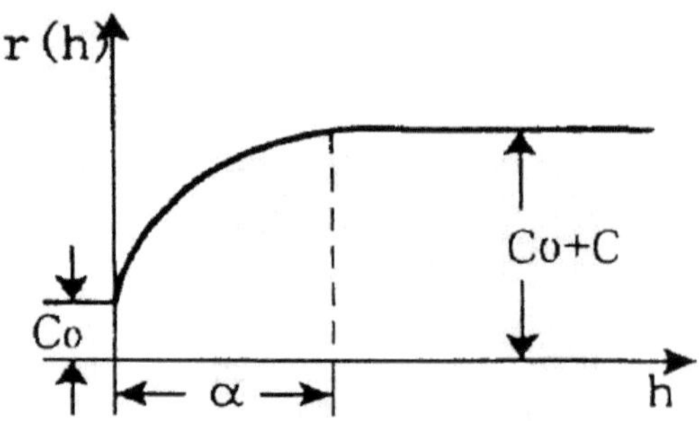

Fig. 3. Semivariogram variation curve

Figure 3 was a ideal semivariogram variation curve[18]. A semivariogram model consists of three important parameters which are C_0, $C+C_0$ and range. C_0 represents the nugget effect, which is the semivariogram value when the distance is zero. The nugget effect is led to random factors, such as sampling error and measurement error. $C+C_0$ is the sill, which represents total variance. We adopted the value of $C_0/(C+C_0)$ to stand for the spatial class. If the ratio of spatial class was < 25%, the variable was considered strongly spatially dependent ; If the ratio of spatial class was > 25% and < 75%, the variable was considered moderate spatially dependent; and if the ratio was > 75%, the variable was considered weakly spatially dependent[19]. Range is also known as the longest correlation distance. If samples separated by distances are smaller than the range, they are related spatially in some extent; if those separated by distances are greater than the range, there were no spatial dependence[20].

The classical statistical analysis of soil parameters were operated with SPSS11.5 software. Variowin 2.2[21] was used to compute the variograms of semivariogram and Arcgis9.0 was used for kriging interpolation. To characterize the spatial variation of the soil properties, experimental variograms were computed using Variowin 2.2. The parameters of the fitted models were then used with the data for ordinary kriging in Arcgis9.0. The values of the soil properties were predicted on grid at un-sampled locations and the predictions were used to produce interpolation maps.

3 Results and Discussion

3.1 Simple Statistics of Soil Nutrients Data

Average value and coefficients of variation for soil NN, TN and AK were listed in Table1. The average value of NN content (19.20mg/kg) in April was greater than in June (3.51mg/kg). The dramatic decrease was mainly caused by the NN absorption of winter wheat during the growth season of elongation stage to mature stage from April to June. Compared with the average value of AK in April (137.60mg/kg), the mean value in June decreased slightly (110.67mg/kg). But there were no change of TN content on average between in April and in June. In April, the variation coefficient (CV) of NN was 38.59%, which was greater than AK (18.72%) and TN (13.44%). In June, CV of NN increased to 61.05%. TN had the smallest CV of 12.36% in June. The CV of AK was 18.72% in April and 14.32% in June. All soil nutrients variables both in April and June were medium-degree variations with CV of 12.36%~61.05%.

Compared with CV of soil nutrients variables in April, CV for all soil nutrients in June changed more or less. Except NN with a marked increase of the CV from 38.59% to 61.05%, the CV of other soil nutrients decreased in different degree from April to June.

Table 1. Descriptive statistical of soil nutrients properties in April and June

Variables	Minimum	Maximum	Mean	SD	CV%	Skewness	Kurtosis
NN-April (mg/kg)	6.85	46.78	19.20	7.41	38.59	0.01	-0.25
TN -April (%)	0.08	0.13	0.10	0.01	13.44	0.32	-0.62
AK-April (mg/kg)	77.30	237.20	137.60	25.77	18.72	-0.20	0.27
NN-June (mg/kg)	0.96	15.13	3.51	2.14	61.05	0.50	0.67
TN –June (%)	0.07	0.13	0.10	0.01	12.36	0.19	-0.65
AK-June (mg/kg)	80.55	171.11	110.67	15.84	14.32	0.25	0.02

Note: NN= nitrate nitrogen; TN = total nitrogen; AK = available potassium; CV = variation coefficient.

Skewness and kurtosis values of soil variables indicated that only the soil TN in April and June was in normally distribution. The other soil properties were in nearly normal distribution after natural logarithm transformation before geostatistics analysis. Skewness and kurtosis values of all soil properties were also given in Table 1.

3.2 Spatial Structure Analysis of Soil Nutrients in Winter Wheat Fields

The semivariogram parameters of soil properties were shown in Table 2. The semivariogram of soil TN in June showed Spherical model, whereas other soil nutrients both in April or June showed Gaussian model.

Table 2. The semivariogram parameters of soil properties

Variables	Model	Nugget(C_0)	Sill($C+C_0$)	Range(m)	Spatial class
NN-April	Gaussian	0.109887	0.17788	170.00	61.78
TN -April	Gaussian	0.000090	0.00019	150.00	48.39
AK-April	Gaussian	0.018496	0.04910	190.00	37.67
NN-June	Gaussian	0.209590	0.27857	220.00	75.24
TN -June	Spherical	0.000040	0.00017	170.00	23.26
AK-June	Gaussian	0.007603	0.01780	61.19	42.71
ΔNN	Exponential	34.285300	52.41203	98.83	65.41
ΔTN	Gaussian	0.000024	0.00004	19.30	66.67
ΔAK	Exponential	270.252000	578.01600	93.59	46.76

Note: NN= nitrate nitrogen; TN = total nitrogen; AK = available potassium; ΔNN= nitrate nitrogen content in June minus in April; ΔTN= total nitrogen content in June minus in April; ΔAK= available potassium content in June minus in April.

The spatial class of TN in June (23.26%) was smaller than 25%, which indicated the existence of strong spatial dependence. The spatial class of NN in June was 75.24%, more than 75%.The other soil nutrients variables had moderate spatial dependence with the spatial class at the range of 37.67%~61.78%.

Except AK with a marked decrease of the range, the ranges of other soil nutrients increased about 20m from April to June. The range of AK in April was 220.00m, whereas that in June decreased to 61.19m.The ranges of all soil variables were bigger than the sampling interval of 30m, which satisfied the effectiveness of geo-statistical analysis.

Table 3 presented the cross validation results of the optimal model for soil variables. Generally, the optimal model should be the one that Root-Mean-Square (RMS) is closer to Average Standardized Error (ASE), Mean Standardized (MS) is closer to 0, and Root-Mean-Square Standardized (RMSS) is closer to 1.

Table 3. The cross validation results of semivariogram model optimal model for each soil variable

Variables	RMS	ASE	MS	RMSS
NN-April	6.661	7.334	-0.016	0.937
TN -April	0.009	0.010	0.004	0.911
AK-April	20.900	20.320	-0.010	1.043
NN-June	2.141	1.845	-0.063	1.193
TN -June	0.007	0.008	0.004	0.870
AK-June	11.240	11.630	-0.007	0.968
ΔNN	6.163	6.730	0.012	0.925
ΔTN	0.006	0.006	-0.001	0.968
ΔAK	21.240	20.890	0.008	1.020

Note: NN= nitrate nitrogen; TN = total nitrogen; AK = available potassium; RMS = Root-Mean-Square, ASE = Average Standardized Error, MS = Mean Standardized, RMSS = Root-Mean-Square Standardized.

3.3 The Spatial Distribution Maps of Soil Nutrients in Winter Wheat Field

Figure 4 indicated that, soils with high NN content in April distributed in the north and east part of the field with downward trend from SW to EN. NN content in most areas of study area was lower than the average value (19.20mg/kg). Different to the map of NN in April, high NN content in June were found in the north part of the field, whereas, low NN content located in the east part of the field. The NN content average value in June was lower than that of in April (19.20mg/kg). It indicated that winter wheat growth strongly influenced the soil NN content and spatial distribution.

The spatial distribution map of TN in April showed that soil with high TN content was distributed in the northeastern of field and it reduced from SW to EN, which was similar to the distribution of TN in June. This implied that the spatial distribution of soil TN had low change during winter wheat growth.

Soil with high AK content in April was located in the north part of the field, and AK content in most areas was lower than the average value (137.60 mg/kg). In June, the soil distributed in the north and east part of field had high AK content, and the content in most areas was lower than the average value of 110.67 mg/kg. The soil AK of research field both April and June had the similar spatial distribution with high content in the EN part and low in the SW part.

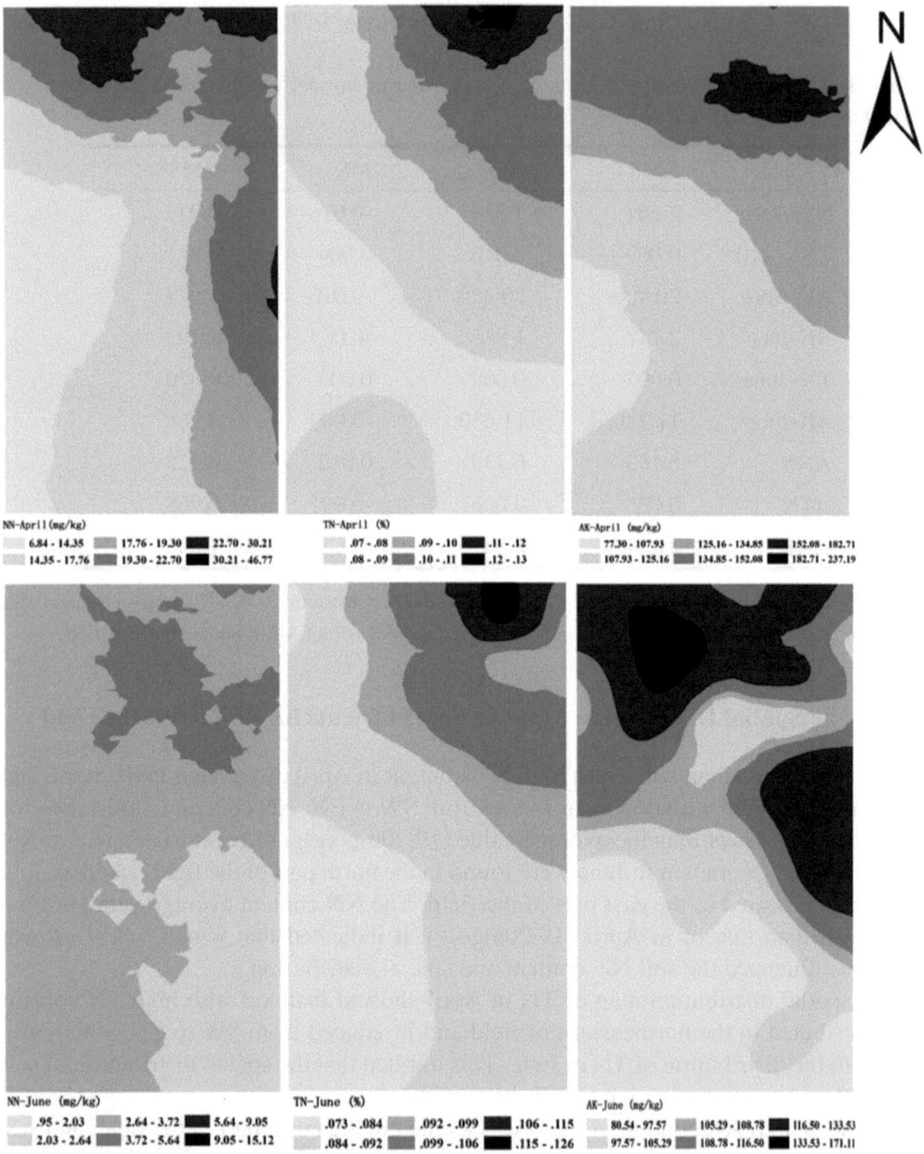

Fig. 4. The spatial distribution maps of nitrate nitrogen (NN), total nitrogen (TN) and available potassium (AK) in April and June

3.4 Temporal Variability Analysis of Soil Nutrients in Winter Wheat Field

In order to clearly reflect the change of soil nutrients during winter wheat growth season, the differences spatial distribution of soil NN, TN and AK between April and June were calculated using the band match function of ENVI software.

Figure 5 showed the differences spatial distribution of NN, TN and AK between April and June. From those, the differences of NN, TN and AK were negative in most areas, which indicated winter wheat not only absorbed original fertilizer but also absorbed the topdressing at the wheat elongation stage. In different places, the changes of nutrients were not same, which indicated the absorptive capacity of winter wheat on soil nutrients were different in this field.

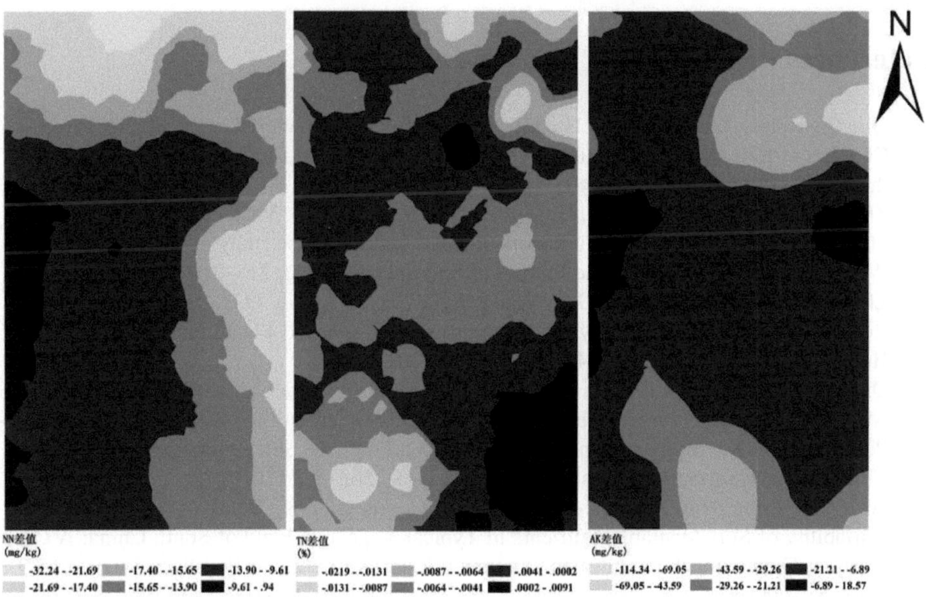

Fig. 5. The spatial distribution maps of differences of nitrate nitrogen (NN), total nitrogen (TN) and available potassium (AK) between April and June

4 Conclusions

Compared with the average value of NN and AK, the average value of NN content decreased obviously in June, and the value of AK in June decreased slightly. But there were no change of TN content on average between April and June. All soil variables both in April and June were medium-degree variations with CV of 12.36%~61.05%.

From the spatial distribution maps of soil nutrient, the NN changed obviously on spatial distribution between April and June. However, the spatial distribution of soil TN and AK changed small, which had the similar spatial distribution that high content in the northeast part and low in the southwest part. Winter wheat growth had great effect on soil NN, whereas little effect on TN and AK.

The differences of NN, TN and AK between June and April were negative in most areas, and the changes of nutrients were not same in different places. These indicated that winter wheat growth need large amount of nitrogen and potassium fertilizer and different absorptive capacity of winter wheat on soil nutrients in different places.

Acknowledgements. This work was financially supported by program of National Natural Science Foundation of China (31071324), program of National Natural Science Foundation of China (41201326) and Shanxi Province Science and Technology Support Program (20110311038), program of State Key Basic Research and Development Program (2011CB311806).

References

1. Qin, J., Kong, X.B., Jiang, G.H., Liu, Y., Li, C.Z.: Characteristics of Spatio-Temporal Changes of Soil Organic Matter in Typical Fringe in Beijing for 25 Years. Transactions of the CSAE 24(3), 124–129 (2008)
2. Zhao, Y.F., Shi, X.Z., Yu, D.S., Huang, B., Wang, H.J., Zhao, Y.C., et al.: Spatial variation of soil nutrients and its affecting factors at small scale - A case study of peri-urban areas in Wuxi Jiangsu Province. Chinese Journal of Soil Science 37(2), 215–219 (2006)
3. Zhang, Q.J., Jiang, H.L., Liu, G.S., Wang, X.J., Liu, Y.Q., Liu, Q.H., Gu, J.G., Liu, H.W.: Research on Spatial Variability of Soil Nutrients in Tobacco Plantation Field. Journal of Henan Agricultural University 45(1), 116–122 (2011)
4. Qi, Z.P., Wei, Z.Y., Li, F.Y., Tang, S.M.: Spatial-temporal Variability of Paddy Soil Nutrient in Hainan Province. Chinese Journal of Soil Science 40(6), 1292–1296 (2009)
5. Peng, T., Lv, X.: Study on Spatio-Temporal Variation of Soil Organic Matter in Oasis Cotton Field. Chinese Journal of Soil Science 41(3), 563–567 (2010)
6. Wang, B., Long, H.Y., ZhuGe, Y.P., Zhang, J.G., Gong, Y.K.: Study on Spatio-Temporal Variability of Soil Available Nutrients in Typical Vegetable Field of South China: A Case Study at Guangxingzhou, Junshan, Yueyang. Plant Nutrient and Fertilizer Science 15(3), 551–558 (2009)
7. Liu, Q.F., Lu, J.L., Li, S.L., Sheng, H.Y., Li, N., Wang, Y.P., Wang, J.M., Zhu, C.L.: The Spatio-Temporal Variation of Nitric Nitrogen in Protected Vegetable Soils in Different Years of Cultivation. Agricultural Research in the Arid Areas 29(2), 159–163 (2011)
8. Huang, Z.G., Li, B.G., Hu, K.L.: Characteristics of the Spatio-Temporal Changes of Soil Organic Matter of Sugarcane Field in Red Soil Hill Areas. Transactions of the CSAE 22(11), 58–63 (2006)
9. Umali, B.P., Oliver, D.P., Forrester, S., Chittleborough, D.J., Hutson, J.L., Kookana, R.S., Ostendorf, B.: The Effect of Terrain and Management on the Spatial Variability of Soil Properties in an Apple Orchard. Catena 93, 38–48 (2012)
10. Hattar, B.I., Taimeh, A.Y., Ziadat, F.M.: Variation in Soil Chemical Properties along Toposequences in an Arid Region of the Levant. Catena 83(1), 34–45 (2010)
11. Huang, B., Sun, W.X., Zhao, Y.C., Zhu, J., Yang, R.Q., Zou, Z., Ding, F., Su, J.P.: Temporal and Spatial Variability of Soil Organic Matter and Total Nitrogen in an Agricultural Ecosystem as Affected by Farming Practices. Geoderma 139, 336–345 (2007)

12. Song, G., Sun, B.: Spatial and Temporal Changes of Soil Inorganic Nitrogen in Rice-wheat Rotation System at County Scale—A Case Study in Yizheng County, Jiangsu Province, China. Journal of Agro-Environment Science 28(2), 636–642 (2008)
13. Zhang, S.J., He, Y., Fang, H.: Spatial Variability of Soil Properties in the Field Based on GPS and GIS. Transactions of the CSAE 19(2), 39–44 (2003)
14. Xie, H.M., Zhu, B., Zhu, Z.L.: Temporal-Spatial Variations of Ammonia and Nitrate under Application of Inorganic and Organic Fertilizers in Purple Soil of Winter Wheat. CJEA 14(1), 118–121 (2006)
15. Chen, F., Pu, L.J., Cao, H., Peng, B.Z., Yang, G.S., Zhou, S.L.: Spatial and Temporal Changes for Soil Nutrients and Their Mechanism in Typical Area of Taihu Lake Valley During the Past Two Decades. Acta Pedologica Sinica 39(2), 236–245 (2002)
16. Nanjing Agriculture College. Analysis Method of Agricultural Soil, pp. 25–69. Agriculture Press, Beijing (1980)
17. Wang, S.Q., Zhu, S.L., Zhou, C.H.: Characteristics of Spatial Variability of Soil Thickness in China. Geographical Research 20(2), 161–169 (2001)
18. Chen, L.P.: Theoretical and Experimental Studies on Variable-Rate Fertilization in Precision Farming. China Agricultural University, Beijing (2003)
19. Iqbal, J., Thomasson, J.A., Jenkins, J.N., Owens, P.R., Whisler, F.D.: Spatial Variability Analysis of Soil Physical Properties of Alluvial Soils. Soil Sci. Soc. Am. J. 69, 1338–1350 (2005)
20. Cambardella, C.A., Moorman, T.B., Novak, J.M., Parkin, T.B., Karlen, D.L., Turco, R.F., Konpka, A.E.: Field-Scale Variability of Soil Properties in Central Iowa Soils. Soil Sci. Am. J. 58, 1501–1511 (1994)
21. Pannatier, Y.: Variowin Software for Spatial Data Analysis in 2D. Springer, New York (1996)

Research of Determination Method of Starch and Protein Content in Buckwheat by Mid-Infrared Spectroscopy

Fenghua Wang[1,*], Ju Yang[1], Hailong Zhu[2], and Zhiyong Xi[1]

[1] Faculty of Modern Agricultural Engineering,Kunming University of Science and Technology, Kunming, Yunnan, China 650500
wangfenghua018@163.com
[2] Engineering Training Center, Kunming University of Science and Technology, Kunming, Yunnan, China 650500

Abstract. Starch and proteins are the vital nutrients in buckwheat, to achieve the fast detection of the buckwheat internal composition has an important theoretical significance and application value for buckwheat breeding, processing and other steps. In the paper, forty buckwheat samples from different origins have been selected. The starch and protein content of buckwheat was determined, and the mid-infrared transmission spectrum of buckwheat has been obtained using Fourier mid-infrared spectroscopy. Forty samples were randomly divided into the prediction set and validation set with 30 and 10 samples respectively. After smoothing preprocessing, the prediction models of buckwheat starch and protein content have been established using the combination method of principal component analysis and artificial neural network, finally the models have been verified. The results showed that the correlation coefficient between the prediction value and measurement value of buckwheat starch content is 0.9029, and the relative error is smaller and its mean value is 2.33%, the method of the buckwheat starch content prediction is feasible. But the prediction for buckwheat protein content is not ideal, need to be further studied.

Keywords: Mid-infrared spectroscopy, Buckwheat, Starch content, Protein content, PCA, BP Neural Network.

1 Introduction

Buckwheat, also known as the triangle wheat, black wheat, belongs to the genus of dicotyledonous Polygonaceae, family of Fagopyrum Mill, mainly have two cultivar of buckwheat is Sweet Buckwheat (Fagopyrum esculentum Moench) and Tartary Buckwheat (Fagopyrum tararicum Gaerth). The buckwheat mainly located in the

* This study was financed by the Application Basis Surface Projects of Yunnan Province (2010ZC028), Kunming University of Science and Technology Analysis and Testing Fund (2010299).

D. Li and Y. Chen (Eds.): CCTA 2012, Part II, IFIP AICT 393, pp. 248–254, 2013.

alpine areas of the Loess plateau and the plateau mountains of Yunnan, Guizhou and Sichuan. It has a high nutritional and health value, known as the food treasures of "Medicine-Food"[1]. Therefore, it is necessary to develop the cultivation industry and further processing technology of buckwheat, to improve the quality of buckwheat has a great significance for increasing farmers' income.

Now, buckwheat is used to buckwheat flour production for making buckwheat food. The level of starch and protein content of buckwheat is an important index for the buckwheat nutritional quality evaluation [2]. Buckwheat starch and protein content measurements is still using the traditional chemical analysis methods, it has the advantage of high measurement accuracy, but the disadvantages are as follows, the detection is slow, the measurement cost is expensive, the sample is destroyed, and it should be detected in the specialize testing organization or laboratory.

With the characteristics of high resolution capability, fast scan time, large radiation flux, low stray radiation and wide spectral range, fast and accurate detection of sample components can be achieved through Fourier Mid-infrared spectroscopy combined with chemometrics. Spectroscopy has been widely used in quality inspection of agricultural products, and the good analysis results have been achieved [3-7]. However, spectroscopy based studies on buckwheat nutrients in domestic and foreign have not been reported. Therefore, eight kinds of buckwheat from different growing areas have been collected. The spectral information of buckwheat have been obtained using spectroscopic techniques, the quantitative relationship model between the spectral information and the buckwheat internal quality have been researched, and a rapid, accurate, stable, easy assessment for buckwheat quality has been established. It has a good theoretical significance and practical value in buckwheat numerous aspects such as breeding, processing and testing.

2 Experimental Instrument and Methods

2.1 Instruments and Equipments

Spectrometer: Germany Bruker EQUINOX 55 Fourier infrared spectroscopy, CS101-2D type electric blast drying oven (Chongqing star joint venture enterprise experimental instrument Co., Ltd.), JLML rice huller machine (Shanghai Jiading Grain and Oil Detection Instrument Factory).

2.2 Experimental Material

Seven kinds of tartary buckwheat and a kinds of sweet buckwheat from Yunnan, Shanxi, Sichuan province in China were selected. They are Kunming local buckwheat, Luxi tartary buckwheat, Zhaotong tartary buckwheat No. 1 and No.2, Shanxi tartary buckwheat, Xichang tartary buckwheat, Dali tartary buckwheat and Dali sweet buckwheat, a total of eight kinds of representative buckwheat varieties, each buckwheat selected five samples, a total of 40 samples. Each sample was air-dried, eliminating Impurities, shelling, crushing over a sieve for the requirement of particle size and consistent uniformity, the diameter of the sieve is 0.2mm. All samples were bagging sealing for spare use.

2.3 Experimental Methods and Steps

Acquisition of the Buckwheat Spectral Features. The transmittance of the buckwheat samples have been obtained Using Germany Brook EQUINOX55 Fourier Mid-infrared spectroscopy, set the spectral resolution of $2cm^{-1}$, the testing range of $4000 \sim 400cm^{-1}$.

At first, 1g buckwheat sample and 1.5g potassium bromide have been mixed in the mortar, and then the mixing powder has been placed in a circular hole to tabletting. Then samples scanning have been carried on, background measurements is required before each scanning. Each sample repeat scanning for three times. The whole operation should be in dark conditions as possible to avoid potassium bromide decomposition.

Buckwheat Starch and Protein Content Measurement. Buckwheat starch content has been measured by enzyme hydrolysis method, and protein content has been measured by the Kjeldahl nitrogen method. The chemical measurement has been accomplished by Analysis and Testing Center of Yunnan Province. Table 1 is the Statistical parameters of the chemical measurement results.

Table 1. Buckwheat chemical constituent's statistical parameters

	Sample number	Minimum	Maximum	Mean	Standard Deviation
Starch content (%)	40	48.97	69.49	62.22	5.73
Protein content (%)	40	8.31	13.04	10.32	1.42

3 Experimental Results and Analysis

3.1 Data Preprocessing

Average smoothing method was used in mid-infrared spectral pretreatment to remove the influence of high-frequency random noise, baseline drift, uneven sample and light scattering. The choice of the smoothing window size was 3. Figure 1 is the buckwheat spectral curves after pretreatment. In the figure, the abscissa is the wave number (cm^{-1}), and the ordinate is the transmission rate (%).

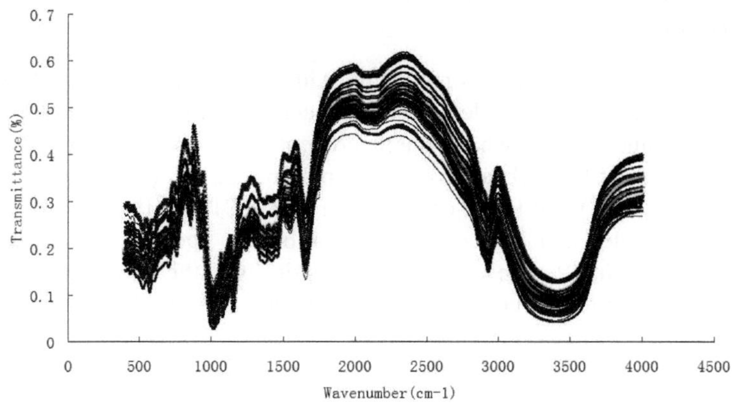

Fig. 1. Transmittance of buckwheat Mid-infrared spectroscopy

3.2 Principal Components Extraction Based on Principal Component Analysis

Forty samples were randomly divided into the prediction set and the validation set with 30 and 10 samples respectively. Sample spectral bands from 4000 to 400 cm^{-1}, there are 1868 data in total. Computational complexity is large when using the full spectrum of computing. And due to the weak spectral information of some regional samples, the composition of the sample or the nature is lack of correlation. The principal component analysis is an effective method for data mining. It can translate the original multiple-wavelength variables into fewer new variables. These new variables, not only are not related to each other, but also can synthetically reflect the information of original multiple-wavelength variables [8, 9]. So the prediction set and the validation set were respectively for principal component analysis. The accumulative reliabilities of the principal components are shown in Table 2.

Table 2. Accumulative reliabilities of the first 6 PCs

Principal component	Accumulative reliabilities of prediction set/%	Accumulative reliabilities of validation set/%
PC01	89.831	92.286
PC02	96.087	98.121
PC03	99.488	99.544
PC04	99.916	99.846
PC05	99.985	99.946
PC06	99.995	99.972

3.3 Prediction Model of Buckwheat Starch Content Based on BP Neural Network

The score of the six principal components which extracted by PCA from the 30 prediction set is the input variables of the BP neural network, and the output layer is the measured value of buckwheat starch content. By order experiments, the best hidden layer nodes of buckwheat starch content is 6, the iteration number of network setting training is 1000, the target error setting is 0.0001, the fitting residual error of the modeling samples is 1.805×10^{-3}. The calibration model of buckwheat starch content was established to predict the 10 validation set, the correlation analysis for prediction and measured values are shown in Figure 2. The error analysis results in Table 3.

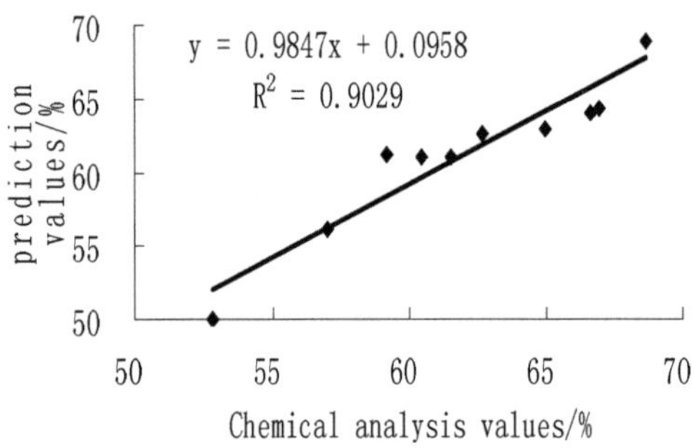

Fig. 2. Relationship between chemical analysis values and prediction values of buckwheat starch content

Table 3. Prediction results for buckwheat starch by BP model

Sample NO.	Chemical analysis value(%)	Prediction value(%)	Absolute error(%)	Relative error(%)
1	60.43	61.06	0.63	1.04
2	62.68	62.61	-0.06	-0.10
3	64.97	62.97	-1.99	-3.07
4	66.62	64.08	-2.54	-3.81
5	68.69	68.94	0.25	0.36
6	66.93	64.39	-2.54	-3.79
7	59.20	61.16	1.96	3.31
8	57.03	56.07	-0.96	-1.68
9	52.80	49.99	-2.82	-5.33
10	61.55	61.08	-0.47	-0.76

3.4 Prediction Model of Buckwheat Protein Content Based on BP Neural Network

The score of the six principal components which extracted by PCA from the 30 prediction set is the input variables of the BP neural network, and the output layer is the measured value of buckwheat protein content. By order experiments, the best hidden layer nodes of buckwheat protein content is 6, the iterations number of network setting training is 1000, the target error setting is 0.0001, the fitting residual error of the modeling samples is 4.173×10^{-3}. The calibration model of buckwheat protein content was established to predict the 10 validation set, the correlation analysis for prediction and measured values are shown in Figure 3. The error analysis results in Table 4.

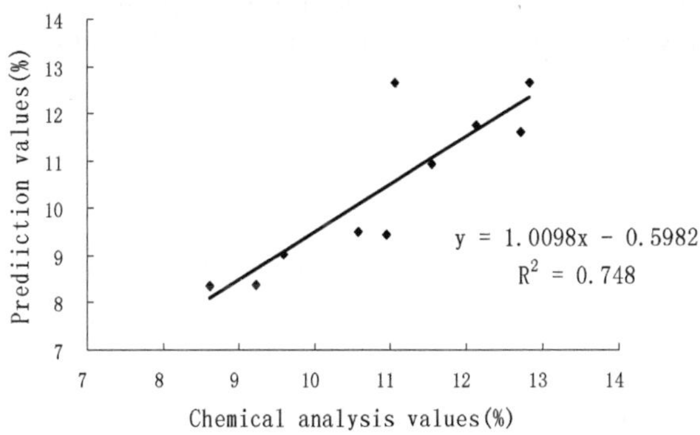

Fig. 3. Relationship between chemical analysis values and prediction values of buckwheat protein

Table 4. Prediction results for buckwheat Protein by BP model

Sample NO.	Chemical analysis value(%)	Prediction value(%)	Absolute error(%)	Relative error(%)
1	12.71	11.61	-1.09	-8.61
2	11.56	10.95	-0.61	-5.25
3	10.58	9.49	-1.09	-10.34
4	9.58	9.02	-0.56	-5.83
5	8.61	8.36	-0.25	-2.92
6	9.23	8.38	-0.85	-9.20
7	11.06	12.66	1.60	14.50
8	10.95	9.45	-1.50	-13.73
9	12.82	12.65	-0.17	-1.33
10	12.13	11.74	-0.39	-3.20

4 Conclusion and Discussion

(1) The neural network prediction model of buckwheat starch and protein content were established respectively using principal component analysis and artificial neural network method, and its predictions results has been verified. The results showed that the predict correlation of the artificial neural network prediction model for buckwheat starch content on the 10 samples validation set is 0.9029, the absolute average of the relative error is 2.33%. So, the method is feasible to predict the starch content in buckwheat.

(2) The buckwheat protein content prediction model was established in our paper. For the 10 samples, the predict correlation of the model is 0. 7480, the absolute average of the relative error is 7.9%. The results showed that the prediction effect for buckwheat protein content is not ideal using infrared spectroscopy, it is necessary to increase the buckwheat varieties and the number of the samples to optimize the modeling methods.

References

1. Guo, Y., Chen, Q., Yang, L., et al.: Analyses of the seed protein contents on the cultivated and wild buckwheat Fagopyrum esculentum resources. Genet. Resour. Crop Evol. 54, 1465–1472 (2007)
2. Choi, S., Ma, C.: Structural characterization of globulin from common buckwheat (Fagopyrum esculentum Moench) using circular dichroism and Raman spectroscopy. Food Chemistry 102, 150–160 (2007)
3. Tu, Z., Zhu, D., Ji, B., et al.: Progress in quality analysis of honey by infrared spectroscopy. Spectroscopy and Spectral Analysis 30(11), 2971–2975 (2010) (in Chinese)
4. Scibisz, I., Reich, M., Bureau, S., et al.: Mid-infrared spectroscopy as a tool for rapid determination of internal quality parameters in tomato. Food Chemistry 125, 1390–1397 (2011)
5. Wang, D., Wang, J., Ma, Z., et al.: Quantitative determination of lime additives in wheat flour by mid-infrared spectroscopy combined with attenuated total reflection. Transactions of the CSAE 28(4), 258–263 (2012) (in Chinese)
6. Engel, R.E., Long, D.S.: Measuring grain protein concentration with in-line near infrared reflectance spectroscopy. Crop Management 10(5), 1–3 (2005)
7. Song, T., Zhang, F., Liu, Y., et al.: Rapid determination of fatty acids in soybean oils by transmission reflection-near infrared spectroscopy. Spectroscopy and Spectral Analysis 32(8), 2100–2104 (2012) (in Chinese)
8. Zhang, S., Wang, F., Zhang, H., et al.: Detection of the Fresh Jujube Varieties and SSC by NIR Spectroscopy. Transactions of the Chinese Society of Agricultural Machinery 40(4), 139–142 (2009) (in Chinese)
9. Zhang, S., Wang, F., Zhang, H., et al.: Near-Infrared Determination of Soil Nutrients Based on Principal component Analysis and BP Neural Network. Journal of Shanxi Agricultural University 29(6), 483–487 (2009) (in Chinese)

Research Progress of Grain Quality Nondestructive Testing Methods*

Fenghua Wang**, Zhiyong Xi, Ju Yang, and Xiaojing Yang

Faculty of Modern Agricultural Engineering, Kunming University of Science and Technology, Kunming, Yunnan, China 650500
wangfenghua018@163.com

Abstract. The application of modern non-destructive testing methods on grain quality testing had important implications for grain production, distribution and sale. In the paper, five test methods of modern electronic information technology, which were near-infrared spectroscopy, electronic nose, machine vision, magnetic resonance and acoustic vibrations, were described, the research progress of near-infrared spectroscopy, electronic nose technology and machine vision technology in grain quality nondestructive testing was reviewed with focuses. The application of modern non-destructive testing methods on grain quality testing could reduce costs and improve detection accuracy, furthermore, make it combined with modern computer technology and data processing, which would play an important role in agriculture.

Keywords: Grain, Quality, Non-destructive testing, Research process.

1 Introduction

Cereal is a traditional diet of Chinese and an important and indispensable food of human survival [1]. Rice annual production of our country is 200 million tons, accounting for 35 percent of world output. Most of the food needs circulation and warehouse storage. However, in domestic and foreign trade and the circulation process of grain, food quality testing methods of many grain purchasing department rely on the person's subjective assessment because the lack of rapid quality detection means, which is not only can not detect a variety of internal quality parameters of food, and there is a great error. Using modern non-destructive testing method to identify the grain quality can reduce costs and improve the detection accuracy, China's grain quality testing technology research and application levels will be improved [2].

---

* This study was financed by the Application Basis Surface Projects of Yunnan Province (2010ZC028), Kunming University of Science and Technology Analysis and Testing Fund (2010299).

** Corresponding author.

D. Li and Y. Chen (Eds.): CCTA 2012, Part II, IFIP AICT 393, pp. 255–262, 2013.

Non-destructive testing (NDT) is general term for all the technical means to test the existence of different degrees of the defects in the detected objects, give out the location, size, number and nature of the defect, determine the technical state (such as eligibility, remaining life, etc) of the detected objects, using sound, light, electricity, magnetism, taste and other characteristics, under the premise that does not damage or affect the performance of the detected objects [3]. Modern non-destructive testing methods (such as near-infrared spectroscopy, the electronic nose and machine vision technology, nuclear magnetic resonance and acoustic vibrations method) are mainly based on the density, hardness, shape, color and other parameters of the detected objects, to determine the quality of agricultural products such as maturity, sugar and acid ratio, moisture content, sugar content, internal lesions and so on, which are both to ensure the integrity of the sample and can effectively determine the internal quality information [4]. In the paper, five test methods of modern electronic information technology, which were near-infrared spectroscopy, electronic nose, machine vision, magnetic resonance and acoustic vibrations, were described. The research progress of the near-infrared spectroscopy, electronic nose and machine vision technology in the non-destructive testing of grain quality were reviewed mainly.

2 Near-Infrared Spectroscopy and the Application Status in the Grain Non-destructive Testing

Near Infrared Spectroscopy (NIRS) is electromagnetic waves that the wavelength is in the 780-2526nm between visible light and infrared light [5].Due to composing frequency and all levels double frequency absorption of containing hydrogen group vibration in organic molecules was consistent with the near infrared region, therefore, vibration information of containing hydrogen group in the sample organic molecules can be obtained by scanning with near infrared spectroscopy [6]. Compared with traditional methods of chemical analysis, the main technical characteristics of the near-infrared spectroscopy are fast analysis, simultaneous determination of multi-component samples without pretreatment, non-destructive analysis, distance measurement and real-time analysis, low analysis cost and simple operation. So the near-infrared spectroscopy is known as the giant of the analyst community [7].

In the 1960s, Norris determined the moisture, protein, and fat content of the grains using near-infrared spectroscopy and multivariate linear regression analysis. Later, Hymowitz uses near infrared technology for the determination of protein, fat, and amino acid content of the cereal crops. Kong Wenwen et al. [8] detected malondialdehyde (MDA) content in the barley leaves using spectroscopy and have high precision. a new method for status monitoring of physiological information about the large scale agriculture of hordeum growth and stress of herbicide has been provided. Xie Xinhua et al. [9] established a near-infrared mathematical model of the single-grain brown rice protein which improves the near-infrared quantitative analysis with high accuracy. Zhu Wenchao et al. [10] applied visible / NIR spectroscopy to realize the identification of the transgenic rice leaves and detection of the chlorophyll content quickly. A new method for the realization of the living identification and continuous monitoring in field has been offered.

The grain protein distribution has been studied by Engel et al. [11]. The predictive model of spring wheat has been established under the laboratory conditions. the online prediction of the protein concentration of spring wheat in farmland by the model, The results showed that the results of distribution of grain protein content by online and laboratory are very similar, the correlation coefficient is 0.93. Lu Hui et al. [12] established the method of the fast species identification and eating quality analysis on rice by near infrared spectroscopy. Near infrared spectrum of 102 crushed rice samples have been collected, the quantitative analysis model of water, protein and amylose on rice by partial least squares (PLS) method, to evaluate the accuracy of the predicted results. The results show that NIRS analysis techniques can be used for rapid non-destructive testing of the quality and variety of rice. Liu Xiaojun et al. [13] found that it had a obvious spatial and temporal variation characteristics of leaf reflectance spectrum in the different soil water and leaves, and sensitive spectral bands of leaf water content located in the near infrared and shortwave infrared region, the spectral index and leaf water content showed a good linear correlation, the linear fitting R2 of 0.80. Therefore, a rapid non-destructive diagnosis of rice leaf water content estimation model has been established to provide the basis for the precise management of rice water. Sun Xiaorong et al. [14] have detected the water content of starch by near infrared reflectance spectroscopy, at first the different types of starch at home and abroad have been collected, then the optimal pretreatment methods and spectral range have been selected, finally quantitative analysis of the water content of starch by partial least squares method, The results showed that rapid non-destructive testing of starch water content using near-infrared spectroscopy is available.

3 Electronic Nose Technology and the Application Status in Grain Non-destructive Testing

The electronic nose is an electronic system of the simulating biological nose to analyze and detect complex odor and volatile chemical composition mainly. The electronic nose is a kind of instrument can identify a variety of odor, which composed of the selective chemical sensor arrays and the appropriate pattern recognition system. Generally it consists of three parts of the gas sensor array, signal processing unit and the pattern recognition unit. On working, the signal is generated after the adsorption of odor molecules by the gas sensor array, then the signal is sent to the signal processing unit for processing, and finally the analysis results is obtained by the pattern recognition unit analyzing the data of signal processing [15,16].

On mouldy detection aspect, Zou Xiaobo et al. [17] developed a rapid detection grain whether moldy electronic nose device, the device can be fast, accurate analysis of the measured grain odor, and its correct rate is 90% or more. Pan Tianhong et al. [18-19] developed a group of thick-film metal oxide tin gas sensors (MOS) array in the identification of moldy corn. The data of the entire reaction process between the sensor and the grain volatile odors, the characteristic value of the sensor and the odor response have been extracted, and have been analyzed by RBF neural network, the result show that the back estimation rate reached 100% and the recognition probability is 92.19%. On the insect detection aspect, Ridgway et al. [20] detected the insect infected wheat samples using the electronic nose, the feasibility of the method has been proved. Jiang Deyun et al. [21] designed test system using the electronic nose to test and compare the gas samples of containing live worms, containing dead insects and standard, then the pattern recognition have been carried on by principal component analysis. The result shows that this method can quickly detect the pests' erosion of food whether or not. Zhang Hongmei et al. [22] detected the pest wheat and the mixture of the pest wheat and normal wheat on the 5storage period by electronic nose. The results show that the pest wheat and normal wheat and mixed with different proportions of insect pests of wheat samples can be detected by this method. On the storage year detection aspect, Pang Lingjiang et al.[23,24] distinguished the five year wheat using portable electronic nose, The results show that the principal component analysis can identify the different years of wheat, the recognition accuracy is up to 85% by BP neural network analysis. Zhang Hongmei et al. [25] identified the wheat of the different storage life using electronic nose, and sensor array signal has been optimized, the BP neural network model has been established using the optimized sensor, and the recognition rate is up to 96%.

4 Machine Vision Technology and the Application Status in Grain Non-destructive Testing

Machine vision technology is a kind of technology using imaging sensor instead of the human eye to obtain the object image, which converts images into digital information to analyze the images and draw a conclusion, and carry on non-destructive testing to internal and external quality of an agricultural products, such as the shape, size, color, surface defects and the maturity [26].

Huang Xingyi et al. [27] identified the rice varieties by the computer vision technology instead of human eye. The method that has combined the shape with the color characteristics of the image to identify the rice varieties has been proposed. The recognition classifier has been established by Bayesian decision method, and the recognition accuracy can reach 88.3%. Casady et al. [28] collected the image of the paddy rice crown using the machine vision systems, the image segmentation and feature extraction have been carried on by gray-scale median values and mathematical morphology. The study found that the correlation of height, width and area of rice crown measured by the system and manual are 0.896, 0.874 and 0.885 respectively, which has provided decision-making support for rational application of nitrogen fertilizer. R. Choudhary et al. [29] extracted the shape, color, texture and wavelet feature of wheat from the color image, the classification accuracy of wheat varieties have been compared using the linear classifier and the quadratic classifier. Combining with the different number of wheat characteristics, the results of the classification show that the detection results combined with wheat shape, color, texture, and wavelet four kinds of characteristics are the best. Aiming at the segmentation problems of a large number of cereal grain adhesion image, Lin Yun et al. [30] improved the erosion and dilation algorithms and watershed segmentation algorithm based on morphological. The relative error is less than 2%.

5 Nuclear Magnetic Resonance and Acoustic Vibration

Nuclear magnetic resonance (NMR) is one of the technologies to detect concentrated hydrogen [31], which is sensitive to respond changes of the agricultural water and lipid in the mixed granulating state. Research object of nuclear magnetic resonance is the nucleus having the magnetic moments. In the static magnetic field, this has a magnetic moment nucleus exist the different energy levels. The samples are radiated by a particular frequency of electromagnetic radiation, the transition between energy level of the nucleus will be carried on, which is nuclear magnetic resonance.

Loveday et al.[32] studied the physical and chemical changes model of the protein bar in the store process of the 50 day in 20°C,the study discover that the chemical works is very small in the protein hardening process, mainly due to moisture migration caused by micro-structural changes and molecular activity of impact. Vanlent et al.[33] determined the small water droplet size and distribution of the six different fabrication of butter using the NMR and confocal laser scanning microscopy to ensure the microbial stability and sensory quality. The results show that the different fabrication of butter put up the noticeably different. In recent years, NMR techniques have been widely used in medicine [34], which plays an important role in the medical examination and biochemical research, but rarely used in agricultural products. With the low field strength, precision, low-cost, high-speed research and development of NMR technology, NMR technology has great potential and will play an important role in the agricultural domain [35].

The acoustic vibration is a means of method to determine the quality of the detected object according to mechanical vibration measurement results that is produced through an incentive to the test specimen. The characteristic of the acoustic vibration is simple, rapid, and inexpensive [36].

Knocking detection is one of the most simple and commonly method used as non-destructive testing techniques relying on the component vibration characteristics [37]. Acoustic vibration detection methods should be further studied from the theory and practice, the acoustic vibration detection methods of easy operation, convenient and reliable would be developed, which will further promote the practical application of acoustic vibration detection technology. On the food safety monitoring aspect, Nassar et al.[38] inspected the dynamics formation process of milk floc using low-energy ultrasonic resonance testing technology, the influence of raw milk temperature and chymosin concentration to milk flocculation have been analyzed. Elvira L et al. [39] studied found that the ultra high temperature sterilization packaged milk has been detected using 800 KHz ultrasonic to determine the changes of the ultrasonic propagation time and amplitude, which can detect the growth of microorganisms in liquid food. Gan T H et al.[40] proved that packaging products can be detect by the non-contact method, even including the detection of liquid with starch as a raw material in microwave packaging contain using air-coupled ultrasonic technology to detect liquid and starch-based raw materials in the container.

6 Summary

Although near-infrared spectroscopy has the advantages of fast, easy, simple, accurate and the simultaneous determination of multi-component samples in the online analysis, it is an indirect analysis technique and still has many shortcomings. The development direction of near infrared spectroscopy technology is the imaging spectroscopy. With the development of two-dimensional array detector, the measurement of composition and component distribution become possible. The electronic nose use as an electronic olfactory sensor system, shown unique features in the food, quality control, quality evaluation and safety testing, Which combined with the parallel coupled system of gas chromatography instrument and flame ionization detector become a new trend. Machine vision technology application in agriculture is still in the initial stage, the agricultural product range is limited, and most used in quality detection of agricultural products. And now, development of the agricultural harvest and processing automation field should be to meet the requirements of increasing of the harvesting and processing automatic degree of the agricultural products and improving of the production efficiency, should be further strengthened to speed up research in the field. NMR techniques played an important role in the medical examination and biochemical fields, but agriculture is still a new field, should increase the strength on cereal and fruit application respect in the future. Acoustic vibration detection methods should be further studied from the theory and practice, to deep study of acoustic vibration testing methods of grain, and establish the relationship between signals of sound vibration and defects. The portable mensuration instrument will become the development trend of the sound vibration technology.

References

1. Wu, S., Cheng, Z.: Grain storage and quality maintenance. Chinese Journal of Animal Science 46(6), 52–54 (2010) (in Chinese)
2. Yu, Z., Hou, Y.: Development of grain ventilation drying quality. Drying Technology and Equipment 7(1), 41–44 (2009) (in Chinese)
3. Li, J., Rao, X., Ying, Y.: Advance on Application of Hyperspectral Imaging to Nondestructive Detection of Agricultural Products External Quality. Spectroscopy and Spectral Analysis 31(8), 2021–2026 (2011) (in Chinese)
4. Sun, T., Xu, H., Ying, Y.: Progress in Application of Near Infrared Spectroscopy to Nondestructive On-line Detection of Products/Food Quality. Spectroscopy and Spectral Analysis 29(1), 122–126 (2009) (in Chinese)
5. Xu, X., Cheng, F., Ying, Y.: Application and Recent Development of Research on Near-Infrared Spectroscopy for Meat Quality Evaluation. Spectroscopy and Spectral Analysis 29(7), 1876–1880 (2009) (in Chinese)
6. Blanco, M., Villarreal, I.: NIR spectroscopy rapid-response analytical tool. Trends in Analytical Chemistry 21(4), 240–250 (2002)

7. Gong, Y., Zhang, W.: Recent Progress in NIR Spectroscopy Technology and Its Application to the Field of Forestry. Spectroscopy and Spectral Analysis 28(7), 1544–1548 (2008) (in Chinese)
8. Kong, W., Liu, F., Hui, F., et al.: Rapid detection of malondialdehyde in herbicide-stressed barley leaves using spectroscopic techniques. Transactions of the Chinese Society of Agricultural Engineering 28(2), 171–175 (2012) (in Chinese)
9. Xie, X., Xiao, X., Li, F., et al.: Mathematic Models for Analyzing Protein Content in Single Brown Rice with Near Infrared Spectroscopy. Transactions of the Chinese Society for Agricultural Machinery 37(8), 120–122 (2006) (in Chinese)
10. Zhu, W., Cheng, F.: Analysis of Transgenic and Non-Transgenic Rice Leaves Using Visible/Near-Infrared Spectroscopy. Spectroscopy and Spectral Analysis 32(2), 370–373 (2012) (in Chinese)
11. Engel, R.E., Long, D.S.: Measuring grain protein concentration with in-line near infrared reflectance spectroscopy. Crop Management 10(5), 1–3 (2005)
12. Lv, H., Zhang, Z., Wang, S.P., et al.: Quality analysis and category identification of rice based on the near infrared spectroscopy (NIRS). Science and Technology of Food Industry 33(3), 322–325 (2012) (in Chinese)
13. Liu, X., Tian, Y., Yao, X., et al.: Monitoring Leaf Water Content Based on Hyperspectra in Rice. Scientia Agricultura Sinica 45(3), 435–442 (2012) (in Chinese)
14. Sun, X., Liu, C., Wu, J., et al.: Rapid detection of water content in starch based on near infrared spectrum. Science and Technology of Food Industry 32(10), 441–442 (2011) (in Chinese)
15. Zhang, H., He, Y.: The Applications of Electronic Nose in the Grain Quality Discrimination. Journal of Agricultural Mechanization Research 15(3), 180–182 (2009) (in Chinese)
16. Wei, L., Zhang, C., Zhao, B., et al.: Electronic Nose Technology and Its Application in the Activity Detection of Wheat Grain. Journal of Agricultural Mechanization Research 9(6), 150–152, 196 (2010) (in Chinese)
17. Zou, X., Zhao, J.: Rapid identification of moldy corn by electronic nose. Transactions of the CSAE 20(4), 121–124 (2004) (in Chinese)
18. Pan, T., Chen, S., Zhao, D.: Application of Electronic Nose Technology in Moldy Foodstuff Recognition. Instrument Technique and Sensor 3, 51–52 (2005) (in Chinese)
19. Zhu, J., Zhao, D., Pan, T., et al.: Identification of moldy foodstuff based on artificial olfactory system. Transactions of the CSAE 21(1), 106–109 (2005) (in Chinese)
20. Hu, J., Guo, M., Zhang, J., et al.: New detection technology and application status of stored-grain insects. Transactions of the CSAE 23(11), 286–290 (2007) (in Chinese)
21. Jiang, D., Kong, X., Li, B., et al.: Research and Application of Electronic Nose for Testing Stored-grain Insects. Journal of Anhui Agricultural University 32(2), 254–257 (2005) (in Chinese)
22. Zhang, H., Jun, W.: Detection of age and insect damage incurred by wheat, with an electronic nose. Journal of Stored Products Research 43, 489–495 (2007)
23. Pang, L.: Application of electronic nose in identification of wheat freshness. Zhejiang University, Hangzhou (2005)
24. Pang, L., Wang, J., Lu, X.: Discrimination of Different Storage Time of the Wheat by Electronic Nose. Chinese Journal of Sensors and Actuators 20(8), 1717–1722 (2007) (in Chinese)
25. Zhang, H., Wang, J., Tian, X., et al.: Optimization of sensor array and detection of stored duration of wheat by electronic nose. Journal of Food Engineering 82(6), 403–408 (2007)

26. Ying, Y., Zhang, W., Jiang, Y., et al.: Application of Machine Vision Technique in Automatic Harvesting and Processing of Agricultural Products. Transactions of the Chinese Society for Agricultural Machinery 31(3), 112–115 (2000) (in Chinese)
27. Huang, X., Li, J., Jiang, S.: Study on identification of rice varieties using computer vision. Journal of Jiangsu University (Natural Science Edition) 25(2), 102–104 (2004) (in Chinese)
28. Casady, W., Singh, N., Costello, T.A.: Machine vision for measurement of rice caopy dimensions. Trans. of the ASAE 39(5), 1891–1898 (1996)
29. Choudhary, R., Paliwal, Jayas, D.S.: Classification of cereal grains using wavelet, morphological color, and textural features of non-touching kernel images. Biosystems Engineering 99(3), 330–337 (2008)
30. Ling, Y., Wang, Y., Sun, M., et al.: Application of Watershed Algorithm to Paddy Image Segmentation. Transactions of the Chinese Society for Agricultural Machinery 36(3), 95–98 (2005) (in Chinese)
31. Zhou, S., Shang, D., Ying, Y., et al.: Detecting Subtle Bruises on Fruits with Nuclear Magnetic Resonance Imaging. Transactions of the Chinese Society for Agricultural Machinery 41(8), 107–110 (2010) (in Chinese)
32. Loveday, S.M., Hindmarsh, J.P., Creamer, L.K., et al.: Physicochemical changes in a model protein bar during storage. Food Research International 42(7), 798–806 (2009)
33. Vanlent, K., Vanlerbergbe, B., Oostreldt, P.V., et al.: Determination of water droplet size distribution in butter: pulsed field gradient NMR in comparison with confocal scanning laser microscopy. International Dairy Journal 18(1), 12–22 (2008)
34. Pang, L., Wang, Y., He, Z., et al.: Application of NMR in Fruit Quality Detection. Journal of Agricultural Mechanization Research (8), 176–180 (2006) (in Chinese)
35. Jin, Z., Zhang, J., Qian, F., et al.: Study on Changes of Grains during Storage by NMR. Food Science 29(9), 66–69 (2008) (in Chinese)
36. Chen, J.: Non-destructive test for the quality of agricultural product. Journal of Changchun University of Technology (Natural Science Edition) 27(3), 262–266 (2006) (in Chinese)
37. Cawley, P., Adams, R.D.: The mechanics of the coin-tap method of non-destructive testing. Journal of Sound Vibration 122(9), 299–316 (1988)
38. Nassar, G., Nongailland, B., Noel, Y.: Monitoring of milk gelation using low-frequency ultrasonic technique. Journal of Food Engineering 48, 351–359 (2001)
39. Elvira, L., Sampedro, F., Montero de Espinosa, et al.: Eight-channel ultrasonic device for non-invasive quality evaluation in packed milk. Ultrasonic 45, 92–99 (2006)
40. Gan, T.H., Hutchins, D.A., Billson, D.R.: Preliminary studies of a novel air-coupled ultrasonic inspection system for food containers. Journal of Food Engineering 53(4), 315–323 (2002)

Maize Disease Diagnosis Model Based on Ontology and Multi-Agent

Liying Cao[1,2], Xiaoxian Zhang[3], Xiaohui San[1], Li Ma[1], and Guifen Chen[1,*]

[1] School of Information Technology, Jilin Agricultural University, Changchun 130118
[2] Anhui Agricultural University, Hefei 230036
[3] Software Institute, Changchun Institute of Technology, Changchun 130012
{caoliying99,zhangxxccit,umbrella,mary,guifchen}@163.com

Abstract. According to the uncertainty, dynamics and dependency of maize diseases, this paper puts forward a kind of maize disease diagnosis model combined ontology with multi-Agent. The knowledge representation of ontology provides a clear formal specification for the onset periods and symptom characteristics during maize disease occurs. Using the Agent's intelligence and the division collaboration of Agent modules, we have solved the bottleneck problem of the acquisition and representation of maize disease knowledge and improved the learning ability of the ontology. Experiments show that the model can diagnose maize diseases accurately and efficiently and has a good application prospect.

Keywords: multi-Agent technology; ontology; maize disease; intelligent diagnosis.

1 Introduction

Maize is the crop holding characters of food, feed and economy concurrently. The maize industry has been one of traditional industries in JiLin Province and its per capita, commodity volume, transfer volume and export volume have been for many years in the first place in the whole country. But in recent years, factors such as maize disease monitoring and evaluation, forecast and warning and timely and accurate diagnosis and treatment to corn have brought some influence to the high, steady and efficient yields. The current maize disease diagnosis and treatment is still basing on experience and more extensive and has the phenomenon such as lack of agricultural experts, untimely and inaccurate diagnosis and unreasonable pesticides. Not only does it waste resources and pollute the environment, but also lead to the maize production can't be high and stable. It is the important link to ensure the high and stable maize yields that researching maize disease diagnosis and treatment and reducing the losing of maize disease.

At present, the expert system applied in crop protection bases on the database expert system mainly, which provides services about the plant disease and caterpillar

* Corresponding author.

D. Li and Y. Chen (Eds.): CCTA 2012, Part II, IFIP AICT 393, pp. 263–269, 2013.

diagnosis, forecast, the comprehensive management, plant quarantine and pesticide management. Even if the expert system based on the model has the bottleneck problem of knowledge acquisition and representation, too. [1] The knowledge model of which is fixed and doesn't have the learning function and can't update real-time according to knowledge changes in the related areas.

This paper takes the maize disease in JiLin Province as the research object, extracts and classifies the knowledge experience in the field and takes knowledge representation, intelligent reasoning and knowledge study by using ontology technology. It adopts multi-Agent to realize ontology learning, task decomposition scheduling, information processing and intelligent diagnosis function respectively [3].

2 Research Data and Related Technology

2.1 Research Data Sources

The research data derives from:
Consulting the domain experts, such as related experts of Jilin Agricultural University and the Institute of Plant Protection, JiLin Academy of Agricultural Sciences.

Relevant knowledge obtained by consulting and concluding from science and technology literature materials, such as the published books, teaching materials, articles published on science and technology journals, proceedings and degree thesis of the expert on the issue.

On the basis of test data accumulated from scientific research project (national 863 project). According to the experiment of the relationship between the yield of corn and harm, we collected maize growth traits index data 29250, such as, maize seedling, huge bellbottom period and mature; the maize disease occurring degree data and the field data 4320 and photos 189 copies; the maize insect damage field data 5760 and photos 306 copies; the maize grass damage field data 4320 and photos 286 copies. We absorbed scientific research and data also.

Collect knowledge and practice experience of all the farmers and agricultural technicians in the agricultural demonstration zone and add source of knowledge.

The source of the picture data is the staff room of plant protection of Jilin Agricultural University.

This paper organizes 13 maize disease common in the JiLin Province, such as, maize dwarf mosaic, maize rough shrink disease, maize leaf blight, corn top rot disease, maize stalk rot, maize stalk rut, maize head smut, maize curvularia leaf spot, corn sheath blight, Bipolaris maydis and northern leaf spot. Corn cercospora sogina, maize stem rot disease, corn.bacterial wilt, corn silk HeiSuiBing, corn curved spore fungus leaf, WenKuBing corn, corn small spot disease and corn circle spot disease. The structures of maize disease organized are shown below:

Table 1. Maize disease organization structure

serial number	disease name	happen period	symptom	prevention method	disease picture

2.2 Ontology

The concept of the ontology comes from the realm of philosophy. In the 1960 s, the field of computer began to use ontology, but until 1993 Gruber gave the ontology standard definition: ontology is the clear specification of the conceptual model[4]. At present, the definition of ontology is the clear formal specification of the sharing the conceptual model. It can be took as four levels, such as the conceptual model, clearity, formalization and share. Specific description as follows:

The concept model, obtained by abstracting out some of the objective world phenomenon related concepts. The meaning showed by concept model is independent of specific environmental state;

Clarity, means that the concept used and constraints used by concepts are clearly defined;

Formalization, with the body language code, ontology can make the computer be read and processed.

Share, ontology embodies knowledge recognized commonly and reflects the concept set recognized publicly in the relevant fields, that is, ontology aimed at groups rather than individuals consensus.

In general, the goal of onto1ogy is to obtain, describe and express knowledge in related fields, and provide the common understanding of the domain knowledge and make the vocabulary knowledge clearly recognized commonly in the field, and give the definition between these words and relationship between these words clearly. General speaking, the onto1ogy is a file which officially define a group of words and the relationship between them. It is the integration of the knowledge representation and the relationship rules for intelligent system. It defines some relevant classes and objects to express the knowledge. The relationship rules express the relationship between knowledge and provide a kind of intelligent reasoning mechanism.

2.3 Multi-Agent Technology

Agent technology is a kind of intelligent technology which develops on the basis of the distributed and artificial intelligence research needs [5]. Agent is a kind of a smart entity with initiative, reactive, sociality and autonomy Characteristics etc. It Can be understood as intelligent computer hardware or software system with three kinds of function of the knowledge, ability and target. The knowledge obtains by users, some Agents or their own learning. The ability is the method and means of Agent to solve some problems. The problem to be solved and the task are the goal of Agent [6]. Due to the system resources quantity and their own capability limits, the tasks which can be accomplished by one single are very limited. Multi-Agent technology takes single Agent as the foundation and takes use of the cooperation and interaction between self-government Agents to solve some kink of problems and complete the complex task together.

3 Construction of Maize Disease Diagnosis Model Based on Multi-Agent and Ontology

3.1 Knowledge Representation and Knowledge Model Based on Ontology

Aiming at the effects brought by complex factors, such as the symptoms of maize disease, pathogenesis regularity complex, vulnerable to climate condition, soil factors and the difference among areas, it is difficult to use a single knowledge representation to obtain accurate expression. Taking use of ontology, some bottleneck problems in the agricultural expert system, including the knowledge representation, knowledge organization and knowledge sharing, can be well solved. The goal of ontology is to obtain, describe and express the knowledge of the related field, provide the common understanding of the domain knowledge, determine the common admissive vocabularies in the field and give a clear definition of the relationship between these vocabulary words and terms from different levels of formal model. Based on above aims, this subject set up the database of maize disease ontology.

We adopt Protege4.0 to construct the maize disease ontology database, use seven footwork in the constructing process, take formalism description on target ontology and generated the owl file [7]. Specific steps as follows:

List the important terms of ontology:

There are many concerned terms in describing maize disease. In order to extract the corresponding information, we choose some necessary terms, such as disease name, onset period, onset place, disease characteristics, disease spot shape, disease spot color and prevention-control method, etc.

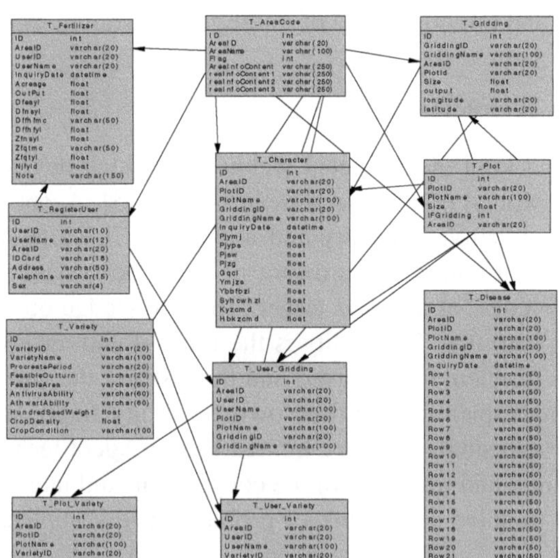

Fig. 1. The knowledge base relationship chart based on ontology

Definite classes and the class level system: make the definition and classification of the class on maize disease ontology.

Definite attributes of classes: aiming at maize disease name, onset period, onset place, disease spot shape, disease spot color and prevention-control method, we establish data attributes to store the specific content of the description.

Define constraint attributes: one attribute may be with multiple constraints. The types of the attribute value include any, integer, string, float, etc.

Create example: make sure the nearest class to the individual and add the individual in the class as an instance and evaluate attributes of this instance. The maize disease instance include 13 common disease name, onset period, onset place, disease spot shape, disease spot color etc. The knowledge base relationship of ontology is shown in fig. 1.

Ontology constructed as shown in figure 2 below.

Fig. 2. concept structure of ontology maize disease

Maize disease ontology constructed can remove large junk information to improve the accuracy of the information retrieval during retrieving and reasoning. It can retrieve contents closely related to retrieval condition and improve the comprehensive information retrieval.

3.2 The Realization of Multi-Agent in the Maize Disease Diagnosis Model

The system can be divided into four layers from function and role. The first level is the user application layer. The second is intelligent processing layer, which is composed of information exploration Agent, intelligent diagnosis Agent, task decomposition and scheduling Agent, ontology evaluation Agent and ontology information processing Agent [8]. The third level, data storage layer, is composed of ontology knowledge base, disease database and ontology learning knowledge database, in which ontology studying completes by ontology studying Agent, respectively. The fourth level is the original data infomation layer.

The process of disease diagnosis can be described as: the user submits remotely diagnosis data through client. after receiving the diagnosis data, the diagnosis system use the diagnosis model in advance of data storage to make preliminary diagnosis analysis. If it can't find the matching model, the diagnosis system studies the corresponding diagnosis model by using ontology studying agent. In data processing layer, the agent of task decomposition and scheduling will breakdown the diagnosis task and use the specific algorithm of this diagnosis module by taking use of intelligent diagnosis agent, and then ontology evaluate agent evaluates the algorithm or ontology information processing agent processes the corresponding diagnosis and finally makes the conclusion and output to the user.

In intelligent diagnosis, we adopt the framework of the reasoning approach, that is, concluding the regular knowledge of agriculture (such as maize disease diagnosis and treatment technology) and depositing at the framework and using different guidance processing according to different situations [9]. The characteristics of this framework reasoning method: divide maize disease into a number of subproblems depending on each other by using frame method and form the knowledge hierarchical structure. Another sub problems focus in one frame to describe and then realize the modular and compact of the knowledge [10]. Basing on improving database, the knowledge base and material database, aiming at the specific disease characteristics needed to be diagnosed, we find out one framework similarest to the disease and match them by "top-down".

We adopt the semantic retrieval in this model. This retrieval can not only diagnose conditions consistent results in grammar form but also diagnose the results which meet the conditions on the semantics. When direct search is without results, the intelligent agent can make semantic expansion and strive to obtain highly relevant results.

4 Summary

This paper proposes a multi-Agent maize disease diagnosis model based on ontology.

In order to verify the validity of the model, we train it by taking use of survey data of the project team. We brought five kinds of common maize disease in JiLin Province to train in the experimental process and then compared with real situation. The inspection results of multi-Agent maize disease diagnosis model based on ontology. It can be seen that the average accuracy of the model is 82.98%.

This paper introduces ontology and multi-Agent technology into maize disease diagnosis for their research, combines it with the implementation of the national "863" maize precise operation system research and application project, verifies the feasibility of the model and has a certain practical value. According to special requirements of the content of maize disease diagnosis, the diagnosis program and the diagnosis result, we select protege4.0, JADE and Eclipse as the based software and construct the processing platform by taking model diagnosis, model learning and diagnosis result output as one system through integrating above three softwares and provide a smooth technical supporting system for the database building of maize disease diagnosis, the learning, diagnosis, inquiry, retrieval and update of the model.

Acknowledgments. This work was funded by the Youth Foundation of Jilin Agricultural University under Grant No. 201136, the National High-Tech Research and Development Plan of China under Grants Nos.2006AA10A309, Changchun Technology Correspondent Project (2009245), Jilin provincial "125" science and technology research subject(No. 201247 and 201248), the research and application of facilities for the safety of vegetables production technology based on Internet of Things(2011-Z20), Jilin province science and technology development program (201101114).

References

1. Liu, D., Yang, K., Chen, J.: The Current Situation and Trend of Agent Research. Journal of Software 11(3), 315–321 (2000)
2. Zhao, J., Yan, P., Chen, G., Hu, L.: The Research on Fault Diagnosis Knowledge Representation of Hobbing Machine based on Ontology. Automation of Manufacturing Industry 8(33), 79–84 (2011)
3. Qin, J., Cao, L., Wang, X.: Fault Diagnosis Method Based on Multi-agent Hybrid Inference. Computer Engineering 9(32), 44–46 (2006)
4. Mu, C., Huang, H., Tian, C., et al.: Intrusion-Detection Alerts Processing Based on Fuzzy Comprehensive Evaluation. Journal of Computer Research and Development 42(10), 1679–1685 (2005)
5. Du, X., Li, M., Wang, S.: A Survey on Ontology Learning Research. Journal of Software 17(9), 1537–1847 (2006)
6. Li, T., Ma, Y., Qiu, Y.: Modeling Emotions for the Intelligent Agent Based on Fuzzy Logic. Computer Science 34(11), 137–139 (2007)
7. Li, C., Chen, Q.: Application of Multi-Agent in Disease Diagnosis System. Computer Engineering 15(33), 182–184 (2008)
8. Xu, Z., Huang, K., Zhang, W.: The Methodology of Ontology Building. Computer Science 29(1) (2002)
9. Feng, W., Li, J., Zhang, Y.: Research of Multi-agent Implemented EAI Model with SOA. Computer Science 36(1), 286–287 (2009)
10. Wang, C., Liu, Z., Fu, H.: Research on knowledge retrieval model based on ontology and multi-agent. Computer Engineering and Design 30(9), 2304–2306, 2310 (2009)

Design of Wireless Sensor Network Middleware for Agricultural Applications

Liang Zhao[1,2], Liyuan He[2], Xing Jin[1], and Wenjun Yu[1]

[1] College of Science, Huazhong Agricultural University, Wuhan 430070, People's Republic of China
[2] Resources and Environmental Science, Huazhong Agricultural University, Wuhan 430070, People's Republic of China
zhaoliang323@mail.hzau.edu.cn

Abstract. Typical WSN (Wireless Sensor Network) applications in precision agriculture include production process management, plant growth optimization, farmland monitoring and so on. Middleware, which can connect hardware, applications, and enterprise systems, is required to construct ubiquitous agricultural environment combining WSN technology with different agricultural applications, but there have been insufficient studies in the field of agricultural-oriented WSN middleware compared to other industries. This paper proposes a middleware to process data collected from agricultural environment by applying WSN technology, analyzes the data and standardizes different data forms, and provides intelligent diagnosis service and event service. One of the advantages of the proposed middleware is to build a model library which will guide the agriculture production process, such as irrigation control and disease prevention. In addition, a web-based remote login is provided to all users around the world to browse agricultural environmental parameters and to reference for intelligent decision.

Keywords: WSN, Agricultural application, Middleware, Intelligent diagnosis, Web service.

1 Introduction

WSN technology is increasingly common in all sectors because of its small size node and low cost advantages. Typical applications include agricultural production process management, precision agriculture, optimization of plant growth, farmland monitoring and so on [1]. In these applications, the acquisition of farmland environmental parameters, such as air temperature, humidity, light intensity, wind speed, soil moisture information, are important foundation of the practice of agriculture and farmland information management.

Due to the inherent diversity, variability, dispersion and continuity of farmland environmental parameters, there are some difficulties to get these parameters continuously and quickly, the same to farmland managers and agricultural decision-makers to make accurate decisions timely. With the maturity and popularity of

D. Li and Y. Chen (Eds.): CCTA 2012, Part II, IFIP AICT 393, pp. 270–279, 2013.

WSN technology, a large number of heterogeneous wireless sensor nodes deployed in the fields, which can be organized into a multi-hop intelligent network to realize the distributed farmland environmental information acquisition timely and continuously.

While WSN can solve the problem of data acquisition, but there are still several contradictions in most applications: First, the data collection and storage of heterogeneous sensor node; Second, how to convert these raw data into meaningful information and decision-making; Third, System maintenance, deployment and development of WSN applications subject to specific environmental constraints, there is no universal solution mode [2]. To solve these problems, the WSN middleware has become a research hotspot in recent years, but the study of WSN middleware for agricultural applications is less due to its specific application-oriented, which makes research more difficult.

2 WSN Middleware Research Review

WSN middleware is a software layer between sensors, gateways and application systems which support the flexible integration of them, in addition to providing services such as distributed computing, network, a main purpose is to shield the heterogeneity of the underlying components and the complexity of the distributed environment, and it's a better solution to the interoperability of applications across different platforms [3],[4],[5].

According to the location in the system, WSN middleware can be divided into the node-side model, server-side mode and mixed mode [2]. Simultaneously, according to the task and design methods, WSN middleware is divided into virtual machine, database, tuple space, event-driven and adaptive middleware [6]. Such as Mate [7], Impala [8], and Agilla [9], all belong to the node-side mode and virtual machine-based middleware. Cougar [10], SINA [11], and MiLAN[12] belong to server-side mode, is also database-based middleware. TinyDB [13] belongs to the database middleware due to its core function, including the COSMOS [14] and DSWare [15] are typical mixed-mode.

Many scholars study WSN middleware for different application areas to provide different services and different architectures[16],[17],[18]. Such as the SOM service-oriented middleware architecture, which consists of the user application layer, the sink node level and the sensor node layer, each provides different services. MiSense, is a service-oriented component-based middleware, which can support different performance needs of distributed sensor applications. In the literature [2], a context-sensitive middleware for greenhouse was proposed, which made the interactions between user application services and the heterogeneous WSN more smoothly, handled sensor data more effectively and provided context-sensitive services more intelligently.

3 Architecture Design of the Proposed Middleware

Agricultural application-oriented middleware (AAOM) systems can collect heterogeneous sensor data through gateways, such as light, humidity, soil moisture, temperature and so on, which stored in database after pretreatment. At the same time, agricultural models and expert knowledge are stored in model database. Based on these data and knowledge, alert service, intelligent diagnosis-making service and feedback control can be conducted to guide agricultural production process. Users can access these services through Web interface, then let users control equipment and monitoring environment parameters manually or automatically more better. The system hierarchy is shown in Fig.1.

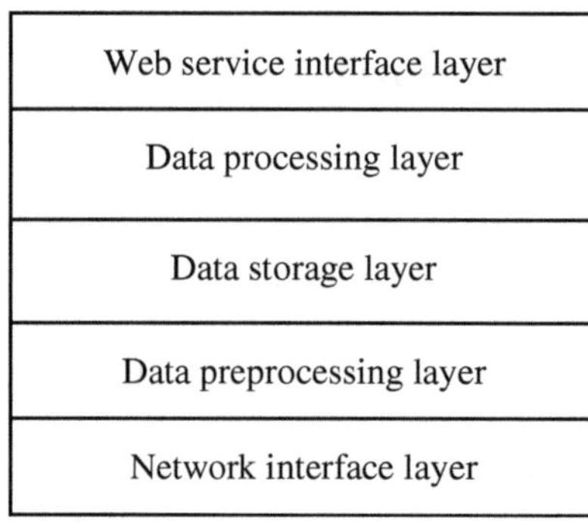

Fig. 1. Hierarchy chart of the proposed middleware

The architecture of the middleware is divided into five layers, namely, the network interface layer, the data preprocessing layer, the data storage layer, the data processing layer, and the Web service interface layer.

The main function of network interface layer is to interact with the heterogeneous WSN, continuously access to sensor data, and at the same time, to manage functions for the states of various sensor networks.

At data preprocessing layer, there are two main functions, that is data filtering and data standardization. The data filtering does not store data delivered from the sensor network interface in the sensor data DB if the data is a duplicate, or stores data only when the data is changed. At the same time, the received sensor data may be not continuous or even missing because of nodes faulty or interference in the transmission process which will impact on the analysis of data. In order to reduce the impacts by the missing data, the corresponding missing value estimation algorithms are used to predict the missing data if the missing data is found when the sensor data is delivered from the data acquisition engine. Data standardization module is called to analyze the missing data and standardize the heterogeneous sensor data forms.

The data storage layer's main task is to store the heterogeneous sensor data preprocessed in data preprocessing layer and then be deposited in the sensor DB. In addition, crop models are stored in crop model DB. Each crop may have more than one crop model, and each model may have one or more rules. A data access controller is placed in the data storage layer to support various forms of queries.

The data process layer plays the role of providing various query processing functions for sensor data and crop model data. In data processing layer, there are two functions, including intelligent diagnosis and real-time alerting. An aggregation engine is called to calculate sensor data, such as the average value, the maximum value and the minimum value during per aggregation cycle, which makes intelligent diagnosis possible by comparing with the rules, in advance to notify the intelligent service management module if it is identical or exceeds some threshold value. In addition, if the data entered has the same value as the level requested by an application, it notifies the real-time alerting service module to process the event services, otherwise the sensor data is stored into the sensor data DB through DB controller.

The Web service interface layer is an open interface, which is the top layer of the proposed middleware, supports connections with the outside users through browse and query processing. The specific data flow chart and structure of the proposed middleware is shown in Fig. 2.

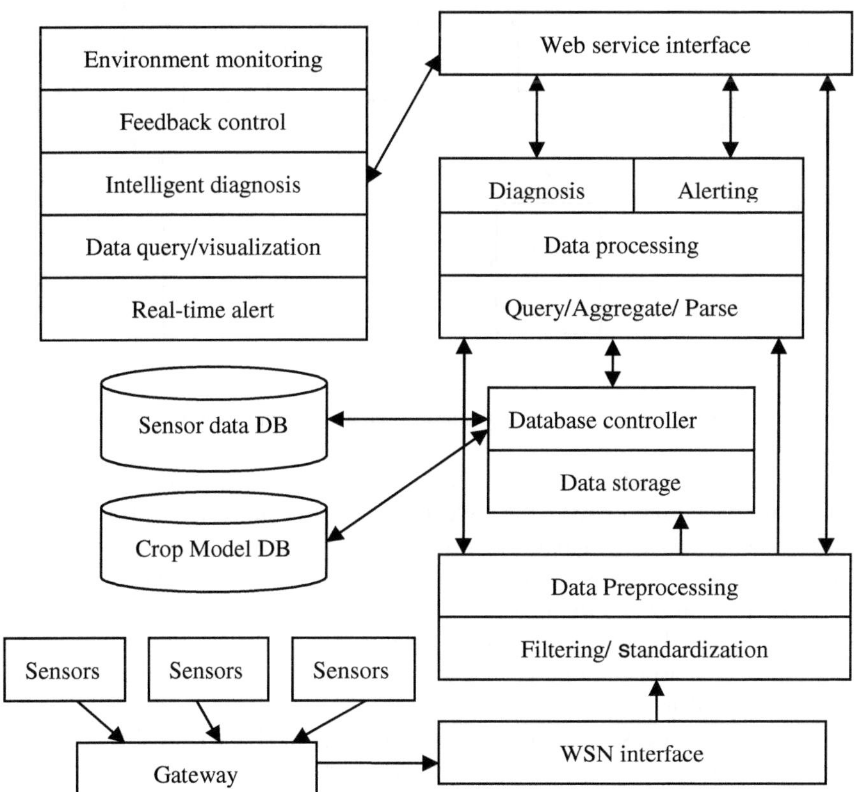

Fig. 2. Specific data flow chart of the proposed middleware

4 Implementation of the Proposed Middleware

4.1 Data Requirement

The main data repository of the proposed middleware is RDBM and is built on MySQL. Fig.3 depicts the entity relationships of the RDBM, composed by 11 tables [19].

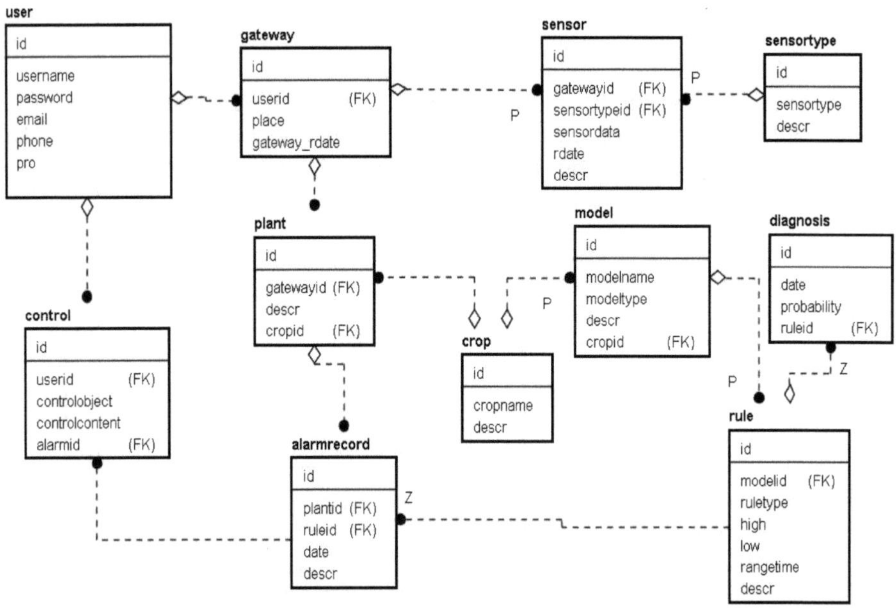

Fig. 3. Entity Relationship chart of the proposed middleware

The main table is called Sensor and it is the most basic device generating data in the network. In the table, sensor type, the collected data value and others such as date, the remaining battery capacity and other information are stored. Each device is also classified according to a user specific management zone. This is accomplished by using the table Gateway, in which the field place stores a relevant location within the zone, and Plant table records some crop planted in this zone. At the same time, there are different types of sensors in one zone, all the sensor types information is stored in Sensortype table. Rules of operation are described through the table Rule, which contains trip point values that are used by the crop model and real-time alert. In addition, crop types are stored in Crop table. There are diagnosis results and real-time alerts in diagnosis table and alarm record differently. The control information is stored in control table if the devices in a zone needed feedback control. Z shows zero or multiple relationship, P expresses one or multiple relationship, and FK is foreign key of a table.

4.2 Intelligent Diagnosis Algorithm

The intelligent diagnosis engine is started to parse crop model and match the rules according to real-time sensor data and the history data in sensor data DB. A diagnosis is arrived if all the rules are successfully matched. Otherwise, the probability of occurrence is calculated if there is a part of rules matched. Intelligent diagnosis algorithm is as follows:

```
(1) Get data;
(2) Start intelligent diagnostic threads;
(3) Determine the current status of the environmental parameters based on the
        rules and sensor data ;
(4) If  The current state and the historical state are different then
(5)        The current status is regarded as historical status, and current time
              is recorded;
(6) Else
(7)       Compute the state duration and determine;
(8)       If  The duration reaches the prescribed time limit  then
(9)            Add a rule item;
(10)           If  All rules were met   then
(11)                Intelligent diagnosis is delivered;
(12)           Else
(13)                Probability of occurrence is computed;
(14)         End if
(15)      Else
(16)          Go to (1)
(17)      End if
(18) End if
```

A simple disease probability monitoring subsystem was developed, when the real-time data is received, each rule is checked against the aggregated sensor data. The example is based on rice blast prediction model. If the temperature has been within 10°C to 35°C, especially within 24 °C and 28 °C, and relative humidity has stayed above 90%, and if this situation lasts 6 to 8 hours, serious rice blast is easily to cause occurrence, and the rules have been consistently triggered throughout a certain number of days, a warning is generated.

4.3 Application of the Proposed Middleware

The proposed middleware in this paper was implemented with Java and JSP languages, and provides services through Web, users can remotely login to the system via internet, browse and query all the information, including historical sensor data and alerting information, real-time sensor data and alerting information, expert knowledge, intelligent diagnosis and other services. Fig. 4 shows the main functions' UI of the integrated system applying the proposed middleware. The system UI (a) shows historical sensor data that can be queried according to users' real requirement. (b) and (d) display real-time sensor data fluctuations such as illumination, temperature, humidity, etc. collected from the environment, at the same time, if an intelligent diagnosis or a real-time alerting is occurred, warning information is respectively displayed in the interface (b) and (d). In addition, (c) was implemented so as to allow managers to directly set a range of environmental parameters and event services so that it could provide intelligent diagnosis service such as automatic control of the environment, etc. and the notification service for dangerous situations.

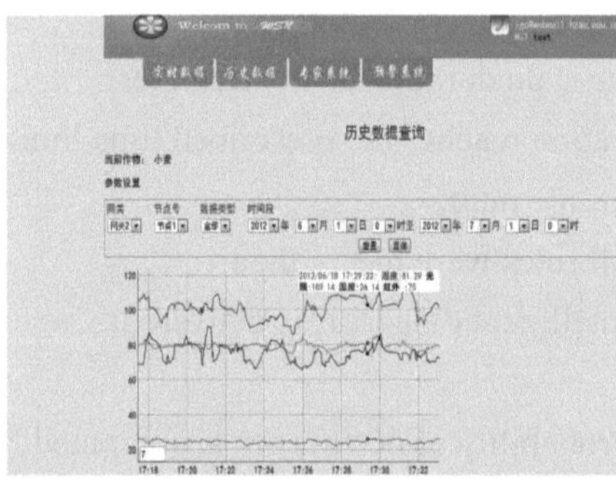

(a) Historical data query

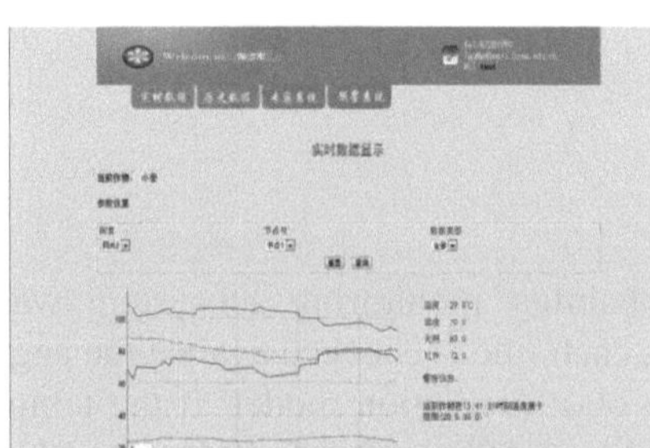

(b) Real-time alerting

Fig. 4. Main functions of the integrated system

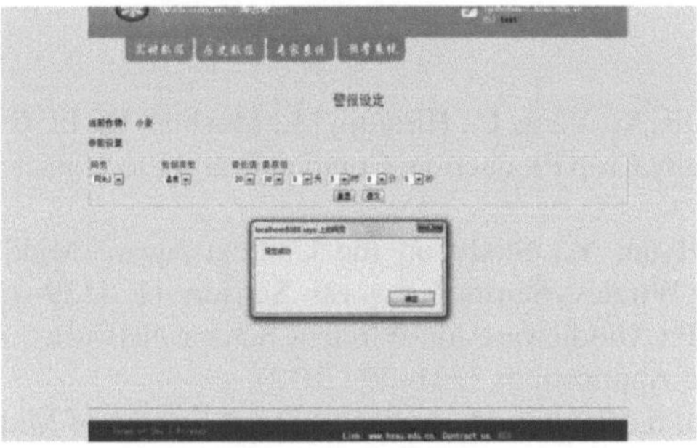

(c) Environmental parameters setting

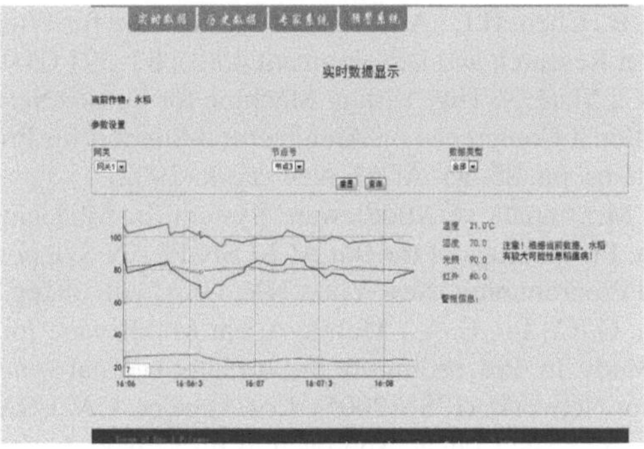

(d) Real-time intelligent diagnosis

Fig. 4. (*Continued*)

5 Conclusion

This paper proposed and developed an agricultural application-oriented middleware to efficiently process data collected from the environment applying WSN technologies and to implement combined services through Web, which was designed to bridge the gap between sensors and intelligent diagnosis. It allows local and remote users to access the middleware through Web service to access sensor data and other information. The middleware runs a data analysis engine and performs intermediate level data processing, through which real-time alerting and intelligent diagnosis are provided. The test results showed that the middleware could access and process heterogeneous sensor data smoothly and provide more accurate intelligence services. In future studies, an expansion of different models is needed to enrich the model DB, which will make the middleware more versatile, and simultaneously expression of crop models and associated expert knowledge needs to be improved. In addition, as middleware, more perfect and optimization functions also needed to access wireless sensor network and network's nodes.

References

1. Lee, W.S., Alchanatis, V., Yang, C., Hirafuji, M., Moshou, D., Li, C.: Sensing Technologies for Precision Specialty Crop Production. Computers and Electronics in Agriculture 74, 2–33 (2010)
2. Jeonghwang, H., Hyun, Y.: Study on the Context-Aware Middleware for Ubiquitous Greenhouses Using Wireless Sensor Networks. Sensors 11, 4539–4561 (2011)
3. Luca, M., Gian, P.P.: Middleware for Wireless Sensor Networks: An Outlook. Journal of Internet Service and Applications 3, 31–39 (2012)
4. Soma, B., Munmun, S., Souvik, M., Subhajit, D.: A Survey of Middleware for Internet of Things. Communications in Computer and Information Science 162, 288–296 (2011)
5. Mauri, K., Marko, H., Timo, D.H.: A Survey of Application Distribution in Wireless Sensor Networks. Journal on Wireless Communications and Networking 5, 774–788 (2005)
6. Li, R., Wei, Y., Fu, B., Chen, H.L.: A Review of Middleware for Wireless Sensor Networks. Journal of Computer Research and Development 45(3), 83–391 (2008)
7. Philip, L., David, C.: Maté: A Tiny Virtual Machine for Sensor Networks. In: Proceedings of the 10th International Conference on Architectural Support for Programming Languages and Operating Systems, pp. 85–95. ACM, New York (2002)
8. Liu, T., Margaret, M.: Impala: A Middleware System for Managing Autonomic, Parallel Sensor Systems. In: Proceedings of the 9th ACM SIGPLAN Symposium on Principles and Practice of Parallel Programming, New York, NY, USA, vol. 38, pp. 107–118 (2003)
9. Fok, C.-L., Roman, G.-C., Lu, C.-Y.: Mobile Agent Middleware for Sensor Networks: An Application Case Study. In: Proceedings of the 4th International Symposium on Information Processing in Sensor Networks (IPSN 2005), Los Angeles, CA, USA, pp. 382–387 (2005)
10. Yong, Y., Johannes, G.: The Cougar Approach to In-Network Query Processing in Sensor Networks. ACM SIGMOD Record 31, 9–18 (2002)
11. Chavalit, S., Chaiporn, J., Shen, C.-C.: Sensor Information Networking Architecture and Applications. IEEE Personal Communication 8, 52–59 (2001)
12. Heinzelman, W.B., Murphy, A.L., Carvalho, H.S., Perillo, M.A.: Middleware to Support Sensor Network Applications. IEEE Network 18, 6–14 (2004)
13. Samuel, R.M., Michael, J.F., Joseph, M.H., Wei, H.: TinyDB: An Acquisitional Query Processing System for Sensor Networks. ACM Transaction on Database Systems 30, 122–173 (2005)
14. Marie, K., Jun, W.L., Yong, J.L., Jae-Cheol, R.: COSMOS: A Middleware for Integrated Data Processing over Heterogeneous Sensor Networks. ETRI Journal 30, 696–706 (2008)
15. Li, S., Lin, Y., Son, S.H., Stankovic, J.A., Wei, Y.: Event Detection Services Using Data Service Middleware in Distributed Sensor Networks. Telecommunication Systems 26, 351–368 (2004)
16. Nader, M., Jameela, A.: A Survey on Service-oriented Middleware for Wireless Sensor Networks. Service Oriented Computing and Applications 5, 71–85 (2011)
17. Kavi, K.K., Subramanian, R.K.: A Service-Oriented Component-Based Middleware Architecture for Wireless Sensor Networks. International Journal of Computer Science and Network Security 9, 174–182 (2009)

18. Li, Q., Li, C., Shi, J.: Research and Application on Application-oriented Event-Based Middleware of WSNs. In: 2010 Second International Conference on Networks Security Wireless Communications and Trusted Computing, pp. 498–501 (2010)
19. Emanuel, P., Miguel, A.F., Raul, M., Carlos, R.C., Juan, A.L., Samuel, R.M., Ferreira, P.J.S.G., Reis, M.J.C.S.: An Autonomous Intelligent Gateway Infrastructure for in-field Processing in Precision Viticulture. Computers and Electronics in Agriculture 78, 176–187 (2011)

Research on the Method of Feature-Based Multi-scale Vector Data Model

Yibing Dong, Jianyu Yang*, Chao Zhang, Dehai Zhu, Xingyue Tu, and Xianzhe Qiao

College of Information and Electrical Engineering,
China Agricultural University, 100083, Beijing, China
ycjyyang@cau.edu.cn, 294726701@qq.com

Abstract. Multi-scale representation of spatial data is a research focus in GIS, while building multi-scale data model is a key to implementing multi-scale representation of vector data. In view of the shortcomings of existing multi-scale data model in geographical cognition and special analysis, this paper puts forward a method of feature-based, and studies on it qualitatively from definition, description and extraction. Compared to conventional single-scale E-R model, in this paper, the key strategies of building multi-scale conceptual model are put forward. Deeply study and analysis are applied on abstraction and expression of multiple geometric characteristics, of multi-attribution, and of semantic relation among different scales, and the design of feature-based multi-scale conceptual model is realized. Finally, the object-oriented multi-scale logic model is researched, which lays a theoretical foundation for building the feature-based multi-scale vector data model.

Keywords: scale, feature-based, multi-scale representation, extended E-R model, object-oriented.

1 Introduction

Multi-scale representation of spatial data refers to different expression forms of one geographical entity in geometry, topology and properties, which arise from different scales (resolutions) of geographical entity in the computer storage, analysis and description [1]. With the continuing expansion of application fields in GIS, there is a growing demand, not only to observe, understand and describe the geographic phenomena in different scales, but also to analyze, manage and express multiple geographic representations.

In recent years, as the development of wavelet compression and image pyramid model, the problems of multi-scale raster data representation and visualization have been solved. However, the relationships among spatial elements are complex, the multi-scale representation and management of vector data have been an open problem [2]. At present, the research of multi-scale vector data representation mainly focus on four aspects, including formal description of the multi-scale data model, design and

* Corresponding author.

D. Li and Y. Chen (Eds.): CCTA 2012, Part II, IFIP AICT 393, pp. 280–289, 2013.

implementation of data structure, link and update mechanism between different levels and consistency maintenance of different expression versions. Specialists and scholars have long committed to improve previous algorithms and data structures, and made considerable achievements in reducing data redundancy, improving the efficiency of data access and maintaining consistency of spatial data [3-7]. In addition, a series of concept were proposed, such as "critical scale", "incremental", providing new ideas and methods for the representation of multi-scale vector data [8-10]. The existing method of multi-scale representation is broadly divided into four categories: multi-scale explicit storage, initial scale changes accumulation, Key scale function evolution and Senior scale automatic generalization. Their basic ideas, advantages and disadvantages are shown in Table 1.

Table 1. Methods and technical strategies of multi-scale representation

Technical Strategy	Basic Idea	Advantage and Disadvantage
Multi-scale explicit storage	Store the pretreated data in different scales explicitly, and when is needed, the data at a certain scale can be extracted immediately	*Advantage:* data structure is simple and easy to access data *Disadvantage:* data redundancy large and scale change discrete, cannot get data at any scale
Initial scale changes accumulation	Store the coarsest data representation and the change data between adjacent scales, therefore, data in any scale can be obtained by accumulated the change on initial data	*Advantage:* save storage spaces and improve access efficiency *Disadvantage:* scale change discrete, cannot get any scales of data and present the difficulty of maintaining consistency within the multiple versions of the maps
Key scale function evolution	Store a series of "key scale" data representation in hierarchical structure, and can derive any data representation between adjacent scales by interpolation or Morphing function	*Advantage:* data redundancy less and can get data at any scale *Disadvantage:* key scale of the different objects is inconsistent and difficult to select
Senior scale automatic generalization	Store the most detailed data representation and using on-line automatic generalization, achieve the export of data representation at any scale	*Advantage:* data redundancy at least and can get data at any scale *Disadvantage:* subject to the automatic generalization, this method cannot be achieved right now

As current GIS lack methods for on-line automated map generalization, the senior scale automatic generalization (the last method in Table 1) remains a challenging area of research, so we cannot get data at any scale [11]. On the other hand, the multi-scale explicit storage (the first method in Table 1) is widely used with simple structure, but the data storage is too large. Therefore, the initial scale changes accumulation (the second method in Table 1) and key scale function evolution (the third method in Table 1) are used as compromise methods for multi-scale vector data representation. Found by previous studies, specialists discuss multiple representation issues from a data-modeling point of view, with emphasis placed hierarchical structure to organize and store data at different scales. But the hierarchical structure has introduced additional problems, including ignore the essential characteristics and intrinsic links of geographical phenomena, educe the efficiency of spatial analysis and complex operation, and bring a certain of error [12, 13].

Based on the above, this paper introduces a "feature-based" thought into the progress of multi-scale modeling, and studies on it qualitatively from definition, description and extraction. It also includes the design of multi-scale conceptual model, which not only overcome the insufficient of previous model, but also achieve the cognitive and description of the geographical world.

2 Qualitative Research of Multi-scale Feature

2.1 Scale Definition

Typically, the scale is a measure of study object, including two parts of space scales and time scales. The spatial scale refers to the size of study scope (geographical scope) or the level of detail (size and level of geographic resolution). The time scale indicates the duration and frequency of study.

In different study areas, the meaning and application of scale is various. In cartography and GIS, scale is often defined as a ratio of the distance on the map and the actual length of a certain object. Among them, the distance is fixed, the relationship between "space scales" in the actual geographical space and "scale" is same as the relationship between denominator and business [10]. In this paper, large-scale refers to larger spatial areas, corresponding to low resolution. On the contrary, small-scale corresponds to smaller spatial areas and high resolution, which can reflect complex phenomena.

2.2 Feature Description

Traditionally, feature was a graphical representation of the real world. By the application of object-oriented methods in GIS, the concept of feature was re-defined as an abstract of real-world phenomena; using these features, all the geographical phenomena can be represented [14]. Geographic feature is an abstract of geographic entity, which includes geographic entity and further descriptions of them.

According to the definition, the feature F in the horizontal is divided into three parts: entity feature EF, attribute feature AF and relation feature RF. Thus, the feature can be shown as:

$$F=<EF, AF, RF> \quad (1)$$

Where EF is a description of the spatial position and geometry of geographical entities, AF describes the composition and characteristics of them, such as name, type, while RF is to describe the relationship between entities, including topological and semantic relations.

Through considering the domain of information representation, the spatial positions and distributions of entities can be represented in the geometric domain, as well as the semantic domain can explain the logic pattern of non-spatial information descriptions. Thus, in vertical, feature F is divided into geometric domain feature GDF and semantic domain feature SDF. Combine of two definitions above, features F can be complete expressed as:

$$GDF=< GEF, GAF, GRF> \qquad SDF=<SEF, SAF, SRF> \tag{2}$$

Where, respectively, GEF, GAF, SRF represents the geometric characteristics of entities, attributes and relationships, and SEF, SAF, SRF is the semantic characteristics of entities, attributes and relationships.

2.3 Multi-scale Feature Extraction

In different scales, a geographic entity will show different features. In this paper, the extraction method of multi-scale features is proposed, which can lay foundations for the establishment of multi-scale vector data model.

According to the definition and description of feature, some features at a variety of scales can be extracted, such as location, geometry, attribute, semantic relationship (shown in Figure 1). The extracted multi-scale features mainly show as the following three aspects:

(1) In different scales, a certain entity shows different geometries or abstract geometric types. For example, with increasing scale, the abstract geometry type of a building will change from surface to point.

(2) The topologic relations of a certain entity in different scales must be consistent, but the semantic relations between different geographical entities will change. For example, as the scale increases, two school buildings and a track field aggregate into a school.

(3) In different scales, the attributes expressed of one certain geometry will be different.

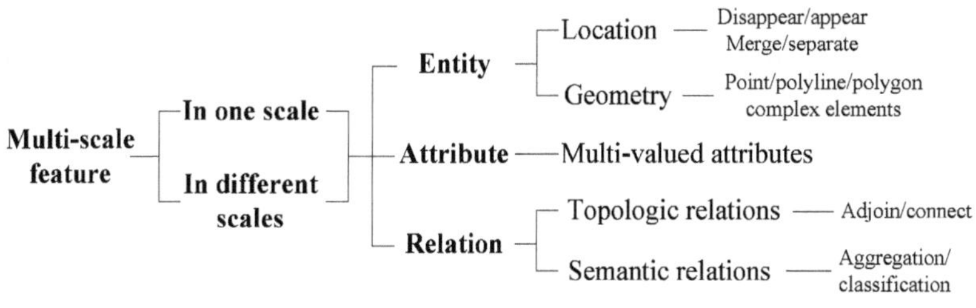

Fig. 1. Extraction of multi-scale features

3 Feature-Based Multi-scales Conceptual Modeling

In GIS, the process of building data model is an abstract to the real world. Like single-scale models, multi-scale models consist of three levels: conceptual model, logical model and physical model. The abstract level of three models is gradually deepened. Now, based on the multiplicity of geometry, attribute and relations in different scales, how to properly abstract, definite and describe the geographic entities and their relations is still a problem, which must face in multi-scale conceptual modeling [15].

3.1 Choice of Conceptual Modeling Tool

The conceptual model is defined as a description of entities in the real world, and there are a lot of methods for conceptual modeling. Entity-Relation (E-R) model is the most famous method, and it can use diagram to represent the relationships between entities and attributes. A basic E-R model, consisted of entities, attributes and relations, can be used to represent geographical features. However, the traditional GIS conceptual model only supported single-level expression that cannot express complex changes between different scales [15].

In this study, the traditional E-R model is been expended to meet the demand of multi-scale representation and application. This paper also introduces a series of abstract concept, like generalization, aggregation, classification, to support multi-valued description of attributes, relations and others.

3.2 Multi-scale Extended E-R Modeling

Multi-scale extended E-R model contains three elements: the entities, attributes and relations. According to the composition and law of multi-scale features, this paper is applied to study the multiple geometric features, multi-valued attributes and the change of semantic relations in different scales.

3.2.1 Abstraction and Expression of Multiple Geometrical Features

Borges said, the focus of building a conceptual model was to make a clear expression of geographical entity, and the type of abstract geometry would affect the spatial relationships between different entities [1]. In this paper, geographic entities were abstracted as point, polyline, polygon and complex elements formed by these simple elements. In any scale, entity can be represented by these elements. In addition, the concept of "abstract feature class" is proposed. So the geometric features of a certain entity in different scales can be unified in Figure 2.

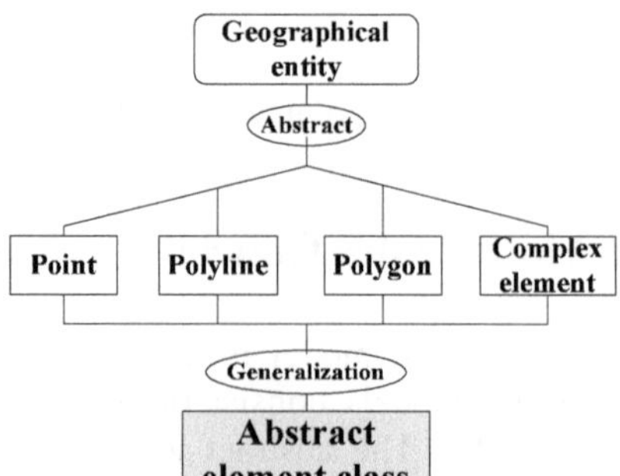

Fig. 2. Abstraction and expression of multiple geometrical features

3.2.2 Extraction and Description of Multi-valued Attributes

In order to reflect the abstract information of entities, in the stage of conceptual modeling, it is very important to extract the relationships between attribute domains or attribute value domains. With the abstraction level dropping in different scales, the attributes of a same entity is considered to be redefined, and the attribute values is consider to be refined, Parent said [16].

(a) (b) (c)

Fig. 3. Multi-scale representation of Haidian District in Beijing. The highlighted part in (a) is Haidian district, the highlighted part in (b) is Sujiatuo Town, and the highlighted part in (c) is Beianhe Village.

For example, in Beijing's Handian District, the gathered characteristics of geographic elements abstracted at different scales is described clearly (shown in Figure 3). Simultaneously, the attribute domains and attribute value domains of highlighted elements in each layer are extracted (Figure 4). Therefore, the fields of attribute domains (XZQM, JDMC, and JFMC) are different and unrelated, which indicates the geographic elements exhibit different attributes in a variety of scales.

Fig. 4. Extraction and description of multi-valued attributes. XZQM, JDMC, JFMC is the name field of highlighted element in (a), (b) and (c) respectively; and QHDM, JDDM, JFDM is the code field of highlighted element in them.

However, the refining fields of attribute value domains (QHDM, JDDM, JFDM), not only reflect the different attributes of geographic elements represented in different scales, but also link multi-scale data together to establish multi-scale hierarchical structure. Besides, the attribute value domains can be used to extract multiple attributes.

3.2.3 Extraction and Representation Semantic Relations

Although the geometric shapes and details in different scales are different, the topologic relations between geographic entities should be consistent. Therefore, the semantic relations should be extracted to express the scaling relations between different scales. The four basic concepts of semantic relations are described as follows:

(1) Aggregation. Aggregation can combine primary objects with different characteristics into a higher level object—composite object. The composited object can derive some new characteristics from primary objects, but cannot own the characteristics inherited from primary objects. The relationships between composite objects and primary objects mainly represent as "part of". For example, the school is a composite object, and school buildings, roads, playgrounds are the parts of it.

(2) Generalization. Similar to aggregation, generalization is an operation that can extract general attributes or characteristics from common class to establish a high-level class. So the relationship between high-level class and common class can be represented as "is a".

(3) Classification. Classification is a method which can form a hierarchic class structure. In this structure, every class is consisting of objects with the same characteristics, and there is a "is instance of" relationship between object and its class.

(4) Association. In one scale, some objects can be combined into an abstract object. The relationship between objects and abstracted object is represented as "is member of".

(a) (b) (c)

Fig. 5. Semantic relation of multiple geographical elements of Haidian District in Beijing

In this paper, some methods are applied to extract multiple semantic relations of Haidian District in Beijing (shown in Figure 5). For example, the elements in neighborhood layer (Figure a) are classified by street code (JDDM). Then the street layer (Figure b) can be achieved by generalizing all classes. In the same way, through classifying and generalizing the elements by zoning code in street layer (Figure b), the block layer (Figure c) can also be gained. Each element in the layer can be simple element or composite element aggregated from geographical entities with different characteristics. Using the four basic semantic relations and making a certain expansion of them, it is possible to establish scaling relation between different scales and thus to connect data representations together.

4 Feature-Based Object-Oriented Multi-scale Logic Modeling

In object-oriented method, the concept of "class" is a collection of objects with common attributes and methods, where the object is a package of its stages and behaviors. For an object, the stage is a collection of attribute values and the behavior is a collection of methods. In this paper, the concept of feature and class are similar, through class the geographical features can described formally. Shown in Figure 6, the object is an instantiation of class, and the feature is an abstraction of entity. Therefore, feature-based multi-scale logic modeling can be achieved by object-oriented data model.

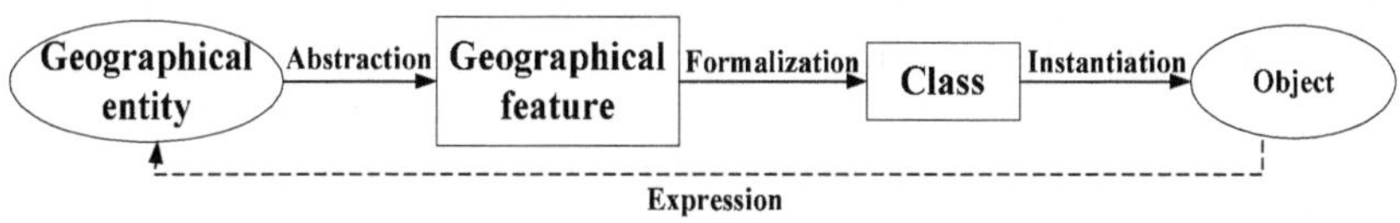

Fig. 6. Formal description of geographical feature

In multi-scale logic modeling, feature is defined as a unity of geographical data and functions, corresponding to the "class" of the object-oriented data model. The "object" described by "feature class" is a geographical entity. Combining the design ideas of feature-based, as well as the method of object-oriented, a unified multi-scale data model can be build from conceptual modeling to logic modeling.

In this study, the feature object FO is divided into geometry object GO and semantic object SO. GO is composed of the spatial locations and relations of geographic entities, which indicates the geometric features. Similar to GO, SO represents the semantic features, which include attributes, non-spatial relations and others (shown in Figure 7). The GO and SO are not independent of each other, and they are encapsulated into FO. Through this method, the features of geographical entities can be represented completely.

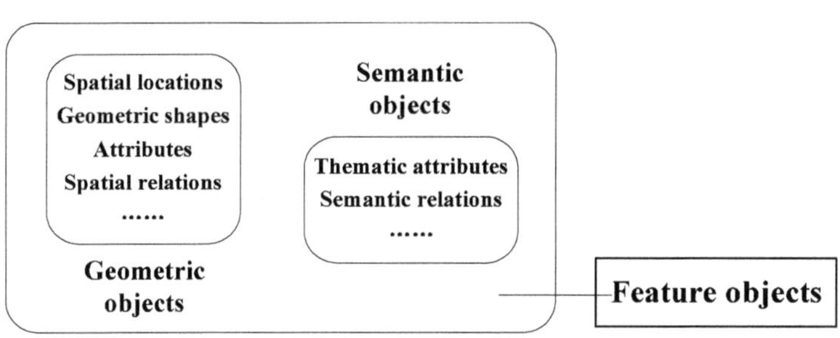

Fig. 7. Composition of feature object

Therefore, the SO are extracted to organize spatial representations in different scales. The semantic relation in SO can be used as the basic description object. Through these semantic relations explained above, we can link a series of spatial representations together, and establish the spatial indexing mechanism efficiently by semantic features. Thus, the multi-scale logic model can be achieved preliminary.

5 Conclusions and Outlooks

The multi-scale vector data model is an important part of multi-scale spatial data representation. However, in terms of geographical cognition and spatial analysis, the existing multi-scale hierarchical data models had some lacks. For the requirements and applications of multi-scale data representation, this paper puts forward the thought of feature-based, and studies on it qualitatively from definition, description and extraction. The design of feature-based multi-scale conceptual model was realized by extended E-R model. Finally, the feature-based object-oriented multi-scale logic model was researched, which lays a theoretical foundation to building the feature-based multi-scale vector data model. From this study, we can draw the following conclusions:

(1) Qualitatively study and deeply analysis of multi-scale features, not only realize the cognition and representation of the real world, but also obtain a more integral abstraction and description of geographical entity to maintain the consistency of topological relations between different scales.

(2) The thought of feature-based and object-oriented is consistent. Using object-oriented method, the cognitive paradigm of geographical space and the representation of geographical data can be unified together naturally. Therefore, a unified multi-scale data model can be build from conceptual modeling to logic modeling.

Next step, the major works are: (1) Combined with experimental data, the quantitative research will be carried out, which will be extract multi-scale features and change rules of geographical entities in different scales. (2) Guided by the conceptual model in this paper, we will deeply study the implement of feature-based object-oriented logic model, including the establishment of object-oriented data structure and spatial indexing mechanism, the maintenance of level connectivity relation and consistency. The establishment of feature-based multi-scale data model will lay foundations for the representation of multi-scale vector data.

Acknowledgments. This research was funded by the National Natural Science Foundation Project "Research on Compression and Progressive Transmission of Vector Data in the Network Environment" (NO: 4171309). The authors would like to thank the anonymous reviewers for the constructive comments and suggestions.

References

1. Li, L., Wu, F.: Multi-scale Spatial Data Representation Model and Its Visualization. Science Press, Beijing (2005)
2. Cheng, C., Lu, F.: A Bi-hierarchical Multi-scale Model and Indexing Approach for Large Vector Data Representation. Journal of Image and Graphics 14(6), 1012–1017 (2009)

3. Ai, T., Van Oosterom, P.: GAP-tree Extensions Based on Skeletons. In: Advances in Spatial Data Handling, pp. 501–514 (2002)
4. Zhou, M., Bertolotto, M.: Efficiently Generating Multiple Representations for Web Mapping. In: Li, K.-J., Vangenot, C. (eds.) W2GIS 2005. LNCS, vol. 3833, pp. 54–65. Springer, Heidelberg (2005)
5. Haunert, J.H., Dilo, A., Van Oosterom, P.: Constrained set-up of the tGAP structure for progressive vector data transfer. Computers & Geosciences 35(11), 2191–2203 (2009)
6. Yang, B., Li, Q.: An Algorithm for Progressive Transmission Vector Map Data over the WWW. Acta Geodaetica Et Cartographic Sinica 34(4), 355–360 (2005)
7. Ren, Y.: The Research of Vector Data Progressive Transmission. Institute of Remote Sensing Applications Chinese Academy of Sciences, Beijing (2007)
8. Zhou, S., Jones, C.B.: Design and Implementation of Multi-scale Databases. In: Jensen, C.S., Schneider, M., Seeger, B., Tsotras, V.J. (eds.) SSTD 2001. LNCS, vol. 2121, pp. 365–384. Springer, Heidelberg (2001)
9. Follin, J.-M., Bouju, A., Bertrand, F., Stockus, A.: An Increment Based Model for Multi-resolution Geodata Management in a Mobile System. In: Li, K.-J., Vangenot, C. (eds.) W2GIS 2005. LNCS, vol. 3833, pp. 42–53. Springer, Heidelberg (2005)
10. Ai, T., Chen, J.: Key Issues of Multi-Scale Representation of Spatial Data. Geomatics and Information Science of Wuhan University 30(4), 1–9 (2005)
11. Ai, T., Li, J.: The Lifespan Model of GIS Data Representation over Scale Space. Geomatics and Information Science of Wuhan University 35(7), 757–781 (2010)
12. Sang Hun, L., Kunwoo, L.: Simultaneous and incremental feature-based multiresolution modeling with feature operations in part design. Computer Aided Design 44(5), 457–483 (2012)
13. Li, W., Li, H., Liang, R., et al.: Research on Feature-based Spatial Data Model. Geomatics & Spatial Information Technology 01, 42–45 (2010)
14. Yin, Z.: Feature-based Data Models in GIS. Management Geological Science and Technology 02, 50–53 (2002)
15. Wang, Y., Chen, J., Jian, J.: Multi-Scale Conceptual Model for GIS Geographical Feature. Journal of China University of Mining & Technology 32(4), 376–382 (2003)
16. Parent, C.: Modeling Spatial Data in the MADS Conceptual Model. In: The Proceedings of the International Symposium on Spatial Data Handling. Springer, Berlin (1998)

The Current and Future Potential Geographical Distribution of the Italian Locust, *Calliptamus Italicus* (Linnaeus) (Orthoptera: Acrididae) in China*

Yujia Qin[1], Zhihong Li[1,**], Li Zhao[2], Glenn Fowler[3], and Yan Fang[1]

[1] Department of Entomology, China Agricultural University, Beijing 100193, P.R. China
[2] College of Agronomy, Xinjiang Agricultural University, Wulumuqi 830052, P.R. China
[3] USDA-APHIS-PPQ-CPHST-PERAL, 1730 Varsity Drive, Suite 300, Raleigh, NC 27606
lizh@cau.edu.cn

Abstract. The Italian Locust, *Calliptamus italicus* (Linnaeus) (Orthoptera: Acrididae), is one of the most harmful species to the grasslands in Xinjiang province, China. It attacks a variety of plants in *Compositae*, *Chenopodiaceae* and *Gramineae*. We used CLIMEX and ArcGIS to determine the potential geographical distribution of *C. italicus* in China under current and future climatic scenarios. Our analysis predicted that under current climatic conditions, *C. italicus* has a wide potential distribution in areas north of the Yangtze River in China. With the climatic changes, the potential range for *C. italicus* was projected to expand, and the suitability levels of most areas would increase, especially in areas with optimal climatic conditions. We therefore suggest that surveys should be enhanced in the areas suitable for *C. italicus* to facilitate early detection and reduce potential damage.

Keywords: *Callipatamus italicus* (L), CLIMEX, ArcGIS, potential geographical distribution, climatic change.

1 Introduction

Callipatamus italicus (Linnaeus) is one of the main insect pests of grasslands in Xin Jiang, and attacks a wide range of host plants including 45 species in 17 families. The nymphs of *C. italicus* prefer *Artemisia frigida* Willd. and *Carex onoei* Franch. *et* Sav., while the adults prefer not only *A. frigida* Willd., but *Salvia deserta* Schang and *Medicago falcate* Linn. [1]. The Italian locust has become the most common and damaging plant pest in Kazakhstan, regularly invading millions of hectares and causing crop damage of great economic importance [2]. In 1999, *Calliptamus italicus* was introduced to from Kazakhstan and Russia into Xinjiang province, China. It subsequently damaged approximately 2 million hectares in 2000 and 2007 in Xinjiang province [1, 3].

* This study received financial supports from the Special Fund for Agro-scientific Research in the Public Interest (201003079).

** Corresponding author.

D. Li and Y. Chen (Eds.): CCTA 2012, Part II, IFIP AICT 393, pp. 290–298, 2013.

The distribution area of *C. italicus* stretches from Western Europe to the meadow-steppes of Central Asia in Siberia and in Kazakhstan and its neighboring countries. Its range includes countries along the northern and eastern edge of the Mediterranean Sea, Central Europe and Central Asia up to Mongolia and western Siberia [4]. These regions include the former USSR, Russia Federation, Spain, Italy, France, Ukraine, northern Africa; Siberia, Kazakhstan, Uzbekistan, Turkey, Mongolia, Iran, and Afghanistan. In China, it occurs in Xinjiang, Gansu, Qinghai, Shaanxi, and Hebei provinces [5, 6].

In this study we modeled the potential range of *C. italicus* in China based on climate suitability. Because pest ranges are likely to shift in response to changes in temperature and precipitation, we also assessed the effects of climatic change on its potential distribution.

2 Materials and Methods

2.1 Overview Of CLIMEX Model and ArcGIS

We used CLIMEX 3.0 to model the potential geographical distribution of *C. italicus* in China under current and future climate scenarios. CLIMEX model is a predictive modeling software that has been used to predict the potential geographical distribution of many pests including: *Ceratitis capitata* (Wiedemann) [7], *Bactrocera zonata* [8], *Frankliniella occidentails* Pergande [9], *Hyphantria cunea* [10], *Solenopsis invicta* [11], *corynespora* Leaf Fall Disease [12], *Nassella neesiana* [13].

CLIMEX is a dynamic model [14] that integrates the weekly responses of a population to climate using a series of annual indices. The model combines the growth index (GI) and the stress indices (SI) into an overall ecoclimatic index (EI). EI describes the climatic suitability of a location for a species as a number between 0 and 100. In this study, we classified the EI into four categories of climatic suitability for *C. italicus*: unsuitable (EI=0), marginal (EI=0-10), suitable (EI=10-20) and optimal (EI>20) [14, 15].

To visualize the results, the CLIMEX output was 'loose-coupled' [15] to a geographical information system (ArcGIS). It can present CLIMEX results in the same map as a species distribution or other information that provides spatial context [16]. Both CLIMEX ver. 3.0 and Arc/INFO ver. 9.3 were purchased by the Plant Quarantine Lab of China Agriculture University.

2.2 Meteorological Databases and Climatic Change

Two climate databases were used in our model. We first used the CLIMEX standard meteorological dataset, based on 30 year averages from 1961 to 1990, to create an initial model output. This dataset contains information for 2500 weather stations worldwide. Secondly, we used a regular gridded dataset of the normal climate for the same period, TYN SC 2.0 (http://www.cru.uea.ac.uk/~timm/grid/TYN_SC_2_0.html) [17] to fine-tune the model output. The dataset consisted of 67,420 weather stations spaced on a 0.5° latitude×0.5° longitude regular grid for significant land areas worldwide.

For the climate change scenario, we modeled an average annual temperature increase of 2.3-3.3℃and, an annual mean precipitation increase of 5-7% by 2050 in China [18].

2.3 Geographic Data Sets

For geographic background datasets, we downloaded the electronic map of China national boundary, province boundary and national grassland distribution from http://nfgis.nsdi.gov.cn/ at a scale of 1:4,000,000.

2.4 Comparing Current and Climatic Change Scenarios

In order to visualize the differences in the potential geographical distribution for *C .italicus* under current and future climate conditions, the respective suitable land and grassland areas for each EI category were calculated in ArcGIS for each scenario.

2.5 Fitting CLIMEX Parameters

To fit the CLIMEX model for *C .italicus*, the parameters were manually and iteratively adjusted until the simulated geographical distribution, as estimated by the EI values, coincided with the species' known native distribution and the reported description of its range. Due to the lack of the data we used the information for the related species *Melanoplus sanguinipes* (Fabricius) [19] to fit the initial values. Parameters used in the CLIMEX model are presented in table 1.

Temperature Index

The lower developmental threshold of *C. italicus* was 15.5°C [20]. In this paper, the minimum temperature for development (DV0) was set at 15.50°C. The lower optimum temperatures for *C. italicus*'s populations were set at 17°Cto correspond to its distribution in northern Africa. Because this species is now established in Xinjiang province, so we set the upper threshold temperature to 35°C [21].

Moisture Index

To fit the known *C. italicus*'s distribution in Xinjiang province, the lower soil moisture limit for development (SM0) was set to 0.01. We used this value because locusts normally require a desiccated environment for embryonic development. They survive well when the soil moisture in May is between 0.03-0.3 [22]. So in this paper, the lower and upper limits for optimal soil moisture (SM1 and SM2) were set to 0.03 and 0.3 respectively. Based on the *C. italicus*'s distribution in the central Europe, the upper soil moisture limit for development (SM3) was set to 0.6.

Table 1. Parameters used in the CLIMEX model for the *C. italicus*

Parameter	Initial value	Final value
DV0 Lower threshold temperature	10.0°C	15.5°C
DV1 Lower optimum temperature	16.0°C	17.0°C
DV2 Upper optimum temperature	28.0°C	28.0°C
DV3 Upper threshold temperature	32.0°C	35.0°C
SM0 Lower threshold of soil moisture	0.02	0.01
SM1 Lower limit of optimum soil moisture	0.05	0.03
SM2 Upper limit of optimum soil moisture	0.30	0.30
SM3 Upper threshold of soil moisture	0.70	0.60
DPD0 Diapause induction day length	11h	10h
DPT0 Diapause induction temperature (average weekly minimum)	11.0°C	11.0°C
DPD1 Diapause termination temperature (average weekly minimum)	3.0°C	6.0°C
DPD Diapause development days	120	90
DPSW Summer or winter diapause ('0' for winter diapause, '1'for summer diapause)	0	0
TTCS Cold stress temperature threshold	-18.0°C	-18.0°C
THCS Cold stress accumulation rate	-0.0004	-0.0008
TTHS Heat stress temperature threshold	35.0°C	35.0°C
THHS Heat stress accumulation rate	0.008	0.0005
SMDS Dry stress soil moisture threshold	0.020	0.010
HDS Dry stress accumulation rate	-0.003	-0.003
SMWS Wet stress soil moisture threshold	0.7	0.6
SMDS Dry stress soil moisture threshold	0.001	0.003

Diapauses Index

Diapause is an important biological characteristic that allows locust to survive harsh environmental conditions. *Calliptamus italicus* overwinters below ground in the egg stage[23]. Based on its distribution in Iran, Afghanistan [4], Tibet and Sichuan province, we set the diapause induction day length to 10. We adjusted the diapauses parameters to identify suitable in areas in Xinjiang province based on *C. italicus*'s distribution, north of the Mediterranean Sea, and north of Africa, based on that distribution, we set the diapause termination temperature and the diapause development days to 6 and 90.

Stress Parameter

We adjusted the stress parameters to account for *C. italicus'* global distribution. This was done to reduce the likelihood that out prediction maps for climatic suitability would not include areas where *C. italicus* cannot survive.

3 Results

3.1 Potential Geographical Distribution under Current Climate

The CLIMEX parameter values were calibrated until the predicted geographical distribution agreed with the observed global distribution for *C. italicus*. With the current climate scenario our model predicted that North America, southern South America, most of the Europe, the edge of the Mediterranean Sea, isolated parts of north and south Africa, Central Asia, West Asia, East Asia and southern Australia were potentially suitable for *C. italicus* establishment (Figure 1). Known regions of occurrence, such as Italy, Spain, and Central Europe, with EIs between 0 and 20 were considered to have marginally suitable to suitable climates for establishment. Locations such as Kazakhstan, western Siberia, and Xinjiang where high population levels occur had EIs>20, and were considered optimally suitable for *C. italicus*.

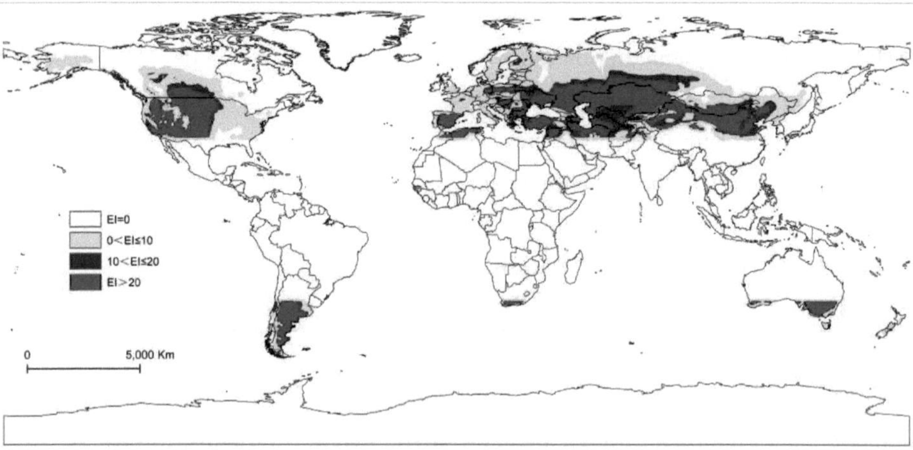

Fig. 1. The global Ecoclimatic Index (EI) values for *C. italicus* under the current climate scenario (1961-1990 averages) (□, unsuitable (0.00); ■, marginal (0.01–10); ■, suitable (10–20); ■, optimal (20.00+).)

Our model predicted that, under current climatic conditions, *C. italicus* could establish north of the Yangtze River in approximately 29% of China (Figure 2a). Optimal climate conditions occurred in most of Beijing, Tianjin, Xinjiang, Inner Mongolia, Gansu, Ningxia, Shaanxi, Shanxi, Hebei and small parts of Qinghai, Jilin, Shandong, and Henan. Climate conditions are projected to be marginal in parts of Heilongjiang, Jilin, Liaoning, Inner Mongolia, Gansu, Xinjiang, Qinghai, Shaanxi, Henan, Shandong, Anhui, Jiangsu, Sichuan and Hubei. The remaining areas south of the Yangtze River were unsuitable for the establishment of *C. italicus.* The grassland GIS layer further refined the model and estimated that (insert at-risk provinces here) were at risk for establishment based on climate and host presence (Figure 2b).

3.2 Potential Geographical Distribution under Future Climate

The climate change scenario indicated that China is likely to become hotter and wetter by 2050. As a result, the potential range for *C. italicus* is projected to expand. Under the climate change scenario, the percentage of China's land area and grassland area at risk for *C. italicus* establishment increased from 29% to 31% and 96% to 98%, respectively. ,

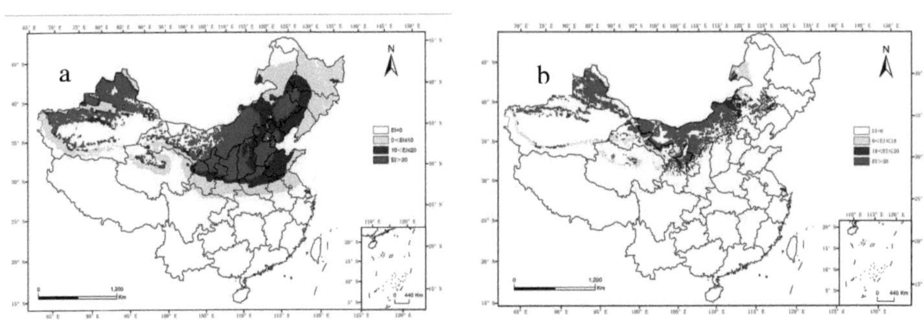

Fig. 2. The Ecoclimatic Index (EI) values for *C. italicus* in China under the current climate scenario (1961-1990 averages) with vegetation distribution (□, unsuitable (0.00); ■, marginal (0.01–10); ■, suitable (10–20); ■, optimal (20.00+).) (a) for land area, b) for grassland area)

Also, the overall climate suitability of China's land area is projected to increase with climate change, especially in those areas with optimal climatic conditions (Figures 3). Our model predicted that by 2050, the proportion of the marginal areas will decrease from 9% to 7%, suitable areas will decrease from 8% to 6%, and optimal areas will increase 11% to 17%. Therefore, the overall suitable area will increase by 2% by 2050. For grassland, the proportion of the marginal areas will decrease from 10% to 4%, suitable areas will decrease from 14% to 7%, and the optimal areas will increase from 71% to 86%, so the overall suitable area will increase by 2% by 2050 (Figure 4).

4 Discussion

In this study, we combined Climex EI values with grassland host presence to increase the precision of the predictive maps. However, the model predictions discussed above surely have some limitations.

Additional factors like dispersal, host density, competition, and the effects of natural enemies [24], soil type, geographical barriers and human activities could be incorporated to generate more robust predictions. Also, the use of model evaluation techniques like the ROC curve [25] could be used to determine which modeling approaches, e.g. Maxent, GARP, DIVA-GIS and CLIMEX, are most appropriate for predicting the establishment potential of pests like *C. italicus* in a particular area.

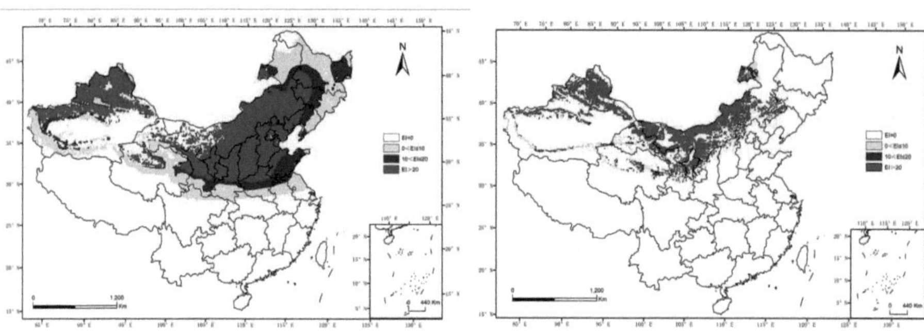

Fig. 3. The Ecoclimatic Index (EI) values for *C. italicus* in China in 2050 with vegetation distribution (□, unsuitable (0.00); □, marginal (0.01–10); ■, suitable (10–20); ■, optimal (20.00+).) (a) for land area, b) for grassland area)

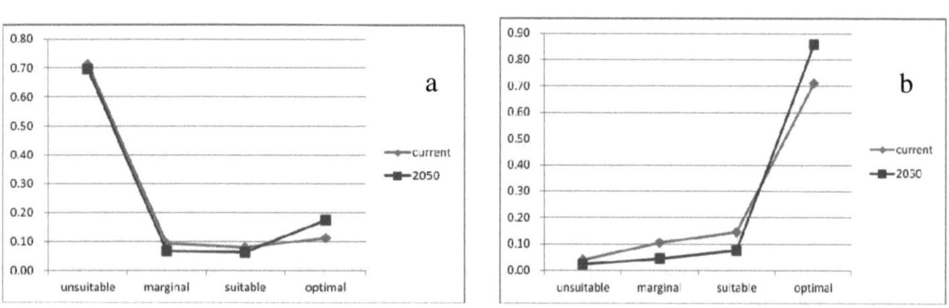

Fig. 4. Changes in area by EI category for total land area and grassland area under current and climate change scenarios (a) for land area, b) for grassland area)

Acknowledgements. The authors would like to thank Zhimei Li and the other members of the Plant Quarantine Laboratory of China Agricultural University (CAUPQL). This study received financial support from the Special Fund for Agro-scientific Research in the Public Interest (201003079).

References

1. Xue, Z.P., Zhang, Q., Yashen, S.L., Wang, G.J., Abudouwaili, Y.M.M., Xiao, H.W.: Food selection of *Calliptamus italicus* and loss assessment for the grassland. Plant Protection 36(1), 95–96 (2010) (in Chinese)
2. Antonow, A.G., Kambulin, V.E.: Forecasting seasonal dynamics of the Asiatic migratory locust using the *Locusta migratoria migratoria–Phragmites australis* forecasting system. In: New Strategies in Locust Control, pp. 81–89 (1997)
3. Zhang, Y., Gao, S., Yashen, S.L., Cao, G.C., Zhang, Z.H.: Numerical analyses on morphometrics gregarious and solitarious phase of *Calliptamus italicus*. Applied Entomology 48(4), 854–861 (2011) (in Chinese)
4. `http://www.fao.org/ag/locusts-CCA/en/1010/1018/1075/`
5. Zhang, S.M., Zhao, Y.X.: The geographical distribution of agricultural and forest insects in China. China Agriculture University Press, Beijing (1996) (in Chinese)
6. Crop Protection Compendium (2007)
7. Worner, S.P.: Ecoclimatic assessment of potential establishment of exotic pests. Journal of Economic Entomology 81, 973–983 (1988)
8. Ni, W.L., Li, Z.H., Chen, H.J., Wan, F.H., Qu, W.W., Zhang, Z., Kriticos, D.J.: Including climate change in pest risk assessment: the peach fruit fly, *Bactrocera zonata* (Diptera: Tephritidae). Bulletin of Entomological Research 10, 1–11 (2011)
9. Cheng, J.F.: The potential geographical distribution of *Frankliniella occidentails* Pergande based on CLIMEX and GIS in China. China Plant Protection 27(1), 16 (2007) (in Chinese)
10. Lin, W.: The potential geographical distribution of *Hyphantria cunea* in China. China Agriculture University, Beijing (1991) (in Chinese)
11. Chen, L.: The potential geographical distribution of *Solenopsis invicta* in China. Chinese Academy of Agricultural Sciences, Beijing (2007)
12. Qu, W.W., Li, Z.H., Huang, G.X., Lin, C.H., Ni, W.L.: The current and future potential geographic range of *Corynespora* Leaf Fall Disease in China. Sensor Lett. 10, 439–446 (2012)
13. Bourdôt, G.W., Lamoureaux, S.L., Watt, M.S., Manning, L.K., Kriticos, D.J.: The potential global distribution of the invasive weed *Nassella neesiana* under current and future climates. Biol. Invasions 6 (2010) 10.1007/s10530-010-9905-6
14. Kriticos, D.J., Sutherst, R.W., Brown, J.R., Adkins, S.W., Maywald, G.F.: Climate change and biotic invasions: a case history of a tropical woody vine. Biological Invasions 5, 145–165 (2003)
15. Kriticos, D.J., Sutherst, R.W., Brown, J.R., Adkins, S.W., Maywald, G.F.: Climate change and the potential distribution of an invasive alien plant: *Acacia nilotica ssp.* indica in Australia. Journal of Applied Ecology 40, 111–124 (2003)
16. Sutherst, R.W., Maywald, G.F., Kriticos, D.J.: CLIMEX Version 3: User's Guide (2009), `http://www.Hearne.com.au`
17. Kriticos, D.J., Alexander, N.S., Kolomeitz, S.M.: Predicting the potential geographic distribution of weeds in 2080. In: Preston, C., Watts, J.H., Crossman, N.P. (eds.) Proceedings of the Fifteenth Australian Weeds Conference, Adelaide, Australia, September 24-28, pp. 27–34. Weed Science Society of Victoria (2006)
18. Qin, D.H., Chen, Y.Y.: Climate and environment evolution in China. Science Press, Beijing (2005) (in Chinese)

19. Olfert, O., Weiss, R.M., Kriticos, D.J.: Application of general circulation models to assess the potential impact of climate change on potential distribution and relative abundance of *Melanoplus sanguinipes* (Fabricius) (Orthoptera: Acrididae) in north American. Psyche, 1–9 (2011)
20. Wumaerbie, K., Xiong, L.: Determination of developmental zero and degree-days to complete one generation of Article black car locust, *calliptamus italicus, Gomphocerus sibiricus*. Xinjiang Animal Husbandry (suppl.), 30–31 (2007) (in Chinese)
21. Xi, R.H., Liu, J.P., Zhang, L., Ma, L.X.: The relationship between locusts lay eggs and climatic factor. Entomology Knowledge 28(2), 76–78 (1991) (in Chinese)
22. Guo, F., Chen, Y.L., Lu, B.L.: The biology of the migratory locusts in China. Shandong Science and Technology Press, Shandong (1991) (in Chinese)
23. Zhang, Q., Qiao, Z., Xiong, L., Bahatiyaer, D.W.T., Zhao, Y., Dang, H.C., Zhang, X.S., Xiao, H.W.: Biological characteristics of *calliptamus italicus*. Xinjiang Agriculture Sciences 6, 256 (1995) (in Chinese)
24. Baker, R.H.A., Sansford, C.E., Jarvis, C.H., Cannon, R.J.C., MacLeod, A., Walters, K.F.A.: The role of climate mapping in predicting the potential geographical distribution of non-indigenous pests under current and future climates. Agriculture Ecosystems and Environment 82, 57–71 (2000)
25. Wang, Y.S., Xie, B.Y., Wan, F.H., Xiao, Q.M., Dai, L.Y.: Application of ROC curve analysis in evaluating the performance of alien species' potential distribution models. Biodiversity Science 15(4), 365–372 (2007)

A New Navigation Line Extraction Method for Agriculture Implements Guidance System

Mingxuan Li, Man Zhang*, Haiyan Huan, and Gang Liu

Key Laboratory of Modern Precision Agriculture System Integration Research, Ministry of Education, China Agricultural University, Beijing, 100083, China
cauzm@cau.edu.cn

Abstract. To make up the shortages of the existing vision navigation algorithms such as the noise interference and the processing speed, a new approach for navigation line detection is presented in this article. This new method was based on image boundary detection. The color images were transformed into binary ones after pre-processing. Then, three stages in this method were presented. At first, the image was divided into several sectors and the boundary of the crops was detected. Then a number of points indicating the centers of the rows were determined. At last, the navigation line was estimated by an improved linear regression algorithm. Compared to other navigation line detection algorithms, this new method could find the navigation lines more accurately. It need only about 89ms to process a picture with 640×320. So it can meet the need of agricultural real-time visual navigation when the speed is 1~2m/s.

Keywords: Navigation line detection, Machine vision, Agricultural implements navigation.

1 Introduction

Steering agricultural machinery within rowed crop fields is a tedious task for producers. Automated guidance of the machinery will not only reduce operator fatigue but also increase both the productivity and safety of the operation[1]. As a machine vision system can detect a path in relation to crop rows, vision-based guidance can be used to guide machinery travelling between crop rows to perform field operations such as cultivating, chemical spraying, and harvesting[2~4].

There are a lot of research have been made in the field of crop line identification. At present, the traditional ways of crop detection method is based on the Hough transformation largely[5~7]. Hough transformation algorithm is a shape matching technology put forward by Paul Hough in 1962. Duda and Hart established the Hough transformation linear testing in 1972. Hough transformation has several advantages such as strong robustness and small noise influence[8]. However, it also has disadvantages such as a high computational complexity[9]. In order to improve the

* Corresponding author.

D. Li and Y. Chen (Eds.): CCTA 2012, Part II, IFIP AICT 393, pp. 299–308, 2013.

speed of Hough, Kultanen P. put forward the Random Hough Transformation (RHT) [6, 10] algorithm. RHT is a technology of multi-to-one. Compared to the traditional Hough transformation, RHT depend on the idea of randomness, reducing the computational complexity greatly. But for the RHT algorithm, random point selection is so important that it is unstable to detect the navigation line[11]. In addition, because the navigation line only depends on the two crops lines nearest to the image center, so it still brings some redundant calculation to extract all the crop lines.

In order to solve these questions above, a new method was introduced in this article. Multistage filtering is put forward to solve the detection problem, and result in a lower computational complexity with the similar detection stability than RHT and traditional Hough transformation. Two sections are contained in this new method, the first one is image pre-processing section and the second one is a new baseline extraction algorithm that designed in this research named Boundary Detection Navigation Algorithm.

2 Materials and Method

2.1 Experimental Materials

The experimental platform, as shown in Fig.1, included several parts, such as vision sensor, computer, controller, hydraulic system and the agricultural implement. In this platform, the camera was used to real-time acquire the image of farmland; the computer was used to identify and position of crop line from the acquired image, and then calculate the lateral deviation of the farm machinery relative to the target path; Based on the lateral deviation, the controller is used to control the hydraulic system work and realize the automatic navigation of fame machinery.

Fig. 1. The experiment platform

The visual sensor used in this paper is OK AC1310 camera made by Beijing Jiaheng Company. The pixel size is 4.65 μm×4.65μm; the resolution is 1300 × 1024; The frame rate is between 8~24Hz; The shortest exposure time of the electronic shutter is 20μs. The data transmission between the video camera and the computer through a USB2.0 interface[12].

The navigation program is developed on the computer of Intel(R) Core (TM2) Quad CPU Q9400 @2.66GHz, 2.96GB in the XP operating system. And the development environment is Microsoft Visual C++ 6.0.

2.2 Methods

The testing method of the navigation line was divided two sections, the pre-processing section and the navigation line extraction section. The pre-processing section included graying, binarization and filtering, and the navigation line extraction section included division and detection (including image segmentation and boundary detection), points determination (including false points removing and navigation points estimation), and navigation line regression, as shown in the Fig.2.

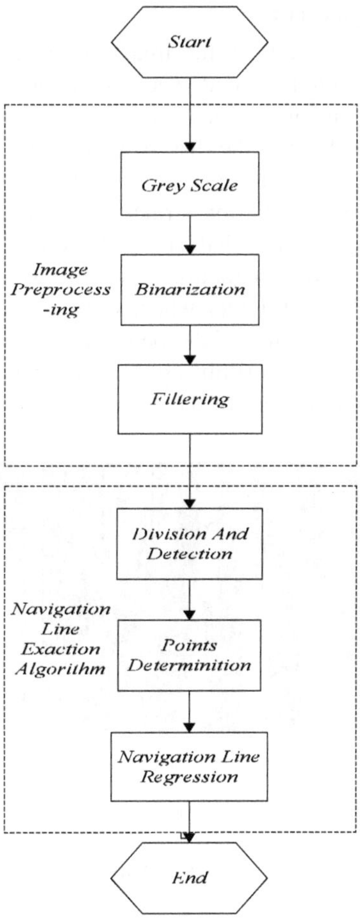

Fig. 2. The flow chart

3 Results and Disscussions

3.1 Image Pre-processing

Considering the complexity of the farmland environment, color images were chosen as research objects in this article because rich color information was contained in it[11]. However, color images also brought some problems such as the redundant information, the noise distribution and the complex nature. Thus, the images needed to be preprocessed before navigation line extraction. The image preprocessing contained three segments, namely as graying, binarization, and de-noising.

Usually, the green component G is much greater than red R and blue B components because the main pigment of the crops was chlorophyll. So the 2GRB method was taken to gray images in order to emphasize green component and restrain the rest two components [12]. The brightness and color will different if the image is acquired under different light condition. The OTSU method was used to transform the gray images into binary images [13]·

There were a lot of noise points in the image after segmentation because of the characteristics of farmland environment, such as weeds, shadow. The method of corrosion-median filter-expansion was selected to reduce the noise according to the characteristics of the weeds. Firstly, the disc structure with 5 pixel radius was chosen to corrode the images. Secondly, the image was filtered through the 5×5 pixel size windows. Finally, the 5×1pixel inflation structure elements was adopted to extrude the vertical part of the images. After that, most of the noise was filtered out and the stability of the navigation baseline extraction algorithm would be improved.

The images of bean, corn and winter wheat were collected to verify the effect of the image preprocessing. The results of the bean images were shown in the Fig.3, it showed that the preprocessing algorithm could effectively reduce the influence of weeds in images and get the boundary of the crops.

a) gray image b) binary image c) filtered image

Fig. 3. The results of image processing

3.2 The Navigation Line Detection Algorithm

Before detecting the row position, a number of points indicating the centre of the rows needed to be determined firstly. In order to obtain these points, the binary image resulting from grayscale image was divided into a series of horizontal strips as shown in Fig.4, and the next step was to estimate where the rows were intersect each individual strip. In principle, the number of strips should equal to the number of pixel rows in the image. In order to reduce the amount of subsequent computation and enhance the stability of subsequent analysis, it was suggested to let each strip involve more than one pixel row. Thus, the number of strips should be kept at a reasonably low value, and this value was denoted as step. Practical experience showed that every 10 pixels divided into a strip was suitable when the system was working on the experimental field. Each image strip was processed as illustrated in Fig. 4. Mathematically, the process could be described as Fig. 5.

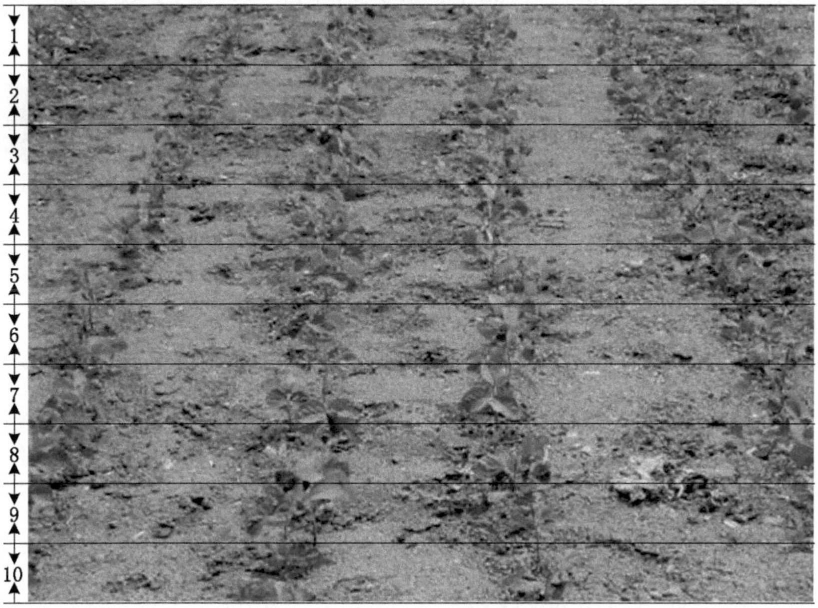

Fig. 4. Color image divided into horizontal strips

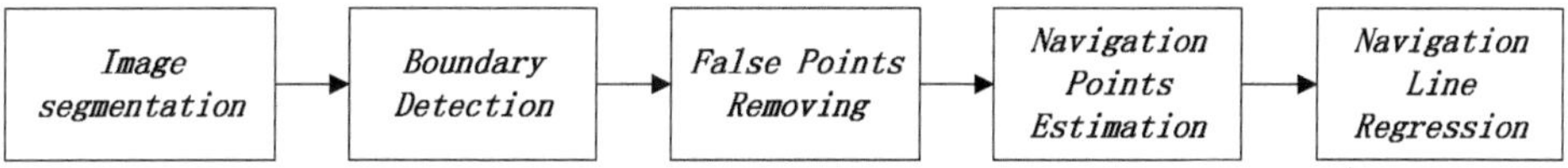

Fig. 5. Steps of the navigation line extraction algorithm

Step 1:Image Segmentation

$V_{i,j}$ was used as the binary value (0-black point and 1-black point) of the pixel in position *(i,j)* of the image strip. The indices i and j denote the pixel row and the pixel column, respectively,

$(i = 1, 2..., N; j = 1, 2..., M; N = Height / step; M = Width)$ .Fig.6 was the first strip of binary image in Fig.4.

Step 2: Boundary Detection

$v = (v_1, v_2 ... v_M)$ was used as a row vector resulting from summing the array V over the index i:

$$v_i = \sum_{j=1}^{N} V_{i,j} \qquad i = 1, 2 ... M \tag{1}$$

The result of this summation is illustrated by the curve in Fig. 7. The position of non-zero value in Fig.7 presents the position of crop in Fig.6. So. Crop position can be determined by detect the point that exchange between 0 and 1. The *ith* non-zero part is represented by a 2D vector like $P_i=(Pl_i,Pr_i)$. Pl_i was the position of the *ith* non-zero part's left border in a strip, Pr_i is the position of the *ith* non-zero part's right border in a strip.

Fig. 6. The first strip of binary image

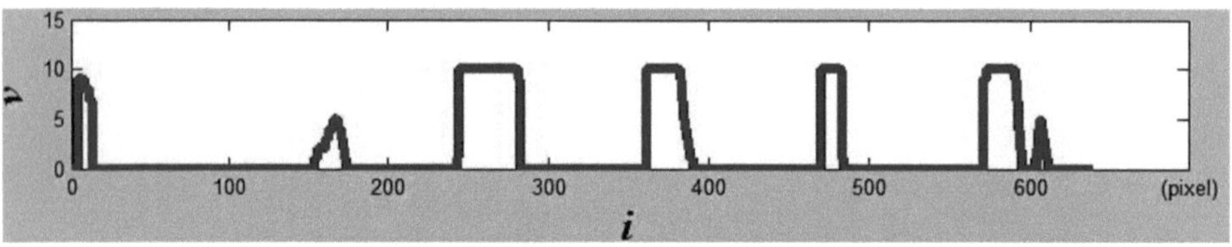

Fig. 7. The number of white points in the first strip

Step 3: False Points Removing

However, not all of these non-zero points represents the real crop because of the influence of weeds. An important difference between crop and weed is their scale in the image. By make use of this feature, a criterion to distinguish the crops and weeds could be set. This criterion describe as follows:

The *ith* non-zero part represent the crop when $(Pl_i - Pr_i) \geq \frac{1}{2K}\sum_{j=1}^{K}(Pl_j - Pr_j)$, (K is the number of non-zero parts in a strip),or it represent the weed.

Step 4: Navigation Points Estimation

After previous step the navigation points (P_i^{navx}, P_i^{navy}) can be estimated by the following formula:

$$\begin{cases} P_i^{navx} = \dfrac{\min_0(|(\frac{Pl_i + Pl_r}{2}) - \frac{width}{2}|) + \min_1(|(\frac{Pl_i + Pl_r}{2}) - \frac{width}{2}|)}{2} \\ P_i^{navy} = step * i - \dfrac{step}{2} \end{cases} \tag{2}$$

In this formula, the $\min_0(S)$ is the smallest value in S , the $\min_1(S)$ is the second smallest value in S , $i = 1, 2...N$, and the navigation point detection result is shown by Fig.8.

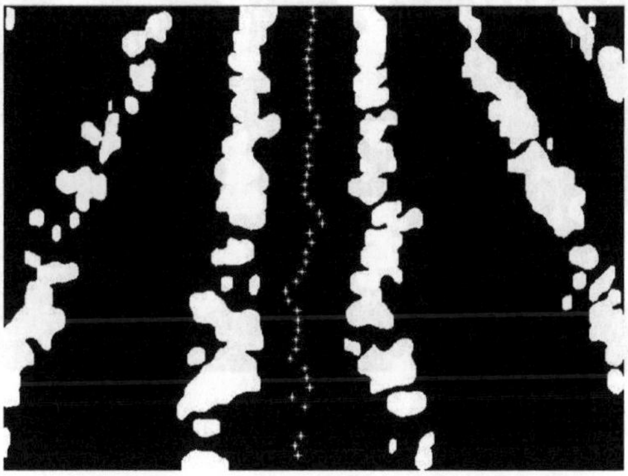

Fig. 8. The detected Navigation Points

Step 5: Navigation Line Regression

Based on these navigation points, the navigation line can be calculated by LSE (least-square estimation) algorithm. However, because of the existing of surplus plant or lack of one or more plant in a row, there are some outliers exist in navigation points, and not all of navigation points should be applied by LSE algorithm. We set another criterion to determine whether a navigation point is a outliers as follows:

The ith navigation point is a outliers when $P_i^{navx} > \alpha\sqrt{\text{var}(P^{navx})}$, and $\text{var}(P)$ is the variance of P . In our experiment the α is 1.5.

Assume the remainder navigation points after removed the outliers are $(P_1^{rx}, P_1^{ry}), (P_2^{rx}, P_2^{ry}), \ldots, (P_{N_r}^{rx}, P_{N_r}^{ry})$. Based on LSE algorithm, the navigation line parameters can be represented as follows (navigation line can be represented as $y = kx + b$):

$$\begin{cases} k = \dfrac{\overline{P^{rx} P^{ry}} - \overline{P^{rx}}\ \overline{P^{ry}}}{\overline{(P^{rx})^2} - \left(\overline{P^{rx}}\right)^2} \\ b = \overline{P^{ry}} - k\,\overline{P^{rx}} \end{cases} \tag{3}$$

Where:

$$\overline{P^{rx}} = \frac{1}{N_r}\sum_{i=1}^{N_r} P_i^{rx},\ \overline{P^{ry}} = \frac{1}{N_r}\sum_{i=1}^{N_r} P_i^{ry},\ \overline{P^{rx}P^{ry}} = \frac{1}{N_r}\sum_{i=1}^{N_r} P_i^{rx}P_i^{ry},\ \overline{(P^{rx})^2} = \frac{1}{N_r}\sum_{i=1}^{N_r}(P^{rx})^2$$

The navigation line detection result is shown in Fig.9.

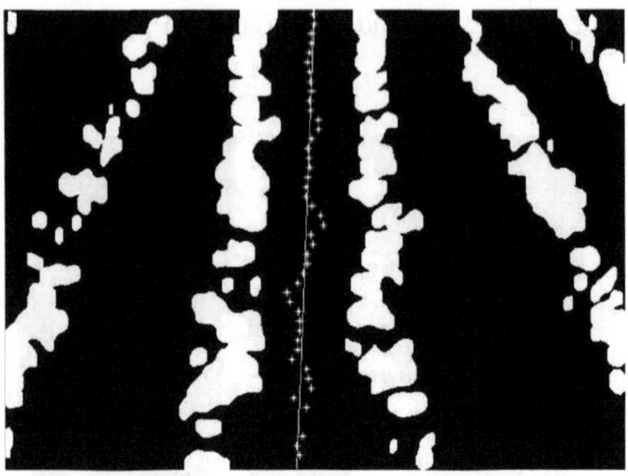

Fig. 9. The detected Navigation Line

3.3 Evaluation Test and Dicussion

This algorithm was named Boundary Detection (BD) algorithm because it partly depended on crop boundary detection. In order to evaluate the speed and accuracy of this BD algorithm, all traditional Hough Transform (HT) algorithm, Random Hough Transform (RHT) algorithm, and BD algorithm were used to analyze the wheat images. The aim of this test was to extract the center line of the image and calculate the parameters of the navigation line. The result was shown in Fig.10. The line parameter results were shown in Table 1. These lines were represented by the polar coordinate, and the ρ resented the polar diameter, the θ represented the polar angle.

According the above analysis, all of these methods could detect the navigation line successfully with acceptable error. From the timeconsuming perspective, HT consumed the longest time and RHT algorithm was the second. BD algorithm needed the shortest time, which was 89ms and less than half of the HT. From the precision perspective, the difference of θ value between BD and HT algorithm was 0.2567°, and 0.43° between BD and RHT algorithm. And the difference of ρ value between BD and HT algorithm was 2.26mm, and 1.88mm between RD and RHT algorithm. The accuracy of all these methods could meet the demand of the agricultural navigation, but the BD algorithm required the least time.

Table 1. The results of linear parameter comparison

Method	θ	R	Time
HT	1.0922	368.32	228ms
RHT	1.2655	372.46	126ms
BD	0.8355	370.58	89ms

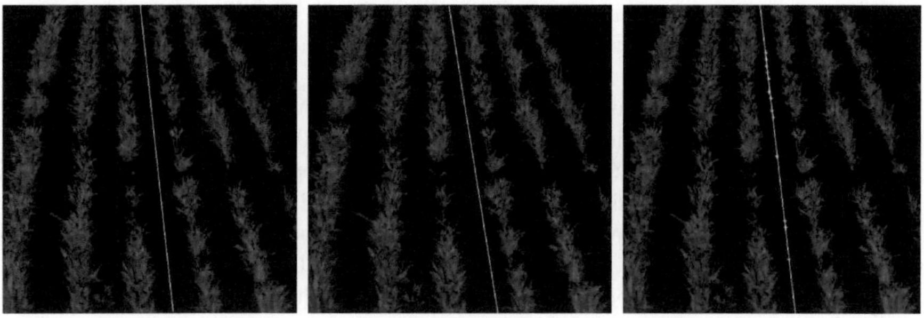

1) HT 2)RHT 3) BD

Fig. 10. The results of navigation line detection

4 Conclusion

A new navigation line extraction algorithm named BD algorithm was proposed to process the farmland environment images in this article. After images acquisition, firstly the color images were transformed into grayscale images by 2G-R-B method according to the characteristics of the farmland crops. Then the grayscale images were transformed into binary images by the OTSU method, meanwhile, a corrosion-median-expansion filter algorithm was adopted to filter the noise. Finally, the parameters of the navigation line were calculated by the BD algorithm. Theoretical analysis and experimental test results showed that the designed BD algorithm was more efficient than HT and RHT algorithm, which only needed 89ms to process an image in this article. The accuracy of the detected line was higher and could be used as the navigation baseline.

Acknowledgements. This work was supported by the National Natural Science Fund (Grant No. 30900869), National Technology Support Program (Grant No. 2011BAD20B06), and 863 Program (Grant No.2012AA101901).

References

1. Marchant, J.A., Brivot, R.: Real time tracking of plant rows using a Hough transform. Real–Time Imaging 1(02), 363–375 (1995)
2. Bakker, T., Wouters, H., van Asselt, K., et al.: A vision based row detection system for sugar beet. Computers and Electronics in Agriculture 60(1), 87–95 (2008)

3. Zhang, H.X., Zhang, T.Z., Chen, B.Q.: Detectation Algorithm for Orientation Lines Based on Pattern Recognition. Transactions of the CSAM 39(2), 107–111 (2008)
4. Jiang, G.Q., Ke, X., Du, S.F., Chen, J.: Detection Algorithm of Crop Rows Based on Machine Vision and Randomized Method. Transactions of the CSAM 39(11), 85–88 (2008)
5. Chen, B., Tojo, S., Watanabe, K.: Detection algorithm for traveling routes in paddy fields for automated managing machines. Transactions of the ASAE 45(1), 239–246 (2002)
6. Fernandes, L., Oliveira, M.: Real - time line detection through an improved Hough transform voting scheme. Pattern Recognition 41(1), 299–314 (2008)
7. Leemans, V., Destain, M.F.: Line cluster detection using a variant of the Hough Transform for culture row localization. Image and Vision Computing 24(5), 541–550 (2006)
8. Jang, G.Q.: Research on Navigation and Localization of Agricultural Machinery Based on Machine Vision. China Agriculture University, Beijing (2009)
9. Cao, Q., Wang, K., Li, H.: Detection algorithm for crop target multi-lines of the field image based on machine vision. Transactions of the CSAE 25(12), 107–113 (2009)
10. Chen, Y., Zhang, M., Ma, W.Q., et al.: Positioning method of integrated navigation based on GPS and machine vision. Transactions of the CSAE 27(3), 126–130 (2011)
11. Liu, Z.X.: Methods of Navigation Control and Path Recognition in Autonomous Agricultural Vehicle. China Agriculture University, Beijing (2011)
12. Chen, J., Jiang, G.Q., Du, S.F., et al.: Crop rows detection based on parallel characteristic of crop rows using visual navigation. Transactions of the CSAE 25(12), 107–113 (2009)
13. Soggaard, H., Olsen, H.J.: Determination of crop rows by image analysis without segmentation. Computers and Electronics in Agriculture 38(2), 141–158 (2003)

Definition and Standardization of Data Elements' Attributes in Land and Resources Management

Lei Cong[1], Yongxia Yang[1,*], Dehai Zhu[1], Min Yin[2], and Jianlin Li[2]

[1] College of Information and Electrical Engineering, China Agricultural University, Beijing 100083, China
[2] Beijing Municipal Bureau of Land and Resources, Beijing 100013, China
yongxia.yang@qq.com

Abstract. The standardization of data element in land and resources management is an important component of the standardization of information technology in the field. It's an essential method to guarantee the interconnection, intercommunication and interoperability, and achieve the information sharing. This study is a case study on Beijing Municipal Land and Resources Management, especially cadastral management and land use management. Considering of the particularity of data in land and resources management from the aspects of language, expression, area, data and application, we study and develop many rules to describe the attributes of data element based on ISO/IEC 11179 and GB/T 19488. This paper shows the specific definition and standardization of data elements' attributes in our study. The whole study can direct the other related working in land and resources management, and give the ideas to carry out the study on the standardization of data element in the other fields.

Keywords: Standardization, Data Element, land and resources, attribute.

1 Introduction

The standardization of data element is an important component of the standardization of information technology. It guarantees the development and running of information system. The success of the system applications is on the basis of it. The standardization of data element is defined as the process of developing and implementing uniform standards, which includes the data elements' general rules, definition, description, classification, expression and registration.

Its importance can be broadly divided into five areas:

- Standardizing and unifying the collection and application of data;
- Building the basis of data sharing for different systems with multiple modes of data;
- Achieving data management in the level of metadata and making the separation of software and data, so the development of software is more flexible.

* Corresponding author.

D. Li and Y. Chen (Eds.): CCTA 2012, Part II, IFIP AICT 393, pp. 309–320, 2013.

- Providing unified standard of data exchanging in business internal areas or between internal and external areas;
- Being the basis of building a unified, integrated and efficient data model in business.

Land and resources management is a multi-disciplinary integrated system relating to land, geography, minerals, etc. Business management needs cooperation and mutual penetration in the area, while it actually lacks of common data models and standards. The present situation causes many problems, such as one word with several equivalents of different meanings and one word with multiple equivalents of the same meaning. The former can make users understand the word with ambiguity like Digital Gap; the later can cause data redundancy and update data inconsistently, and further, it goes ill with the integration of information system, which is called Information Island. These problems add the confusion of the system, and limit data exchanging and sharing. In summary, the standardization of data element in land and resources management is a vital foundation in this field, especially for the goals of "A Map" and "Golden Land Project".

ISO (the International Organization for Standardization) and IEC (the International Electrotechnical Commission) has released "ISO/IEC 11179-Information techology-Specification and standardization of data elements" in 1999, and updated it to "Information technology--Metadata registries (MDR)" in 2004. It's an international standard for defining, describing, standardizing and registering data element or metadata for an organization. For the standard's importance, China National Institute of Standardization translated it into "GB/T 18391-Information techology-Specification and standardization of data elements" in 2002 and "GB/T 18391- Information technology--Metadata registries (MDR)" in 2009 for Chinese organizations' use. In addition, CNIS also released "GB/T 19488 Date Elements for e-government" for better understanding and more practical, which refines and embodies the contents of principle and theories in GB/T 18391.

This study is on the standardization of data element in Beijing Municipal Land and Resources Management, especially cadastral management and land use management. The above standards are the basis of the study, whereas, because of language, expression, area, data and application, there are still some problems that need further research. In this paper, we record and summarize the issues and our solutions that focus on the standardization of data element's attributes.

2 Definition of Data Element' Attributes

Data element is described by a set of attributes. The basic attributes of data element can be organized into five categories, including a total of 22 attributes as following [3].

- **Identifying:** Name (mandatory), Identifier (conditional), Version (conditional), Registration Authority(conditional), Synonymous Name (optional), Context (conditional)
- **Definitional:** Definition (mandatory)

- **Relational:** Classification Scheme (optional), Keywords (optional), Related Data Reference (optional), Type of Relationship (conditional)
- **Representational:** Representation Category (mandatory), Form of Representation (mandatory), Data Type of Data Element Values (mandatory), Maximum Size of Data Element Values(mandatory), Minimum Size of Data Element Values (mandatory), Layout of Representation(conditional), Permissible Data Element Values (mandatory)
- **Administrative:** Responsible Organization (optional), Registration Status (conditional), Submitting Organization (optional), Comments (optional)

The above attributes are just recommended in the standards. In the actual study, they can be added, deleted and modified with many factors. First, we can identify priorities on attributes considering of the application of data elements. For example, if the

Table 1. Attributes of Data Element in Land and Resources Management

Category	Attribute	Obligation
Identifying	Chinese Name	M
	Chinese Spelling	M
	Identifier	M
	Version	M
	Registration Authority	M
	Synonymous Name	O
	Context	C
Definitional	Definition	M
	Object Class Term	M
	Property Term	M
Relational	Keywords	O
	Relationship	O
Representational	Representation Term	M
	Data Type	M
	Data Format	M
	Value Domain	M
	Unit of Measurement	O
Administrative	Status	M
	Submitting Organization	O
	Date of Approval	O
	Comments	O

standardization of data element is for the internal needs of an organization, the attributes of Administrative should be completed and detailed for future use in business management; while the standardization of data element is for data sharing among different organizations, the attributes of Administrative may have no much sense. Second, because of the difference of languages and areas, we can add some specialized attributes adapt to the current environment, such as Chinese name, English name, Chinese spelling. In addition, we can also merge and resolve some attributes in order to make them clearer and more concise.

Considering of the actual need in business areas of Beijing Municipal Land and Resources Management, we determine the following factors to choose the attributes with the aim of less attributes but also a clear expression.

1) Requires attributes, such as name, identifier, definition, data type, etc.
2) Attributes for management, such as version, registration authority, submitting organization.
3) Attributes for inquiring and reading by user, such as keywords, synonymous name.

By the above factors, we define the attributes which are used to describe data elements in land and resources management (see table 1).

3 Standardization of Data Element's Attributes

This section shows the key points of the standardization of data element's attributes in our study fields, which are cadastral Management and land use management. We study and develop many proprietary and practical rules to solve the problems caused by language, expression, area, data and application. This study complements and improves the processes and methods for the standardization of data element's attributes in specific field, which are abstract and conceptual in the standards.

3.1 Chinese Name

Chinese name is the name of data element with single or multiple Chinese words. It's generally composed of object class term, property term, representation term and sometimes also qualifier term, which are related to the data model of data element. In order to make data element's concept, classification and representation more clearly, we add the first three terms in the attributes of data element, and the methods to identify them are showed in the ensuring chapters. In this part, the development of naming rules is the key, even in our whole research. A standardized name is the basis to solve "Digital Gap" and "Information Island" in the system.

Besides the basic naming rules in the standards as following [5] [7]:

Semantic rules:

- *One and only one object class term shall be present.*
- *One and only one property term shall be present.*

■ *One and only one representation term shall be present.*

■ *Qualifiers will be added as needed to make the name unique within a specified context. The order of the qualifier terms is not significant. Qualifier terms are optional.*

EXAMPLE

宗地法人代表姓名 Name of Parcel Legal Representative

法人代表	Legal Representative	(object class term)
姓名	Name	(property term)
名称	Name	(representation term)
宗地	Parcel	(Qualifier term)

NOTE "姓名"and "名称" are used in different situations in Chinese, while they are a same word in English. We explain it later in the article.

Syntactic rules:

■ *The object class term shall occupy the first (leftmost) position in the name.*

■ *The property term shall occupy the next position.*

■ *The representation term shall occupy the last position. If any word in the representation term is redundant with any word in the property term, one occurrence will be deleted.*

EXAMPLE

宗地法人代表姓名 Name of Parcel Legal Representative

The property term is "姓名", and the representation term is "名称". They make the meaning repeated, so the representation term is deleted in the name.

■ *Qualifier terms shall be prefixed to the part qualified. The order of qualifiers shall not be used to differentiate names.*

EXAMPLE

"变更前宗地地籍号" and "宗地变更前地籍号" both mean parcel number before changing. The position of Qualifier term "变更前（Before Changing）" have no effect.

Uniqueness rule:

All names in each language shall be unique within one context.

We also develop the additional lexical rules to adapt to the particularity of this field, considering of the use of Chinese words.

Lexical rules:

■ *If the object class of data element is human, its name should use "姓名" to represent. While the object class of data element is thing or organization, its name should use "名称" to represent.*

EXAMPLE

宗地法人代表姓名 Name of Parcel Legal Representative

宗地权属单位名称 Name of Parcel Ownership Unit

Comparing to a word of name in English, it seems complex in Chinese. More than one word can correspond to one meaning but also have a little difference. So rules are much needed to prevent the confusion.

- *International abbreviations and acronyms are allowed to use in the name.*

EXAMPLE

DEM Digital Elevation Model

DTM Digital Terrain Model

Although they are the abbreviations and acronyms of English words, almost everyone knows and consistently uses them in this field. On the contrary, if they are translated to Chinese, someone may have doubts about them and the name of data element would be too long.

- *Arabic numerals can be used in the name to distinguish data elements.*

EXAMPLE

线状地物权属单位 1 Name of Linear Feature Ownership Unit

线状地物权属单位 2 Name of Linear Feature other Ownership Unit

This is a means to make the name of data element simple and brief, but it's on the premise of a clear definition.

These are all of rules we have studied and developed for naming data elements. As there aren't unified English name for data elements in this field, we don't consider English name as one attribute of data element.

3.2 Chinese Spelling

Chinese spelling is a specific attribute for Chinese data element, which is composed of the spellings of Chinese characters in the name of data element. All the letters should be lowercase, and connect with a hyphen"-".

There are two points needed to pay attention to:

NOTE 1:

"u" and "v" should be distinguished properly and used in the right place, such as "吕(lv)" and "区(qu)".

NOTE 2:

If the Chinese name of data element with Arabic numerals, the Chinese spelling should retain the Arabic numerals.

For example, 线状地物权属单位 1 (xian-zhuang-di-wu-quan-shu-dan-wei-ming-cheng-1).

3.3 Identifier

Each data element should have an identifier, which is defined by the registration authority. This process includes the study of classification of data elements. There we suggest that the authority just provides the idea of classification instead of changeless classification number, because the standardization of data elements is an increasing long-term work.

Since there are so many horizontal and vertical business areas in land and resources management, we use the method of classification combining surface and line. First, we divide the field of land and resources management to three areas of land, resource and geological environment by the means of surface classification. Second, we use the combinative method to get business class and business sub-class. After that, the classification can be refined by the object class that data elements belong to. Fig.1 shows the classification result of land and resources management.

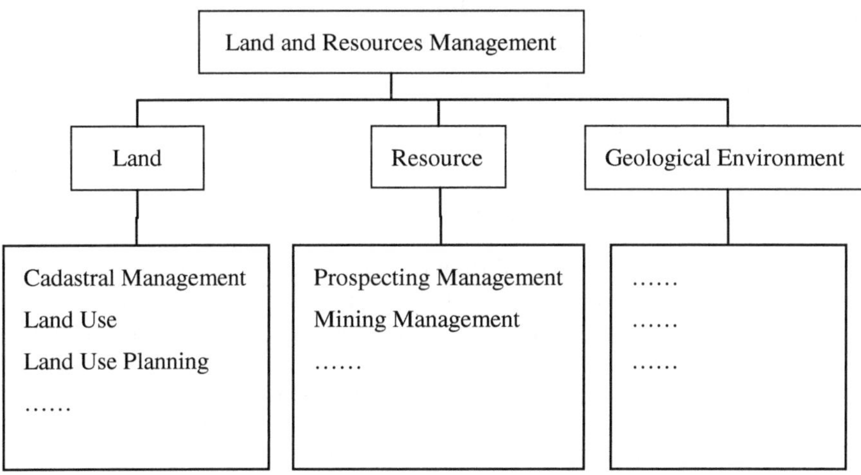

Fig. 1. The classification result of land and resources management

Then we can define identifier of data element based on the above result. Fig.2 shows the frame of identifier.

XX	XX	XX	XX	XXX
|	|	|	|	|
Professional Code	Business Code	First-level Classification Code	Second-level Classification Code	Third-level Classification Code

Fig. 2. The frame of identifier in land and resources management

1) Professional Code is the code of three professional areas in land and resources management, land, resource and geological environment. It is represented by two letters.

2) Business Code is the code of business types. It is represented by two numbers.

3) First-level Classification Code is about the object class that data elements belong to.

4) Second-level Classification Code can refine the object class in first-level classification.

5) Third-level Classification Code is the sequential number.

EXAMPLE

宗地地籍号 cadastral number of parcel

Identifier TD020101001

Professional Code: TD

Business Code: 02

First-level Classification Code: 01

Second-level Classification Code: 01

Third-level Classification Code: 001

3.4 Synonymous Name and Context

Synonymous name refers to the different names in the different application environment for one data element. A data element may have none, one or more synonymous name, which are separated by semicolon. The collection of synonymous name can further solve "Digital Gap" and "Information Island" in the system, considering of the traditional habits in the business management.

As synonymous name's corresponding, context records the application environment and process generating and using the data element. A data element may have none, one or more context, which are separated by semicolon, that's decided by the quantity of synonymous names.

3.5 Definition

Definition is the statements used to express the essential characteristics of data element. It can distinguish data element from all other ones. We should make them brief and accurate. Besides, definitions should be unique in a field and use singular form. Reading various terminology standards for reference is the best way to make definition.

3.6 Object Class Term

Object class term can show the objects and concepts that data elements belong to, and represent an activity or object in a context. The extraction of object class term is a problem in our study, because it usually confuses with qualifier term. To solve this problem, we introduce the thought of modeling, such as Entity Relationship Diagram (ERD).

An object class is a set of ideas, abstractions or things in the real world that are identified with explicit boundaries and meaning, and whose properties and behavior follow the same rules. It corresponds to the entity or relation in the data conceptual model. Through rebuilding the ERD, We can sort out all the entities and relations in the whole business areas. The results are the object class and their attributes are data elements.

In order to prevent object class term confused with qualifier term, the entities and relations must be minimal. The object class that data element directly belongs to, which isn't able to be subdivided, generates the object class term of data element, while the object class at the higher level may be the qualifier term. Sometimes the object class is too small to present in the ERD, so it uses the object class at the higher level instead of the direct one.

3.7 Property Term

A property is a characteristic common to all members of an object class. Each property has a name. Property term represents the characteristics to distinguish the object classes of data elements significantly. It generally corresponds to the attributes in the ERD, such as address, name, price, etc.

3.8 Representation Term

Representation term describes the form of the set of valid values of an administered item which includes representation. Often, the representation term may be redundant with part of the property term. When this occurs, one term or part of one term may be eliminated in a structured name. This can be established as a naming rule.

Representation term is helpful to achieve the functions of data element naming, analysis, analog and inquiry. The international used representation terms include date, code, name, sum and so on. They can also be added by the demand in the study field.

3.9 Relationship

Relationship describes the connection between the present data element and other ones. We can know the deep-rooted network of relationship in this business area by the attribute. It includes the following basic relationships: derive-from, compose-of, replace-of, link-with [7].

EXAMPLE

Parcel compose-of Boundary lines

3.10 Data Type and Date Format

Data type is the type of data element value. The basic data type includes character, number, Date, Date Time, Boolean and binary-type. Because of the particularity of data in land and resources management, we should define the additional data types for spatial data. Based on the standards of OPENGIS, we add seven data types of spatial data, which are point, linestring, polygon, multipoint, multilinestring, multipolgon and geometrycollection.

Data format orders the format requirement for data elements in a business perspective. It combines three basic attributes which are data type of data element values, maximum size of data element values and minimum size of data element values. There are the exact basic requirements in the existing standards. Besides them, we rule that the data format of spatial data is represented by "O", and the data format of binary data is represented by "B".

3.11 Value Domain and Unit of Measurement

Valued domain shows the allowable value set of data element. It's given by four ways as following [7]:

1) By name, pointing out the name of valued domain directly.
2) By reference.
3) By list, representing every possible value with its instance or meaning.
4) By rules indirectly.

Unit of measurement should refer to the related international standards. It needs to be represented in Chinese to avoid users' not understanding the abbreviations.

3.12 Status

Status shows the current state of data element for registration during its whole lifetime. We define four types of status for data element in land and resources management.

1) Draft: the relevant departments and units can give widely comments and suggestions during this stage.
2) Trial: the result can be used for a try after a certain technical procedures. The feedbacks in this stage are some practical advices.
3) Implementation: the content of data element has become the standards to be complied with.
4) Abolished: the content of data element would be deleted from the standards.

4 Discussion and Conclusion

This study is based on the theories and concepts provided by the international standards about the standardization of data element; in addition, we consider of the particularity of data in land and resources management from the aspects of language, expression, area, data and application. The proper attributes of data element in this field are defined and the specific rules to describe them are developed in our study. They can direct the other related working in land and resources management, and give the ideas to carry out the study on the standardization of data element in the other fields.

Although the study has been basically completed, there are some uncertain issues needed for further study.

1) Chinese name is too long. That's because of the semantic rules. The rules make the name standardized, but also different from the previous used name. Can we add an attribute called "short title" to record the previous used name?
2) Synonymous name and context are two corresponding attributes. Context shows the synonymous name's application environment. How to show the application environment of Chinese name?
3) We define the data type and data format of spatial data. But they can be only used for small range. There should be an international standard to define them.

By this study, we have achieved more than 400 data elements' standardization in the areas of cadastral management and land use management. We complete and improve the whole process and rules during the case study. Besides, our research results also include a work guideline to direct the related work in the future. Further, we are making an effort to develop it up to a local standard in Beijing. For the unified management of the standardized data element in the future, a management information system is needed. It ensures the interoperability of data elements in the different departments and business, and achieves data sharing successfully. So the design, development and maintenance of the MIS will be the important work.

This study is only about two areas of cadastral management and land use management. There may be still many problems in the case study of other fields. So the results of this study needs to be improved constantly.

Acknowledgements. This research was funded by China Agricultural University Young Teacher Innovative Project "study on the standardization of data element in Beijing Municipal Land and Resources Management". Young Teacher directed and contributed to the whole study. The authors would like to thank the anonymous reviewers for their constructive comments and suggestions.

References

1. ISO/IEC 11179-1 Information technology—Metadata registries (MDR)—Part 1: Framework (2004)
2. ISO/IEC 11179-2 Information technology—Metadata registries (MDR)—Part 2: Classification (2005)

3. ISO/IEC 11179-2 Information technology—Metadata registries (MDR)—Part 3: Registry metamodel and basic attributes (2003)
4. ISO/IEC 11179-2 Information technology—Metadata registries (MDR)—Part 4: Formulation of data definitions (2004)
5. ISO/IEC 11179-2 Information technology—Metadata registries (MDR)—Part 5: Naming and identification principles (2005)
6. ISO/IEC 11179-2 Information technology—Metadata registries (MDR)—Part 6: Registration (2004)
7. GB/T 19488 Date Elements for e-government (2004)
8. Liu, W., Bao, F.: Representation Method of Intelligent Transport Systems (ITS) Data Elements. Transport Standardization (2011)
9. Nogueras-Iso, J., Zarazaga-Soria, F.J., Lacasta, J., Béjar, R., Muro-Medrano, P.R.: Metadata standard interoperability: application in the geographic information domain. Computers, Environment and Urban Systems (2004)
10. Liu, Y.: Study on the Standardized model of data element. Science & Technology Information (2010)
11. Hüner, K.M., Otto, B., Österle, H.: Collaborative management of business metadata. International Journal of Information Management (2011)
12. Lin, Y.: Traffic Scientific Data Design Based on Data Elements Technology. Science Technology and Engineering (2011)
13. Wu, B.: Study on ecological metadata and its standard. Journal of Central South University of Forestry & Technology (2010)
14. Kim, T.J.: Metadata for geo-spatial data sharing: A comparative analysis (1999)
15. Ouyang, Y., Yao, L., Dai, C.: Data Element Distill Method on Demand-oriented of Information System. Computer Engineering (2006)
16. Wei, H., Zhang, J.: The Definition and Application of Data Element in E-government Affair Standards System. Information Technology & Standardization (2004)

Design of Monitoring Network for Cultivated Land Quality in County Area Based on Kriging Estimation Variance*

Sai Tang[1], Jianyu Yang[1,**], Chao Zhang[1], Dehai Zhu[1], and Wenju Yun[2]

[1] College of Information and Electrical Engineering,
China Agricultural University, 100083 Beijing, China
[2] Land Consolidation and Rehabilitation Center,
Ministry of Land and Resources, 100035 Beijing, China
757165996@qq.com, ycjyyang@cau.edu.cn

Abstract. Through the monitoring network for cultivated land quality in county area, distribution and changing trend of the quality should be reflected. Besides, the quality of non-sampled locations should also be estimated with the data of sampling points. Due to the correlation among spatial samples, traditional methods such as simple random sampling, stratified sampling and systematic sampling are inefficient to accomplish the task above. Thus, spatial sampling method based on Kriging estimation variance is presented in this paper. The method considers the spatial structure characteristics of cultivated land and is configurable to restrict the expense or estimation accuracy, making it more flexible to determine the number of samples. What's more, tessellation polygon is used to densify the network and this helps Kriging interpolation to achieve higher efficiency than traditional methods. Furthermore, inference for the population mean of natural quality indices stays in a very precise level at the same time.

Keywords: cultivated land quality, spatial sampling design, monitoring network, Kriging, estimation variance.

1 Introduction

China is an agricultural country with a large population but not enough cultivated land. Until 2011, the cultivated land per capita in China was 1.38 mu (0.09ha), only 40% of the world average. With the rapid development of economy, industrialization and urbanization, this situation becomes even worse [1]. Therefore, the rate of change of cultivated land is concerns for Chinese governments. At present, methods for monitoring the quantity of cultivated land have become reliable; some researchers analyzed and interpreted remote sensing images to keep track of the quantity condition [2-6]. However, unreasonable land exploitation in China leaves a negative influence on both quantity and quality of cultivated land. When tracking quality condition, the

* The research was funded by the development of the information management system for Farmland grading monitoring (201011006-5).

** Corresponding author.

D. Li and Y. Chen (Eds.): CCTA 2012, Part II, IFIP AICT 393, pp. 321–333, 2013.

method of remote sensing is not applicable any more so that on-the-spot sampling and investigation become necessary.

As the dynamic monitoring of cultivated land quality is the next task of Ministry of Land and Resources[7], we have sampled seven hundred counties in China and in each county, a network would be set up to monitor the change of cultivated land quality. With the basis of current Land Resource Survey and Cultivated Land Evaluation, the paper centers on how to design the monitoring network in county area.

2 Study Area and Data

The study area Daxing district, located in the south of Beijing (longitude 116°13′-116°43′E, latitude 39°26′-39°51′ N), is a river alluvium plain with a sub-humid warm temperate continental monsoon climate. The terrain slopes gradually from east to west and the span of altitudes is generally between 14 and 52 meters. The average annual temperature and precipitation are 11.6℃ and 556mm.

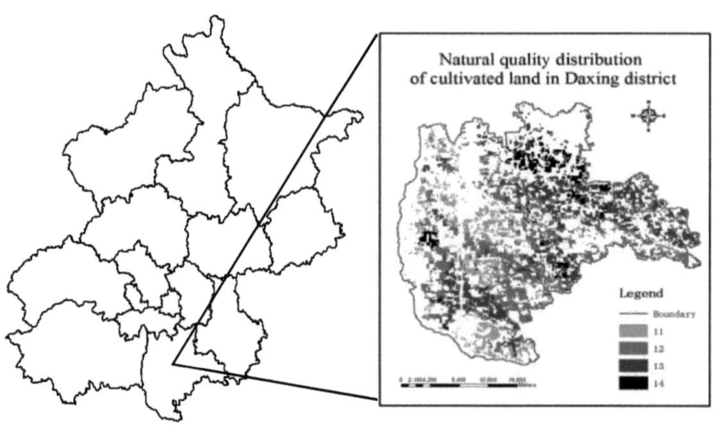

Fig. 1. Study area. Daxing district, located in the south of Beijing, has four natural quality grades (from 11 to 14), showing the ladder-like distribution from northeast to southwest. Higher quality land of 14th grade mainly in northeast makes up 7.72%, that of 13th grade 36.19%, that of 12th grade 35.35% and the rest mainly located in middle and south area belongs to 11th grade.

The data used in our study was the layer of cultivated land quality of Daxing district from last investigation and the corresponding scale was 1: 10000. It included 5182 map spots, with the total area of 416.97km^2. Daxing district belongs to the region of Huanghuanghai and the sub-region of piedmont plain of Yanshan and Taihang Mountains [8]. When designing the network, only natural quality was concerned and natural quality index was used as the attribute variable. The reasons not to take utilization and economy quality into consideration are as follows: a) Natural quality index could reflect the natural conditions and agricultural facilities of the region. b) After calculating the natural quality index, the corresponding utilization and economy quality indices will be obtained by introducing the coefficients which are assigned by

Chinese provinces [8]. c) Utilization and economy quality indices are associated with input cost and grain output, so they are largely affected by human factors and spatial interoperation cannot be applied directly.

3 Methods

The design of monitoring network is to sample from 5182 blocks of cultivated land and place monitoring points on each sampled block, then infer the quality of non-sampled blocks based on investigation of monitoring points. Currently, most designs are originated from classical sampling models, which mainly include simple random sampling, stratified sampling, systematic sampling and cluster sampling [9].These methods ignore the spatial correlation [10] so that they are not suitable for our study because cultivated land belongs to spatially distributed object and there are correlations among samples. Therefore, sampling method based on Kriging estimation variance is presented in this paper.

Kriging method was first raised by Matheron [11, 12], originally applied to researches of geology mineral resources. Oliver used it in the interpolation of GIS [13] and later on the method was widely spread in many fields. It is built upon the theoretical analysis of semi-variance function, considering the distance between non-sampled and sampled points. Kriging method belongs to optimal unbiased estimation so its prediction is better than that of traditional interpolation methods. By interpolating the data of monitoring points, we can get the relationship among estimation variance, sampling distribution, spatial correlation and number of samples [14].

3.1 Examination of the Data Distribution

Data examination is necessary before Kriging interpolation because Kriging is based on the assumption of stationarity, which demands all data values come from distributions that have the same variability. Secondly, certain Kriging methods require the data to be approximately normally distributed. Normal QQ Plot was used to examine the data in the study. In the diagram below, Data points fall close to the 45-degree reference line, approximating normal distribution. Therefore, the application of Kriging interpolation is allowed.

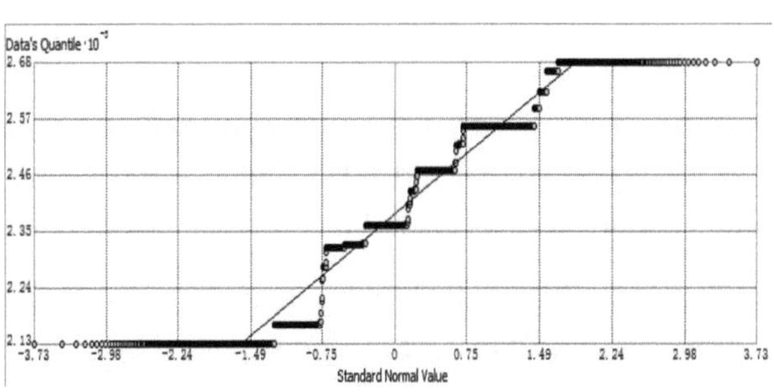

Fig. 2. Normal QQ Plot: the quantile values of the standard normal distribution are plotted on the x-axis, and the corresponding quantile values of the dataset are plotted on the y-axis

3.2 Measures of Monitoring Network Performance

The performance of a monitoring network can be measured using different indices. The following lists three common measures [6]:

Measure 1: *MSE* (mean square error) of the mean estimation

$$\sigma_n^2 = E\left[\frac{1}{n}\sum_{i=1}^{n} f(x_i) - \frac{1}{A}\int_A f(x_i)dx_i\right]^2 \tag{1}$$

Where E denotes the mathematical expectation; n is the total number of samples; $f(x_i)$ is the real value of point x_i in the region A; and term $\frac{1}{A}\int_A f(x_i)dx_i$ is the true mean of A. In practice, this mean is estimated by the arithmetic mean of n sampling points and the corresponding formula is $\frac{1}{n}\sum_{i=1}^{n} f(x_i)$. This measure is quite useful for estimating population mean [15] or total [16].

Measure 2: *MSE* of discrepancy

$$MSE\ \hat{f}(x_i) = E\left[\left(\hat{f}(x_i) - f(x_i)\right)^2 | A\right] \tag{2}$$

Where $f(x_i)$ denotes the real value at point x_i and $\hat{f}(x_i)$ is the estimated value of $f(x_i)$, which is calculated from interpolation of monitoring points. This measure is suitable to estimate other points by spatial interpolation on existing sampling points [17].

Measure 3: Maximum site error

$$\left\{\max_i \left\|\hat{f}(x_i) - f(x_i)\right|,\ \forall i \in A\right\} \tag{3}$$

The applicable condition of this measure is similar to that of measure 2.

The quality of non-sampled cultivated land should be predicted by interpolation in our study. As administrators only concern the global accuracy so that measure 2 is appropriate, which places sampling points in suitable positions and has the most accurate estimation through Kriging interpolation [18]. Because natural quality was the prior knowledge, formula of measure 2 was redefined like this:

$$MSE = \frac{1}{n}\sum_{i=1}^{n}\left(\hat{f}(x_i) - f(x_i)\right)^2 \tag{4}$$

Where n means the total of cultivated land map spots, which was 5182. By the definition, the higher the estimation accuracy of interpolation is, the lower the *MSE* is.

3.3 Design of Monitoring Network

The steps to set up the monitoring network are as follows: a) Lay out the basic network according to regular grids; the choice of grid size depends on both estimation variance and number of sampling points. b) Adjust the centers of grids which do not fall in the area of cultivated land to form the global monitoring network. c) Densify the global network by adding sampling points in tessellation polygons whose *MSE*s are relatively high. The workflow is displayed in figure 3.

Basic Monitoring Network. In geographical space, using regular grids to place samples is the common way to quantitatively describe spatial variation and a square grid is the most convenient in practical applications [19]. The reasons to introduce regular grids are listed in the following: a) Sampling points are distributed uniformly so that the entire district could be represented. b) When applying Kriging interpolation, if numbers of point pairs of different lag classes vary greatly, the estimation efficiency of semi-variation function will fall [20]. c) Generally, spatial correlation declines when the distance between monitored objects increases so that the grid sampling is more effective than random sampling in decreasing estimation variation [14].

Referring the bottom left corner of the layer, sampling points are placed according to square grids of different sizes. Using the data of sampling points, a Kriging interpolation is applied to estimate the value of non-sampled cultivated land. Compare the grid sizes and the corresponding *MSE*s; then a suitable grid size can be chosen for the basic network.

Global Monitoring Network. Becasue the center of a grid may be not located inside a map spot of cultivated land, we create a searching circle around this center. Choose the centroid of a map spot, which is closest to the center and inside the searching circle be a candidate point. As these candidate points may raise the estimation variance, a further observation is needed to decide whether to add them or not. The observation should be: apply Kriging interpolation and calculate the *MSE* after adding each candidate point; add the one that decreases *MSE* by the least to the basic network and take the points which make *MSE* decline as candidates for the next calculation. The loop stops when no point reduces the global *MSE*.

Densification of the Network. Adding sampling points in complex areas where natural quality of cultivated land changes intensely can not only increase the accuracy of interpolation but also be quite meaningful because there may be a high chance of quality grades change.

At present, the main approach to densify the network is to use existing points to get weights of non-sampled locations and then add the point with largest weight. A. G. Journel's method was to apply Kriging interpolation at first and get the Kriging variance of every non-sampled point, then set the point with largest Kriging variance as the new one to add [21]. Kriging variance is used to evaluate the uncertainty of Kriging interpolation [22]. But in this study, real values of cultivated land blocks were known so that error of estimation, which shows the deviation of real and estimated value, was more reliable than Kriging variance. Hence, estimation variance was used as the weight in the study.

However, points with large estimation variance may stand as outliers; to solve this problem, tessellation polygon was introduced. Tessellation polygon was proposed by Netherland meteorologist A. H. Thiessen to predict average rainfall based on data of weather stations in discrete distribution. He linked all adjacent weather stations as triangles and drew perpendicular bisector of each line to produce tessellation polygon. At last, the rainfall intensity of polygon region was presented by the only weather station inside the polygon. The tessellation polygon has the following features: a) There is only one discrete point inside each polygon. b) All objects inside a polygon are closest to the corresponding discrete point. c) The objects on the line of a polygon have the same distance to the two points on both sides of the line.

When using monitoring points to build tessellation polygon, every polygon is the control area of one point. If *MSE* of some polygon is relatively high, the corresponding estimation accuracy is low. To some degree, this approach can exclude individual outliers to be added. The details are listed as follows: Selecting the point of maximum variance in a tessellation polygon with a high *MSE*. Add it to the network only if it makes the global *MSE* smaller. The procedure of densification stops until the count of sampling points reaches the specific value or the *MSE* of tessellation polygon is lower than the threshold. Till then, the work to set up the monitoring network is complete.

3.4 Estimate the Population Mean

After monitoring network is finished, the population mean of natural quality indices of cultivated land will be calculated by following formula:

$$\overline{f}(x_i) = \frac{1}{n}\sum_{i=1}^{n} \hat{f}(x_i) \tag{5}$$

Where n means the total of cultivated land map spots, which is 5182; $\overline{f}(x_i)$ is the population mean; $\hat{f}(x_i)$ is the estimation value.

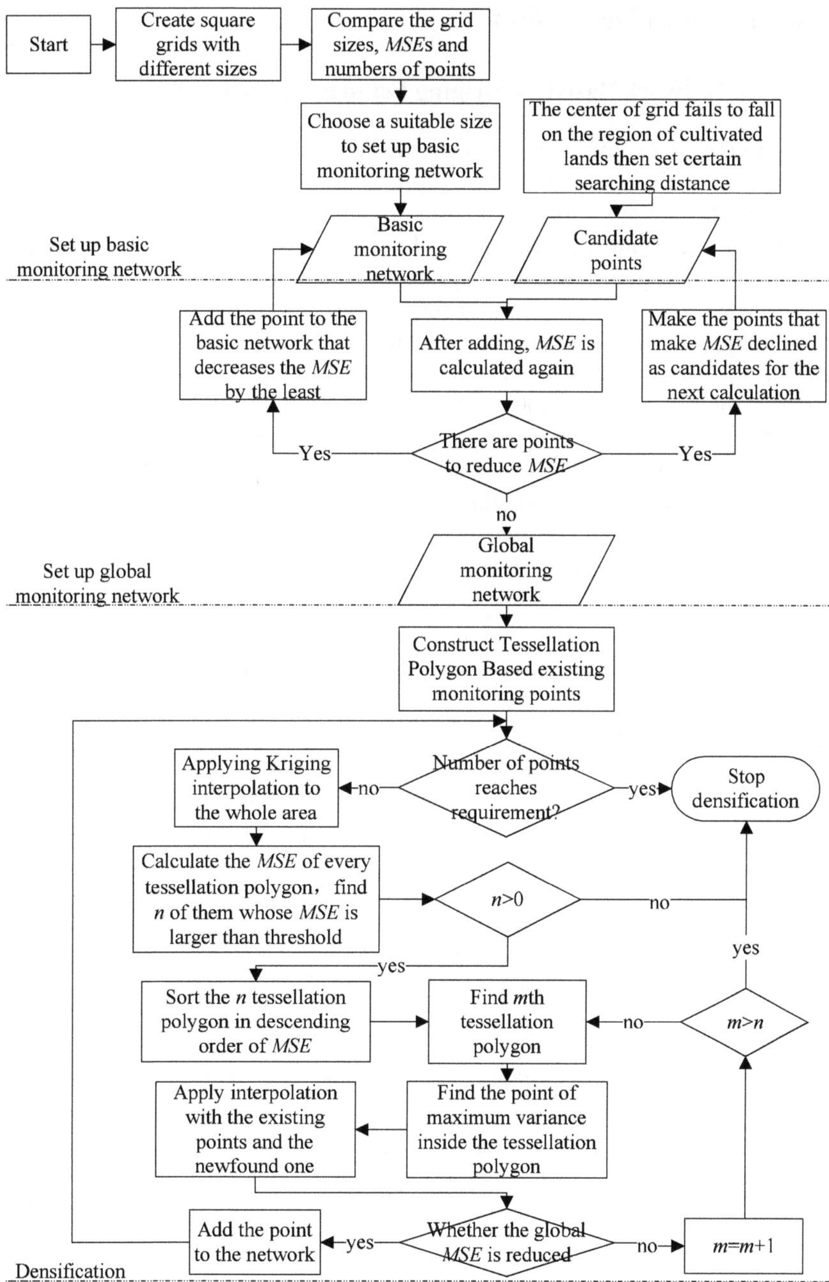

Fig. 3. Workflow of setting up the network: *N* could be set to the final count of samples according to the cost (*N*=5182by default). A threshold could be set and new points are added to *n* regions whose *MSE*s are bigger than the threshold (*n* equals the number of global sampling points by default). Variable *m* is initially set to 1.

4 Results and Discussion

4.1 Set Up Network Based on Kriging Estimation Variance

According to the procedure mentioned above, different sizes of square grid were used and searching circle's radius was set as 100m. In principle, the smaller the size of grid is, the larger the count of points is. But in light of discontinuous distribution of cultivated land, the count was affected by the position relation of grids and map spots. So, it is legal that sometimes smaller grids contain fewer points. Table 1 shows the number of points and *MSE* of each grid size.

Considering both *MSE* and number of points, we found it reasonable to choose the 4000m grid for three reasons: (a) The *MSE* of 4000m grid with 39 points was 13130, smaller than that of 3900m, 3800m, 3600m, 3500m, 3300m and 3100m. This meant that the 4000m grid contained fewer points but had higher accuracy. (b) Although the *MSE*s of 3700m, 3400m, 3200m or sizes below 3000m were smaller, the number of points increased significantly along the way, followed by the augmentation of sampling cost. (c) To leave basic monitoring network for further densification, we made the number of points as small as possible.

Table 1. Number of points and *MSE* of each grid size

Grid size	Number	*MSE*	Grid size	Number	*MSE*	Grid size	Number	*MSE*
4500	34	17077.8	3600	53	13258.8	2700	88	12288.0
4400	34	15527.0	3500	59	13288.6	2600	94	10338.6
4300	36	13763.5	3400	49	12336.4	2500	108	12510.7
4200	37	13995.1	3300	49	13635.9	2400	102	11123.3
4100	31	15932.5	3200	67	12913.5	2300	126	10857.5
4000	**39**	**13130.1**	3100	74	14568.0	2200	141	10602.5
3900	48	13930.6	3000	69	12693.5	2100	154	9505.5
3800	48	14675.4	2900	74	12402.4	2000	161	10046.4
3700	51	12413.1	2800	82	11072.8			

Even though the searching radius was set up as 100m, from the figure 4(a), we can see that sampling points did not cover every grid. To fix this, we set the radius as 500m and got 11 candidate points, as showed in figure 4(b) with the symbol of triangle. To follow the procedure above, three points was selected with the purpose to increase the estimation accuracy, which came down to a 42 points network. The *MSE* was 12198.37, less than former network with the *MSE* of 13130.05.

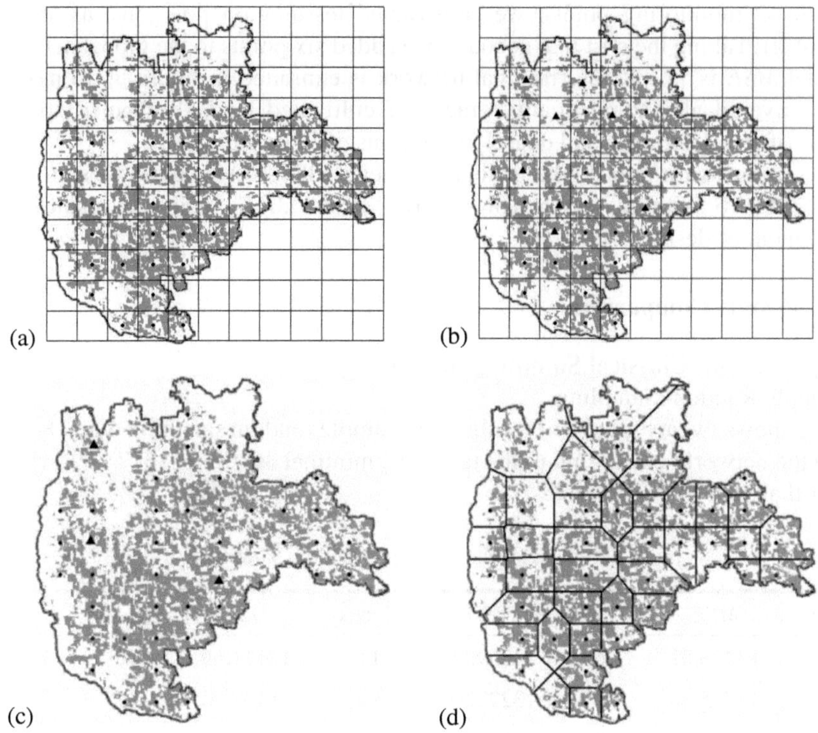

Fig. 4. (a) is the basic network of 39 sampling points by 4000m grids. (b) shows the locations of candidate points. (c) is the global network of 42 points. (d) shows the tessellation polygons.

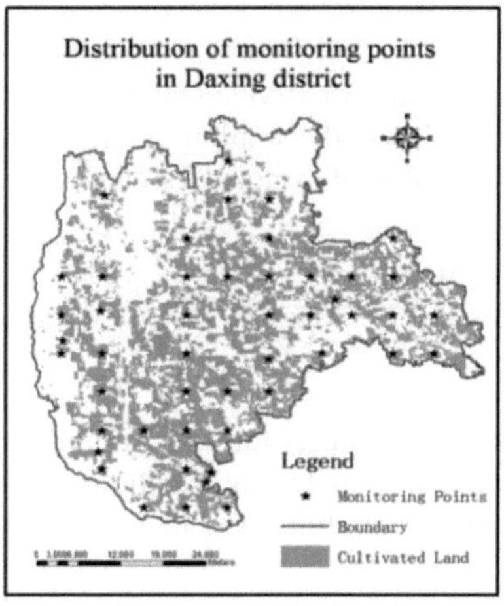

Fig. 5. Distribution of monitoring points in Daxing district

With 42 monitoring points, we constructed tessellation polygons, as showed in figure 4(d). Taking the cost into account, we added six points to the global network and the final *MSE* is 11238.32. The final network is exhibited in figure 5; 48 monitoring points covered all natural quality grades of cultivated land, distributed sparsely in regions of small variance and densely in regions of large variance.

According to formula 5, the estimation mean 2384.16 was quite close to the real mean 2385.76 with a very small relative error 0.07%, and the result met the requirement of design.

4.2 Results Comparison

Compared with Classical Sampling Models.
(a) Simple Random Sampling
Table 2 shows twenty groups of results using simple random sampling with 48 points to set up the network. According to the table, the minimal *MSE* was 13171.69, which was bigger than that of our method.

Table 2. Results of simple random sampling

Times	*MSE*	Times	*MSE*	Times	*MSE*	Times	*MSE*
1	13423.81	6	13759.86	11	**13171.69**	16	14000.15
2	14775.30	7	13222.51	12	15108.00	17	14467.08
3	15999.14	8	14419.24	13	15345.55	18	14091.05
4	14154.64	9	14595.11	14	16037.21	19	13274.75
5	13180.20	10	**18165.41**	15	14811.47	20	15200.25

(b) Stratified Sampling
When applying stratified sampling, we divided study area into 4 stratums (from 11 to 14 natural quality grade) and placed 48 sampling points as well. The formula followed [23] was used to assign the number of samples for each stratum:

$$n_h = \frac{W_h S_h / \sqrt{C_h}}{\sum W_h S_h / \sqrt{C_h}} \times n \text{ , } W_h = \frac{N_h}{N} \tag{6}$$

Where h means the hth stratum; N stands for total number of cultivated land; N_h is the number of land in hth stratum; W_h denotes the weight of hth stratum; S_h means standard deviation of natural qualities of h stratum; And C_h is the cost for setting up a sampling point.

Based on the assigned numbers in table 3, the study used random method to place them. Results of twenty experiences are contained in table 4, with the minimal *MSE* being 13990.60 and maximal one being 18546.03, far worse than 11238.32 of our method.

Table 3. Standard Deviation and number of samples in each stratum

Natural quality classification	11	12	13	14
Standard Deviation	19.37	26.57	49.16	19.49
Number of sampling points	7	13	25	3

Table 4. Results of stratified sampling

Times	*MSE*	Times	*MSE*	Times	*MSE*	Times	*MSE*
1	15107.41	6	17697.43	11	14791.12	16	**18546.03**
2	**13990.60**	7	18089.52	12	14441.22	17	17209.28
3	16501.53	8	15171.63	13	14660.60	18	14002.29
4	17101.16	9	16482.26	14	14697.30	19	16447.92
5	14936.49	10	14722.58	15	16422.40	20	16319.17

The results indicated classical sampling models lead to low estimation accuracy of interpolation because of the correlations among samples.

Compared with Traditional Square Grids. From table 1, the count of points of 3800m and 3900m equaled that of our method, but the estimation accuracy was significantly lower than ours. Besides, until sampling points reached one hundred, traditional grids can provide sufficient accuracy. In summary, the densification of our method is valid.

Compared with Adding Points without Tessellation Polygon. If we did not use tessellation polygon as a limit and directly added points of maximum variance, the *MSE* was 11680.68.Due to the fact that point of maximum variance could be quite different from surrounding values, directly adding may decrease the estimation accuracy and *MSE* did not monotonically decrease during the procedure of densification. This experience illustrated that accuracy could be improved when we limited the new points to the area of tessellation polygon with relatively high *MSE*.

5 Conclusion

By comparing our approach to traditional sampling methods in cost (count of sampling points), surface fitting (measured by *MSE*) and inference of the population mean, we can conclude that:

(a) Because spatial samples are correlated, traditional sampling methods will reveal their shortages like low efficiency and accuracy. We came up with a method based on Kriging estimation variance, which takes the spatial variation and distribution characteristics into account, and involves interpolation to calculate the value of non-sampled cultivated land.

(b) When densifying the network, points of maximum variance cannot be added directly in case that some outliers appear to decrease accuracy. Introducing tessellation polygon can deal with the problem to some degree.

(c) The paper reports the way to build the monitoring network for cultivated land quality in Beijing Daxing district. By interpolating the whole area, the estimation accuracy is better than traditional methods and the statistical inference of population mean stays in an accurate level as well.

References

1. Shi, S., Chen, Y., Yao, Y., Li, Z., He, Y.: Methodology for Impact Assessment of Regional Cultivated Land Resources Change Based on 3S Technology. Transactions of the CSAE 24(7), 91–96 (2008)
2. Liu, J., Zhu, W.: Accuracy and Efficiency of Different Spatial Sampling Schemes for Cropland Change Monitoring. Transactions of the CSAE 26(10), 331–336 (2010)
3. Zhang, J., Pan, Y., Hu, T., Chen, L., Dong, Y.: Analysis of Influence Factors about Space Sampling Efficiency of Winter Wheat Planting Area. Transactions of the CSAE 25(8), 169–173 (2009)
4. Hu, T., Zhang, J., Pan, Y., Song, G., Dong, Y., Jia, B.: Researches on Remote Sensing Area Measurement Based on Different Sampling Methods. Remote Sensing for Land & Resources 3, 37–41 (2008)
5. Jiao, X., Yang, B., Pei, Z.: Paddyrice Area Estimation Using a Stratified Sampling Method with Remote Sensing in China. Transactions of the CSAE 22(5), 105–110 (2006)
6. Wang, J., Liu, J., Zhuan, D., et al.: Spatial Sampling Design for Monitoring the Area of Cultivated Land. International Journal of Remote Sensing 23(2), 263–284 (2002)
7. Wu, Y., Yun, W., Li, W.: Research on Standard-plot Based Monitoring and Early-warning of Arable Land Quality. China Land Science 20(4), 40–45 (2006)
8. Ministry of Land and Resources: Regulations for Classification on Agricultural Land (TD/T1004 - 2003). China Standards Publishing House, Beijing (2003)
9. Kish, L.: Survey Sampling. John Wiley and Sons, USA (1985)
10. Wang, J., et al.: Spatial Analysis. Science Press, Beijing (2006)
11. Matheron, G.: The Intrinsic Random Functions and Their Applications. Advances in Applied Probability 5(3), 439–468 (1973)
12. Matheron, G.: Kriging or Polynomial Interpolation Procedures – a Contribution to Polemics in Mathematical Geology. Canadian Mining and Metallurgical Bulletin 60 (1967)
13. Oliver, M.A., Webster, R.: Kriging: a Method of Interpolation for Geographical Information Systems. International Journal of Geographical Information Systems 4(3), 313–332 (1990)
14. Wang, J., Robert, H., Steve, W.: Spatial Sampling Design for Monitoring Drought, Flood and Earthquake Disaster in China. Progress in Natural Science 9(4), 50–59 (1999)
15. Griffith, D.A., Haining, R., Arbia, G.: Heterogeneity of Attribute Sampling Error in Spatial Data Sets. Geographical Analysis 26(4), 300–320 (1994)
16. Barnett, V.: Elements of Sampling Theory. Edward Arnold, London (1974)
17. Griffith, D.A., Bennett, R.J., Haining, R.P.: Statistical Analysis of Spatial Data in the Presence of Missing Observations: a Methodological Guide and an Application to Urban Census Data. Environment & Planning A 21(11), 1123–1151 (1989)
18. Simbahan, G.C., Dobermann, A.: Sampling Optimization Based on Secondary Information and Its Utilization in Soil Carbon Mapping. Geoderma 133, 345–362 (2006)

19. Guo, R.: Spatial Analysis. Higher Education Press, Beijing (2001)
20. Jiang, C., Wang, J., Cao, Z.: A Review of Geo-Spatial Sampling Theory. Acta Geographica Sinica 64(3), 368–380 (2009)
21. Journel, A.G.: Nonparametric Geostatistics for Risk and Additional Sampling Assessment. In: Principles of Environmental Sampling, pp. 45–72. American Chemical Society (1988)
22. van Groenigen, J.W., Siderius, W., Stein, A.: Constrained Optimisation of Soil Sampling for Minimisation of the Kriging Variance. Geoderma 87, 239–259 (1999)
23. Wang, J., Jiang, C., Li, L., Hu, M., et al.: Spatial Sampling and Inference. Science Press, Beijing (2009)

The Potential Geographical Distribution of *Bactrocera cucurbitae* (Diptera: Tephritidae) in China Based on Eclosion Rate Model and ArcGIS*

Zhimei Li[1,3], Ningbo Wang[1,2,3], Jiajiao Wu[4], Jay Richard Stauffer[5], and Zhihong Li[1,**]

[1] Department of Entomology, College of Agronomy and Biotechnology, China Agricultural University, Beijing 100193, P.R. China
[2] Beijing Science and Technology Publishing Co., Ltd, Beijing Academy of Science and Technology, Beijing 100035,P.R. China
[3] These authors contributed equally to this work.
[4] Guangdong Entry-Exit Inspection and Quarantine Bureau, Guangzhou, 510623, P.R. China
[5] Ecosystem Science and Management, Penn State University, University Park, PA 16802, USA
lizh@cau.edu.cn

Abstract. The melon fruit fly, *Bactrocera cucurbitae* Coquillett (Diptera: Tephritidae), is one of the important insect pests of fruits and vegetables. In order to monitor and control it effectively, it is necessary to know the potential geographical distribution of this pest. The ER (Eclosion rate) model was constructed from empirical biological data, and analyzed with ArcGIS. Based on the soil temperature and moisture data of Chinese meteorological stations, the potential geographical distribution of *B. cucurbitae* from January to December in China was predicted. Six categories were used to describe different levels of suitability for *B. cucurbitae* in China. The potential geographical distribution and suitable levels for every month in China were obtained and showed that almost all locations were suitable from May to September. Further analysis showed that monitoring measures should be taken in Guangdong, Guangxi, Yunnan, and Hainan provinces throughout the year.

Keywords: *Bactrocera cucurbitae*, potential geographical distribution, ArcGIS, eclosion rate model, plant quarantine.

1 Introduction

The melon fruit fly, *Bactrocera cucurbitae* Coquillett (Diptera: Tephritidae), originated from India [1], spread throughout Southeast Asia in the 1990s, and then spread to other regions. At present, *B. cucurbitae* is found in some areas of Asia, Africa, and the Pacific Islands, including more than 30 tropical and subtropical countries and regions [2-3]. In China, *B. cucurbitae* was mainly distributed in Guangdong, Guangxi, Hainan, Fujian, Yunnan (the geographic distribution in Yunnan can be divided into year-round

* This study was supported by the Program of Ministry of Agriculture, China (No. 2012-Z15).
** Corresponding author.

D. Li and Y. Chen (Eds.): CCTA 2012, Part II, IFIP AICT 393, pp. 334–342, 2013.

distribution and seasonal distribution. The year-round distribution area was located at the South of latitude 24°), Taiwan, and Hong Kong. In addition, *B. cucurbitae* was found in Sichuan, Chongqing, Guizhou, Zhejiang, Jiangsu, Hunan, and other places in recent years [4-10]. *Bactrocera cucurbitae* has been reported to attack more than 100 species of fruits and vegetables, and has been listed as an important quarantine pest by many countries including China [11-12]. *Bactrocera cucurbitae* often caused losses of 30% to 90%. According to the reports, *B. cucurbitae* had caused a reduction in production of watermelon of India, bitter gourd of Nauru, and towel gourd and pumpkin of Solomon by 28.55%, 95%, 90% and 87%, respectively [1]. Japan had spent 20.4 billion yen to eradicate *B. cucurbitae* in Okinawa islands [13].

Population fluctuation of *B. cucurbitae* not only has been correlated with the temperature, but also with the relative humidity and rainfall [14].Temperature had a significant impact on *B. cucurbitae* [15], under laboratory conditions; when the temperature was below 14℃, all instars of *B. cucurbitae* did not develop normally [16-17]. The low temperature of developmental threshold for pupae was 11.14℃, developmental temperature range was 11-36℃ [17], and the optimum developmental temperature range was 22-30℃ [18]. When the temperature was greater than 32.2℃, the mortality rate to increased, and under the condition of 34°C pupae mortality rate was 95.27%±0.35% [18]. Soil moisture also affected the pupae greatly. In India, the report showed two peak period of melon flies attacking the fruits -- the one is in June with the relative humidity 72.0% -84.0 % and rainfall 53.3-61.5 mm, the other is the second and third weeks in July with the relative humidity 85.0% and rainfall 114.1-247.5 mm [19]. Mature larvae of *B. cucurbitae* pupated in the soil of 0.5-15 cm depth [20-21], and it is favorable for the mature larvae to pupate with higher eclosion rate when the soil moisture is less than 25%. When the soil moisture is greater than 30%, both the pupation rate and eclosion rate are lower [18].

The potential geographical distribution study of fruit flies started with, climate maps, ecological niche modeling software, technical methods, etc. and involved several species (e.g., *Ceratitis capitata* Wiedemann, *B. dorsalis* (Hendel), *B.* (*Daculus*) *oleae*). In 1924, Cook (1925) used climate maps predict the pest potential geographical distribution, and then this technology was further developed [22]. Messenger (1972) used artificial climate chambers to simulate a dozen typical weather conditions of the United States, to study the growth and development of the *C. capitata* Wiedemann, *B. dorsalis* (Hendel) and *B. cucurbitae* under different climatic conditions. Combined with climate analysis, the suitable geographical distribution of the three fruit fly species in the United States was described [23]. Yonow applied CLIMEX to study the potential geographical distribution of the *B.* (*B.*) *tryoni* in Australia [24].

The potential geographical distribution study of fruit flies started later in China, but there are many papers in recent years, including the Multiple Indicator fuzzy comprehensive evaluation technology, agro-climatic similarity distance technology, geographic information systems (GIS) technology and ecological niche modeling software, among others. Many fruit flies were involved such as *B. dorsalis* (Hendel), *B. cucurbitae* Coquillett, *Anastrepha ludens* Loew and *Rhagoletis pomonella* Walsh. Zhou (2005) built the lethal temperature model and the effective accumulated temperature model of *B. cucurbitae* Coquillett in 2005, and used 30 years' climate data of the 670 meteorological stations in China to run the model. The prediction results showed that the melon fly could occur in 36.27% of China, the northern boundary of the distribution is about 31°±2°N. It could produce 2-10 generations/year, and mainly produced 4-7 generations/year [25]. Kong et al. (2008) analyzed the potential geographical distribution of *B. dorsalis* (Hendel) and *B. cucurbitae* Coquillett in China and world by combining the CLIMEX and DIVA-GIS in 2008. The highly suitable areas of *B. cucurbitae* Coquillett in China included Guangdong, Guangxi, Hainan, southern of Fujian, southern of Yunnan, western of Taiwan, and the Sichuan Basin, while moderate to low suitable areas include parts of Jiangxi, Hunan, Guizhou, Chongqing, Shanghai and Sichuan, Yunnan, Fujian, Zhejiang, Jiangsu, Anhui, Hubei, Shaanxi, Henan, and Gansu [26].

The purpose of this study was to establish the ER (eclosion rate) model of B. cucurbitae, and analyze the potential geographical distribution and suitability levels using the ER model and displaying distribution patterns using soil temperature and soil moisture of past years in China by ArcGIS.

2 Materials and Methods

The *B. cucurbitae* samples collected from Huangpu in Guangdong province were selected for the eclosion experiment. Eggs of *B. cucurbitae* were obtained from adults that had been reared for four generations on an artificial diet. Mature larvae (6 days after egg hatch) were placed in moist sand (75% relative humidity) at 29°C for pupation. All 6480 pupas were gathered after 24h under and held at 25°C.

The eclosion rate (ER) data were collected by placing pupae in a plastic box (high 7cm, diameter 12.5cm) containing medium soil (Guangdong, DaHan) in an Artificial Climate box (Germany Binder Kbwf240). Data were analyzed using SPSS13.0 (http://www.seekbio.com/soft/1492.html) and ArcGIS 9.0 (Environmental Systems Research Institute, ESRI). Soil temperature and moisture (2001-2003) were obtained from China Meteorological Administration.

The ER model, based on soil temperature and relative humidity was obtained from a crossover design experiment conducted at the plant quarantine laboratory of Guangdong Entry-Exit Inspection and Quarantine Bureau. The design specified six soil temperature grades: 9°C, 14°C, 19°C, 24°C, 29°C, and 34°C. Six relative humidity grades included 0%, 20%, 40%, 60%, 80%, and 100%. Experiments consisted of 36 treatments and 3 replications during about 25 days and every box had 60 pupas. All pupas were placed 2 cm under the soil and held in artificial climate boxes. Water was added as needed [27].The ER of *B. cucurbitae* was obtained and the ER model was derived by stepwise regression (SPSS 13.0)

Geographic distribution and suitability was calculated for each month of the year for all of China and displayed as maps by ArcGIS. Soil temperature and moisture data were limited, but at data for a minimum of 10 days per month were obtained. Since values were obtained from different locations in different months, the full data set is presented in Table 4. The suitability of *B. cucurbitae* was analyzed on the basis of these data and ER model results. The ER of *B. cucurbitae* for every location in China was obtained.

For each location, suitability was categorized into 6 levels; negligible (where the *B. cucurbitae* is unable to occur and survive) (ER=0), extremely low (0<ER≤0.1), low (0.1<ER≤0.2), moderate (0.2<ER≤0.3), high (0.3<ER≤0.4) and extremely high (0.4<ER≤1). Suitability maps were plotted for each month using the inverse distance weight (IDW) raster interpolation.

3 Results

3.1 ER Model

SPSS was used to build the modelas it is convenient, fast, and reliable. The binary model was not significant (p>0.05) and the binary cubic model is too complicated to explain, so we chose binary quadratic regression model as the eclosion rate model of *B. cucurbitae*. After XY was dropped, the model chosen for eclosion rate was as follows:

$$Z=-0.0041X^2-0.00005Y^2+0.17976X+0.01044Y-1.618$$

Z is the ER (eclosion rate) of *B. cucurbitae*

X is the soil temperature

Y is the soil moisture

Table 1. Regression model summaries

Model	R	R Square	Adjusted R Square	Std. Error of the Estimate
1	0.840 (a)	0.706	0.692	0.206

a Predictors:(Constant), xy, xx, yy, y, x

Table 2. Analysis of variance of regression

Model		Sum of squares	Df	Mean square	F	Sig.
1	Regression	10.396	5	2.079	48.991	1.31E-25
	Residual	4.329	102	0.042		
	Total	14.725	107			
a	Predictors: (Constant) , xy, xx, yy, y, x					
b	Dependent Variable: z					

Table 1 shows the regression model summary, and four constants are available including XX, X, Y and YY. Table 2 shows the analysis of variance of regression. Stepwise regression ordered the independent variables according to their explanatory power. Analysis showed that XY did not contribute to the explanatory power of the model significantly (p>0.05), so it was dropped. By regression analysis, after XY was dropped, t=0.107, P>0.01(Table 3). The model was reliable because of R^2=0.706, F=48.991>F0.05 and P<0.01. The regression analysis of the ER showed that P was significant on 0.01 level.

Table 3. Excluded variables of regression model

Model		Unstandardized Coefficients		Standardized Coefficients	t	Sig.
		B	Std. Error	Beta		
1	(Constant)	-1.618	0.158		-10.218	2.69E-17
	x	6.112	0.485	4.157	12.593	1.68E-22
	y	1.004	0.253	0.928	3.961	0.0001
	xx	-4.743	0.367	-4.138	-12.906	3.59E-23
	yy	-0.539	0.199	-0.519	-2.713	0.0078
	xy	0.025	0.231	0.018	0.107	0.9152
a	Dependent Variable: z					

3.2 Potential Geographical Distribution of B. Cucurbitae in China

After the suitable range and suitable level analysis of *B. cucurbitae* in China monthly the results showed that the percent of suitable locations was about 11% in January and February, then it increased to 37.14% quickly in March, and it was 76.19% in April. Almost all locations were suitable from May to September, but it was 61.76% in October, and then decreased to 19% in November, and it was 12.28% in December. For the most part, it was basically the same in December, January and February, and there was a turning point in March and April with the suitable locations increasing. May to September was the most suitable period, and there was another turning point in October and November with the suitable locations decreasing. Finally, it accomplished an alternating cycle of the four seasons. Combined with the change of the different suitable level locations, May and September took the largest proportion of extremely high and high suitable level locations, followed by them was April, June, August and October, and then it was July, November, December, February and January. The percentage of extremely high level locations was the lowest and the percent of extremely low locations was the highest in January. The percentage of extremely high and high level locations increased linearly from January to May, and then it decreased from May to July. It increased from July to September, and then decreased from September to December. It was the lowest in July because the soil temperature exceeded the eclosion threshold temperature of *B. cucurbitae*.

The distribution of *B. cucurbitae* diffused towards north in March. Guangdong, Guangxi, Yunnan and Sichuan came to be the extremely high or high suitable level locations. We must pay attention that Shannan prefecture of Tibet, Juxian and Laiyang of Shandong came to be the extremely high or high suitable level locations in March. Ba county of Hebei Province (latitude 39.07°, longitude 116.23°) was the northern boundary of the suitable regions in March. The eclosion of *B. cucurbitae* occurred at Ba county in March when monthly average soil temperature was 10.89°C, the monthly average soil relative humidity was 16.90% (Fig. 1).

As a turning point, we can clearly see the division of the potential geographical distribution of *B. cucurbitae* in October. Extremely high and high suitable area concentrated in the southern provinces of China, in which we must pay attention to the Nuomuhong of Qinghai and Tianjin. Yili of Xinjiang (latitude 43.57°, longitude 81.20°) was the northern boundary of the suitable regions in October. Monthly average soil temperature was 10.93°C, and the monthly average soil relative humidity was 17.01% (Fig. 2).

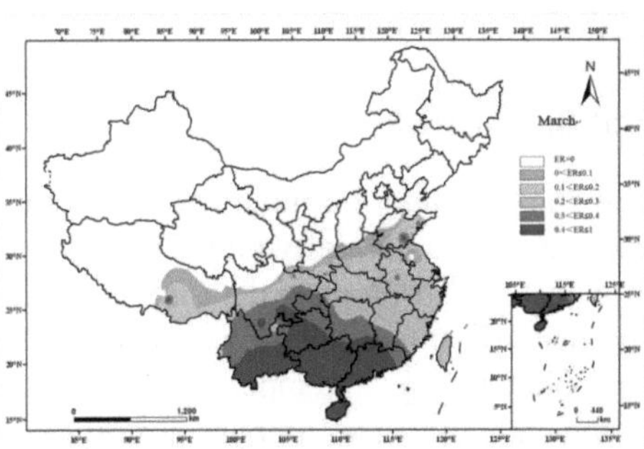

Fig. 1. The potential geographical distribution of *B. cucurbitae* in March

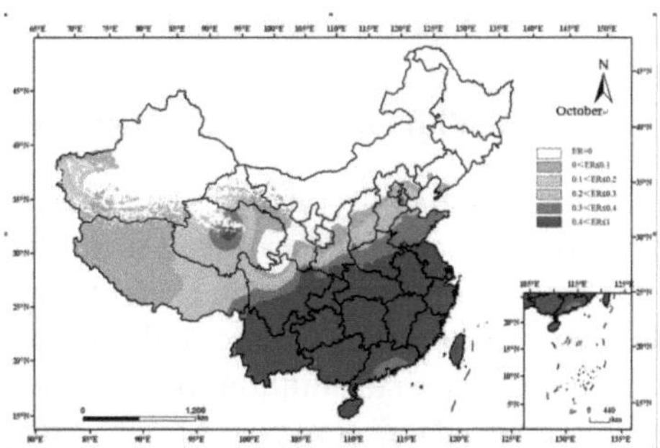

Fig. 2. The potential geographical distribution of *B. cucurbitae* in October

3.3 Conclusions

The results indicated that soil temperature and soil moisture affected the potential geographical distribution of *B. cucurbitae* in China and suitable areas changed with the seasons. From the coefficients of the variables in the model, the soil temperature affected the eclosion of *B. cucurbitae* more than soil moisture. In winter, the potential geographical distribution of *B. cucurbitae* was concentrated in the south of China, such as Yunnan, Guangxi, and Guangdong. In March, *B. cucurbitae* started to spread from south to north. Almost all locations were suitable from May to September.

By the analysis month by month, the suitable range and level of *B. cucurbitae* of the 12 months in China were obtained, and the causes of its distribution were analyzed in detail. It is useful for the quarantine departments to determine the distribution of the density of monitoring locations, the monitoring time according to the local conditions and the pest management time.

4 Discussion

4.1 Monitoring and Trapping Periods

This study indicated four monitoring and trapping periods for fruit fly eclosion in China. If continuous two months are not suitable, monitoring should be suspended. Monitoring measures should be taken throughout the year in Guangdong, Guangxi, Yunnan, and Hainan. The second period was from March 1 to November 30 and included Hubei, Hunan, Sichuan, Chongqing, Guizhou, Jiangxi, and Fujian. The third period, from March 1 to October 31 included Shandong, Jiangsu, Shanghai, Zhejiang, Anhui, Tibet. and Taiwan. The last period was from April 1 to October 31, and monitoring areas included Liaoning, Heilongjiang, Jilin, Inner Mongolia, Qinghai, Beijing, Tianjin, Hebei, Henan, Shaanxi, Shanxi, Gansu, Ningxia, and Xinjiang.

In summary, the two important natural factors (soil temperature and moisture) determined the potential geographical distribution of *B. cucurbitae* over 12 months in China were reported, and the ER model can be used to predict where and when these conditions are suitable for eclosion. This model also can be applied to other countries and will be improved with the development of the future research.

4.2 Potential Geographical Distribution of B. Cucurbitae in China

Almost all suitability analysis methods of fruit flies include the main natural factors that affect the occurrence of fruit flies, and temperature and humidity often is the key to determine fruit fly outbreaks. Under the situation that meteorological data are relatively limited, how to make full use of these data is particularly important; however, we often encounter data substitution or conversion problems (e.g., temperature and soil temperature data conversion), as well as the relationships among atmospheric humidity, rainfall, and soil moisture, which are very important to the adaptability analysis research of fruit flies.

In addition, the mode of 'biological experiments + biological model +GIS analysis' was suitable for suitability analysis of invasive fruit flies. To date, pest suitability analysis techniques use models combined with software analysis as its main analysis methods; thus, models and software for medical and military fields can be used to analyze the pest suitability. Once related environmental meteorological data are available, we can analyze the suitability of any instars of the invasive species year by year, month by month, even ten days by ten days and day by day, and a detailed study can increase the accuracy of suitability analysis and forecast. Detailed and completed information obtained from the literatures and biological experiments is very important for the potential geographical distribution research. We also need to explore how to combine the results of quantitative analysis and graphics, as the analysis of the species turns from qualitative methods towards quantitative methods.

4.3 Management Countermeasures

Quarantine measures should be strengthened domestically as well as at the borders (inspection and quarantine). The critical tasks of border inspection and quarantine are port inspection and quarantine from May to September, because many areas of China are suitable for *B. cucurbitae* during this time. Especially from imports of these continents in which the *B. cucurbitae* distributes: Asia (Oman, Bangladesh, Brunei, Cambodia, India, Indonesia, Japan's Ryukyu Islands, Laos, Malaysia, Myanmar, Nepal, Pakistan, Philippines, Sri Lanka, Thailand, Vietnam), Africa (Egypt, Kenya, Tanzania, Mauritius, some areas of Reunion), and Solomon Islands, Mariana Islands, Papua Island, Hawaii of USA and other parts of the Pacific Islands[1,3,28]. In addition, domestic inspection and quarantine should monitor the transport of fruits and vegetables from south to north in China, such as Guangdong, Guangxi, Hainan, Fujian, Yunnan, Taiwan, Hong Kong, Sichuan, Chongqing, Guizhou, Zhejiang, Jiangsu, Hunan, and other places [4-10]. Quarantine measures ought to be reinforced to prevent the infestations of fruits and vegetables.

Acknowledgments. The authors would like to thank Liang Guangqin, Mo Renhao, Hu Xuenan, Liu Haijun, Li Chunyuan and all other officers of Guangdong Inspection and Quarantine Technology Center to support the experiment. Thanks for all members of the Plant Quarantine and Invasion Biology Laboratory of China Agricultural University (CAUPQL). This study was supported by the Program of Ministry of Agriculture, China (No. 2012-Z15).

References

1. Dhillon, M.K., Singh, R., Naresh, J.S., et al.: The melon fruit fly, *Bactrocera cucurbitae*: A review of its biology and management. Journal of Insect Science 5 (2005)
2. Animal and Plant Quarantine Administration of the People's Republic of China and Plant Quarantine Institute of Ministry of Agriculture: A Selection of the entry plant quarantine pests of China. China Agriculture Press, Beijing (1997) (in Chinese)
3. EPPO. Distribution Map of Quarantine Pests for Europe *Bactrocera cucurbitae* (2006), `http://pqr.eppo.org/datas/DACUCU/DACUCU.pdf`
4. Wang, X.J.: Overview about important fruit fly pests of fruits and vegetables. Plant Quarantine 9, 20–30 (1995) (in Chinese)
5. Wang, X.J.: Diptera Tephritidae insects of East Asia. Acta Zootaxonomica Sinica, 331–338 (1996) (in Chinese)
6. Zhou, Y.S., Zhao, H.P.: Biology and other comprehensive prevention and control of *B. dorsalis* for mangos. Journal of Southwest Agricultural University 3, 210 (1996) (in Chinese)
7. Jiang, X.L., Ren, L.Q., Xiao, S., He, W.Z., Sun, B.Z.: Study on the biology of *Bactrocera cucurbitae*. Plant Quarantine 17, 74–76 (2003) (in Chinese)
8. Jiang, C.M., Ai, H.M., Zhao, S.X.: Pest Risk Analysis of *Bactrocera cucurbitae*. Entomological Journal of East China 14, 256–259 (2005) (in Chinese)
9. Jiang, C.M., Ai, H.M., Zhao, S.X.: Life tables of the melon fly laboratory population reared on various host fruits. Journal of Fujian Agriculture and Forestry University (Natural Science Edition) 35, 24–28 (2006) (in Chinese)
10. Zhu, Z.H., Ye, H., Zhang, Z.Y.: Genetic relationships among four *Bactrocera cucurbitae* geographic populations in Yunnan Province. Chinese Journal of Applied Ecology 16, 1890–1892 (2005) (in Chinese)

11. Chen, H.D., Zhou, C.Q., Yang, P.J., Liang, G.Q.: On the seasonal population dynamics of melon and oriental fruit flies and pumpkin fly in Guangzhou area. Journal of Plant Protection 22, 348–354 (1995) (in Chinese)
12. Li, R.K.: Damage and Control of *Bactrocera cucurbitae*. China Vegetables, 26–27 (1997) (in Chinese)
13. Koyama, J., Kakinohana, H., Miyatake, T.: Eradication of the melon fly, *Bactrocera cucurbitae*, in Japan: importance of behavior, ecology, genetics, and evolution. Annual Review of Entomology 49, 331–349 (2004)
14. Zaman, M.: Assessment of the male population of the fruit flies through kairomone baited traps and the association of the abundance levels with the environmental factors. Sarhad Journal of Agriculture 11, 657–670 (1995)
15. Manzar, A., Srivastava, J.P.: Population fluctuation of fruit flies, *Bactrocera* spp. infesting bitter gourd in Central Uttar Pradesh. Progressive Horticulture 36, 146–149 (2004)
16. Keck, C.B.: Effect of temperature on development and activity of the melon fly. Journal of Economic Entomology 44, 1001–1002 (1951)
17. Liang, G.Q., Liang, G.Z., Lin, M., et al.: Fruit fly and its Control. Guangdong Science and Technology Press, Guangzhou (1993) (in Chinese)
18. Yuan, S.Y., Kong, Q., Xiao, C., Chen, B., Yang, S.S., Zhang, J.B., Sun, W., Li, Z.Y.: Influence of Temperature on Development, Survival and Fecundity of *Bactrocera dorsalis* Hendel. Journal of Huazhong Agricultural 24, 588–591 (2005) (in Chinese)
19. Ingoley, P., Mehta, P.K., Verma, K.S.: Seasonal occurrence of fruit fly, *Bactrocera cucurbitae* Coq., on cucumber in midhills of Himachal Pradesh. Himachal Journal of Agricultural Research 28, 48–53 (2002)
20. Jackson, C.G., Long, J.P., Klungness, L.M.: Depth of pupation in four species of fruit flies (Diptera: Tephritidae) in sand with and without moisture. Journal of Economic Entomology 91, 138–142 (1998)
21. Pandey, M.B., Misra, D.S.: Studies on movement of Dacus cucurbitae maggot for pupation. Shashpa 6, 137–144 (1999)
22. Cook, W.C.: The distribution of the alfalfa weevil Phytonomus Posticus Gyll. A study in physical ecology. Journal of Agricultural Research 30, 479–491 (1925)
23. Messenger, P.S.: Bioclimatology and prediction of population trends. In: Pro. FAO Conf. Ecology in Relation to Plant Pest Control, Rome, pp. 21–45 (1972)
24. Yonow, T., Sutherst, R.W.: The geographical distribution of the Queensland fruit fly, *Bactrocera (Dacus) tryoni*, in relation to climate. Australian Journal of Agricultural Research 49, 935–953 (1998)
25. Zhou, W.C.: Research on geographic distribution model of *Bactrocera cucurbitae*. Entomological Research Progress, 36–40 (2005) (in Chinese)
26. Kong, L.B., Lin, W., Li, Z.H., Huang, G.S., Liang, Y.B.: Effects of climatic factors on the growth and geographical distribution of Oriental fruit fly, *Bactrocera dorsalis*. Chinese Bulletin of Entomology 45, 528–531 (2008) (in Chinese)
27. Wu, Q.H., Shao, Z.X., Su, D.M.: Insect ecology experiments. Fudan University Press, Shanghai (1991) (in Chinese)
28. Stonehouse, J.M., Stravens, R., Dacambra, P.: The Melon Fly *Bactrocera cucurbitae* in the Republic of Seychelles. In: Proceedings of the Indian Ocean Commission, Regional Fruit Fly Symposium, pp. 211–213 (2000)

The Potential Geographical Distribution of *Locusta migratoria tibetensis* Chen (Orthoptera: Acrididae) in Qinghai-Tibet Plateau*

Xiongbing Tu[1,2], Zhihong Li[1,**], Zehua Zhang[2,*], Zhigang Wu[1], Wenlong Ni[1], Liao Fu[1], and Yasen Shali[2]

[1] Department of Entomology, College of Agriculture and Biotechnology, China Agricultural University, Beijing 100193, China
[2] Key Laboratory for Biological Control of Ministry of Agriculture, Institute of Plant Protection, Chinese Academy of Agricultural Sciences, Beijing, 100193, China
txb1208@163.com, lizh@cau.edu.cn, lgbcc@263.net

Abstract. *Locusta migratoria tibetensis* Chen (Orthoptera: Acrididae) is one of the most harmful species of Acrididae. It causes extensive damage to gramineous crops. The climate mapping program, CLIMEX 3.0, and the GIS software, ArcGIS 9.3, were used to model the current and future potential geographical distribution of *L. m. tibetensis*. The model predicts that, under current climatic conditions, *L. m. tibetensis* will be able to establish itself throughout much of Qinghai-Tibet Plateau in China, including Qinghai , Tibet and Sichuan province. A total of 94.23% of Qinghai-Tibet Plateau's land mass or 2.22×10^6 km^2, is climatically suitable under current climatic conditions. In this study, the potential distribution of *L. m. tibetensis* under future climate scenarios was projected and quantified climatically suitable areas. In Qinghai-Tibet Plateau, the total suitable climate land mass was increased by 0.16×10^6 km^2 or 0.70% until 2100. However, compared to 2020 and 2050, the suitable land mass was decreased by 0.47×10^6 km^2 and 0.17×10^6 km^2. The most optimal area would increase 0.16×10^6 km^2 until 2100, the most of Qinghai-Tibet Plateau would be expected to offer optimal climatic conditions for the possible spread and establishment of the pest. The analysis of this information indicated a high degree of accuracy for the CLIMEX model. The significant increases in the potential distribution of *L. m. tibetensis* projected under the climate change scenarios considered in this study suggest that biosecurity authorities should consider the effects of climate change when undertaking pest risk assessments.

Keywords: *Locusta migratoria tibetensis* Chen, CLIMEX, potential geographic distribution, ArcGIS, climate change.

* This study received financial supports from the Special Fund for Agro-scientific Research in the Public Interest (201003079).

** Corresponding author, China Agricultural University and Chinese Academy of Agricultural Sciences contributed equally to this paper.

D. Li and Y. Chen (Eds.): CCTA 2012, Part II, IFIP AICT 393, pp. 343–351, 2013.

1 Introduction

Locusta migratoria tibetensis Chen is one of the most harmful pests and causes large amounts of damage to gramineous crops. Since 1990s, the occurrence of *L. m. tibetensis* has showed a rising trend. Especially in 2003 to 2006, there were 1000 to 3000 individuals per m^2 along the valley areas of Jinsha River, Yarlung Zangbo River and Yalong River, which had threaten the production of highland barley and grass [1-2].

The distribution area of *L. m. tibetensis* is only in China at present, including northern of Qinghai, Tibet and west of Sichuan province[3]. Studies showed that there was a significant correlation between increasing of potential distribution area and global warming. The annual surface temperature increased at the speed of 0.0301°C, while the potential distribution area increased 504.38 km^2 per year. This result indicated that a little increment in temperature would bring a great extension in distribution of *L. m. tibetensis*[2].

The CLIMEX model has been used to describe the potential distribution of some pests, such as *Ceratitis capitata* (Wiedemann)[4-5], *Bactrocera tryoni*(Froggatt) [6-7], *Carpomya vesuviana* Costa[8], *B. tsuneonis*(Miyake)[9], *B. scutellata* (Hendel)[10], *Toxotrypana curvicauda* Gerstaecker[11], *B. zonata*(Saunders)[12], *Melanoplus sanguinipes* (Fabricius) [13]. This model has also previously been used to predict the effects of climate change on species' potential distributions using either a regional global climate model [14] or synthetic climates[15-16]. Species examined using this approach include *B. dorsalis* (Handel)[17], *Melaleuca quinquenervia*[18] and *Melanoplus sanguinipes* (Fabricius)[13]. To know the potential distribution of the *L. m. tibetensis* under climate change assumptions due to potential climate change impact, the CLIMEX 3.0 model was used to assess the response of *L. m. tibetensis* to climate and to make predictions for the 2100s.

2 Materials and Methods

2.1 Overview of the CLIMEX Model

CLIMEX is a dynamic model[19] that integrates the weekly responses of a population to climate using a series of annual indices. The growth and stress indices are calculated weekly and then combined into an overall annual index of climatic suitability, the Ecoclimatic Index (EI). This index is expressed on a scale of 0 to 100. Some suggested guidelines are provided[20-21] based on the EI. An EI of 0.00 indicates that the climate is unsuitable for the species and that the species cannot persist in an area under average environmental conditions. An EI of 0.01–9.99 indicates marginal conditions, an EI of 10.00–19.99 indicates suitable conditions, and an EI of 20+ indicates optimal conditions. An EI of 100 indicates that conditions are perfect throughout the year. However, few environments are sufficiently stable to provide perfect habitat all year[17].

To visualize the results, the CLIMEX output was 'loose-coupled'[16] to a geographical information system (ArcGIS). This system was purchased from ESRI to create thematic maps. ArcGIS was also used to make projections of the land areas that would have a suitable climate for the pest species.

2.2 Meteorological Databases and Climate Change

Two climate databases were used in this modeling exercise. By default, CLIMEX uses 30-year averages of climate data to estimate climatic suitability. The CLIMEX standard meteorological dataset was first used to create an initial fit. This dataset accompanied CLIMEX version 3.0 and consists of 30-year averages from 1961 to 1990 for an irregularly spaced set of 2500 climate stations. Subsequently, a regular girded dataset of the normal climate for the same period TYN SC 2.0 (http://www.cru.uea.ac.uk/~timm/grid/TYN_SC_2_0.html)[22] was used to fine-tune the parameter fit. The dataset consisted of 67,420 points spaced on a 0.5° latitude×0.5° longitude regular grid for significant land areas worldwide.

The future average annual temperature will increase 1.3-2.1°C, 2.3-3.3℃, 3.9-6℃ until 2020, 2050, 2100 respectively. While the future annual mean precipitation will increase 2-3%, 5-7%, 11-17% until 2020, 2050, 2100 respectively in China[17,23]. In this study, we use the average data of temperature and annual mean precipitation for the potential geographical distribution analysis in the future.

2.3 Fitting CLIMEX Parameters

To fit the CLIMEX model for *L. m. tibetensis*, the parameters were manually and iteratively adjusted until the simulated geographical distribution, as estimated by the EI values, coincided with the species' known native distribution. Parameters used in the CLIMEX model are presented in table 1.

Temperature Index

In this paper, the minimum temperature for development (DV0) was set at 14.6°C (Li *et al.*, 2007). The lower and upper optimum temperatures for *L. m. tibetensis* populations were set at 15–28°C[24]. So the lower and upper temperature optima (DV1 and DV2) were set at 15 and 28°C, respectively[24-25]. As a result of the species is established in Tibet, and the day average temperature is no more than 32°C (http://cdc.cma.gov.cn/), so the upper threshold temperature (DV3) was set at 32°C.

Moisture index

In commonly, Locust like desiccated environment for embryonic development stage. They survive well when the soil moisture in May at 0.03-0.3[26]. However, we have found that the soil moisture can be reach to 0.40 in October, this would be related with there was a long snow-covered time in Qinghai-Tibet Plateau. So in this paper, the lower soil moisture limit for development (SM0) was set to 0.02, the lower and upper limits for optimal growth (SM1 and SM2) were set to 0.03 and 0.5 respectively[25,27]. According to one year soil moisture changes of Tibet, the upper soil moisture limit for development, SM3, was set to 0.7 (http://cdc.cma.gov.cn/).

Table 1. Parameters used in the CLIMEX model for the *L. m. tibetensis*

Parameter	Value
DV0 Lower threshold temperature	14.60
DV1 Lower optimum temperature	15.00
DV2 Upper optimum temperature	28.00
DV3 Upper threshold temperature	32.00
SM0 Lower threshold of soil moisture	0.02
SM1 Lower limit of optimum soil moisture	0.03
SM2 Upper limit of optimum soil moisture	0.50
SM3 Upper threshold of soil moisture	0.70
DPD0 Diapause induction day length	11
DPT0 Diapause induction temperature (average weekly minimum)	11.0
DPD1 Diapause termination temperature (average weekly minimum)	3.0
DPD Diapause development days	120
DPSW Summer or winter diapause ('0' for winter diapause, '1'for summer diapause)	0
TTCS Cold stress temperature threshold	-17.00
THCS Cold stress accumulation rate	-0.0003
TTHS Heat stress temperature threshold	33.00
THHS Heat stress accumulation rate	0.005
SMDS Dry stress soil moisture threshold	0.002
HDS Dry stress accumulation rate	-0.003
SMWS Wet stress soil moisture threshold	0.70
HWS Wet stress accumulation rate	0.001

Diapause Parameter

Diapause is an important biology characteristic for locust to endure bad environment. In this paper, we take *Melanoplus sanguinipes* (Fabricius) for setting diapause parameter of *L. m. tibetensis*[13]. Compared to *Melanoplus sanguinipes*, the diapause day length and development days of *L. m. tibetensis* are very close to *Melanoplus sanguinipes* (http://cdc.cma.gov.cn/), also *L. m. tibetensis* is winter diapauses[26].

However, there was no related report about diapause induction and termination temperatures.

Stress Parameter

All stress parameters were adjusted by the real distributions in China, to ensure that the predictions of areas suitable for the insect were consistent with the insect's current absence.

2.4 Estimating the Size of Land Areas of Suitable Climate Projected by the CLIMEX Model

The land areas projected to have a climate suitable under current and future conditions for *L. m. tibetensis* (EI>0) were quantified for Qinghai-Tibet Plateau (table 2).

Table 2. Projected area that is climatically suitable (EI>0) for *L. m. tibetensis* under the current climate, expressed as an area ($10^6 km^2$) and as a percentage of the total land of Qinghai-Tibet Plateau, and the percentage changes in Qinghai-Tibet Plateau for 2020, 2050 and 2100 in comparison with those obtained under the current climate. The climate model was run for the future climate changes in China[23].

	Area with EI>0 under current climate	Mean changes in areas with EI>0 under future climate in comparison with those obtained under current climate		
		2020	2050	2100
Total area ($10^6 km^2$)	2.22	-0.47	-0.17	0.16
% total land area	94.23	-1.99	-0.72	0.70

3 Results

3.1 The Potential Distribution of L. m. Tibetensis under Current Climatic Conditions

The projections using the model suggest that under current climatic conditions, *L. m. tibetensis* can potentially establish itself to the Qinghai-Tibet Plateau (fig. 1, a). The results of the analysis indicate that Qinghai-Tibet Plateau would be expected to offer optimal climatic conditions for the possible spread and establishment of the pest. After CLIMEX modeling, the extent of the land area that is climatically suitable for *L. m. tibetensis* under the current climate is quantified by ArcGIS in table 2. A total of 94.23% of Qinghai-Tibet Plateau's land mass or 2.22×10^6 km^2, is climatically suitable.

3.2 The Potential Distribution of L. m. Tibetensis under Future Climate Scenarios

In this study, the potential distribution of *L. m. tibetensis* under future climate scenarios was projected and quantified climatically suitable areas. In Qinghai-Tibet Plateau, the total suitable climate land mass was increased by 0.16×10^6 km^2 or 0.70% until 2100 (table 2). However, compared to 2020 (fig.1, b) and 2050 (fig.1, c), the suitable land mass was decreased by 0.47×10^6 km^2 and 0.17×10^6 km^2 (table 2). The most optimal distribution area (EI: 20.00+) would increase 0.16×10^6 km^2 until 2100 (fig.1, d).

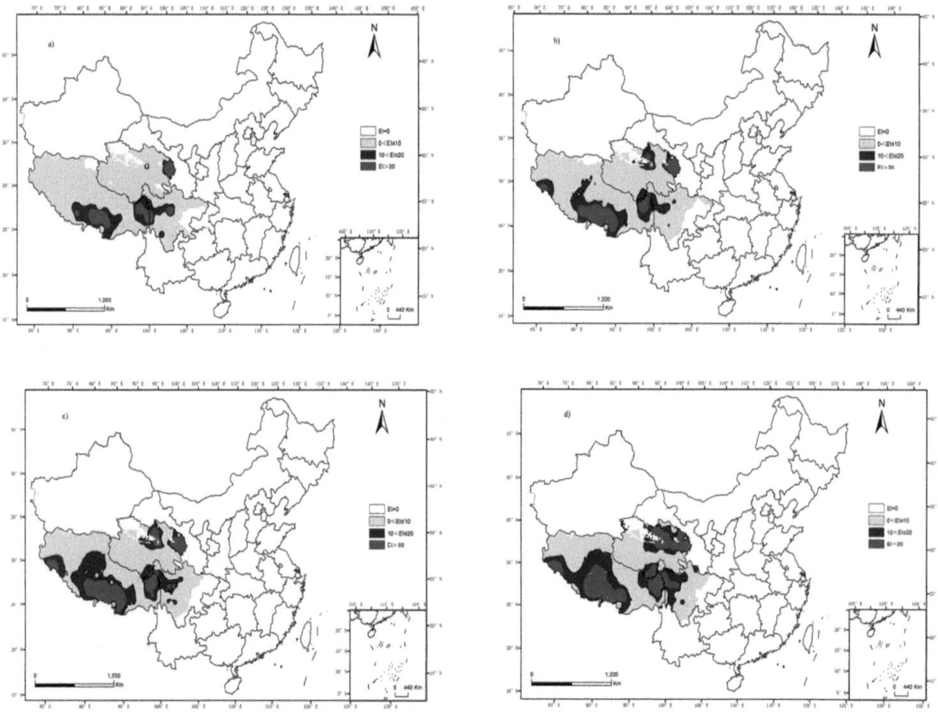

Fig. 1. Climatic suitability (EI) for the locust in the current and future climate scenarios projected using CLIMEX™. a)Qinghai-Tibet Plateau, current climate; b)Qinghai-Tibet Plateau, 2020; c)Qinghai-Tibet Plateau, 2050; d)Qinghai-Tibet Plateau, 2100. □, unsuitable (0.00); ▢, marginal (0.01–10); ■, suitable (10–20); ■, optimal (20.00+).

4 Discussion

L. m. tibetensis as the independent subspecies which has been nominated by Chen[3]. The taxonomy of *L. m. tibetensis* has been studied recently; Zhang *et al.*[28-29] had illustrated the potential importance of phylogeographical history in shaping the divergence and distribution patterns of *L. m. tibetensis* and other subspecies, he thought that *L. m. tibetensis* had independent genetic evidence for strong population substructure, as a result of geographical blocking for there was no gene flow between the inner and outside of Qinghai-Tibet Plateau. This is the reason for the current distribution of *L. m. tibetensis*, and we only predicted the potential distribution in Qinghai-Tibet Plateau in this paper.

The results of this study indicate that *L. m. tibetensis* is expected to be able to establish itself under temperate zones. Such conditions are to be found in most of Qinghai-Tibet Plateau. Moreover, the results suggested that as climate changes, the potential distribution of *L. m. tibetensis* would expand into cold regions, as it would expand into northward of the Plateau.

We have known that *L. m. tibetensis* is only in Qinghai-Tibet Plateau at present[3]. Studies showed that a little increment in temperature would bring great extension in distribution of *L. m. tibetensis*[2]. This study presented an example of influence of global warming on plateau ecology. So how climatic changes influence the potential distribution of insect in Qinghai-Tibet Plateau will need more attentions[30-31].

Potential distributions cannot be predicted based on climate alone. There is also a need to consider dispersal and species interactions, such as host availability. For example, mangoes, *Mangifera indica* Linn, are distributed in over 70 countries, more than 90% are grown in Asia (India, Pakistan, Bangladesh, Myanmar, Malaysia, China, Thailand, etc.), while Tanzania, Zaire, Brazil, Mexico, the United States, etc., also cultivate mangoes[32]. Competition and the effects of natural enemies[33] should also be considerations. Moreover, the limiting factors considered to affect the geographical distributions of pests should include soil type, geographical features, natural and geographical barriers and human activities. For example, the host plant *B. tsuneonis* limited citrus[9], so the prediction of the potential distribution of *B. tsuneonis* in China considered the factor of host distribution. Additionally, the irrigation and land use map were also taken into consideration for *Carpomya vesuviana*[8] and *Rhagoletis pomonella*[34]. The land use map removed the desert, river and lake, and urban land use type, etc. Therefore, the predictions based on the model and the results discussed above surely have some limitations; these additional factors should also be considered, and more detailed risk assessments should be useful for obtaining more scientific predictions.

Acknowledgements. The authors thank Prof. Stauffer at Pennsylvania State University for his valuable suggestions and modifying the draft. We would like to thank Yan Fan, Zhimei Li and all other members of the Plant Quarantine Laboratory of China Agricultural University (CAUPQL); Also we would like to thank Dezhi Li and other members of Key Laboratory for Biological Control of Ministry of Agriculture, Institute of Plant Protection, Chinese Academy of Agricultural Sciences.

References

1. Chen, Y.L., Zhang, D.E.: Historical evidence for population dynamics of Tibetan migratory locust and the forecast of its outbreak. Insect Science 6, 135–145 (1999)

2. Feng, C.H., Shan, X.N., Guo, C., Luo, L.M.: The change of potential distribution of *Locusta migratoria tibetensis* Chen (Orthoptera: Acrididae) from 1961 to 2005. Acta Ent. Sin. 54, 694–700 (2011) (in Chinese)
3. Chen, Y.L.: A new subspecies of *Locusta migratoria*-Tibetan migratory locust (Locusta migratoria tibetensis subsp. n.). Acta Ent. Sin. 12, 463–475 (1963) (in Chinese)
4. Worner, S.P.: Ecoclimatic assessment of potential establishment of exotic pests. J. Econ. Ent. 81, 973–983 (1988)
5. Vera, M.T., Rodriguez, R., Segura, D.F., Cladera, J.L., Sutherst, R.W.: Potential geographical distribution of the Mediterranean fruit fly, *Ceratitis capitata* (Diptera:Tephritidae), with emphasis on Argentina and Australia. Environ. Ent. 31, 1009–1022 (2002)
6. Yonow, T., Sutherst, R.W.: The geographical distribution of the Queensland fruit fly, Bactrocera tryoni, in relation to climate. Aus. J. Agr. Res. 49, 935–953 (1998)
7. Sutherst, R.W., Collyer, B.S., Yonow, T.: The vulnerability of Australian horticulture to the Queensland fruit fly, *Bactrocera tryoni*, under climate change. Aus. J. Agr. Res. 51, 467–480 (2000)
8. Lv, W.G., Lin, W., Li, Z.H., Geng, J., Wan, F.H., Wang, Z.L.: Potential geographic distribution of Ber fruit fly, *Carpomya vesuviana* Costa, in China. Plant Quarantine 6, 343–347 (2008) (in Chinese)
9. Wang, J.W., Li, Z.H., Chen, H.J., Geng, J., Wang, Z.L., Wan, F.H.: The potential geographic distribution of *Bactrocera tsuneonis* (Diptera: Tephritidae). Plant Quarantine 1, 1–4 (2009) (in Chinese)
10. Ni, W.L., Li, Z.H., Wan, F.H., Wang, Z.L.: Potential geographic distribution of *Bactrocera scutellata* (Hendel), in China. Chin. J. Biol. Control (add) 2, 61–67 (2009) (in Chinese)
11. Qu, W.W., Li, Z.H., Wan, F.H., Wang, Z.L.: Potential geographic distribution of papaya fruit fly, in China. Chin. J. Biol. Control (add) 2, 68–73 (2009) (in Chinese)
12. Ni, W.L., Li, Z.H., Chen, H.J., Wan, F.H., Qu, W.W., Zhang, Z., Kriticos, D.J.: Including climate change in pest risk assessment: the peach fruit fly, *Bactrocera zonata* (Diptera: Tephritidae). Bull. Ent. Res. 10, 1–11 (2011)
13. Olfert, O., Weiss, R.M., Kriticos, D.: Application of general circulation models to assess the potential impact of climate change on potential distribution and relative abundance of *Melanoplus sanguinipes* (Fabricius) (Orthoptera: Acrididae) in north American. Psyche, 1–9 (2011)
14. Kriticos, D.J.: The role of modelling in weed management. In: Shepherd, R.C.H. (ed.) Proceedings of the Eleventh Australian Weeds Conference, Melbourne, Australia, September 30-October 3, pp. 560–569. Weed Science Society of Victoria (1996)
15. Kriticos, D.J., Sutherst, R.W., Brown, J.R., Adkins, S.A., Maywald, G.F.: Climate change and biotic invasions: a case history of a tropical woody vine. Biol. Invasions 5, 145–165 (2003a)
16. Kriticos, D.J., Sutherst, R.W., Brown, J.R., Adkins, S.A., Maywald, G.F.: Climate change and the potential distribution of an invasive alien plant: Acacia nilotica ssp. indica in Australia. J. Appl. Ecol. 40, 111–124 (2003b)
17. Stephens, A.E.A., Kriticos, D.J., Leriche, A.: The current and future potential geographical distribution of the oriental fruit fly, *Bactrocera dorsalis* (Diptera: Tephritidae). Bull. Ent. Res. 97, 369–378 (2007)
18. Watt, M.S., Kriticos, D.J., Manning, L.K.: The current and future potential distribution of *Melaleuca quinquenervia*. Eur. Weed Research Society Weed Research 49, 381–390 (2009)

19. Sutherst, R.W., Maywald, G.F., Kriticos, D.J.: CLIMEX Version 3: User's Guide. Hearne Scientific Software Pty Ltd., `http://www.Hearne.com.au` (accessed November 12, 2008)
20. Sutherst, R.W., Maywald, G.F., Bottomley, W., Bourne, A.: CLIMEX v2 CD and User's Guide. Hearne Scientific Software Pty Ltd., Melbourne (2004)
21. Sutherst, R.W., Maywald, G.F.: A climate model of the red imported fire ant, *Solenopsis invicta Buren* (Hymenoptera:Formicidae): implications for invasion of new regions, particularly Oceania. Environ. Ent. 34, 317–335 (2005)
22. Kriticos, D.J., Alexander, N.S., Kolomeitz, S.M.: Predicting the potential geographic distribution of weeds in 2080. In: Preston, C., Watts, J.H., Crossman, N.P. (eds.) Proceedings of the Fifteenth Australian Weeds Conference, Adelaide, September 24-28, pp. 27–34. Weed Science Society of Victoria (2006)
23. Qin, D.H., Chen, Y.Y.: Climate and environment evolution in China. Science Press, Beijing (in Chinese)
24. Li, Q., Feng, C.H., Zhang, M., Jiang, F., Yang, G., Luo, L.M.: The biological characteristics of *Locusta migratoria tibetensis* (Chen). Chin. Bull. Ent. 44, 210–213 (2007) (in Chinese)
25. Jia, M.: The research about the life habit, classes and biological features of locust at Ali Prefecture of TAR. Tibet Agr. Tech. 29, 11–16 (2007) (in Chinese)
26. Guo, F., Chen, Y.L., Lu, B.L.: The biology of the migratory locusts in China. Shandong Science and Technology Press, Shandong (1991) (in Chinese)
27. Chen, Y.L.: The main locust and ecological management of locust plague in China, pp. 68–70. Science Press, Beijing (2007) (in Chinese)
28. Zhang, D.X., Yan, L.N., Kang, L., Ji, Y.J.: Some unorthodox views on the classification and evolution of the migratory locusts in China prompted by molecular population genetic study. Acta Zool. Sin. 49, 675–681 (2003) (in Chinese)
29. Zhang, D.X., Yan, L.N., Kang, L., Ji, Y.J., Hewitt, G.M., Huang, Z.S.: Unexpected relationships of substructured populations in Chinese *Locusta migratoria*. BMC Evol. Biol. 9, 144–155 (2009)
30. Huang, F.S., Huang, C.M., Liu, J.P.: Studies on the acridoid fauna of Xizang and it's evolution. Acta Ent. Sin. 3, 157–170 (1981) (in Chinese)
31. Chen, Y.L.: A new record of the Desert locust, *Schistocerca Gregaria* (Forskal) in Tibet, China. Acta Ent. Sin. 25, 67 (1982) (in Chinese)
32. Ye, Q., Huang, X.J.: World mango production and trade. Chin. J. Trop. Agr. 22, 44–49 (2002)
33. Baker, R.H.A., Sansford, C.E., Jarvis, C.H., Cannon, R.J.C., MacLeod, A., Walters, K.F.A.: The role of climate mapping in predicting the potential geographical distribution of non-indigenous pests under current and future climates. Agr. Ecosyst. Environ. 82, 57–71 (2000)
34. Geng, J., Li, Z.H., Rajotte, E.G., Wan, F.H., Lu, X.Y., Wang, Z.L.: Potential geographical distribution of *Rhagoletis pomonella* (Diptera: Tephritidae) in China. Insect Science, 1–8 (2011)

Study on Application of Scale Invariant Feature Transform Algorithm on Automated Geometric Correction of Remote Sensing Images*

Hui Deng[1,2], Limin Wang[1,2], Jia Liu[1,2], Dandan Li[1,2], Zhongxin Chen[1,2], and Qingbo Zhou[1,2]

[1] Key Lab. of Resources Remote Sensing & Digital Agriculture, Ministry of Agriculture, 10081, Beijing, P.R. China
[2] Institute of Agriculture Resources and Regional Planning, Chinese Academy of Agricultural Sciences, 10081, Beijing, P.R. China
denghuizxy@126.com.

Abstract. In recent years, data collected from remote sensing satellite and aerophotography have been showing a geometric sequence increase. A method of Scale Invariant Feature Transform (SIFT) algorithm could be employed for the automatic geometric fine correction. This method could avoid the impact of the rotation and zooming of template matching during the image matching process, and it can also save the labor during the image processing operation. Based on the SIFT algorithm, this paper proposes a two-step method, which firstly conducts coarse match on feature points, and then further conducts fine correction on the coarsely matched feature points by using the least squares technique. The result indicates that, this method is an effective automatic matching method for remote sensing images.

Keywords: remote sensing image, automatic geometric correction, SIFT algorithm, least squares.

1 Introduction

Along with the rapid increase of huge amount of remote sensing data, the workload of geometric fine correction on manual image matching for the image data collected from different sources and in different time period is huge. Image matching is the preliminary work of a lot of remote sensing applications (such as multi-source data fusion, variation detection, image mosaic, motion detection, and object identification, etc, and its result directly affects the quality of the follow-up works[1]. Currently, the matching technology for the satellite remote sensing images from the same sensor, at different time period, with small angular variation has been relatively mature, and

* Foundation: The National High Technology Research and Development Program ("863" Program) of China (No. 2012BAH29B02， No. NC2010FB0002) and National Non-Profit Scientific Institution, Ministry of Finance of China(202-15).

D. Li and Y. Chen (Eds.): CCTA 2012, Part II, IFIP AICT 393, pp. 352–358, 2013.

some classical feature extraction algorithms, such as arithmetic operators[2] , arithmetic operators[3], are extracted. They proposes a multiscale registration technique using robust Scale Invariant Feature Transform (SIFT) features in Steerable-Domain[4]. The matching the images from multi-sensor with different resolutions is the focus and the difficulty of current researches[5]. Scholars both at home and abroad have conducted extensive in-depth research on this topic, but no perfect method, which could correctly match the multi-source image under different radiation distribution, dimension, rotation, and translation, has been established.

In recent years, scale invariant Feature Transform (SIFT) feature has achieved great success in image matching. The matching method based on the SIFT descriptor has been successfully applied in many fields, such as object identification [6] , motion panorama[7], and Structure From Motion (SFM)[8]. SIFT algorithm was introduced [6] and [9] in 1999 and was well established in 2004. They also conducted experiments on the many representative descriptors under six circumstances, such as different scenes, change in illumination, and geometry deformation of images, resolution difference, rotation, blurring, and image compression. The results indicate that SIFT descriptor has the best performance [10].

2 Introduction to the Algorithm

The extraction and matching algorithm based on SIFT point features mainly includes three steps: extraction of point features, computation of feature descriptors, and feature matching from [6] and [11].

The Gaussian smoothing is applied on original image with different standard deviation σ, to obtain the Gaussian differential image. The maximum or minimum gray-scale values on differential images are taken as the feature points. Formula (1) is gaussian smoothing formula, and the extraction process of feature points is illustrated in the Figure 1. In the Figure, each pixel on the Gaussian differential image (expressed by crosshair) is compared with 26 adjacent pixels (expressed by dot) within the 3X3 area of current and upper and lower adjacent dimensions. If the average values of Difference of Gaussian of a pixel is larger or smaller than the adjacent pixels, the pixel is taken as the local extreme point. Then, some local extreme points are eliminated according to the preset threshold value of Difference of Gaussian and edge effect standard, and remaining extreme points are taken as the final SIFT feature points.

$$G(x,y,\sigma) = \frac{1}{2\pi\sigma^2} e^{-(x^2+y^2)}/2\rho^2 \quad (1)$$

By taking feature point as the center, the image area with specified height and width is taken, to calculate the gradient direc0 tion and gradient strength of each pixel within the area. The total gradient strength of different gradient directions are calculated and are taken as the eigen vector. The calculation process of the eigen vector is illustrated in the Figure 3. In the Figure, the arrow in the left stands for the gradient strength and direction. The total gradient strengths of 8 directions within the 2×2 sub-areas are calculated separately, so as to obtain an eigen vector with 32 dimensions. The calculation of Lowe's study (Lowe, 2004) was on 8 directions within a 4×4 sub-area, so the feature descriptor of each SIFT point feature is the eigen vector of 4×4×8=128 dimensions.

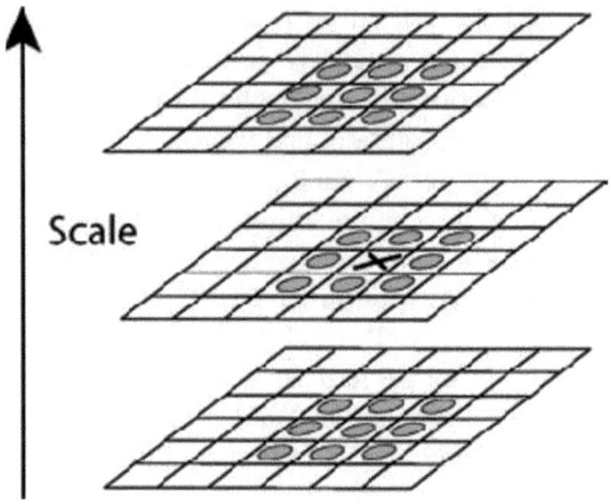

Fig. 1. SIFT point feature extraction process (Lowe, 2004)

Euclidean distance of the eigen vector of different feature points of image to be matched and Reference image is calculated, and the feature point with minimum distance is taken as the initial matching point. The error matching points are eliminated according to the nearest neighbor and second nearest neighbor Euclidean distance ratio.

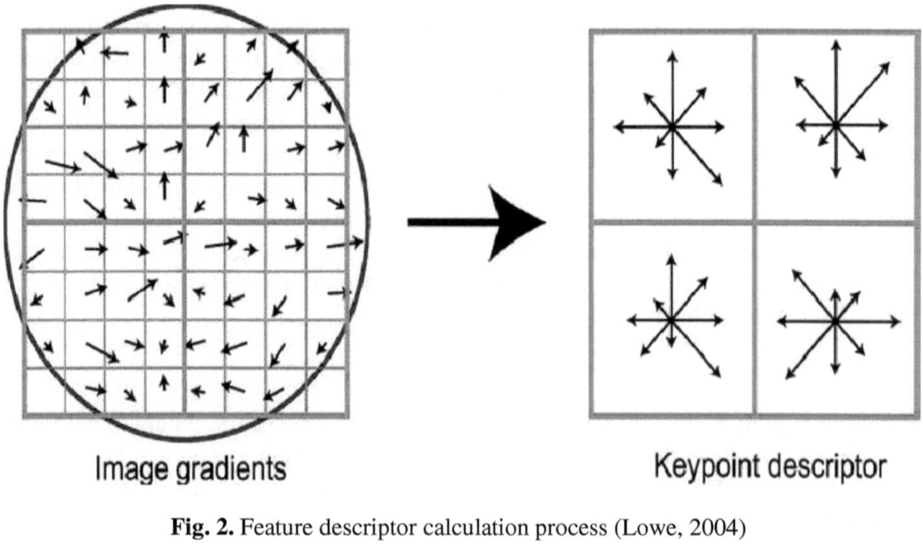

Fig. 2. Feature descriptor calculation process (Lowe, 2004)

3 Analysis and Conclusion

Firstly, feature point and eigen vector information of images to be corrected and Reference images is calculated. The partitioning technology is used to locate feature point pairs, so as to improve the feature point search efficiency, and then the coarse matching is conducted based on both images to find out the matched feature point pairs on both images. After the coarse matching, the fine matching on feature points by using least squares algorithm, to ensure that the total error of the feature points after fine matching is less than 0.5 pixel. Finally, the geometry fine correction based on finely matched feature points is conducted.

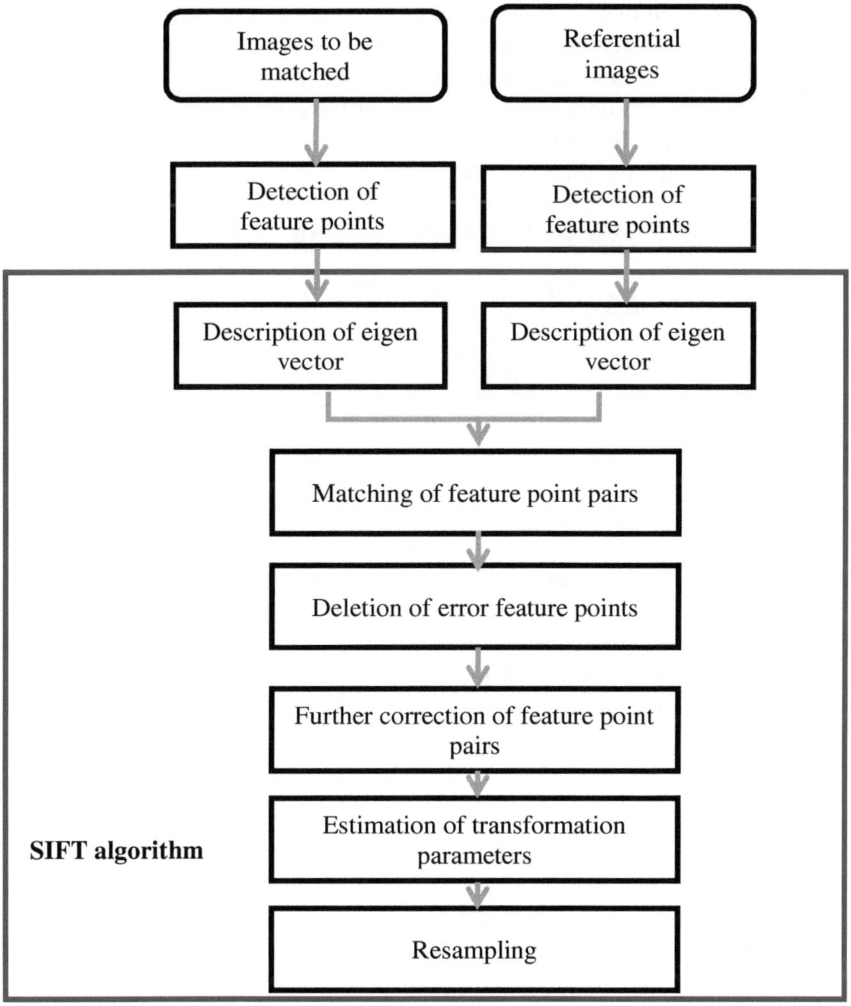

Fig. 3. Flow Chart of geometry fine correction

Feature points and eigenvector refer to the local features of detected images (e.g. edge, angular point, outline, etc). Then it is necessary to combine and convert the features based on the need of matching targets, so as to form eigen vectors which are easy to be matched with good stability, to convert the image geometry correction issue into feature matching issue, and further to convert the feature matching issue into clustering issue of eigenvector of feature space. Eigen vector information is a set of vectors with 128 dimensions. The following Figure illustrates part information of feature points and eigen vector of the Reference image.

The matching is conducted according to the similarity of eigen vector. Usually, various distance functions are taken as the similarity measurement of features, such as Euclidean distance and Mahalanobis distance, etc. The following figure is the schematic diagram of feature point location of two images after coarse matching. The SIFT algorithm is used to extract the 435 feature point pairs of two images.

A total of 50 pairs of feature points are randomly selected from 435 pairs of feature points and they are taken as the ground control points input into the ERDAS geometric correction procedure, to eventually get the error result, which is illustrated in the following Figure. In the figure, the error of the X direction is 4.9251, error of Y direction is 5.0334, and the total error is 7.0422.

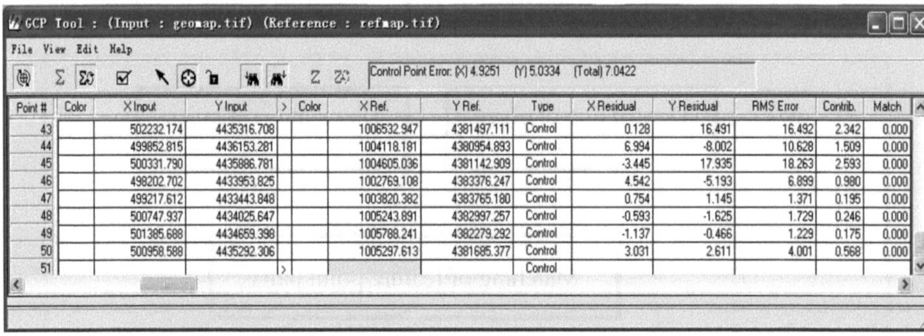
GCP Tool : (Input : geomap.tif) (Reference : refmap.tif)

File View Edit Help

Control Point Error: (X) 4.9251 (Y) 5.0334 (Total) 7.0422

Point #	Color	X Input	Y Input	>	Color	X Ref.	Y Ref.	Type	X Residual	Y Residual	RMS Error	Contrib.	Match
43		502232.174	4435316.708			1006532.947	4381497.111	Control	0.128	16.491	16.492	2.342	0.000
44		499852.815	4436153.281			1004118.181	4380954.893	Control	6.994	-8.002	10.628	1.509	0.000
45		500331.790	4435886.781			1004605.036	4381142.909	Control	-3.445	17.935	18.263	2.593	0.000
46		498202.702	4433953.825			1002769.108	4383376.247	Control	4.542	-5.193	6.899	0.980	0.000
47		499217.612	4433443.848			1003820.382	4383765.180	Control	0.754	1.145	1.371	0.195	0.000
48		500747.937	4434025.647			1005243.891	4382997.257	Control	-0.593	-1.625	1.729	0.246	0.000
49		501385.688	4434659.398			1005788.241	4382279.292	Control	-1.137	-0.466	1.229	0.175	0.000
50		500958.588	4435292.306			1005297.613	4381685.377	Control	3.031	2.611	4.001	0.568	0.000
51				>				Control					

Fig. 4. Error result of coarse matching feature points as GCP

The fine matching on the feature points of the coarse matching is conducted by using least squares algorithm, and the final matching accuracy is optimized. The specific operation process is as follows:

- To establish the dualistic quadratic regression equation between the coarsely matched feature points.
- The feature points are calculated based on the regression equation of each feature point of image to be corrected.
- The error between simulation feature points and the feature points on Reference image is calculated and the maximum error is calculated.
- When the maximum errors are larger than the system preset default parameter of 0.5, the feature point is deleted and the procedure return to step 1 to recalculate; otherwise, the fine matching of feature point pairs is completed. Then exit the calculation.

The following figure is the schematic diagram of feature point location after fine matching. The number of the feature point pairs after fine matching is 134.

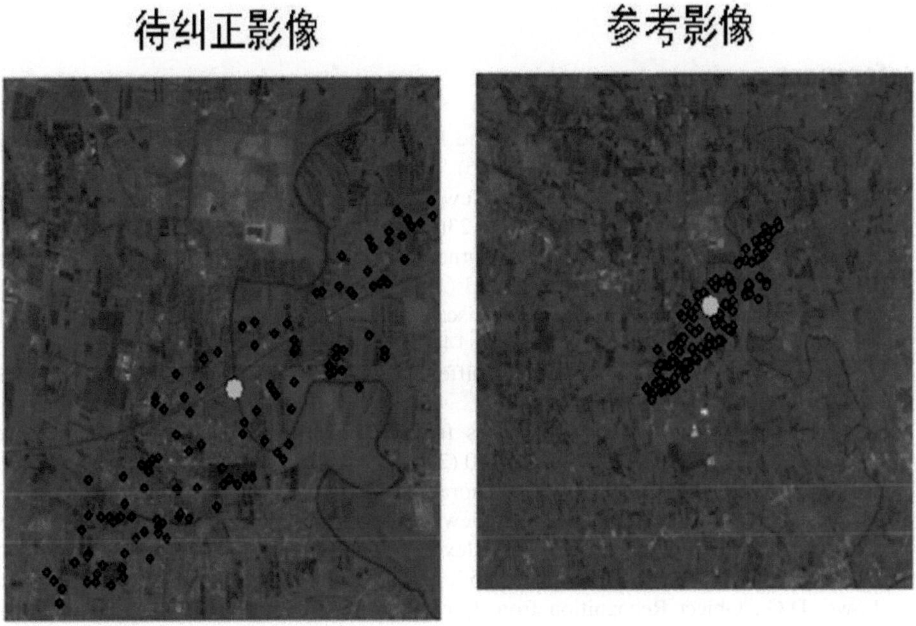

Fig. 5. Schematic diagram of feature point pair location after fine matching

A total of 134 pairs of feature points are input into the ERDAS geometric correction module, to eventually get the error result, which is illustrated in the following Figure. The error of X direction is 2.4689, error of Y direction is 2.1418, and the total error is 3.2685.

Resampling of corrected image is conducted by using the feature points after fine matching, to finish the final geometry fine correction.

The following figure is the schematic diagram of the location of the image after fine correction and the original Reference image.

A total of 20 pairs of control points are randomly selected from the image after fine correction and the original Reference images, to conduct the accuracy analysis, and the final error result is obtained, which is illustrated in the following Figure.The error of the X direction is 1.1018, the error of Y direction is 0.8426, and the total error is 1.3870.

The paper analyses the limitation of the conventional SIFT algorithm in the matching of remote sensing images, and the extraction time and matching accuracy of feature points is optimized. Optimization algorithm can extract more matched feature points with higher accuracy. Based on the SIFT algorithm improved by this paper, and by firstly conducting coarse matching and then fine correction, the error is reduced from 7.0422 to 1.3870. The accuracy of the automatic matching of remote sensing images is improved, which is suitable for the business operation. However, the expenses for the system calculation time is relatively high. The further optimization on algorithm in the aspect of parallel computation could be conducted so as to reduce the operation time.

References

1. Zhang, M.Q.: Research on the Line-Based Registration of Multi-Source Remote Sensing Imagery. Hohai University, Nanjing (2006)
2. Smith, S.M., Brady, J.M.: SUSAN-A New Approach to Low Level Image Processing. International Journal of Computer Vision 23(1), 45–78 (1997)
3. Harris, C., Stephens, M.: A Combined Corner and Edge Detector. In: Fourth Alvey Vision Conference, Manchester, UK, pp. 147–151 (1988)
4. Liu, X., Tian, Z., Chai, C., Fua, H.: Multiscale registration of remote sensing image using robust SIFT features in Steerable-Domain 14(2), 63–72 (2011)
5. Dufournaud, Y.: Matching Images with Different Resolutions. Theory in Computer Vision. Kluwer Academic Publishers (2004)
6. Lowe, D.G.: Distinctive Image Features from Scale-Invariant Keypoints. International Journal of Computer Vision 60(2), 91–110 (2004)
7. Brown, M., Lowe, D.G.: Recognising Panoramas. In: Proceedings of the 9th
8. Schaffalitzky, F., Zisserman, A.: Multi-view Matching for Unordered Image Sets, or How Do I Organize My Holiday Snaps? In: Heyden, A., Sparr, G., Nielsen, M., Johansen, P. (eds.) ECCV 2002, Part I. LNCS, vol. 2350, pp. 414–431. Springer, Heidelberg (2002)
9. Lowe, D.G.: Object Recognition from Local Scale-Invariant Features. In: International Conference on Computer Vision, Corfu, Greece (1999)
10. Mikolajczyk, K., Schmid, C.: A Performance Evaluation of Local Descriptors. IEEE Trans Pattern Analysis and Machine Intelligence 27(10), 1615–1630 (2005)
11. Li, Q.L., Wang, G.Y., Liu, J.G., Chen, S.B.: Robust scale-invariant feature matching for remote sensing image registration. IEEE Geoscience and Remote Sensing Letters 6(2), 287–291 (2009)

Research on Identifying Method of Freezing-Thawing Soil Hydraulic Properties

Zilong Wang[1,2], Qiuxiang Jiang[1], Qiang Fu[1], and Tianxiao Li[1]

[1] College of Water Conservancy & Architecture,
Northeast Agricultural University, P.R. China 150030
[2] Postdoctoral Mobile Research Station of Agricultural and Forestry Economy Management,
Northeast Agricultural University, P.R. China 150030
wzl1216@163.com

Abstract. The identification of freezing-thawing soil hydraulic properties is the basis and key problem of studying soil water movement. Current identifying methods of the properties have many limitations, such as big error, high uncertainty and low operability. Thus, the identifying method of freezing-thawing soil hydraulic properties was researched in the research. The estimating parameters of soil freezing characteristic curve were calculated by classical statistical method. Targeting at the maximum absolute difference between measured and simulated value of unsaturated soil hydraulic properties, the particle swarm optimization algorithm was used to estimate the optimal values of soil water characteristic curve, unsaturated hydraulic conductivity and specific water capacity. The numerical simulation of freezing-thawing soil water movement proved that the measured and simulated value of soil unfrozen water content had high degree of agreement. The identifying method proposed in the research is effective and feasible.

Keywords: Freezing-thawing soil, Soil hydraulic properties, Identifying method, Particle swarm optimization.

1 Introduction

Mathematical simulation is the method of researches on seasonal freezing-thawing soil water movement. The freezing-thawing soil hydraulic properties are the necessary data and the basis of quantification study precision whether in the analytic or numerical solutions. The hydraulic properties include soil freezing characteristic curve, soil water characteristic curve, unsaturated hydraulic conductivity (*K*), unsaturated diffusivity (*D*) and specific water capacity (*C*). For the later three properties can be related by the relational expression $D = K / C$, generally, two of them have to be identified.

The identifying method of freezing-thawing soil hydraulic properties includes direct manner and indirect manner [1]. The direct manner involves laboratory test and field experiment [2-3]. The indirect manner is the priority research area and can be found in many interrelated studies, such as pedo-transfer functions (PTFs), fractal,

D. Li and Y. Chen (Eds.): CCTA 2012, Part II, IFIP AICT 393, pp. 359–366, 2013.

soil morphology, numerical inversion and empirical formulae [4]. The current means have great error due to the complexity of freezing-thawing soil water movement [5-11]. After comprehensive analysis of current means, a new set of identifying methods of freezing-thawing soil hydraulic properties was proposed and its validity was verified in the study.

2 Basic Theories

2.1 Soil Freezing Characteristic Curve

Soil liquid water is not completely frozen when soil temperature is lower than 0℃ and a certain amount of soil unfrozen water exists all along. The content of unfrozen water in frozen soil depends on earthiness, environmental condition and freezing-thawing history. Soil freezing characteristic curve reflects a dynamic balance relation between unfrozen water content and negative temperature in frozen soil and can be expressed as follows:

$$\theta_u = A T_{abs}^{-B} \tag{1}$$

where θ_u is the soil unfrozen water content, T_{abs} is the absolute value of negative temperature, A and B are the empirical coefficients related with earthiness.

Soil freezing characteristic curve is a normal straight line and can be obtained by regression analysis of experimental data.

2.2 Van Genuchten-Mualem Model

van Genuchten-Mualem model is an empirical model of calculating unsaturated soil hydraulic properties. It is widely used in soil water movement research for it can properly fit filed experimental data, although its mathematical form is complex [12-13]. The formulae of van Genuchten-Mualem model used in the study are as follows:

$$|\psi| = \{[(\theta-\theta_r)/(\theta_s-\theta_r)]^{\frac{n}{1-n}} - 1\}^{\frac{1}{n}} / a \tag{2}$$

$$K = K_s\{1-(a|\psi|)^{n-1}[1+(a|\psi|)^n]^{\frac{1-n}{n}}\}^2 / [1+(a|\psi|)^n]^{\frac{n-1}{2n}} \tag{3}$$

$$C = (n-1)(\theta-\theta_r)\{1-[(\theta-\theta_r)/(\theta_s-\theta_r)]^{\frac{n}{n-1}}\} / |\psi| \tag{4}$$

where θ_s is the saturated soil water content, θ_r is the soil residual water content, ψ is the soil matric potential, a and n are the empirical parameters.

The two empirical parameters in van Genuchten-Mualem model can be obtained by reversely solving the relation curve of the measured soil unfrozen water content data and the corresponding soil matric potential data. However, the curve is complicated and its parameter estimation is a nonlinear fitting problem, therefore the particle swarm optimization algorithm is performed in the study to solve the problem.

Particle swarm optimization (PSO) algorithm was proposed by Kennedy and Eberhart in 1995 [14-15]. It is a kind of self-adapting evolutionary computation technology and swarm intelligence algorithm on the basis of population search. Each solution of optimization problem is considered as a particle among the searching space in PSO algorithm. Each particle has its own position, speed and adaptive value (*fitness*) decided by a certain optimization problem. The best position that a particle has experienced during the flight is the best solution found by the particle (called the personal extremum, *pbest*). The best position that the whole group has experienced is the best solution found by the whole group (called the global extremum, *gbest*). Every particle constantly updates itself by using *pbest* and *gbest* and then creates a new population. At last, the whole population can comprehensively search the solution region and find the optimal values [16].

Let N be the particle population size, x_i be the position of the i th particle ($i=1,2,\cdots,N$), v_i be the speed and f_i be the adaptive value of the particle. After the initial position and speed generate randomly, a particle updates itself by following the personal extremum $pbest_i(t)$ and the global extremum $gbest(t)$ in each iteration. At the moment of t+1, any particle i can update its own position and speed by the following formulae:

$$v_i(t+1)=wv_i(t)+c_1r_1(t)(pbest_i(t)-x_i(t))+c_2r_2(t)(gbest(t)-x_i(t)) \tag{5}$$

$$x_i(t+1)=x_i(t)+v_i(t+1) \tag{6}$$

where c_1 and c_2 are the learning factors, r_1 and r_2 are the random numbers evenly distributing in the interval of (0, 1), w is the inertia weight.

Then the personal extremum of each particle and the global extremum of all particles can update by the following formulae:

$$pbest_i(t+1)=\begin{cases} x_i(t+1) & f_i(t+1)\geq f(pbest_i(t)) \\ pbest_i(t) & f_i(t+1)< f(pbest_i(t)) \end{cases} \tag{7}$$

$$gbest(t+1)=x_{\max}(\mathrm{t}+1) \tag{8}$$

where $f_i(t+1)$ is the adaptive value of the ith particle at the moment of t+1, $f(pbest_i(t))$ is the best adaptive value of the ith particle in its searching history, $x_{\max}(t+1)$ is the position of the particle that corresponds with the biggest $f(pbest_i(t))$ among all the particles at the moment of t+1.

The two parameters (a and n) are considered as the particle of PSO. The objective function of PSO can be expressed by the following formula:

$$\min Q = \max(|Y_1 - y_1|, |Y_2 - y_2|, \cdots, |Y_m - y_m|) \tag{9}$$

where Y_i and y_i are the calculated and measured value respectively, m is the number of the measuring point in soil profile.

The algorithm will stop running when the difference between the former and later iteration value of the global optimal particle is less than error threshold. Thus, the convergence criterion of PSO can be expressed as follows:

$$\left|a_{gbest}^{p} - a_{gbest}^{p-1}\right| \le e_1 \,\&\, \left|n_{gbest}^{p} - n_{gbest}^{p-1}\right| \le e_2 \tag{10}$$

where a_{gbest} and n_{best} are the global optimal values of the model parameters (a and n) in each iteration, p and $p-1$ are the iteration number, e_1 and e_2 are the convergence thresholds of the model parameters (a and n).

3 Results and Analysis

3.1 Fixing Soil Freezing Characteristic Curve

The soil profile in the study was divided into two layers, one extended from 0 to 100cm and the other one is under 100cm. The functional relations of soil unfrozen water content and negative temperature were built separately on the basis of the measured data from two different soil layers. The curve fitting situations were showed in Fig. 1.

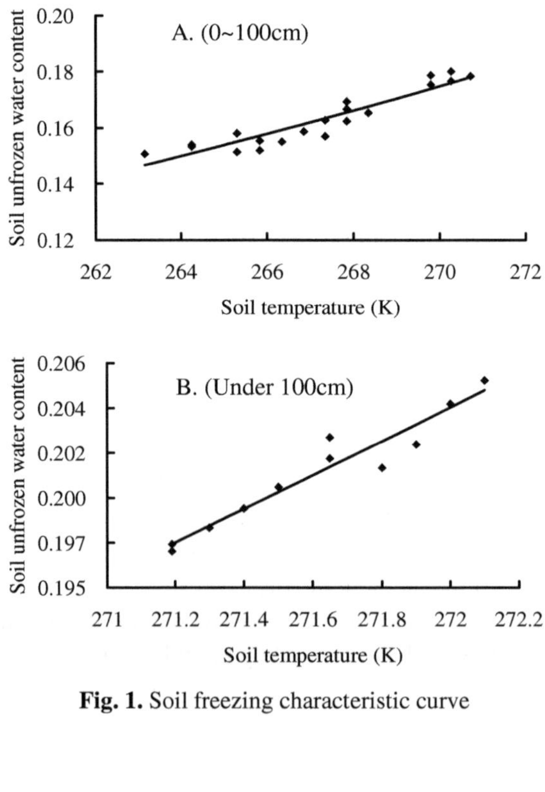

Fig. 1. Soil freezing characteristic curve

The expressions of soil freezing characteristic curves of the two soil layers are as follows:

$$\theta_u = 2E - 18(T + 273.16)^{6.935} \quad (0\text{~}100\text{cm}) \tag{11}$$

$$\theta_u = 6E - 32(T + 273.16)^{12.535} \quad (\text{Under } 100\text{cm}) \tag{12}$$

where θ_u is the soil unfrozen water content, T is the negative temperature.

3.2 Estimating Unsaturated Soil Hydraulic Properties

The PSO algorithm was programmed in Matlab 7.0. Set population size of particle swarm is 200; the maximum iteration time is 10000; the convergence thresholds as 1×10^{-8} and 1×10^{-4}; the learning factors as 2 and the inertia weight changing from 0.9 to 0.1. After running the PSO algorithm program, the optimal solution of the objective function was found when the program ran to the convergence criterion and the model parameters of the two soil layers were listed in Table 1.

Table 1. Results of estimating model parameters

Depth (cm)	θ_s	θ_r	a (cm^{-1})	n
0~100	0.46735	0.02	0.000086	1.6939
Under 100	0.458	0.03	0.0684	1.1271

The soil water characteristic curves of the two layers were fixed by putting the parameters from Table 1 back in the van Genuchten-Mualem model, and showed in Fig. 2.

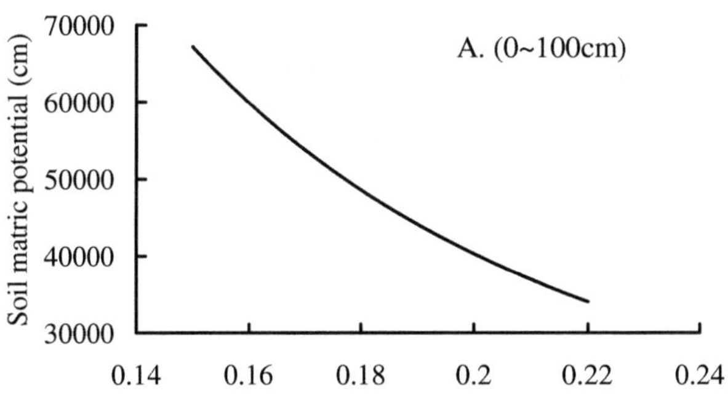

Fig. 2. Soil water characteristic curve

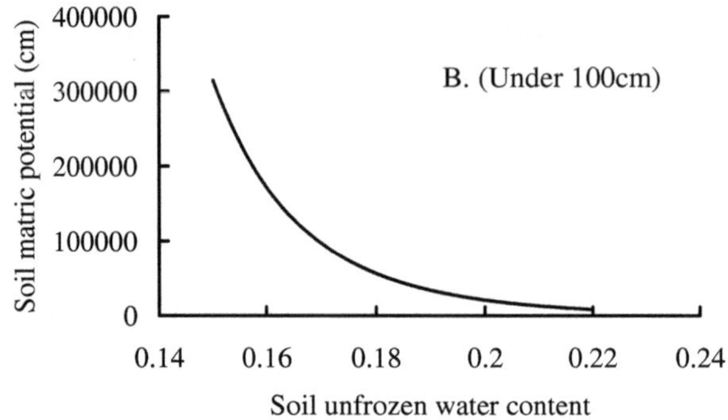

Fig. 2. (*Continued*)

The expressions of unsaturated hydraulic conductivity and specific water capacity in each soil layer were as follows:

$$\begin{cases} K = \dfrac{K_s\left\{1-(8.6\times10^{-5}|\psi|)^{0.6939}\left[1+(8.6\times10^{-5}|\psi|)^{1.6939}\right]^{-0.4096}\right\}^2}{\left[1+(8.6\times10^{-5}|\psi|)^{1.6939}\right]^{0.2048}} \\ C = \dfrac{0.6939(\theta-0.02)\left\{1-\left[(\theta-0.02)/0.44735\right]^{2.4411}\right\}}{|\psi|} \end{cases} \quad (0\sim100\text{cm}) \tag{13}$$

$$\begin{cases} K = \dfrac{K_s\left\{1-(0.0684|\psi|)^{0.1271}\left[1+(0.0684|\psi|)^{1.1271}\right]^{-0.1128}\right\}^2}{\left[1+(0.0684|\psi|)^{1.1271}\right]^{0.0564}} \\ C = \dfrac{0.1271(\theta-0.03)\left\{1-\left[(\theta-0.03)/0.428\right]^{8.8678}\right\}}{|\psi|} \end{cases} \quad (\text{Under } 100\text{cm}) \tag{14}$$

3.3 Verifying Validity of Identifying Method

In order to verify the validity of the identifying method of freezing-thawing soil hydraulic properties, the comparison between numerical simulated value and filed measured value is necessary. Thus, the numerical simulation of freezing-thawing soil water movement was conducted on a period in 2008.

The soil unfrozen water content distribution in December 19, 2008 was simulated and the freezing status was that the frozen depth had developed to 55cm. The comparison between simulated and measured values was showed in Fig. 3.

According to Fig. 3, the simulated and measured soil unfrozen water content in December 19, 2008 coincided with each other very well and had the same trend. The average relative error of the numerical simulation of soil unfrozen water content was 20.18%, therefore, the simulation was successful.

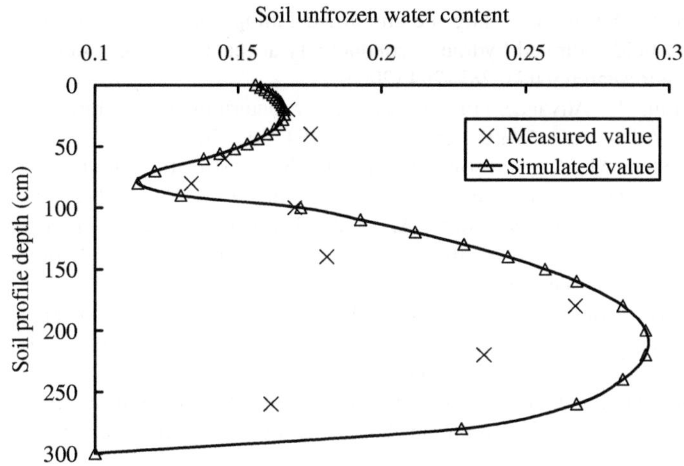

Fig. 2. Comparison between the measured and simulated soil unfrozen water content (2008-12-19)

4 Conclusions

A new set of identifying methods of freezing-thawing soil hydraulic properties was proposed in the study. It includes two part, one part is fixing soil freezing characteristic curve by a regression and fitting analysis of measured data, the other part is using particle swarm optimization algorithm to estimate the empirical parameters of van Genuchten-Mualem model and then to obtain soil water characteristic curve, unsaturated hydraulic conductivity and specific water capacity. At last, a comparison and error analysis was performed on the simulated and measured data of freezing-thawing soil water movement for the purpose of verifying method validity. The results showed that the identifying method of freezing-thawing soil hydraulic properties proposed in the study had good effectiveness and suitability.

Acknowledgements. The study was supported by the National Natural Science Foundation (51209039, 51279031), China Postdoctoral Science Foundation (2011M500632, 2012T50321), Heilongjiang Postdoctoral Fund (LBH-Z11231), Foundation of Heilongjiang Educational Committee (12521016), Doctoral Foundation of Northeast Agricultural University (2012RCB13) and Foundation of Harbin S&T Bureau (2010RFXXN043).

References

1. Lei, Z., Yang, S., Xie, S.: Soil Water Dynamics. Tsinghua University Press, Beijing (1988)
2. Wang, L., Fan, J., Wang, Q.: Measurement of soil conductivity by automated disc tension infiltrometer. Transactions of the CSAE 23(9), 72–75 (2007)

3. Bagarello, V., Sgroi, A.: Using the simplified falling head technique to detect temporal changes in field-saturated hydraulic conductivity at the surface of a sandy loam soil. Soil and Tillage Research 94(2), 283–294 (2007)
4. Xu, S., Liu, J.: Advances in approaches for determining unsaturated soil hydraulic properties. Advances in Water Science 14(4), 494–501 (2003)
5. Chen, H., Shao, M.: A review on indirect methods for estimating unsaturated soil hydraulic properties. Journal of Basic Science and Engineering 10(2), 103–109 (2002)
6. Augeard, B., Assouline, S., Fonty, A., et al.: Estimating hydraulic properties of rainfall-induced soil surface seals from infiltration experiments and X-ray bulk density measurements. Journal of Hydrology 341(1/2), 12–26 (2007)
7. Minasny, B., Hopmans, J.W., Harter, T., et al.: Neural networks prediction of soil hydraulic function for alluvial soils using multistep outflow data. Soil Science Society of America Journal 68(2), 417–429 (2004)
8. Yu, J., Guo, H., Fang, Q.: Confirmation of the parameters of the unsaturated soil moisture movement: A case study of Loam in Zhumadian of Henan Province. Ground Water 28(4), 25–27 (2006)
9. Wei, Y., Liu, Z., Kang, L.: Derivation and verification of soil hydrodynamic parameters in cinnamon soil. Journal of Hydraulic Engineering (3), 81–86 (2004)
10. Shao, M., Wang, Q., Robert, H.: A simple infiltration method for estimating soil hydraulic properties of unsaturated soils I: Theoretical analysis. Acta Pedologica Sinica 37(1), 1–8 (2000)
11. Shao, M., Wang, Q., Robert, H.: A simple infiltration method for estimating soil hydraulic properties of unsaturated soils II: Experimental results. Acta Pedologica Sinica 37(2), 217–224 (2000)
12. Mualem, Y.: A new model for predicting the hydraulic conductivity of unsaturated porous media. Water Resources Research 12, 513–522 (1976)
13. van Genuchten, M.T.: A closed-form equation for predicting hydraulic conductivity of unsaturated soils. Soil Science Society of America Journal 44, 892–898 (1980)
14. Kennedy, J., Eberhart, R.C.: Particle Swarm Optimization. In: IEEE International Conference on Neural Networks, pp. 1942–1948 (1995)
15. Eberhart, R.C., Kennedy, J.: A New Optimization Using Particle Swarm Theory. In: Proceedings of the Sixth International Symposium on Micro Machine and Human Science, pp. 39–43 (1995)
16. Wang, Z., Fu, Q., Jiang, Q.: Soil nutrient management zones based on particle swarm optimization algorithm. Transactions of the CSAE 24(10), 80–84 (2008)

Dynamic Simulation of Water Resources Sustainable Utilization of Kiamusze Based on System Dynamics

Qiuxiang Jiang[1,2], Zilong Wang[2], and Qiang Fu[2]

[1] Key Laboratory of Water-saving Agriculture, College of Heilongjiang Province, Northeast Agricultural University, P.R. China 150030
[2] College of Water Conservancy & Architecture, Northeast Agricultural University, P.R. China 150030
jiangqiuxiang914@163.com

Abstract. In order to determine the sustainable supporting capacity of current Kiamusze water utilization situation to future society and economy, a dynamic simulation model of water resources sustainable utilization was built based on system dynamics (SD). The simulation results indicated that current Kiamusze water resources could not satisfy future demand of industrial and agricultural production and also restricted socioeconomic development. In view of the situation, the water resources sustainable utilization schemes were designed and simulated by targeting water supply and demand balance and adjusting sensitive parameters of the model. The analysis showed that the coordinated scheme not only realized water resources sustainable utilization but also enhanced the sustainable supporting capacity of water resources to society and economy within the fixed number of simulating years. Thus, the coordinated scheme is the best one for water resources sustainable utilization of Kiamusze and can provide policy guidance for further exploitation of regional water resources.

Keywords: Water resources, Sustainable utilization, System dynamics, Dynamic simulation, Kiamusze.

1 Introduction

Kiamusze is located in northeast Heilongjiang Province and in the heartland of Sanjiang Plain surrounded by Songhua River, Heilongjiang River and Wusuli River. The region with flat terrain, fertile soil and abundant water is suitable for agricultural development. After years of exploitation and construction, it has become an important grain production base of China. In recent years, the high interference of human activities and unreasonable utilization of water resources have caused serious imbalance between surface and ground water, low utilization efficiency and supporting capacity of water resources [1-3]. These problems have seriously threatened the sustainability of regional resources utilization and socioeconomic development. Therefore, on the premise of guaranteeing national food security and regional ecological security, rational allocation of water resources, enhancing

D. Li and Y. Chen (Eds.): CCTA 2012, Part II, IFIP AICT 393, pp. 367–375, 2013.

supporting capacity of water resources to socioeconomic development and realizing regional water resources sustainable utilization become especially important.

In the study, a dynamic simulation model of water resources sustainable utilization was built firstly. Then the current situation of Kiamusze water resources utilization was simulated, and the key problems restricting regional socioeconomic development were found out. Lastly, the water sustainable utilization schemes were designed and simulated by system dynamic (SD) model, and the best scheme of water resources sustainable utilization was sought to provide scientific guidance for further exploitation and optimal allocation of regional water resources.

2 SD Model of Water Resources Sustainable Utilization

2.1 Model Boundaries

The spatial boundary of SD model is the administrative boundary of Kiamusze including four districts, six counties and seventeen state-owned farms. The temporal boundary of SD model is from the year 2000 to 2030. The former nine years (2000 to 2008) provide historical statistical data and the latter twenty-two years (2009 to 2030) is the forecasting period. The time step is one year in order to reduce error. The content boundary of SD model includes water resources, land resources, population, national economy, grain production, etc.

2.2 Model Building and Testing

Before model building, Kiamusze water resources sustainable utilization system was analyzed and the feedback mechanism of all elements in the system was primarily understood. The model building stood by practical, problem-oriented and systemic principles [4-5], and included structure design, variable selection and quantitative expression of the relationship between variables. The dynamic simulation model of water resources sustainable utilization with 90 variables was built and showed in Fig. 1. The details of variable structure equations and data sources in the model can be found in reference [6] and [7].

The model validity and history testing results showed that the model had rational structure and high precision and could quantitatively characterize a real system operation [8-10]. Sensitivity analysis of the model found out nine sensitive parameters that had great influences on model operation results [11-13]. The parameters included reclaimed water reuse rate, paddy field irrigation net quota, dry land irrigation net quota, water transport index in canal system, industry water quota, industrial water recycling rate, surface water utilization level, ground water utilization level and transit water resources utilization.

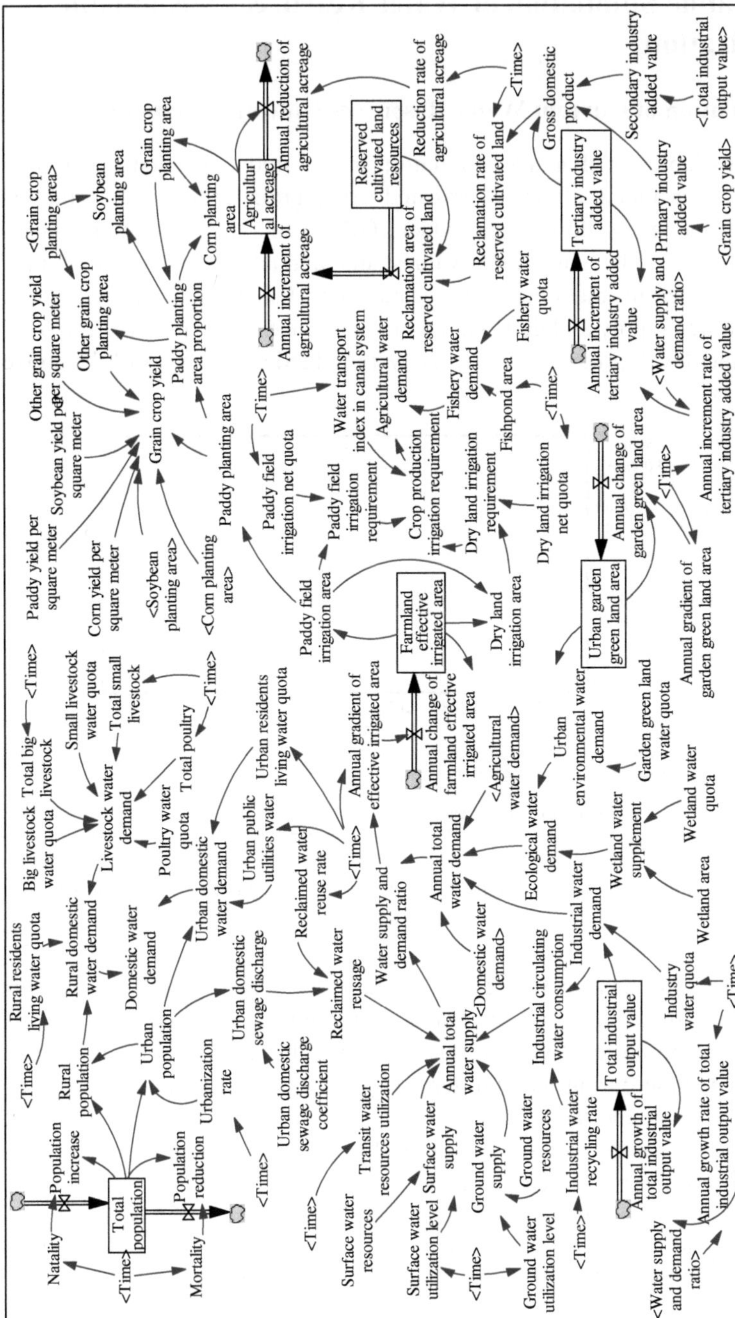

Fig. 1. SD simulation model of water resources sustainable utilization

3 Dynamic Simulation of Water Resources Sustainable Utilization

3.1 Simulation of Current Water Resources Utilization Situation

The output variables of dynamic simulation in the study included water supply and demand ratio, farmland effective irrigated area, grain crop yield, total industrial output value and gross domestic product (GDP), and their variation diagram was showed in Fig. 2. The simulation values and annual growth rates of the output variables in different planning-level years were listed in Table 1. The balance situation of water supply and demand, and its influence on agricultural production and socioeconomic development in Kiamusze were judged on the basis of the above data.

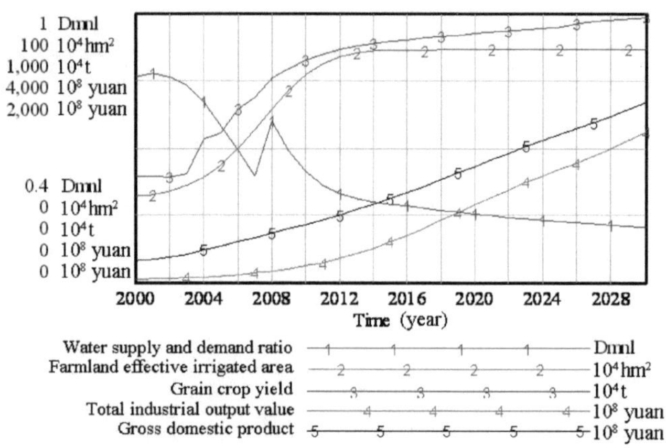

Fig. 2. Dynamic simulation diagram of current water resources utilization situation

According to the data in Fig. 2 and Table 1, the water supply and demand ratio reduced from 0.76 to 0.57 and the balance of supply and demand was $35.5\times10^8 m^3$ during recent planning-level years (2009 to 2015). The water supply and demand ratio continued to decline during medium-term (2016 to 2020) and long-term (2021-2030) planning-level years and the ratio reduced to 0.52 in 2030. Water supply and demand ratio had great influence on agricultural production. Farmland effective irrigated area increased slowly even stopped when the ratio was less than 0.6. The development stagnation of farmland effective irrigated area would directly induce the stagnation of paddy planting area. In this case, regional grain crop yield would be seriously affected for paddy was a high yield and good adaptability crop in Kiamusze.

Table 1. Dynamic simulation values of current water resources utilization situation

Level year		2000	2008	2010	2015	2020	2030
Water supply and demand ratio		0.86	0.76	0.65	0.57	0.55	0.52
Farmland effective irrigated area	Simulation value (10^4 hm^2)	31.98	64.396	77.204	86.136	86.285	86.285
	Annual growth rate (%)	-	13.23	8.07	0.32	0.00	0.00
Grain crop yield	Simulation value (10^4 t)	395.19	757.53	824.16	894.61	925.46	982.58
	Annual growth rate (%)	-	10.10	3.87	0.86	0.61	0.32
Total industrial output value	Simulation value (10^8 yuan)	47.12	170.16	248.40	602.15	1153.96	2240.51
	Annual growth rate (%)	-	20.51	20.85	17.71	11.23	6.36
Gross domestic product	Simulation value (10^8 yuan)	159.74	368.52	433.26	623.87	865.27	1342.81
	Annual growth rate (%)	-	10.77	8.25	7.35	6.13	4.25

As the same as agricultural production, water supply and demand ratio also restricted the developments of total industrial output value and GDP. Kiamusze has abundant mineral resources, so its industry has large development space. Annual growth rate of total industrial output value maintained more than 20% before 2010, but during medium-term and long-term planning level years its growth speed slowed along with the shortage aggravation of water resources. Kiamusze with a low GDP per capita (1.58×10^4yuan) is a relatively poorer region. For industrial and agricultural production in the region restricted by water resources grew slowly, the annual growth rates of recent, medium-term and long-term planning level years were 7.35%, 6.13% and 4.25% respectively.

The analysis of dynamic simulation results indicated that imbalance between water supply and demand was the key contradiction of water resources system. Water resources supporting capacity can not satisfy industrial and agricultural production demand and restricts their development simultaneously, and finally causes slow economic development. In addition, a backward economy will reduce investment in water conservancy construction and water resources exploitation level, and finally aggravate contradiction of water supply and demand. This is a vicious circle.

3.2 Scheme Design of Water Resources Sustainable Utilization

The primary object of scheme design of water resources sustainable utilization is to solve the imbalance problem of water supply and demand in future and to eliminate restriction of water resources to socioeconomic development. On the basis of above analysis, five schemes, including natural continuation scheme, increasing income scheme, decreasing expenditure scheme, comprehensive scheme and coordinated scheme, were designed and their regulation strategies were as follows:

(1) Natural continuation scheme: Sensitive parameters maintained at current year level (2008).

(2) Increasing income scheme: The scheme takes enhancing surface water supporting capacity and exploiting transit water resources as primary measures, and

improving sewage disposal and industrial water recycling capacity as supplementary measures. Its sensitive parameters include surface water utilization level, transit water resources utilization, reclaimed water reuse rate and industrial water recycling rate. Based on the water conservancy and socioeconomic development plan of Kiamusze and the reclaimed water reuse rate and industrial water recycling rate of developed area, the annual gradient of sensitive parameters in different periods were designed and listed in Table 2.

(3) Decreasing expenditure scheme: The scheme considers lowering ground water utilization level and improving water-saving of industrial and agricultural production. Its sensitive parameters include industry water quota, paddy field irrigation net quota, dry land irrigation net quota, water transport index in canal system and ground water utilization level. On the basis of the water conservancy development plan of Kiamusze and all industries water quota of developed area, the annual gradient of sensitive parameters in different periods were designed and listed in Table 2.

(4) Comprehensive scheme: The scheme is comprised by increasing income scheme and decreasing expenditure scheme. Its sensitive parameters and values are as the same as the two schemes.

Table 2. Regulation strategy of water resources sustainable utilization scheme

Sensitive parameter	Value in 2008	Annual gradient of sensitive parameter (%)											
		Increasing income scheme			Decreasing expenditure scheme			Comprehensive scheme			Coordinated scheme		
		R[a)]	M	L	R	M	L	R	M	L	R	M	L
Surface water utilization level	0.4289	2.5	2	1.5	0	0	0	2.5	2	1.5	2.5	0	0
Transit water resources utilization (10^4m^3)	8950	50	15	1.5	0	0	0	50	15	1.5	50	0	0
Industrial water recycling rate	0.3756	3	2.5	2	0	0	0	3	2.5	2	3	2	1
Reclaimed water reuse rate	0.025	20	15	10	0	0	0	20	15	10	20	15	10
Industry water quota ($m^3/10^4$yuan)	161	0	0	0	-5	-4.5	-4	-5	-4.5	-4	-5	-4	-3
Dry land irrigation net quota (m^3/hm^2)	550	0	0	0	-2.5	-2.25	-2	-2.5	-2.25	-2	-2.5	-2.25	-2
Paddy field irrigation net quota (m^3/hm^2)	4500	0	0	0	-4.5	-4	-3.5	-4.5	-4	-3.5	-4.5	-4	-3.5
Water transport index in canal system	0.55	0	0	0	2	1.5	1	2	1.5	1	2	1	0.5
Ground water utilization level	0.886	0	0	0	-2	-0.5	0	-2	-0.5	0	-2	-1.5	-1

a) R, recent years; M, medium-term years; L, long-term years.

(5) Coordinated scheme: A certain economic inapplicability and incipient fault of resources sustainable utilization exist in increasing income scheme and decreasing expenditure scheme, for sensitive parameter design of the two schemes is to enhance water supply or induce water demand furthest. In order to guarantee resources sustainable utilization in coordinated scheme, the exploiting level of surface, ground and transit water, the restriction of industry water quota and the increment rate of

industrial water recycling rate and water transport index in canal system are reduced properly. The annual gradient of all sensitive parameters, except for reclaimed water reuse rate, dry land irrigation net quota and paddy field irrigation net quota, are adjusted and listed in Table 2.

3.3 Dynamic Simulation of Water Resources Sustainable Utilization Scheme

In order to investigate the influence of water resources sustainable utilization scheme to regional water supply and demand balance, the water supply and demand ratios of different schemes were simulated and showed in Fig. 3.

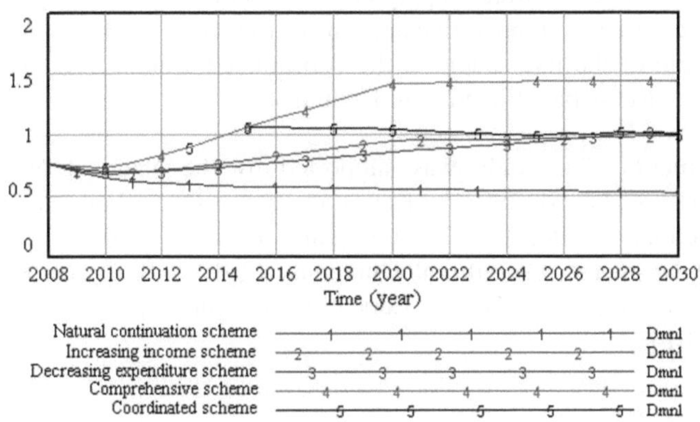

Fig. 3. Simulation of water supply and demand ratios of different schemes

According to Fig. 3, the water supply and demand ratio of natural continuation scheme showed a decreasing trend year by year and the water supply and demand did not balance during the simulated years. The water supply and demand ratio of increasing income scheme showed an increasing trend year by year, but the water supply and demand did not balance till the year 2030. The decreasing expenditure scheme realized balance in 2029. The comprehensive scheme realized balance in 2015, but the ratio increased year by year which signified a great waste of water resources. The coordinated scheme realized balance in 2015, however, the ratio decreased year by year and maintained at 1 which signified a dynamic balance. Thus, the coordinated scheme, that realizes balance and ensures resources sustainable utilization, is the best scheme for water resources sustainable utilization in Kiamusze.

4 Conclusions

A dynamic simulation model of regional water resources sustainable utilization was built in the study on the basis of water resources system analysis and feedback mechanism analysis of all elements. The model included 90 variables from many aspects, e.g. water resources, land resources, society, economy and grain. Nine sensitive parameters were found out by sensitivity analysis of the model and used as regulation variables of subsequent water resources sustainable utilization scheme.

Current water resources utilization of Kiamusze was simulated by using the dynamic simulation model of water resources sustainable utilization. The simulating results showed that the balance of regional water supply and demand was not realized during simulating years and the difference between supply and demand increased yearly. The imbalance of water supply and demand restricted industrial and agricultural development and become the key problem of Kiamusze water resources system.

For the purpose of regional water resources sustainable utilization, five schemes were designed by targeting water supply and demand balance and adjusting sensitive parameters. According to comparative analysis of dynamic simulation of water supply and demand ratio of the five schemes, the coordinated scheme was the best one for water resources sustainable utilization in Kiamusze.

Acknowledgements. The study was supported by the National Natural Science Foundation (51209038, 51179032), Key Laboratory Open Foundation of Water-saving Agriculture, College of Heilongjiang Province (2011KFJ03), Doctoral Foundation of Northeast Agricultural University (2012RCB58) and Research Fund for the Doctoral Program of Higher Education of Ministry of Education of China (20092325110014).

References

1. Hou, C.: Water resources carrying capacity assessment of Kiamusze. Heilongjiang Science and Technology of Water Conservancy 39(3), 20–22 (2011)
2. Li, L., Yang, C., Zhang, Y.: Problems and countermeasures of Kiamusze ground water exploitation and utilization. Ground Water 33(4), 36–37 (2011)
3. Dong, Y., Song, W.: Problem analysis of Kiamusze water resources exploitation and utilization. Heilongjiang Science and Technology of Water Conservancy 35(4), 102 (2007)
4. Cai, L.: The Application of System Dynamics in The Research of sustainable development. China Environment Science Press, Beijing (2008)
5. Hao, Y., Fan, J.: Method and Application of Systems Engineering. Science Press, Beijing (2007)
6. Jiang, Q.: Study on carrying capacity evaluation and dynamic simulation of sustainable utilization of water and land resources in Sanjiang Plain. Northeast Agricultural University, Harbin (2011)
7. Heilongjiang Provincial Bureau of Statistics: Statistical Yearbook of Heilongjiang Province. China Statistics Press, Beijing (2009)
8. Chen, N., Wang, Y.: The Water Resource Sustainable Utilization of Henan Province Based on System Dynamics. Journal of Irrigation and Drainage 29(4), 34–37 (2010)
9. Chen, C., Yan, G.: Research on system dynamics model of water resources sustainable development in China. Journal of University of Shanghai for Science and Technology 22(2), 154–159 (2000)

10. Huang, L., Cao, Y., Zhao, N., et al.: The Sustainable Development of Water Resources in Shandong based on System Dynamics Simulation. Water Power 34(6), 1–3 (2008)
11. Zhong, Y., Jia, X., Li, X., et al.: System Dynamics. Science Press, Beijing (2009)
12. Jia, R., Ding, R.: System Dynamics: Analysis of Feedback Dynamic Complexity. Higher Education Press, Beijing (2002)
13. Wang, Q.: System Dynamics. Tsinghua University Press, Beijing (1994)

Visual Space Research and Application of the Data Mining in Soil Fertility Evaluation*

Hang Chen and Guifen Chen**

College of Information and Technology, Jilin Agricultural University, Changchun, 130118, China
{chenhang0811,guifchen}@163.com

Abstract. China is a populous agricultural country, the arable land is the most basic resources which is needed for agricultural production, the quality of arable land has a direct impact on the entire agricultural production and development ,so to carry out the Evaluation of Farmland is a very important foundation work which is advantage to know china's arable land resources and to improve the efficiency of use for arable land and promote the development of modern agricultural , and which is helpful to further identify china fertility status of cultivated soils, the distribution of nutrients, soil fertility and agricultural land degradation, etc., so as to achieve the realization of agricultural modernization and to improve the comprehensive capacity of agricultural production. This paper use visualization of spatial data mining technology which is used to research and apply on soil fertility evaluation, which is put the technology of spatial data mining and visualization technology, GIS, technology tightly integrated, that makes a research to the elements of soil fertility, yield and other data mining, and achieve the application of visualization of spatial data mining techniques in the evaluation of soil fertility, this application is more conducive to promotion and application of precision agriculture for variable rate fertilization, variable spraying, at the same time it provides real-time, accurate decision making basis for agricultural production management department, to promote the transformation of traditional agriculture to modern agriculture, which has important theoretical significance and practical value.

Keywords: Soil fertility evaluation, GIS, spatial data mining.

1 Introduction

In resent years, with the development of visualization techniques and data mining technology, in order to solve the visible and interactive problem of the data mining

* Foundation project: Supported by National "863" High-tech Project (2006AA10A309), National Spark Project (2008GA661003) and Changchun Technology Project (09KT14), The research and application of facilities for the safety of vegetables production technology based on Internet of Things(2011-Z20).

** Corresponding author.

D. Li and Y. Chen (Eds.): CCTA 2012, Part II, IFIP AICT 393, pp. 376–385, 2013.

process, under the combination of the two information visualization techniques and data mining technology to form a visual data mining ,Visual data mining technology make full use of the advantages of data mining and visualization technology to solve the problem of the "black box" data mining operation, and gradually being valued by the people, which is an important aspect of data mining research and has a very high research value.to carry out the evaluation of cultivated land which is use spatial data mining technology and agricultural technology can reduce the manpower, material and financial resources of arable land evaluation work which is required, and it can improve the efficiency of the evaluation, but also can improve the accuracy of the evaluation results, and conducive to the evaluation results summary, management, updates, cartographic output and promote the application. So as to identify the production capacity of China's cultivated land base, soil fertility status, to achieve the purpose of a strong impetus to the modernization of agriculture and sustainable development.

2 Soil Spatial Data Acquisition

In support of GIS and related technologies, we can quickly obtain a series of basic spatial information about soil and its influencing factors, such as topography, land use, crops, soil fertility, etc.. These information is the basis of the analysis and evaluation of soil space. This paper uses global positioning system (GPS) and3G wireless sensor technology for data acquisition.

2.1 Global Positioning System (GPS)

Global Positioning System (GPS) for information acquisition and implementation of accurate positioning. Widely used in order to improve the accuracy DGPS (Differential Global Positioning System) technology, the so-called "differential correction global satellite positioning technology." In order to improve the accuracy,, widely used DGPS (Differential Global Positioning System) technology, which is called" the differential correction of global satellite positioning technology". It is characterized by high positioning accuracy, the accuracy of the GPS system are free to choose according to different purposes.

2.2 3G Wireless Sensor

3G wireless sensor can be used to collect information on soil plots and be used to environmental monitoring on soil topography, crop of soil, and soil fertility, etc.. The user can monitor, process the data and show the data chart on soil environmental through INTERNET and 3G networks, to provide users with efficient and stable environment parameters, while addressing the defects of traditional long-distance wireless transmission barriers, and the true realization of remote soil environmental monitoring.

3 GIS Spatial Analysis

Spatial analysis is one of the important functions of the GIS system is based on the geographic location of the object and form of spatial data analysis techniques, with the aim of the extraction and transfer of spatial information. The object of spatial analysis is a series of spatial location data, which include two spatial coordinates and professional attributes.

3.1 Overlay Analysis

Overlay analysis in GIS is to refer to the same region, the unified scale, two sets of data files or polygon feature more than two groups superimposed by the simulation of multiple properties in the region, to find and identify multiple attributes polygon or statistical analysis of the property characteristics within the scope of the polygon. In this paper, which the district overlay analysis of the District to be about the level of thematic maps superimposed to produce a new data aspects of the operation, the results of a combination of the original two-tier or multi-layer elements with attributes.

3.2 The Interpolation of Spatial Data

Spatial data according to their own requirements for sampling point observations, such as land type, land nutrient sampling points, ground elevation, the distribution of these points are often irregular, discontinuous, it is impossible to fill the entire study area. Thus formed by the polygon internal changes can not be more accurate, more specific, but can only reach the average level expression, However, at some point, the user might like to know the exact value of the observation point certain characteristics of interest, which resulted in the spatial interpolation technology. The purpose of the interpolation

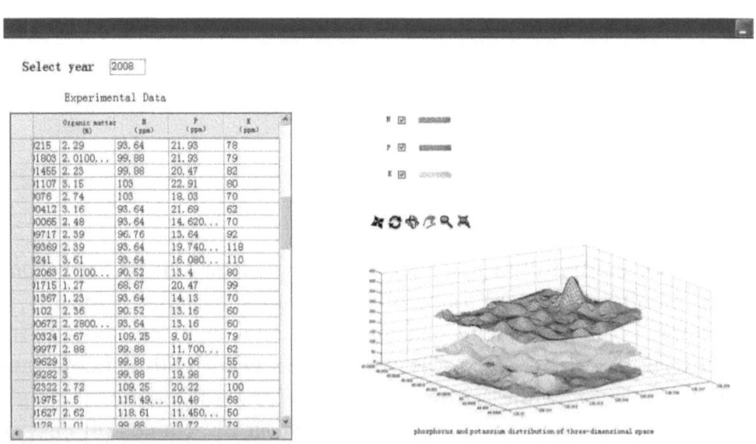

Fig. 1. Phosphorus and potassium distribution of three-dimensional space

is according to the known point attribute reasonable deduction and predict the position near the attribute values. The main methods: inverse distance difference, spline interpolation and kriging. In this paper, the Kriging interpolation method to establish the soil of nutrients, moisture and crop yields three-dimensional spatial variation diagram.The conditions of application of the Kriging interpolation method is regionalized variable spatial correlation.

4 The Integrated Application of GIS and Technology of Spatial Data Mining in Agriculture

Spatial data mining is a process of extracting general knowledge, laws and rules from the spatial database or Spatial Data Warehouse. The knowledge, laws and rules are implied in them, unclear in advance, potentially useful and spatial or non-spatial which ultimately easy to understand[3]. The purpose of spatial data mining is that help people to improve the accuracy and reliability of the decision-making, and enable people to maximize the efficient use of data.

4.1 The Classification of Spatial Data Mining Technology with the Support of GIS

The spatial data mining technology can be divided to describe, explain and predict by function. Descriptive model makes the spatial distribution characterized, such as spatial clustering. Explanatory model deals with spatial relationships, such as the relationship between a space object and the factors affecting their spatial distribution.

4.2 The Evaluation of Soil Fertility Based on Spatial Weighted Fuzzy Clustering Algorithm

4.2.1 The General Situation of Test Area

Using GPS, GIS, RS, and sensor technology, access information of corn farmland from experimental ground, and then divided the ground into grid cell (specifications size is 40m*40m), A1~L10 are sampling points. To sample in this grid cell, the sampling depth is 25 cm and the sampling method is the five points plum sampling, which the soil samples of the four corners of the mesh and grid center were mixed as the grid soil samples. After sampling, select 63 sampling points as the sample data, record the attribute values of spatial coordinates in the soil, organic matter, available nitrogen, available phosphorus and available potassium.

4.2.2 Algorithm Application

First of all, according to the analytical data of the soil nutrients from pilot area, soil nutrient spatial variability law analysis. Analysis results show that, the available phosphorus in soil spatial variability, the coefficient of variation was 31.12%, available potassium spatial variation coefficient is 21.51%, organic matter spatial variation coefficient is 29.31%, available nitrogen spatial variability coefficient of 11.69%. Be evaluated in accordance with the results and the local characteristics of the soil and construct paired comparison matrix B.

$$B=\begin{bmatrix} 1 & 1.333 & 0.4 & 2 \\ 0.75 & 1 & 0.3 & 1.5 \\ 2.5 & 3.333 & 1 & 5 \\ 0.5 & 0.667 & 0.2 & 1 \end{bmatrix}$$

Thus obtain nutrients weight vector $A = \{0.2105, 0.1579, 0.5263, 0.1053\}$.

The fuzzy equivalent matrix $t(R)$ is calculated from original data matrix D using calculate weighted, and then adjust the value of λ by decreasing to establish the dynamic diagram of fuzzy clustering.

When values of λ are different, calculating the Quantity statistics of F-, Results such as shown in table 1.

Table 1. F-value calculations

Num.	λ	Class number	Value of F-
1	1	100	0.339
2	0.997	92	0.652
3	0.996	86	1.305
4	0.995	71	1.371
5	0.994	48	1.469
6	0.993	40	1.543
7	0.992	32	2.304
8	0.992	28	3.445
9	0.990	23	4.594
10	0.987	15	4.875
11	0.986	11	4.898
12	0.984	10	4.787
13	0.983	5	2.069
14	0.981	4	1.121

Compared F with $F_{0.05}(r-1, n-r)$, then we know when $F > F_{0.05}(r-1, n-r)$,the values of λ are 0.981, 0.983, 0.984, 0.986, 0.987, 0.990, 0.991, 0.992, 0.993, 0.994, 0.995, calculating $F - F_{0.05}(r-1, n-r)$, the bigger value of λ is 0.986, From the table also can be seen the class number is 11. Total divided into 11 classes. Shown in Table 2.

Table 2. Classification table

category of Classes	Number of sample points	higher of the soil Nutrient Content	Lower of the soil Nutrient Content	The soil nutrient tend to average	Note
category 1	43				The soil nutrient is average
category 2	7	Available N	Available P	organic matter, Available K	
category 3	1	Available K	Available N, Available P	organic matter	
category 4	1	Available K, Available N	Available P	organic matter	
category 5	4		Available P	Available K, organic matter Available N	
category 6	2	Available N	Available P	Available K, organic matter	
category 7	1			Available P, Available N, Available K, organic matter	

Table 2. *(Continued)*

category 8	1	Available N	Available P	below-average
category 9	1	organic matter		below-average
category 10	1	Available K	Available P	Tend to average
category 11	1			Artificial add noise

The table 2 shows that space fuzzy clustering results based on weighting than unweighted space fuzzy clustering results are obviously different. Category 1, Category 2, Category 3, Category 4, Category 5 and Category 6 are divided into Category 1 in the space fuzzy clustering results of traditional. Available P tends to average in the soil, and available N, available K, organic matter are higher than average by analysis. The kinds of nutrient content in the soil tend to average what are divided into category 1 after weight, however the space fuzzy clustering results of traditional in the category 1 is divided into category 2 to category 6 under the influence of available P which has the Maximum Weight. As the same time, we find that adjacent grids are adjacent; it is mean that spatial continuity is in soil nutrient content, above characteristics benefit the precise task of fertilizer applicator and convenient to soil sampling and field management.

Calculation of the factor of each index weight is by AHP. And according to the reality, making a comprehensive evaluation to organic matter in the soil , nutrition of available N, available P and available k and reflecting the fact is in soil nutrition.

5 Implementation of Visual Spatial Data Mining in Soil Fertility Evaluation

Under Visual spatial data mining support to achieve automation and quantitative evaluation arable land in fertility by mathematics method and mathematical model of systematic clustering method, analytic hierarchy process, fuzzy evaluation , to get the figure of Fertility of arable land and level distribution, to analysis limiting factor of each level of cultivated land. fertility evaluation of the farmland、 the cultivated land fertility dynamic analysis and farmland improvement use application of the feasibility and the scientific nature to provide feasible technology for accurate in the cultivated land fertility evaluation dynamic analysis and improvement, and it has the practical application value of scientific management and sustainable utilization of the cultivated land resource.

5.1 Comprehensive Evaluation Index System of the Cultivated Land Fertility

5.1.1 The Choice of Evaluation Index System

According to the national general 55 index frameworks, to select climate , terrain parts, Ground slope, soil thickness, into soil organic, section configuration, soil erosion, top layer texture, farmland infrastructure, the soil physical and chemical properties, traffic and economic development level and the level of production, etc. According to the selection rules of which has great influence on cultivated land, has a larger variation in YuShu city demonstration area and which is stable in time series, to select the profile character, the site conditions, physical and chemical characteristics, nutrient, soil management five types and 14 indexes as evaluation factors, which reflects objectively YuShu city productivity grade difference and has the very strong objectivity.

5.1.2 Sample Library Established

Based on the Second soil survey results, Representative face of Average each point in the center of Northeast China is 333.33～666.67 hm2. Cultivated land area in Yushu city 331100 hm2, the glebe is about 250000 hm2, Paddy fields is about 80000 hum2, and other is Awning vegetables about 1100 hm2. In YuShu city, the amount of the test-bed for sample points is determined and is the control number, superposition of painted for figure of land used status and the compiling of soil map to confirm the final sample points for 574.

5.2 Realized of Visual Spatial Data Mining on Soil Fertility Evaluation

Soil classification	Integrated fertility index(IFI)	Yield(kg/667m2)	Cultivated land area(Million hm2)	Percentage to total area(%)
Grade 1	>0.85	>600	4.14	12.50
Grade 2	0.81-0.85	551-600	11.25	33.99
Grade 3	0.76-0.80	501-550	8.37	25.29
Grade 4	0.71-0.75	451-500	4.17	12.60
Grade 5	0.65-0.70	400-450	3.24	9.79
Grade 6	<0.65	<400	1.94	5.83

From the figure we can see that the soil of 1~2 level is mainly distributed in the ridge of hilly platform, low terraces of the valley plain and high flood area and placer, diluvia plain, which the terrain is flat and basically no erosion or mild erosion; The land of 3~4 level most located in undulating terraces or the slope of alluvial flood plain, in

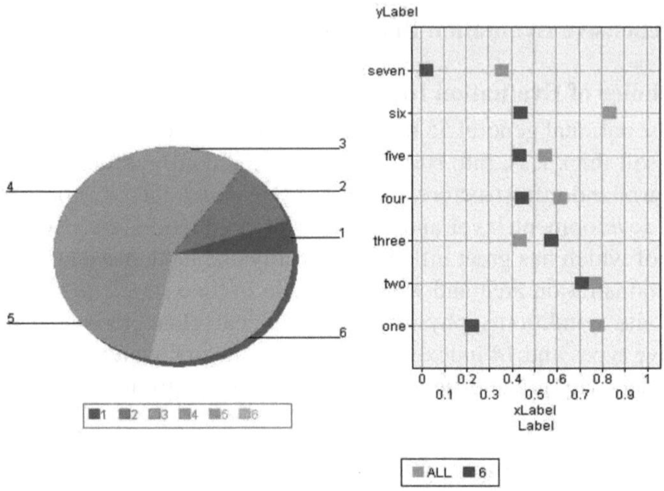

Fig. 2. Results visualization of data mining

undulating terraces or alluvial flood plain slope, in which the terrain is gentle, mild and moderate erosion; The 5~6 level most in low mountains and hills, the top of Undulating tableland and low river floodplain, which terrain is hilly or steep slope greater than 7°. Soil erosion is generally moderate to severe or the low-lying, easy to form a flood.

6 Conclusion

In this paper, based on spatial data mining, site conditions, soil management, soil nutrient ,profile traits and physical and chemical properties, the five factors as the sample input. Using AHP hierarchical analysis the cultivated land fertility soil comprehensive evaluation, with a strong ability of the evaluation and expansion ability. The research results of the cultivated land fertility soil comprehensive evaluation and soil the cultivated land fertility management, provides the theoretical basis and technology. However, the analytic hierarchy process applied to the soil cultivated land fertility is still a new field, its theoretical foundation to be further improved, such as the classification error of the sample data, and the type of kernel function and parameter selection, how to simplify the discriminate formula to reduce the computational volume and improve the discrimination efficiency. The main conclusions are as follows:

① According to the analysis of the spatial clustering of the actual needs of the status and soil fertility decisions, the use of fuzzy clustering algorithm based on the weighted space of soil fertility evaluation, propose a new method for the accurate evaluation of soil fertility.

② Use of spatial analysis methods and GIS in the statistical analysis methods and data mining visualization combined. Conducted in-depth analysis of the various factors that affect soil fertility, and evaluated, to establish an evaluation system to improve the administrative departments for agriculture take full advantage of the ability of the various properties and spatial information data to make decisions.

③ Spatial data mining technology has penetrated the field of agriculture, but because of the complexity and variability of the agriculture itself, making the spatial data mining techniques in dealing with these large, multi-dimensional, dynamic, incomplete, uncertainties and other characteristics of data, data processing and mining tools need to further research and exploration.

So, the diversity of the data entry forms based on the uncertainty and ambiguity of spatial data mining, data warehousing and spatio-temporal data mining, knowledge of the visual representation of user participation and intelligent data mining and other future agricultural areas of application of data mining technology development trends and hot spots.

References

1. Zhang, S.H.: Based on GPS, GIS for precision agriculture variable rate fertilization theory and technology research, pp. 1–3. Jilin University, Jilin (2003) (in Chinese)
2. Miao, X.K., Xia, K.J., Wang, X.: Precision farming variable rate fertilization intelligent decision support system research. Computer Applications 24(11), 153–155 (2004) (in Chinese)
3. Li, Y., Shi, Z., Wu, C.F., Li, F., Cheng, J.L.: Definition of management zones based on fuzzy clustering analysis in coastal saline land. Scientia Agricultura Sinica 40(1), 114–122 (2007) (in Chinese)
4. Chen, G.F., Cao, L.Y., Wang, G.W., Wan, B.C., Liu, D.Y., Wang, S.S.: Application of spatial fuzzy clustering algorithm in precision fertilization. New Zealand Journal of Agricultural Research 50, 1249–1254 (2007)
5. Zhang, S.H., Ma, C.L., Li, W., Xu, Y.: Experimental study on the influence of variable rate fertilization on maize yield and soil nutrients. Transactions of the Chinese Society of Agricultural Engineering 22(8), 64–67 (2006) (in Chinese)
6. Chen, L.P.: Precision agriculture variable rate fertilization theoretical and experimental research, pp. 98–104. China Agricultural University, Beijing (2003) (in Chinese)
7. Sparling, G., Schipper, L.: Soil quality monitoring in New Zealand: trends and issues arising from a broad-scale survey. Agriculture, Ecosystem and Environment 104, 545–552 (2004)
8. Turner, B.L., Meyer, W.B.: Land use and land cover in global environmental change: considerations for study. Int. Soi. Sci. J. 130, 669–680 (1991)

Research on Construction and SWRL Reasoning of Ontology of Maize Diseases

Li Ma, Helong Yu, Guifen Chen, Liying Cao, and Yueling Zhao

College of Information and Technology Science, Jilin Agricultural University, Chang Chun, Jilin, China
mary19801976@sohu.com

Abstract. In this paper, according to the characteristics of maize disease knowledge, OWL DL language was used to build maize diseases ontology, and the reasoning rule of maize diseases was defined by using the expressive ability of SWRL rule language. The author introduced several realizable reasoning functions, and achieved the diagnostic reasoning of maize disease knowledge by Jess inference engine. The results indicated that constructing the maize diseases ontology, and introducing SWRL rule into maize disease ontology provided an effective way for the construction of high-intelligent, shareable and reused maize disease knowledge database and diagnostic rule database.

Keywords: ontology of maize diseases, SWRL, SQWRL, Rule reasoning.

1 Introduction

Corn is the main grain crop, cultivate area and total output is next to wheat, rice and reside the third. Disease influencing corn production of major natural disasters, annual loss of 6% ~ 10%.In recent years, with the replacement of varieties and cultivation techniques of change, the serious diseases happened in the past, have not been solved thoroughly, new disease occur ceaselessly, bring great harm to maize production, prevention and treatment of diseases in maize has become the guarantee of sustainable yield is the key link of Maize[1,2].

At present, the diagnosis and treatment of diseases, mostly based on expert experience, building the database, expert system is used to complete the diagnosis. The traditional knowledge representation methods are difficult to realize complex integrated expert system for the rapid development and the sharing and reusing of demand, recall and precision rate can not be guaranteed. Resulting in the need for a new knowledge representation method to construct the knowledge base, to realize the sharing and reuse between heterogeneous systems[3].

Ontology is the shared conceptual model explicit specification [4], and the definition of ontology for knowledge sharing and reusing the body being applied to artificial intelligence and other related fields become the good knowledge representation tools relatively. However, semantic expression ability of ontology is limited to description logic, can't say the rules of general form, knowledge reasoning ability is limited. The

D. Li and Y. Chen (Eds.): CCTA 2012, Part II, IFIP AICT 393, pp. 386–393, 2013.

SWRL can make up of the shortcomings for ontology modeling technology and ontology language of the OWL, realize the perfect combination of the rules and knowledge[5].

In this paper, use OWL DL language as ontology building language, construction of maize disease ontology database, using SWRL to construct the maize disease related SWRL rules, using the SQWRL query and diagnosis for establishment, to provide reference for the building of sharing and reuse maize disease knowledge base.

2 The Establishment of Maize Disease Ontology

2.1 The OWL Description of Ontology

Ontology is a description of the relationship between the concept and concept model, through the relationship between the concepts to the semantic description. Ontology contains five basic modeling primitives; those are concepts, relations, functions, axioms, instances. OWL (Web Ontology Language) is the standard of language W3C recommendation to describe Ontology. The description of the OWL for knowledge mainly from the following several aspects, (1) Concept: on behalf of the Concept of field set, Concept of each Concept is a object class, (2) Instances: and that individuals, is a special case of the concept, or the bottom of the object; (3) Properties: attribute hierarchical relationships between the concept is set, used to describe and definition and concept of ontology and examples, it including data attributes and object attributes, object attributes value range is the definition of the class, and the data attribute values can be Boolean, byte, double etc data types[6,7].

2.2 The Concept and Attribute of Maize Disease Ontology

Maize diseases mainly include: Top rot, leaf blight, helminthosporium maydis, brown spot, stalk rot, dwarf mosaic disease, rough dwarf disease, Gray Leaf Spot, black silk ear disease, bacterial wilt, General describe as "leaf blight disease onset of parts are in leaves, leaf sheaths or bracts, the early symptoms of leaves on a brown spindle spot" ," Corn Gray Leaf Spot happens in growth period, leaf symptoms early maturity for small spot necrosis, hazel, has faded green halo, Disease spot is not transparent", and so on. Terminology Description of Maize Diseases has a lot of, for example, diseases related terms have names, site of disease onset, symptoms, onset period, and incidence portions include the root, stem and leaf, ear, Onset period can be divided into seedling stage, growth period, and maturation.

In this paper, extract maize disease related terms, including the name of disease, pathogenic sites, the symptom and onset period, Form 4 top-level classes, including: disease class, describe the name of disease; time class, describe the onset period; symptoms class, describing onset of symptoms; class part, describing onset of parts. After each top level class determined, according to knowledge of top level class, subclasses are built. In class symptoms establishing site symptoms subclasses, including 4 parts: class root, class stem, class ear, class leaf, to describe root, stems, ear, leaf disease symptom. In the ontology, the specific name of disease, pathogenesis, symptoms of disease, disease period, are added in the form of an example. Corn Disease ontology concept system as shown in Figure 1.

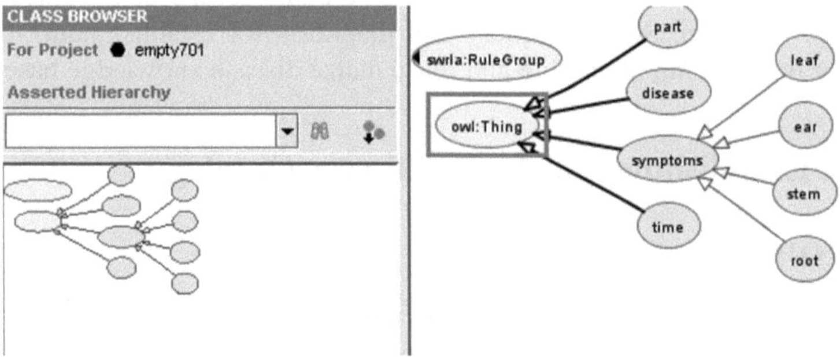

Fig. 1. Corn Disease ontology concept system

After definition of the concept of hierarchical structure, in order to describe the semantic relation between concepts, need to define the attributes of concepts. The object property is used to describe the concepts and concept or instance and instance relations. We also need to define the domain, range, property type and inverse properties as constraint conditions. In this paper, the maize disease ontology set up a total of 5 object properties, results as shown in table 1.

Table 1. The properties of Maize disease ontology

Object properties	Domain	Range	description	Inverse properites
is_time	time	disease	Onset period	
is_part	part	symptoms	Incidence of parts	
hassyptom	disease	symptoms	Onset of symptoms	issyptomof
is_disease	thing	disease	The name of the disease	
issyptomof	symptoms	disease	The symptoms of the disease	hassyptom

In the ontology, the specific name of disease, pathogenesis, symptoms of disease, disease period, is added with the form of examples. Adding example of maize disease 10, example of the root symptoms 3, examples of the stems root symptoms 3, examples of the stem symptoms 6, examples of the ear 5, examples of the leaf symptoms 16, examples of onset period 3, examples of sites 4, and determine the constraint axiom. Corn Disease ontology instances and constraints are as shown in figure 2.

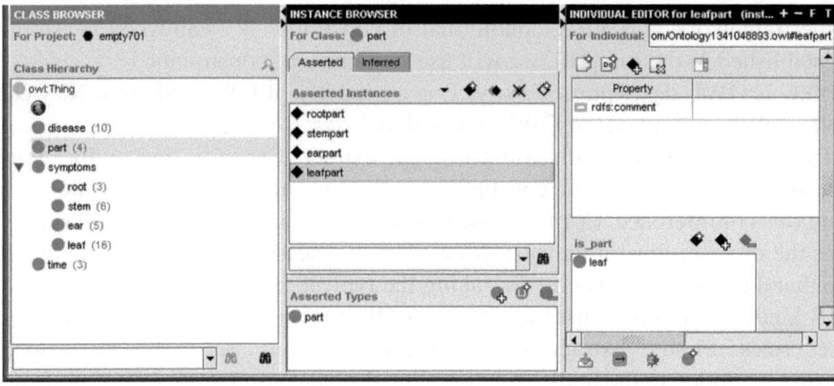

Fig. 2. The ontology constraints and examples

2.3 The Establish of SWRL Rules

SWRL (Semantic Web Rule Language) is a strong ability to express the rule description language based on OWL, Which is based on OWL language OWL DL and OWL Lite.

SWRL extends the OWL axiom set, joined the horn clause form rules, rules and the OWL knowledge base can be combined. SWRL consists of Imp, Atom, Variable and Build-in, totally four parts. Imp SWRL rules,Imp is the SWRL rules part, Consists of body and head. Head and body are made up of atom conjunctive composition, variables of atom record in the variable. Built-in is a modular component of SWRL, which records the SWRL can refer to logical relations, namespace is swrlb[8].

SQWRL (Semantic Query-Enhanced Web Rule 13 Language) is based on the SWRL ontology query language, with build-in predicate thesaurus as expansion point; use build-in to build search rule [9].SQWRL includes Core Operators and Collection Operators two parts, Core Operators query through sqwrl: select, Collection Operators support more advanced search functions, to provide advanced grouping and aggregation function. Because SQWRL use standard SWRL syntax, does not change the semantics of SWRL, so in the SWRL editor SQWRL rules can be added and modified, through SQWRL Query Tab convert the SWRL to a query effectively. SQWRL supports OWL class description, by inference can be constrained examples, can be used to retrieve the ontology contains the implicit information, and expanded the build-in predicate lexicon, enriched the numerical, logic comparison predicate and statistical functions, very convenient to use.

3 Maize Disease of Reasoning Based on SWRL

3.1 Reasoning Rules Construction

SWRL Editor is the expansion of Protégé -OWL, so that the user can edit rules at Protégé, more convenient and intuitive. SWRL Editor can verify SWRL rules of basic syntax and semantics test, to ensure that each rules in grammar is correct.

In the framework of SWRL, conditional of constraints is established in Atom, the rules established in the Imp[10], we will use two kinds of constraints:

C (x):C is OWL description P (x, y):P is a property of OWL, wherein X, y can be variables, OWL individuals or OWL data value.

In the established maize disease ontology, extracted specific reasoning elements disease involved according to concepts of ontology, and according to the element forming rules of inference. Corn Disease ontology rules are mainly based on symptoms to infer the disease, thus in the extraction of corn disease symptom knowledge, must ensure that the knowledge is correct, taking the typical, main symptoms. In this paper, take 10 kinds of common diseases in maize diseases as an example, writing maize diseases rules. Firstly, extract maize disease symptom. Secondly, the associated symptoms are added to the body, to ensure that the rules used in all atoms are from the body or property, that is to say not introduce new variables in the rules, this is SWRL defined by a safety limit. The extracted portions of the rules are as follows:

Rule1- Top rot: disease (? X) ∧ hassymptom (? X, Upper Leaf Chlorosis) ∧ hassymptom (? X, the blade edge or tip withered yellow stripe generated) ∧ hassymptom (? X, internode transverse visible Brown lesion) ∧hassymptom (? X, ear top is changed to pink _) ∧ is_time (? X, chengshuqi)∧ is_time (? X, shengzhangqi) ∧is_time (? X, shengzhangqi) ∧ is_part (? X, Jing) ∧ is_part (? X, SUI) ∧ is_part (? X, ye) → is_disease (x, dingfubing?)

Rule2- helminthosporium maydis:disease (? X) ∧hassymptom (? X, early immersion shape tan or reddish brown spots then)∧ hassymptom (? X, severe ear rot seed black mildew) ∧is_time (? X, chengshuqi) ∧ is_time (? X, miaoqi) ∧is_time (? X, is_part (shengzhangqi)? X, ye) ∧ is_part (? X, SUI) → is_disease (?x, helminthosporium maydis)

In the diagnostic rules, "disease" is the class in the ontology, "is_disease", "hassymptom", "is_time", "is_part" is the object attribute.

3.2 SWRL Rule Reasoning Process

Reasoning rules can be carried out reasoning and searching after building the ontology, but SWRL Editor itself does not perform any reasoning, need the help of other inference tool for reasoning, this paper uses Jess (Java expert system shell) inference engine to do the reasoning[11]. Jess is the CLIPS (C Language Integrated Production System) inference engine based on Java language, its core is composed of 3 parts of fact base, rule base, inference engine and the use of production rules as basic knowledge expression model. Jess supports the forward and backward reasoning, is a good performance of the open type inference machine. Base on advantages above, use Jess as the reasoning engine in the SWRL reasoning rules.

Maize Disease ontology is build using OWL language, then to construct the corresponding rules of SWRL to make OWL reasoning ability inadequacy, thereby forming an ontology knowledge base, including rule library and fact library two aspects. Through Jess and SWRL editor is integrated, the diseases of Maize OWL ontology and rules of SWRL converted to Jess fact library and Jess rule library, calling the Jess engine for reasoning, and push reasoning result into the knowledge library, and format new knowledge. The reasoning mechanisms of SWRL are as shown in Figure 3.

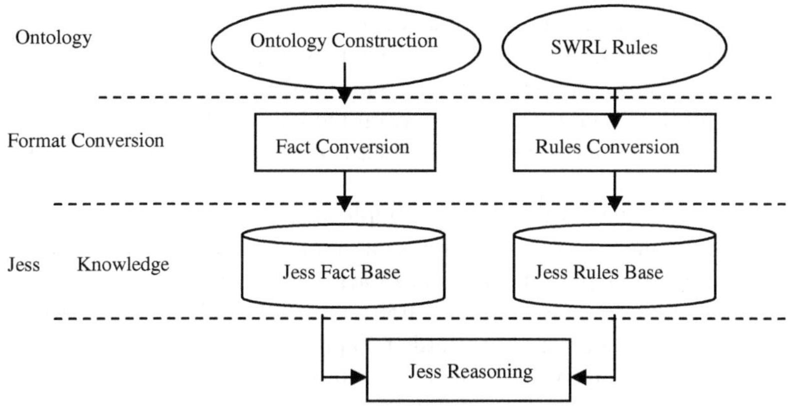

Fig. 3. The reasoning mechanism of SWRL

3.3 Maize Disease Ontology SWRL Diagnostic Reasoning

This paper uses ontology as a knowledge organization method, using SWRL Editor to preparation rules, through the SQWRL query, to mainly achieve the following function of maize diseases diagnosis:

known symptoms of the disease, diagnosis disease name

For example, rule: disease (? X) ∧ hassymptom (? X, Upper Leaf Chlorosis) ∧ hassymptom (? X, the blade edge or tip withered yellow stripe generated) ∧ hassymptom (? X, internode transverse visible Brown lesion) → sqwrl: select (? X)

Known symptoms of Upper Leaf Chlorosis and leaf edges, yellow stripes or tip withered, internodes transverse visible brown lesion tissue, through the SQWRL query, receive the diagnosis conclusion: may be suffering from top rot.

Known disease name, to infer the possible symptoms

For example, rule: hasSymptom (top rot,? Y) ∧is_part (?x, leaf) → sqwrl: select(? y)

Known as the top rot disease on site for leaf, through the SQWRL query, receive leaf symptoms

(3) According to the different condition, realize the query function. For example, rule:

① is_part (? x, leaf) → sqwrl:select (?x) Through the SQWRL query, the query of the ontology all may occur in the leaf disease of.

② is_time (?x, Seedling stage) → sqwrl: select(? x) Through the SQWRL query, get possible disease of corn seedling.

4 Conclusion

This paper studied the maize disease ontology construction, and use SWRL and SQWRL to describe the rules of maize disease diagnosis, completes the ontology in a specific subject application, and practice it. Many domestic scholars carried out through SWRL disease diagnosis [12], but they are using SWRL known disease symptoms to diagnose the disease name, and must meet all of the symptoms can be diagnosed with the disease. In this paper, on the basis of SWRL, using the SQWRL extension reasoning function, rich in disease diagnosis function, improve disease diagnosis.

But corn disease diagnosis is a very complicated process, which involves elements of multiple and complex, the same disease at a different position and different period have been demonstrated, so if you want to achieve accurate maize disease diagnosis function, will add more judgment conditions, and will continue to refine the concept. The next step work include: (1) to further improve maize disease ontology, concept refining; (2) research on SWRL semantic extension, solving predicate logic operation and between a predicate logic or operational problems, further improve the semantic rules legibility and universality, reducing the maintenance cost and rules, to simplify the process of reasoning, facilitates users to complete reasoning.

Acknowledgments. Funding for this research was provided by youth fund of Jilin agricultural university (NO.201133,) and the development of science and technology plan projects of Jilin province (NO. 201101114), Jilin provincial projects(NO.201248),The word bank project(No. 2011-Z20), Hall of Jilin province science and technology science and technology support program(20110237,20120282).

References

1. Jie, C.: Maize disease diagnosis and prevention and control. Shield Press (2002)
2. Chen, L., Wang, L.: Probabilistic neural network in the maize leaf department of the application of the disease recognition. Farm Machinery Research (2011)
3. Ma, L., Yu, H., Wang, Y., Chen, G.: The Knowledge Representation and Semantic Reasoning Realization of Productivity Grade Based on Ontology and SWRL. In: Li, D., Chen, Y. (eds.) CCTA 2011, Part I. IFIP AICT, vol. 368, pp. 381–389. Springer, Heidelberg (2012)
4. Studer, R., Benjamins, V.R., Fensel, D.: Knowledge engineering: principles and methods. Data & Knowledge Engineering (1998)
5. Liu, C.: The Research of Corn Disease and Pest Prevention and Cure Semantic Searching System Based on Ontology. Jilin University (2007)
6. Li, J., Su, X.-L., Qian, P.: The methodology of developing domain ontology. Computer and Agriculture, 7–10 (2003)
7. Wang, X., Wei, Y.-Y.: Research on Construction and SWRL Reasoning of Swine Disease Noumenon. Acta Agriculturae Jiangxi (2012)

8. Zongnan, S., Zheng, Y.L., Qian, P.: Based on the GuiYu SWRL disease diagnosis knowledge representation and semantic reasoning realize. Journal of Intelligence (2010)
9. Wang, H., Fan, L., Li, Z.-Z.: Semantic Web Service Discovery Based on SQWRL. Microelectronics & Computer (2010)
10. Ji, Z.-H., Li, C.-H.: Constructing and Strategy Analyzing of Rule System for the Semantic Web Based on SWRL and Jess. Journal of Huaihai Institute of Technology (Natural Sciences Edition), 26–29 (2009)
11. Sandia National Laboratories. JESS, the Rule Engine for the Java Platform (EB/OL) (November 11, 2008), `http://Herzberg.ca.sandia.gov.JESS` (July 12, 2009)
12. Mou, D.M., Fan, Y.: SWRL reasoning mechanism based on the ecg ontology design and implementation. Journal of Intelligence (2010)

An Object-Oriented Binary Change Detection Method Using Nearest Neighbor Classification

Jie Liang[1], Jianyu Yang[1], Chao Zhang[1], Jiabo Sun[1], Dehai Zhu[1], Liangshu Shi[2], and Jihong Yang[2]

[1] College of Information and Electrical Engineering, China Agricultural University, Beijing 100083, P.R. China

[2] China Land Surveying and Planning Institute, Beijing 100035, P.R. China

{Jianyu Yang}ycjyyang@cau.edu.cn

Abstract. Threshold selection is a critical step in using binary change detection methods. The threshold determines the accuracy of change detection results but is highly subjective and scene-dependent, depending on the familiarity with the study area and the analyst's skill. Nearest neighbor classification is a non-parametric classifier, which was applied to remove the threshold. In order to find the most suitable feature to detect construction and farmland changes, a variety of single and multiple variables were explored. They were regional similarity (RSIM), brightness difference images (BDIs), multi-band difference images (MDIs), multi-band ratio difference images (MRDIs), a combination of RSIM and BDIs (RSIMBD), a combination of RSIM and a optimum band difference and a optimum band ratio difference (RSIMDR), MDIs and MRDIs multiple variable groups. All were tested for two study sites of the bi-temporal SPOT 5 imagery, the results indicated that RSIM, RSIMDR, RSIMBD were significantly better than other single and multiple variables.

Keywords: Object-Oriented, Binary Change Detection, Nearest Neighbor Classification, RSIM.

1 Introduction

Change detection is the process of indentifying differences in the state of an object or phenomenon by observing it at different times.[1] It is a core problem in resource and environmental monitoring, disaster monitoring, land management, city expansion, geographic information update and dynamic monitoring of the military battlefield. In the past years, researchers have put forward a large number of change detection techniques which can be categorized into two groups: (1) those detecting binary change information (i.e., change vs. no change); and (2) those detecting detailed 'from-to' change information.[2]

'From-to' change detection algorithms (e.g., post-classification comparison) provide detailed information about the location and type of land_use/cover change (e.g., from farmland to building) for every pixel or object under investigation. The key in using these methods is to create accurate thematic classification images. The errors of individual-date

D. Li and Y. Chen (Eds.): CCTA 2012, Part II, IFIP AICT 393, pp. 394–406, 2013.

thematic images will affect the final change detection accuracy. [3] Even though the ideal goal of change detection is often to extract 'from-to' change information, however, sometimes rapidly obtaining change and no change information is valuable. A variety of methods have been applied to binary change detection, for example, image differencing, image ratioing, vegetation index differencing and PCA.[4-6]

One critical step in using binary change detection approaches is to select the appropriate thresholds to generate binary change mask. Empirical threshold method is usually used for selection of thresholds, which is an interactive procedure or manual trail-and-error procedure. For example, the optimum threshold can be determined by using the mean and standard deviation of the training samples.[7] However, the threshold is highly subjective and scene-dependent, depending on the familiarity with the study area and the analyst's skill. In order to improve the change detection results, many semi-automatic or automatic threshold determination methods are developed. Im et al. used a calibration approach by autonomously testing thousands of thresholds and selected the optimum threshold(s) showing the highest accuracy.[8] The systematic search technique and a Moving Threshold Window (MTW) were proposed to improve the efficiency and accuracy, respectively.[9, 10] Wang et al. put forward adaptive spatial neighborhood analysis and Rayleigh-Gauss distribution fitting for change detection in multi-temporal remote sensing images, while the optimal threshold was calculated automatically and effectively by the improved Kittler-Illingworth (KI) threshold selection algorithm.[11] Although some advanced threshold approaches have been developed, these are still only applied to specific applications for a specific environment due to its complexity. Nearest neighbor classification as a non-parametric classifier, is only determined by the quality of samples and the selection of features, which is independent of threshold.

Change detection is a complicated and integrated process.[12] Successful remote sensing change detection requires careful attention to: (1) remote sensor system considerations, such as radiometric resolution, spatial resolution, look angle, and (2) environmental characteristics of the study areas, including atmosphere condition, complexity of the landscape and topography.[13] After image data and study areas have been determined, selection of an appropriate change detection method assumes considerable significance. Traditional binary change detection methods usually use a single variable (e.g., band 1 difference). However, the use of multiple bands simultaneously might produce better results (e.g., band 1 difference and band 2 ratio difference). Thus, how to customize the optimum multiple variables becomes another problem. In addition, now there is great interest in indentifying change in high spatial resolution multi-temporal data, such as that provided by GeoEye, Inc., DigitalGlobe, Inc., Leica Geosystems, Inc. Unlike medium- to coarse-resolution imagery, there are ample spectral information, textural information, structure information in high spatial resolution imagery. Many traditional change detection methods based on per-pixel analysis may not function successfully in the high-resolution domain. A serious problem found in per-pixel approaches is the 'salt-and-pepper' phenomenon in change detection results, which is meaningless for application. Therefore, objected-oriented change detection methods have been put forward where homogeneous polygons found in two dates of imagery are compared.[14, 15]

Hence, the objectives of this study included: (1) development of an object-oriented binary change detection method using nearest neighbor classification to replace the thresholds, (2) customization of a variety of single and multiple image variables to extract change information, and (3) validation of binary change classifications with the variables, especially focusing on a comparison between the detection results using single versus multiple image variables.

2 Materials and Methods

2.1 Study Area and Data

Two study sites representing new construction in Changping District, Beijing were selected to demonstrate the nearest neighbor classification for object-oriented binary change detection (Fig.1). The first study site (Site A) is located in southeast Changping District, and exhibits new buildings and change between different growth farmlands. The second study site (Site B) is located 1km east of Site A, and exhibits change from farmlands and water to buildings.

SPOT 5 imagery acquired over the two study sites in Changping District on October 01, 2002 and October 16, 2007 was used for validating the binary change detection. The area obtained spectral information in four multispectral bands (i.e., near-infrared (NIR), red, green, and short waved-length infrared (SWIR)) with 10-meter spatial resolution and one panchromatic band with 2.5-meter spatial resolution (wavelength ranges from 0.48-0.71μm).

In the preprocessing, the bi-temporal images were geo-referenced to a Transverse Mercator (UTM) projection in WGS 84. Rectification errors were less than 0.5 pixel (RMSE). Even though they were obtained on near-anniversary dates, it was still necessary to radiometrically normalize the two dates of remote sensor data since the subsequent analysis (e.g., multi-band differencing, multi-band ratio differencing) directly compared the Digital Number (DN) values from the bi-temporal imagery. The radiometric normalization method using histogram matching was applied. Then Principal Component Analysis method was used to do image merge to get high spatial resolution and fine detail.

2.2 Methods

This object-oriented change detection research aimed to identify the single or multiple variable(s) that is/are most suitable for extracting buildings and farmlands change regions through a comparison of the customized features using the nearest neighbor classification. In order to generate one–to-one mapping relationship of the object with the bi-temporal imagery, two dates of imagery were used together in multi-resolution segmentation procedure. Then, several training/test data were randomly collected within each study site. The binary change information for each sample location (i.e., change vs no change) was identified through visual interpretation of the bi-temporal SPOT 5 imagery. A variety of variables associated with change information were

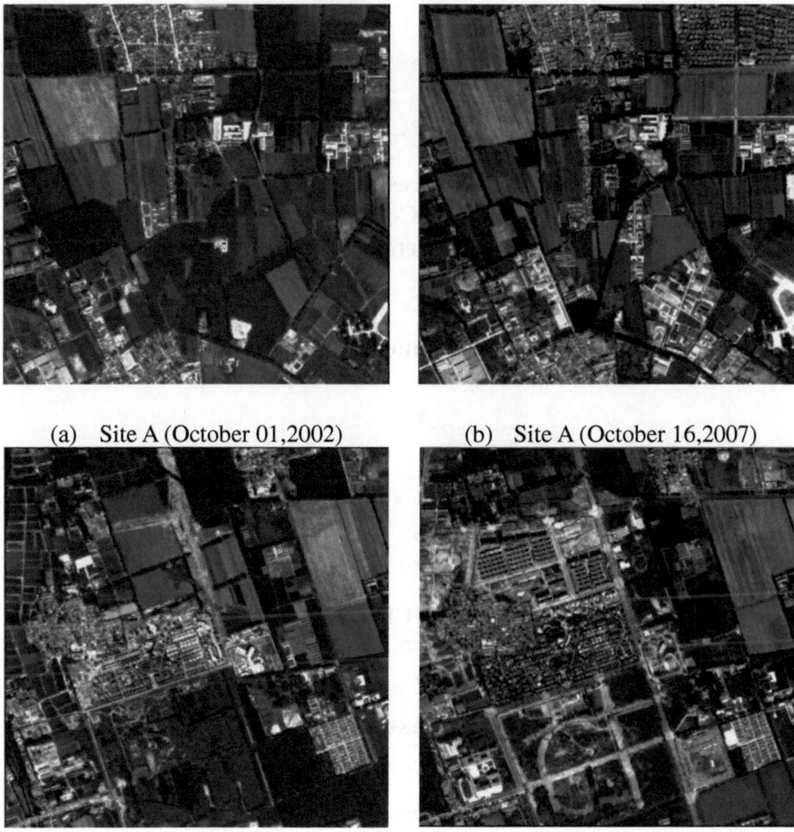

(a) Site A (October 01,2002) (b) Site A (October 16,2007)

(c) Site B (October 01,2002) (d) Site B (October 16,2007)

Fig. 1. (a)-(d) Bi-temporal SPOT 5 imagery for two study sites (A, B), respectively

customized and applied to the nearest neighbor classification. A brief summary of the change detection flow was found in Fig.2.

Nearest Neighbor Algorithm to Binary Change Detection Method

Nearest neighbor classifier is a popular non-parametric classifier used in many field.[16] When input several samples for each class, the nearest neighbor classifier computes the distance from the object to be classified to the nearest training object and assigns it to the class of the training object. Due to its simplicity and avoiding threshold problem, it is selected as a technique for change classification. However, there are two mainly disadvantages of nearest neighbor classifier. One is that its classification accuracy depends on the training data. Therefore, 24-26 objects as training samples were collected for each study site through visual inspection with at

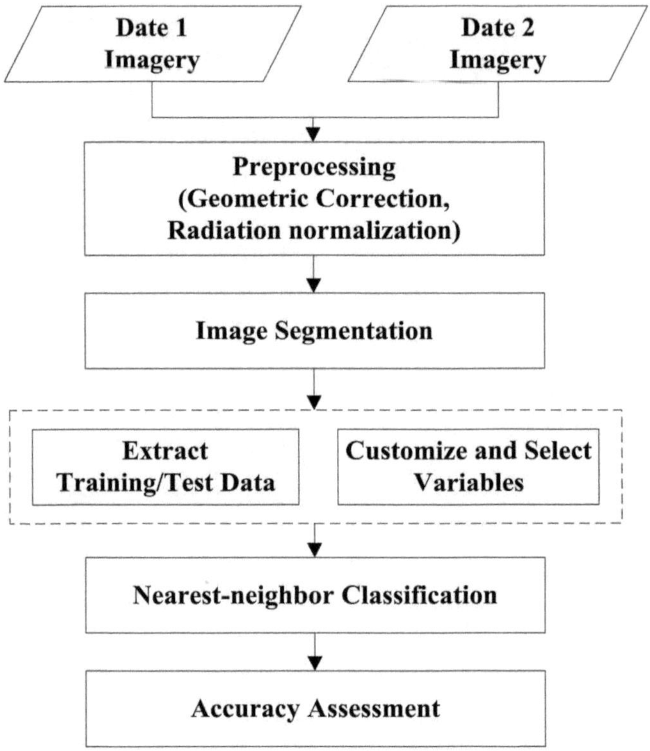

Fig. 2. Flow diagram of the binary change detection method

least 12 objects per class. For a reasonable comparison between the change detection results, the same training data for each site were used in the classification. And the other is called the 'curse of dimensionality'. Dimensionality means the number of features. Usually when the dimensionality is more than 20, the calculation is very slow. What's worse, the accuracy of the classification tends to deteriorate after the dimensionality up to a specific value. So it is necessary to optimize the feature space.

Single and Multiple Variables

In order to identify binary change detection methods that were most appropriate for new construction, a variety of variables were explored. These variables included regional similarity (RSIM), brightness difference images (BDIs), multi-band difference images (MDIs), multi-band ratio difference images (MRDIs) generated from the bi-temporal imagery were used as input variables. Every variable was introduced as follows:

RSIM[17] is used to detect change by evaluating the degree of regional similarity in the same region in two registered remote sensing images obtained on different dates. If the spectral changes of the object between two image dates are significant, the RSIM will fall off to a lower value. There exists three parts of RSIM, which are correlation, brightness ratio and contrast. See the equations:

$$RSIM=\frac{4\sigma_{12}\overline{BV_1}\,\overline{BV_2}}{(\sigma_1^2+\sigma_2^2)[(\overline{BV_1})^2+(\overline{BV_2})^2]}=\frac{\sigma_{12}}{\sigma_1\sigma_2}\cdot\frac{2\overline{BV_1}\,\overline{BV_2}}{(\overline{BV_1})^2+(\overline{BV_2})^2}\cdot\frac{2\sigma_1\sigma_2}{(\sigma_1^2+\sigma_2^2)} \quad (1)$$

Where

$$\sigma_{12}=\frac{1}{B-1}\sum_{i=1}^{B}(BV_{i1}-\overline{BV_1})(BV_{i2}-\overline{BV_2}) \quad (2)$$

and σ_{12} is the covariance between object mean spectral values found in all bands of the two date imagery, σ_1, σ_2 are the standard deviation of the object mean spectral values found in all bands of each date imagery, respectively. While BV_{i1} is the i th object mean spectral value of image 1, BV_{i2} is the i th object mean spectral value of image 2, B is the total number of the bands for each image, and $\overline{BV_1}$, $\overline{BV_2}$ are the means of object mean spectral values found in all bands of the two date imagery, respectively.

Brightness difference images (BDIs) is used to detect change by computing the brightness difference of the two date imagery. As some classes are quite different in brightness, such as buildings and farmland, this variable is suitable to extract those change regions.

$$Diff\_Brightness=\frac{1}{B}\sum_{i=1}^{B}BV_{i2}-\frac{1}{B}\sum_{i=1}^{B}BV_{i1} \quad (3)$$

Multi-band difference images (MDIs) are the differences of object mean spectral value in each band for the two date imagery, and multi-band ratio difference images (MRDIs) are the ratio differences of object mean spectral value in each band for the two date imagery.

$$Diff\_Band(\mathrm{i})=BV_{i2}-BV_{i1} \quad (4)$$

$$Diff\_Ratio(\mathrm{i})=\frac{BV_{i2}}{\sum_{j=1}^{B}BV_{j2}}-\frac{BV_{i1}}{\sum_{j=1}^{B}BV_{j1}} \quad (5)$$

A total of 20 single variables were tested with the nearest neighbor change classification for the two study sites. Each variable group (i.e., MDIs, MRDIs) was used as the inputs for multiple variables. The another group of multiple variables that consisted of one variable from the BDIs and MDIs, each showing the best performance in the experiments with the single variables, and RSIM were selected and tested, namely RSIMDR. In order to demonstrate BDIs provide change information that can be used as ancillary data to facilitate change detection with RSIM, a combination of RSIM and BDIs (RSIMBD) was also selected and tested. Thus, a total of 8 groups of multiple variables were evaluated with the binary change detection method for two study sites.

Accuracy Assessment

An accuracy assessment was accomplished for the binary change classifications (i.e., change and no change) of Site A and Site B. In each case, the accuracy assessment included overall accuracy (OA), the Kappa coefficient, the user's accuracy (UA) and the producer's accuracy (PA). And test samples for each class were randomly collected within each study site.

3 Results and Discussion

3.1 Object-Oriented Binary Change Detection Using Single Variables

A variety of single variables associated with change information were customized for the binary change detection, including regional similarity (RSIM), brightness difference images (BDIs), multi-band difference images (MDIs), multi-band ratio difference images (MRDIs). The image objects used in generating those variables were created from the multi-resolution segmentation technique in eCognition, one of the widely used object-oriented image processing software. A scale parameter of 70, a spectral factor of 80%, a shape factor of 20%, which was subdivided into smoothness (6%) and compactness (14%), and equal weights for the all eight bands were used in the image segmentation process. While the RSIM and BDIs had only one variable each, the MDIs and MRDIs contained four individual variables each, using the four bands of the SPOT 5 imagery.

The overall accuracy and Kappa coefficients of each single variable for the two study sites were shown in Table 1(a). RSIM achieved the best overall accuracy and Kappa performance for both two study sites, while the band 1(i.e., band NIR) difference resulted in the worst overall accuracy and Kappa performance for both two study sites. In addition, the single variable of the same band from the MRDIs was better than this from the MDIs.

The nearest neighbor change classification using the band 4 difference variable produced the highest overall accuracy and Kappa coefficient among the single variables from the MDIs for Site A. The band 3 ratio difference variable yielded the best performance (OA=85.19%, Kappa=0.707) among the single variables from the MRDIs for Site A. The single variables from the MDIs resulted in larger variation in the OA and Kappa than those from MRDIs (Δ OA=14.83% vs. 7.41%,

Δ Kappa=0.295 vs. 0.147) for Site A. Dissimilarly to Site A, the band 3 difference variable from the MDIs, the band 2 and band 3 ratio difference variable from the MRDIs produced the same good performance (OA=76.92%, Kappa=0.538) for the Site B. However, the band 1 difference variable yielded the worst change detection result (Kappa=0.080) among all single variables for the Site B. In addition, the single variables from the MDIs resulted in larger variation in the OA and Kappa than those from MRDIs (Δ OA=23.07% vs. 11.54%, Δ Kappa=0.458 vs. 0.230) for Site B.

3.2 Object-Oriented Binary Change Detection Using Multiple Variables

Based on the change detection results using the multiple variables for the two study sites, as shown in Table 1(b), RSIMBD multiple variable group was slightly better or equal to that based on the RSIMDR group, which produced the highest classification accuracies. However, the change detection results using MDIs, MRDIs multiple variables resulted OA of 74.07% and 77.78% for Site A, which were even lower than the accuracy achieved by the single band 4 differencing and band 3 ratio differencing. Similarly to Site A, the MDIs, MRDIs multiple variables also did not produced higher accuracies than those yielded using single band 3 differencing and band 3 ratio differencing. Therefore, the multiple variables may not always do good performance. On the other hand, as each variable had its own characteristics, representing different capabilities for distinguishing land-use/cover types, a suitable combination may improve the final accuracy of the change classification. For example, the BDIs single variable which only detected the change objects existing large brightness differences, resulted a bad accuracy, however, it facilitated change detection with RSIM. The OA of RSIMBD (RSIM+BDIs) multiple variables increased 11.14% and 3.84% compared with RSIM for Site A and Site B, respectively.

Table 1. OA and Kappa using the single and multiple variables for Site A, Site B

(a) Single variables

Site	Single varible	OA (%)	Kappa
Site A	RSIM	85.16	0.702
	Diff_Brightness	74.07	0.485
	Diff_Band(1)	66.67	0.331
	Diff_ Band(2)	70.37	0.410
	Diff_ Band(3)	77.78	0.560
	Diff_ Band(4)	81.50	0.626
	Diff_Ratio(1)	81.48	0.628
	Diff_Ratio(2)	77.78	0.560
	Diff_Ratio(3)	85.19	0.707
	Diff_Ratio(4)	85.19	0.703

Table 1. *(Continued)*

Site B	RSIM	88.46	0.769
	Diff_Brightness	61.54	0.231
	Diff_Band(1)	53.85	0.080
	Diff_Band(2)	73.08	0.462
	Diff_Band(3)	76.92	0.538
	Diff_Band(4)	69.23	0.385
	Diff_Ratio(1)	73.08	0.462
	Diff_Ratio(2)	76.92	0.538
	Diff_Ratio(3)	76.92	0.538
	Diff_Ratio(4)	65.38	0.308

(b) Multiple variables

Site	Multiple variables	OA (%)	Kappa
Site A	MDIs (band 1 through 4 difference)	74.07	0.485
	MRDIs (band 1 through 4 ratio difference)	77.78	0.560
	RSIMDR (Diff_Band(4) + Diff_Ratio(3) + RSIM)	96.15	0.923
	RSIMBD (RSIM+ Diff_Brightness)	96.30	0.926
Site B	MDIs (band 1 through 4 difference)	73.08	0.462
	MRDIs (band 1 through 4 ratio difference)	76.92	0.538
	RSIMDR (Diff_Band(3) + Diff_Ratio(3) + RSIM)	92.30	0.846
	RSIMBD (RSIM+ Diff_Brightness)	92.30	0.846

As the performance of change detection results should not only be based on the indicators of OA and Kappa, but also consider other accuracy statistics such as user's and producer's accuracies. Thus, the PA and UA comparison using the single and multiple variables for the two study sites were shown in Fig.3. Although the OA and Kappa values of single variables were lower than the accuracy using RSIMBD multiple variable group, it was found that some single variables were very suitable to identify certain change class. For example, the band 4 difference was good at identifying the change class and had small errors of omission as the producer's accuracy reached to 0.929 for Site A, while the band 3 difference was suitable to identifying the change class as the producer's accuracy reached to 0.923 for Site B. In addition, the band 3 difference, band 2 ratio difference and band 3 ratio difference all produced a higher producer's accuracy in extracting no change class than RSIM for Site A, and band 3 difference also yielded a higher producer's accuracy in extracting no change class than RSIM for Site B. Thus, the RSIMDR multiple variable group was good at identifying both change and no change class.

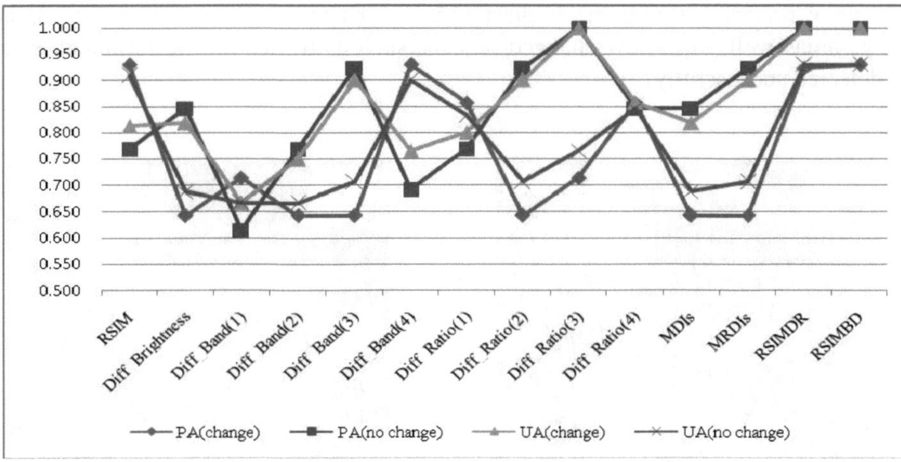

(a) Site A

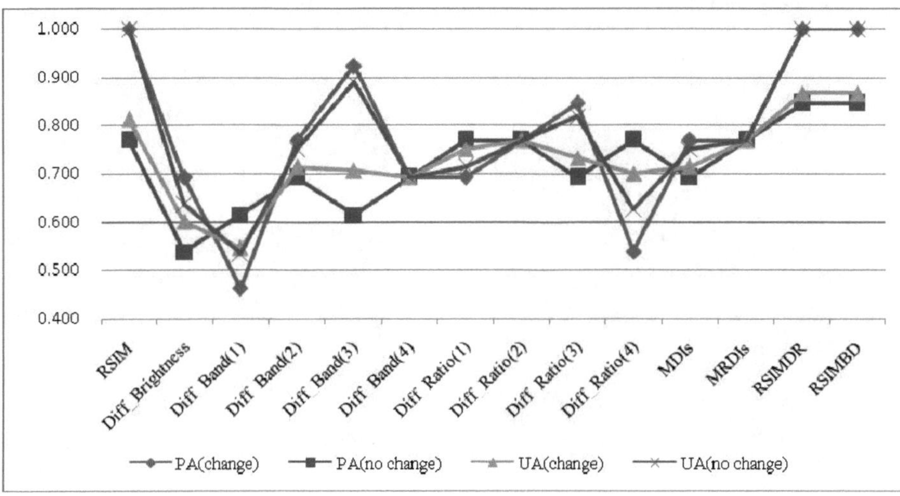

(b) Site B

Fig. 3. PA and UA comparison using the single and multiple variables for Site A, Site B

Comparison of binary change detection for the best and the worst performance using the single and multiple variables for each site were shown in Fig.4. The band 1 difference variable misidentified many unchanged buildings as change objects in the upper middle part and down left part of Site A (Fig.4.(a)), and also misidentified some unchanged buildings as change objects in the middle part of Site B (Fig.4.(c)). Besides, the band 1 difference variable did not detect some new construction as change in the middle down part of Site B (Fig.4.(c)). On the other hand, the RSIMBD multiple variables yielding the highest OA and Kappa for each site detected the change areas well, but a few changed objects, such as from water to farmland, were not detected successfully.

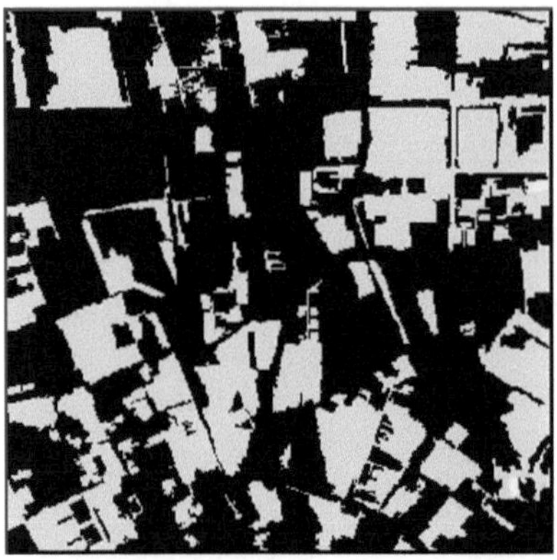

(a) Band 1 difference single variables for Site A
Worst

(b) RSIMBD multiple variables for Site A
Best

(c) Band 1 difference single variables for Site B
Worst

(d) RSIMBD multiple variables for Site B
Best

change   no change

Fig. 4. (a)-(d) The worst and best change detection results for Site A, Site B

4 Conclusion

Binary change detection methods have many advantages and disadvantages. Generally, the binary change detection method is a threshold-based calibration process, which the optimum threshold was found through trial and error. However, such an approach is time-consuming and highly subjective. The research successfully employed the binary change detection method with nearest neighbor classification technique, removing the threshold problem existing in the traditional binary change detection method.

Additionally, a variety of single and multiple variables were explored to extract change information for two study site. The results showed that not all multiple variables produced better performance than the single variables. However, a combination of the best performance in the experiments with the single variables, or those achieved good performance in a certain class, usually produces significantly better accuracy than others. The RSIM single variable yielded the best performance in the nearest neighbor change classification when using single variables. What's more, the combinations of RSIM and other single variables from MDIs and MRDIs, or BDIs, produced the highest overall accuracy and Kappa value than any other multiple variables. The fact that the RSIM combined the correlation, brightness ratio and contrast of all eight bands simultaneously explained the superiority.

The drawback of the proposed method concerned only spectral features of high spatial resolution SPOT 5 imagery, so objects of the same class with different spectral value is misidentified changed class easily. Future works could focus on developing other useful texture or structure features to binary change detection method and applying the method with different types of remote sensor data, such as QuickBird, WorldView.

Acknowledgments. The research was funded by the Pre-Development of the High Resolution Application Demonstration System on Land Resource "Research on the Land Use Type Quickly Extraction with High Resolution Data"(E0202/1112/0104), and "Potential Study on Land Resource Management Applications of Archived Remote Sensing Data" (201109109188). The authors would like to thank the anonymous reviewers for their constructive comments and suggestions.

References

1. Singh, A.: Digital Change Detection Techniques Using Remotely-Sensed Data. International Journal of Remote Sensing 10, 989–1003 (1989)
2. Chan, J.C.W., Chan, K.P., Yeh, A.G.O.: Detecting the Nature of Change in an Urban Environment: A Comparison of Machine Learning Algorithms. Photogrammetric Engineering and Remote Sensing 67, 213–225 (2001)
3. Lu, D., Mausel, P., Brondizio, E., Moran, E.: Change Detection Techniques. International Journal of Remote Sensing 25, 2365–2407 (2004)
4. Lunetta, R.S., Ediriwickrema, J., Johnson, D.M., Lyon, J.G., McKerrow, A.: Impacts of Vegetation Dynamics on the Identification of Land-Cover Change in a Biologically Complex Community in North Carolina, USA. Remote Sensing of Environment 82, 258–270 (2002)
5. Bruzzone, L., Prieto, D.F.: Automatic Analysis of the Difference Image for Unsupervised Change Detection. IEEE Transactions on Geoscience and Remote Sensing 38, 117–1182 (2000)
6. Ridd, M.K., Liu, J.: A Comparison of Four Algorithms for Change Detection in an Urban Environment. Remote Sensing of Environment 63, 95–100 (1998)

7. Lu, D., Mausel, P., Batistella, M., Moran, E.: Land-Cover Binary Change Detection Methods for Use in the Moist Tropical Region of the Amazon: a Comparative Study. International Journal of Remote Sensing 26, 101–114 (2005)
8. Im, J., Rhee, J., Jensen, J.R., Hodgson, M.E.: An Automated Binary Change Detection Model using a Calibration Approach. Remote Sensing of Environment 106, 89–105 (2007)
9. Im, J., Rhee, J., Jensen, J.R.: Enhancing Binary Change Detection Performance Using A Moving Threshold Window (MTW) Approach. Photogrammetric Engineering and Remote Sensing 75, 951–961 (2009)
10. Im, J., Jensen, J.R., Hodgson, M.E.: Optimizing the Binary Discriminant Function in Change Detection Applications. Remote Sensing of Environment 112, 2761–2776 (2008)
11. Wang, G.T., Wang, Y.L., Jiao, L.C.: Adaptive Spatial Neighborhood Analysis and Rayleigh-Gauss Distribution Fitting for Change Detection in Multi-Temporal Remote Sensing Images. Journal of Remote Sensing 13 (2009) (in Chinese)
12. Li, D.R.: Remotely Sensed Images and GIS Data Fusion for Automatic Change Detection. International Journal of Image and Data Fusion 1, 99–108 (2010)
13. Jensen, J.R.: Introductory Digtial Image Processing: A Remote Sensing Perspective, 3rd edn. Prentice Hall, Upper Saddle River (2005)
14. Im, J., Jensen, J.R., Tullis, J.A.: Object-Based Change Detection Using Correlation Image Analysis and Image Segmentation. International Journal of Remote Sensing 29, 399–423 (2008)
15. Walter, V.: Object-Based Classification of Remote Sensing Data for Change Detection. ISPRS Journal of Photogrammetry and Remote Sensing 58, 225–238 (2004)
16. Viswanath, P., Murty, M.N., Bhatnagar, S.: Partition Based Pattern Synthesis Technique with Efficient Algorithms for Nearest Neighbor Classification. Pattern Recognition Letters 27, 1714–1724 (2006)
17. Zhang, C., Yue, A.Z., Yang, J.Y., Su, W., Zhao, D.L., Du, X.J., Chen, M.J., Zhang, Y.N.: Object-Oriented Change Detection Method based on Regional Similarity. China Patent (2009) (in Chinese)

Quantifying the Type of Urban Sprawl and Dynamic Changes in Shenzhen

Ruifang Hao[1,2], Wei Su[1,*], and Deyong Yu[2]

[1] College of Information and Electrical Engineering, China Agricultural University, Beijing 100083, China
[2] State Key Laboratory of Earth Surface Processes and Resource Ecology, Beijing Normal University, Beijing 100875, China

Abstract. Urban sprawl is a hot topic of global concern with increasing dramatically and impact on ecological environment. But there is not a mature method to explain the detail of process of urban sprawl and dynamic changes. In this paper, the dynamic changes of urban is studied based on 1980, 1988, 1994, 2000 and 2005 five remote sensing images of Shenzhen. It is distinguished three types including infilling, edge-expansion and outlying using six landscape metrics related to character the urban pattern. Combining urban sprawl type and urban pattern makes deepen understanding of urban expansion. The results show that the urban area of Shenzhen increases rapidly with growth rate and it experiences urban sprawl phase. In the early stage, it is at the diffusion period with the outlying area in dominant place in sprawl type and landscape fragmentation aggravation showed by NP, LPI increasing in pattern. With the infilling and edge-expansion area increasing rapidly, the urban sprawl turns into coalescence period with good landscape connectivity and landscape fragmentation remission. The method of combining type of urban sprawl and landscape metrics is useful to determine the time of urban sprawl phase. The results can help urban plan and design.

Keywords: Urbanization, urban sprawl, land use/land cover, landscape metric, Shenzhen.

1 Introduction

Urbanization is defined as rural land is converted to urban land use or other construction sites (Karen et al., 2005). The dynamic process of urbanization has led to fundamental changes with resulting in various impacts on the structures, functions, and dynamics of ecological systems at a wide range of scales (Sun et al., 2012).

Reasonable urban spatial pattern is of great significance to promote the sustainable development of cities. In recent studies, urban landscape index has been combined with remote sensing to explain the urban sprawl and dynamic changes, and another focus of researches is how the urban landscape indicators impact on results when it is

* Corresponding author.

D. Li and Y. Chen (Eds.): CCTA 2012, Part II, IFIP AICT 393, pp. 407–415, 2013.

applied to the multi-scale and multi-temporal data sets (Sun et al.,2012; Y et al. 2012).

However, there are considerable controversies on the definition of "urban sprawl". The most universally accepted one is the characteristics of urban sprawl by Ewing. Forman (1995) thought that there were mainly three types in the process of urban sprawl including infilling, edge-expansion and outlying. Infilling means the non-urban area surrounded by urban being converted to urban. Edge-expansion refers to the newly developed urban area spreading out from the fringe of existing urban patches. The outlying growth tends to be distributed at a larger distance from existing developed areas (Sun *et al.*, 2012; Y *et al.* 2012). Technology about quantifying different urban growth types from remote sensing image has not been adequately investigated (Sun *et al.*,2012).

Hoffhine *et al.* (2003) first quantified three urban growth types from Landsat classification imageries. Xu *et al.* (2007) observed that the ratio between the length of common edge and patch perimeter could be used to distinguish them, but no spatial visualization of different urban growth types has been provided. Pham and Yamaguchi (2011) used a percentage of a like adjacency metric to generate the types. Liu *et al.* (2010) proposed Landscape Expansion Index (LEI) with buffer zone analysis. Sun *et al.* developed a quantitative method in order to identify and visualize different types. Each of the types is generated by a specific urban growth process, which may lead to different environmental impact.

In our study, the process of urbanization of Shenzhen is quantified in three types, and some landscape indicators that are relate to the sprawl types are calculated. Combining the type of urban sprawl and the landscape metrics to analyse the urbanization characteristics of Shenzhen between 1980 and 2005 is used.

2 Material and Method

2.1 Study Area

Economic development of Shenzhen was very slow in the three decades before the reform and opening up. But in 1979, the population and Gross Domestic Product (GDP) of the city have been 31.41 million and 196 million yuan respectively. In august of 1980, Shenzhen special economic zone officially proclaimed.

2.2 Data Pre-processing

In our study, five remote sensing images of October 13, 1980(MSS), December 10, 1988(TM), November 8, 1994(TM), November 1, 2000(ETM+) and November 23, 2005(ETM+) are selected. The data pre-processing includes geometric rectification and extraction of study area which base the vector data of Shenzhen.

After the processing of classification, the land use system includes build-up area, arable land, forest, water body, orchard, shrub, wet land, vacant land. Since the focus of this study is urban growth, the classified land use types are further converted into the two classes: urban and non-urban. The rest operation aim at the urban areas only.

2.3 The Type of Urban Sprawl

Based on Hoffhine *et al.* (2003) and Sun *et al.* (2011), a field called T is defined whose value equals the ratio of lc to l, and the lc is the common lengths between new-urban patches and old-urban patches and the l is the perimeter of the corresponding new-urban patches. The value of T is between 0 and 1. If T>0.5, it means that at least 50% of the new-urban patches is surrounded by the old-urban patch, and it represents the infilling type (Fig.1(a)); if 0<T<0.5, the new-urban patches develop from the edge of the old-urban patches, and the common length is less than 50% of its boundary. This type is edge-expansion (Fig.1(b)); if T=0, it means that the new-urban patches have no spatial association with the old-urban patches, and this is outlying type (Fig.1(c)).

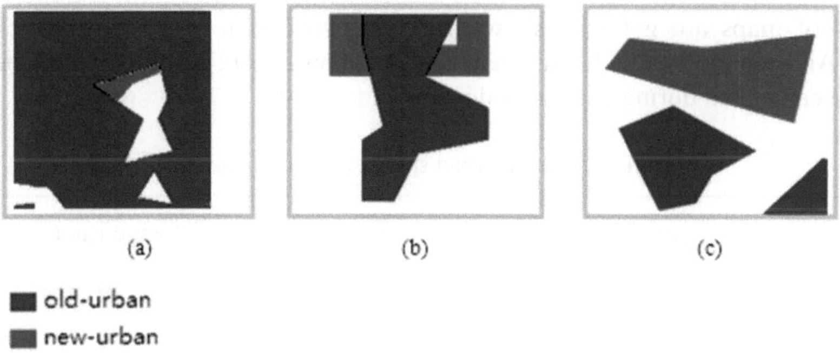

Fig. 1. The three types of urban sprawlAbstract

2.4 Landscape Spatial Metrics

In our study, landscape metrics are selected based on the physical meaning of the indicators themselves, as well as its relationship with the type of urban sprawl.

Number of patches (NP) is one of the Area/Density/Edge metrics, and a measure of landscape fragmentation. Largest patch index (LPI) is also one of the Area/Density/Edge metrics, and a measure of patch dominance. Mean Euclidean nearest-neighbor distance (ENN_MN) is a metric of Isolation/Proximity and it describes landscape connectivity. Contagion index (AI) is a physics aggregation index. Mean patch size (AREA_MN) and landscape shape index can describe landscape fragmentation.

The spontaneous growth will result in increase of the number of patches (NP) and decrease of the contagion index (AI). Edge-expansion causes that largest patch index (LPI), mean patch size (AREA_MN) and contagion index (AI) increasing and mean Euclidean nearest-neighbor distance (ENN_MN) decreasing. Infilling will result in largest patch index (LPI), mean patch size (AREA_MN) and contagion index (AI) increasing. They are calculated about five raster data in Shenzhen based on landscape analysis software Fragstats 3.3.

3 Results

3.1 Monitoring Urban Sprawl Area

The Tab.1 shows that the urban area base in 1980 of Shenzhen is only 12.3km^2. During 1988-1994, the wide urban sprawl area has reached 234.9 km^2 (Tab.1) growing at an average annual rate of 39.15km^2/y. Between 1994 and 2000, the urban sprawl area increased to 268.8km^2 (Tab.1) at an annual rate 44.8km^2/y. The size of sprawl area continues to fall with discrete distribution during 2000-2005. Tab.1 shows that the annual change rate has been rising during study periods.

3.2 Quantifying the Urban Sprawl Type

A series of maps are got who show the type distribution of urban areas changes (Fig.2). And for the detailed characteristics of urbanization, sprawl area of the three types are calculated during each period as the Fig.3 shows.

Table 1. Specific urban sprawl area in Shenzhen

Time	Sprawl urban area (km^2)	Annual change rate(km^2/y)
1980	12.3	
1980-1988	163	20.38
1988-1994	234.9	39.15
1994-2000	268.8	44.8
2000-2005	265	53

As Fig.2(a) and Tab.2 show that during 1980-1988, there is no infilling area, and the dominant type is edge-expansion which is collective. The edge-expansion area mainly gathered in the south. The outlying area is discrete distribution in each zone (Fig.2(a)). Comparing to last time period, the area distribution of outlying type decreases slightly between 1988 and 1994(Fig.2(b)). The infilling and edge-expansion area both increase (Fig.3). The edge-expansion type is dominant and the outlying area is discrete as usual (Fig.2(b) and Fig.3). On the contrary, the rest of two types are gathered. As we can see in the Fig.2(b), the coalescence of urban area of the south Shenzhen is happening. The infilling area is the largest one (Fig.3). The outlying area appeared always in a small area of shape (Fig.2). Between 2000 and 2005, the urban area continues growing (Fig.3). The infilling area is still dominant and larger than the last period, whereas the rest two types both decreased (Fig.3). Shenzhen is merging to a whole large urban (Fig.2(d)).

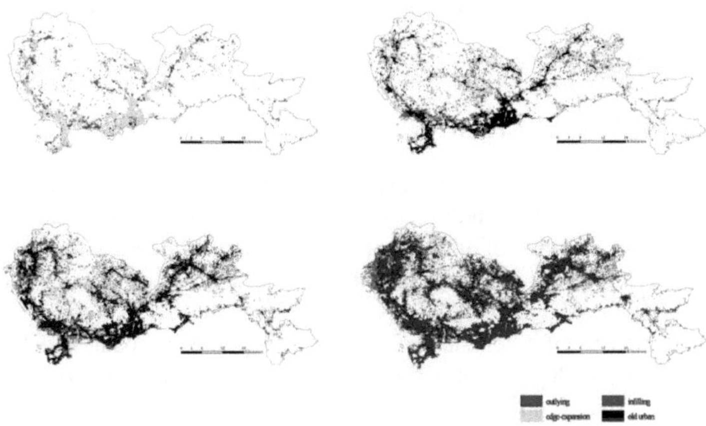

Fig. 2. Spatial distribution of three sprawl types in Shenzhen during 1980-1988, 1988-1994, 1994-2000 and 2000-2005 four periods respectively

3.3 Changes in the Spatial Pattern of the Whole Landscape

The spatial metrics can show the spatial pattern during urbanization. The resolution of 30m×30m is only selected which is certified as the finest one by Wu et al.(2010).The spatial metrics changing graph in Tab.2show the characteristic of the urban growth dynamics patterns.

Table 2. The trend of three types area in each period

Time	LPI	LST	NP	AREA_MN	ENN_MN	AI
LPI	5.6226	18.6107	291	4.5875	797.2751	85.3953
1988	29.4629	31.837	723	23.5697	401.5861	92.891
1994	40.4032	47.2713	1136	33.1193	287.1715	93.7575
2000	50.3377	43.1191	1145	49.9471	262.1326	94.7072
2005	68.5713	41.5652	922	77.3067	288.5362	95.4362

LST rises steadily during 1980-2000 and decreases slightly after 2000(Tab.2). LPI and AREA_MN increase exponentially during urbanization (Tab.2). AI increases steadily for the two stages of the boundary in 1988(Tab.2). With the emergence of the new urban center, urban patches merge and boundaries of urban areas dissolve gradually.

4 Discussion

4.1 Characteristics of Urban Sprawl Type

Fig.4 shows patch percents of three expansion types during Shenzhen urban growth. The phenomenon shows that the process of edge-expansion growth indicates the area rising weather the number of urban patch increasing or decreasing, whereas if the patches of outlying and infilling growth types rise, the urban area for the two types increase.

The urban growth of Shenzhen verifies the hypothesis postulated by Dietael et al.(2005). Between 1980 and 1988, the patch percent of outlying was 90% (Fig.4). In this period, the urban is in course of diffusion. However, the patch percent of edge-expansion is 10% (Fig.4). During the second period, there are infilling patches and the patch percent of outlying is still the biggest (Fig.4). During the rest periods, the patch percents of outlying are dominant (Fig.4), but the area sprawl is mainly edge-expansion and infilling especially infilling type whose patch percent is the least but area is the largest. The urban are alternating from diffusion to coalescence.

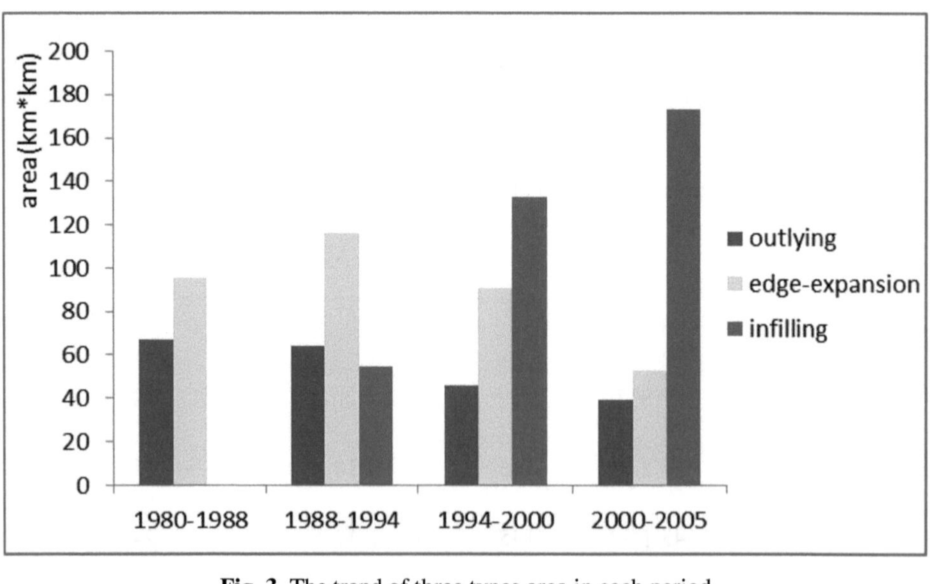

Fig. 3. The trend of three types area in each period

4.2 Dynamic Changing of Urban Growth Types Along with Landscape Pattern

During 1980-1994, the rising of NP is very fast, and this is consistent with outlying patches increasing. The LPI and AREA_MN are also rising, and this is due to edge-expansion area sprawling. The change of LST, ENN_MN and AI indicates landscape fragmentation. However, the AI turns changing at 1988 when the infilling type grows, so the AI is relate to infilling growth.

During 1994-2000, the trends of LST, NP and ENN_MN change more slowly, while the infilling area is increasing faster. So it means the landscape fragmentation ease and the urban area become less dispersed in this period.

Between 2000 and 2005, the slope of LPI and AREA_MN are greater. The LST and NP are decreasing which show that the urban is experiencing coalescence. The ENN_MN increases slightly, and this indicates the aggregation of urban patches.

Shenzhen sprawl rapidly relates to policy of government and its own conditions in many respects. Our research provides the foundation for understanding the development of city driving force and impact on the environment.

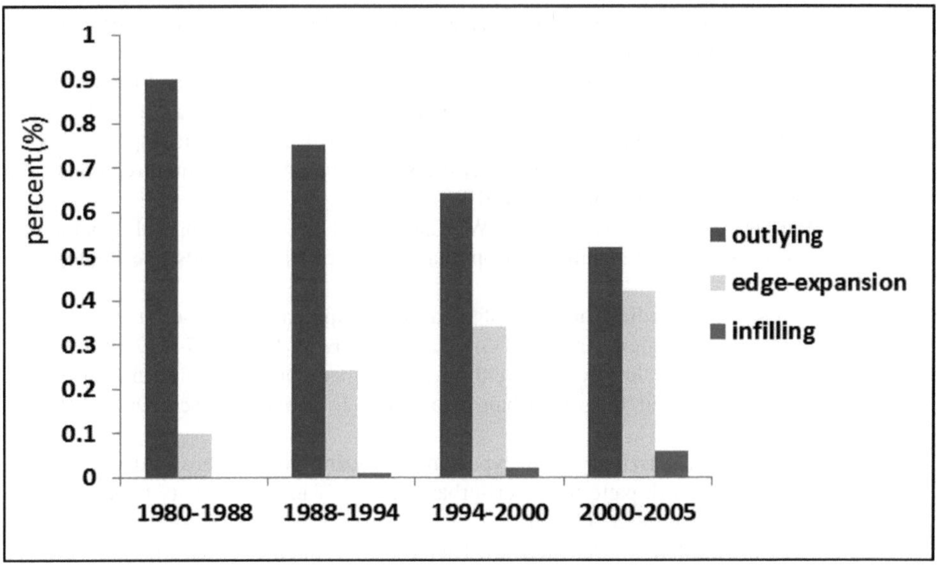

Fig. 4. Patch percents of three types during Shenzhen urban growth

5 Conclusion

In our study, the type of urban sprawl is quantified and the dynamic changes of urban pattern is analysed. The method of distinguishing the types of urban sprawl is effective and Landscape metrics verified it. Combination the urban sprawl type with landscape metrics produces stronger persuasion.

At the early stage of urbanization, Shenzhen is experiencing diffusion with dispersion urban area. The area of outlying is large and there is little infilling area. With the gradually decreasing outlying area and increasing rapidly infilling area, the urban turns into coalescence stage. And it is consistent with trend of landscape metrics. The fragmentation of the urban landscape is serious at diffusion stage, while in the stage of coalescence, the urban patches gather gradually to be more compact.

In the future research, we hope to study couple urban growth phases and find the regular spatial distribution of patches of different sprawl types. The results may play an important role in urban planning and designing in Shenzhen.

Acknowledgements. This research is funded by the Program of National Natural Science Foundation of China (41171404)and National Key Technology R&D Program of the Ministry of Science and Technology(2012BAK19B04-03).

References

1. Sun, C., Wu, Z.-F., Lv, Z.-Q., Yao, N., Wei, J.-B.: Quantifying different types of urban growth and the change dynamic in Guangzhou using multi-temporal remote sensing data. Int. J. Appl. Earth Observ. Geoinf. (2012)
2. Seto, K.C., Fragkias, M.: Quantifying spatiotemporal patterns of urban land-use change in four cities of China with time series landscape metrics. Landscape Ecology 20, 871 (2005)
3. Wu, J., Darrel Jenerette, G., Buyantuyev, A., Redman, C.L.: Quantifying spatiotemporal patterns of urbanization: The case of the two fastest growing metropolitan regions in the United States. Ecological Complexity (2010)
4. Xu, C., Liu, M., Zhang, C., An, S., Yu, W., Chen, J.M.: The spatiotemporal dynamics of rapid urban growth in the Nanjing metropolitan region of China. Landscape Ecology 22, 925 (2007)
5. Luck, M., Wu, J.: A gradient analysis of urban landscape pattern: a case study from the Phoenix metropolitan region, Arizona, USA. Landscape Ecol. 17(4), 327–339 (2002)
6. Charles, D., Martin, H., Jeffrey, H., Keith, C.: Spatio-temporal dynamics in California's Central Valley: empirical links to urban theory. Int. J. Geogr. Inf. Sci. 19(2), 175–195 (2005)
7. Shi, Y., et al.: Characterizing growth types and analyzing growth density distribution in response to urban growth patterns in peri-urban areas of Lianyungang City Landscape and Urban Planning 105, 425–433 (2012)
8. Jaeger, J.A.G., Bertiller, R., Schwick, C., Kienast, F.: Suitability criteria for measures of urban sprawl. Ecological Indicators 10, 397–406 (2009)
9. Dietzel, C., Herold, M., Hemphill, J.J., Clarke, K.C.: Spatio-temporal dynamics in California's central valley: Empirical links to urban theory. International Journal of Geographical Information Science 19(2), 175–195 (2005)
10. Forman, R.T.T.: Land mosaics: The ecology of landscapes and regions. Cambridge University Press, Cambridge (1995)
11. Liu, X.P., Li, X., Chen, Y.M., Tan, Z.Z., Li, S.Y., Ai, B.: A new landscape index for quantifying urban expansion using multi-temporal remotely sensed data. Landscape Ecology 25(5), 671–682 (2010)

12. Xu, C., Liu, M.S., Zhang, C., An, S.Q., Yu, W., Chen, J.M.: The spatiotemporal dynamics of rapid urban growth in the Nanjing metropolitan region of China. Landscape Ecology 22(6), 925–937 (2007)
13. Hoffhine Wilson, E., Hurd, J.D., Civco, D.L., Prisloe, M.P., Arnold, C.: Development of a geospatial model to quantify, describe and map urban growth. Remote Sens. Environ. 86(3), 275–285 (2003)
14. Ewing, R.: Is Los Angeles-style sprawl desirable. J. Am. Plann. Assoc. 63(1), 107–126 (1997)

Extraction of Water Body Based on LandSat TM5 Imagery – A Case Study in the Yangtze River

Zhongshi Tang, Wenhao Ou, Yue Dai, and Yu Xin*

3S Center, Tsinghua University, Beijing 100084, China
gis@tsinghua.edu.cn

Abstract. A improved method named vegetation normalized difference water index (VNDWI) has been proposed in this paper based on NEW of Xiao Yun-fang(2010), which uses Band 1 and Band 7 and the normalized difference vegetation index (NDVI) in TM5 imagery to construct the VNDWI. This method has been tested in the Yangtze River Basin, and compares the results with other methods is the best. Due to the reflection characteristics differences between water and cloud, NEW method cannot distinguish the differences effectively. However, joins the NDVI parameter, VNDWI can enhance the difference between water and cloud, so it can remove clouds and its shadow noise from water body information without using complicated procedures, which is particular difficult work should be done in pretreatment.

Keywords: Extraction of Water Body, Imagery Processing, Binarization.

1 Introduction

The water body research is inseparable from the water scope and accurate extraction from the boundaries, it is particularly important to the coastline change monitoring, flood disaster monitoring, and disaster evaluation. In macro level, remote sensing monitoring of large scope area is having incomparable advantage. This paper extraction research based on the TM5 image of Yangtze River basin, and put forward the improved water extraction method.

Based on the research of water extraction from TM5 imagery, domestic and foreign scholars have made a lot of research, puts forward a variety of algorithm. In the foreign scholars' researches, 1990 Moller Jensen used the TM4 and TM5 imageries and set threshold to extract water[1], 1996 Mc Feeter using normalized water index NDWI = (TM2-TM4) / (TM2 + TM4) by inhibiting vegetation and soil information to extract water[2]. And domestic scholars Xu HanQiu proposed improved normalization water index MNDWI = (TM2-TM5) / (TM2 + TM5) in 2005, and set certain threshold to extract water [3]. In 2004 Chen HuaFang, put forward the method: (TM2 + TM3)-(T4-TM5) > k, which can be well separated by mountain shadow and water [4]. Xiao YanFang using the Band 1 and Band 7 in TM5 imagery for water information

* Corresponding author.

D. Li and Y. Chen (Eds.): CCTA 2012, Part II, IFIP AICT 393, pp. 416–420, 2013.

extraction [5], this method not only can extract water body like NDWI and MNDWI, but also has advantage in urban areas than other methods.

The above method has been widely used in practical work, but there are also have some disadvantages, such as the extraction of water information may contain cloud, vegetation and residents influences which may cause the inaccurate results, Salt and Pepper phenomenon as well.

2 Materials and Methods

In this paper, experimental data is TM5 imagery of the Yangtze River basin on July 5, 1988 from http://glcf.umiacs.umd.edu website. This image contains a complete water body, probably suit for verifying the precision of extraction of the water body by improved method.

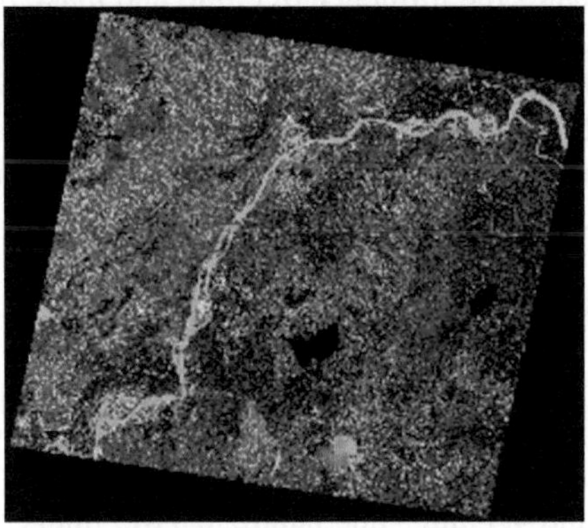

Fig. 1. Yangtze River Basin

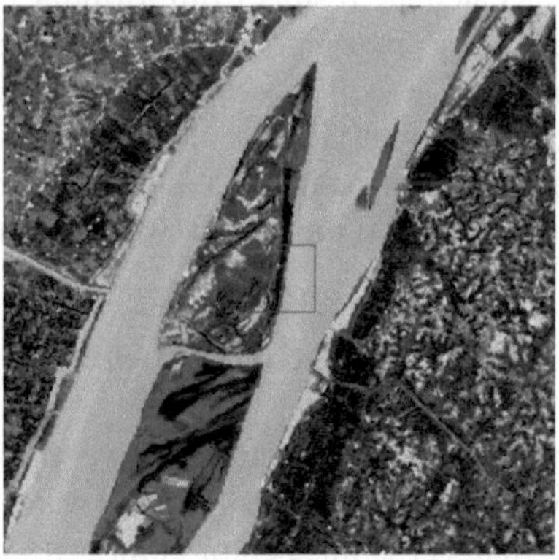

Fig. 2. Part of Yangtze River

Within the most scope of the spectrum from remote sensing sensors, water has a low reflection. For clear water, in the blue and green band reflectance reaches 4% ~ 5%, from the red part of the reflectivity reaches 2% ~ 3%, and in the near infrared and short wave infrared, almost all of the energy absorbed, because the characteristics of vegetation and soil formation is obviously, so often use red light band to recognize water body[6]. A normalized difference index of water named NDWI puts forward by Mc feeters, Chinese scholars Xu HanQiu put forward the modified normalization water index (MNDWI) and XiaoYanFang's NEW method is based on the principle from the differences in energy absorbing by different bands of sensor. However, Xiao YanFang's research indicates that band 1 and band 7 of TM5 imagery makes the minimum difference between water and buildings, and their sum is far higher than water, so it can be used to eliminate buildings and roads influence to the water body extraction.

But in actual study operation, the study area in the Yangtze River basin for this paper mentioned, buildings information may less than others, but cloud and its shadow are the main factors which may affect the accuracy of water extraction. For the cloud reflection characteristics of the bands in TM imageries, a statistic result table can be listed:

Table 1. The Average Value of Water and Cloud

	Band	Band2	Band3	Band4	Band5	Band7	NDVI	NDWI	MNDWI	NEW
Water	106.5	56.9	35.9	25.1	14.6	9.6	-0.391	0.346	0.986	0.997
Cloud	235.9	124.5	142.1	105.9	162.6	89.8	-0.0366	0.0429	0.753	0.909

From the table 1 we can see the largest difference of water and cloud is the NDVI value, so the proper way to distinguish between water and cloud is from NDVI value. This paper improves the NEW algorithm, join NDVI parameters, which named VNDWI here, the formula is:

$$\text{VNDWI} = (\text{TM1-TM7}) / (\text{TM1} + \text{TM7}) > k * \text{NDVI} \qquad \textbf{(Formula 1)}$$

The calculate results from formula 1 eliminates the influence of cloud and make the water body extraction more precisely, and buildings and subsequent shadow elimination is well preserved. Statistical analysis showed that the new method in this paper can extract the water body like NDWI, MNDWI and NEW, it also indicates more accurate extraction in the Yangtze River basin water information from filter out the part of the clouds and the clouds' shadow influences.

3 Results and Discussion

From the imagery of Yangtze River basin, we can see some clouds and the shadow over the water body. In order to show water body more clearly in remote sensing imagery, RGB imagery is false color composite with band 4, band 3 and band2 in TM5 imagery, also make the NDVI imagery through band 3 and band4 for next step.

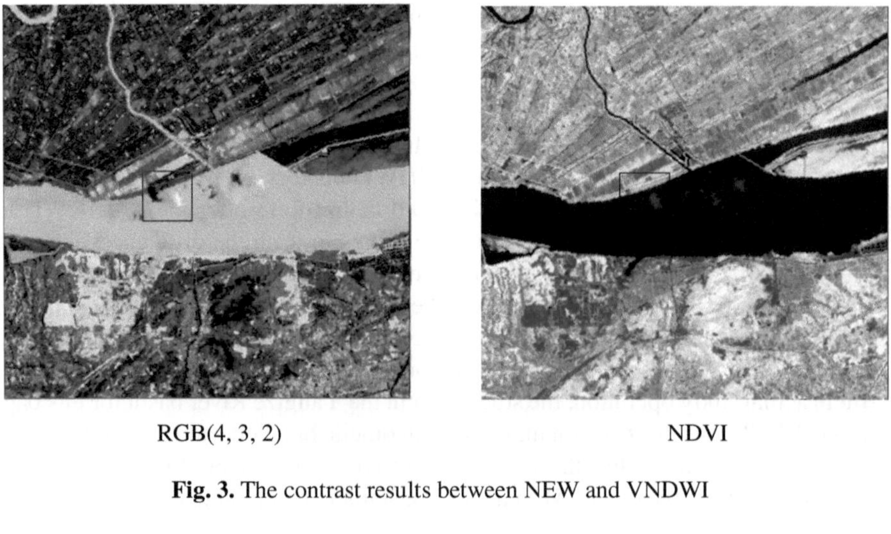

RGB(4, 3, 2) NDVI

Fig. 3. The contrast results between NEW and VNDWI

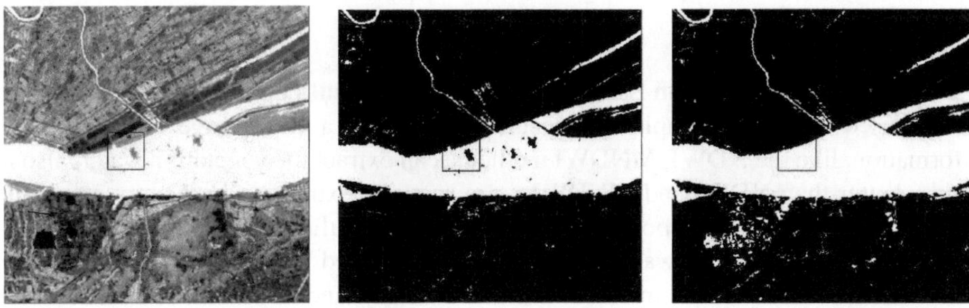

NEW Binarization results of NEW Binarization results of VNDWI

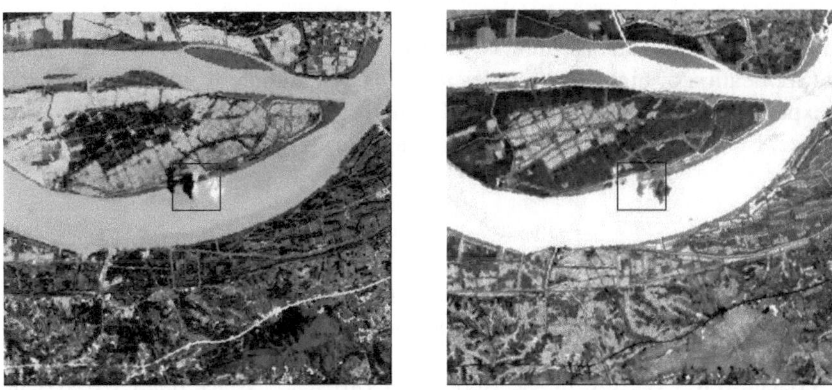

RGB(4, 3, 2) NDVI

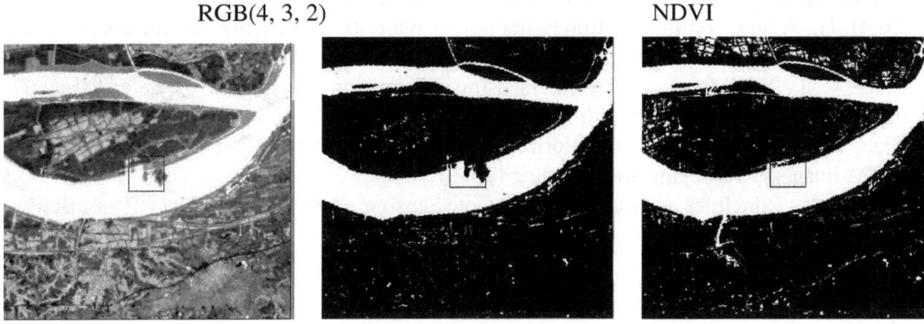

NEW Binarization results of NEW Binarization results of VNDWI

Fig. 3. *(Continued)*

After process from the NDVI imagery, we can see it has eliminate a little influence from the RGB imagery, then let k equals 100 in this particular study area, and from formula 1, we can get the binarization results of NEW and VNDWI from Figure3, which compares NEW and VNDWI imagery, it shows that VNDWI eliminates the influences of cloud and clouds' shadow from the binarization imagery, and buildings and subsequent shadow elimination is well preserved, make the extraction more precisely in the cloud cover area.

4 Conclusion

Use Band 1 and Band 7 from Landsat TM5 imagery to build up the water index NEW, can be used for rapid, simple and accurate extraction a wide range of water body information, like the NDWI, MNDWI methods, can extract the vegetation water, also it can be better than NDWI and MNDWI more accurate extraction in the city area; But the NEW algorithm is not pointed to consider clouds influences in water extraction. However, this paper put forward a improved method called VNDWI algorithm, to some extent, it filter out the cloud influence from water body, and also keep its buildings and mountain shadows filtering characteristics, greatly improving the water body information extraction accuracy.

Acknowledgements. The research is funded by the "Eleventh Five-Year" national satellite application of high technologies in China. All authors would like to appreciate many experts from Office of the State Council Three Gorges Project Construction Committee, Beihang University and Chongqing Normal University.

References

1. Jensen, O.: Knowledge based classification of an urban area using texture and context information in Landsat TM imagery. Photogrammetric Engineering and Remote Sensing 56 (1990)
2. Mc Feeters, S.K.: The use of the Normalized Difference Water Index (NDWI) in the delineation of open water feature. International Journal of Sensing (1996)
3. Xu, H.-Q.: A Study on Information Extraction of Water Body with the Modified Normalized Difference Water Index. Journal of Remote Sensing (2005)
4. Yang, C.-J., Xu, M., et al.: Remote sensing information mechanism of water extraction method. Geography Research, 86–89 (1998)
5. Xiao, Y.-F., Zhu, L.: A study on information extraction of water body using band1 and band7 of TM imagery. Surveying and Mapping (2010)
6. Zhao, S.-Y.: Principles and methods of Remote Sensing Applications. Science Press, Beijing (2003)

Validation and Application of Model ISAREG in a Typical Semiarid Sand-Meadow Area of Horqin Sandy Land*

Yao Wu[1], Tingxi Liu[1], Luis Pereira[2], Paula Perards[2], and Haiyan Wang[1]

[1] College of Conservancy and Civil Engineering, Inner Mongolia Agricultural University, Hohhot 010018, China
[2] Agricultural Engineering Research Centre, Institute of Agronomy, Technical University of Lisbon, Portugal
{wuyao.will,txliu1966}@163.com

Abstract. Win ISAREG is a model for performing soil water balance and estimating impacts of water stress on yields for crops, concentrating on the evaluation of irrigation scheduling. The soil water balance of different type of soil and crop were study in the typical semiarid Horqin Sandy land area in this paper. With the data from Agula experimental station and concentrate on site A3 with Artemisia and site D1 with leymus, the calibration and validation for both sites were performed by using field observations of 2008 and 2009, aiming to adjust the crop coefficients Kc and soil water depletion fractions for non-stress conditions p to the local conditions. Results show that the coefficients of regression was close to 1.0, the relative errors to estimate the soil water content was 0.09for Artemisia and 0.10 for leymus. Results prove the practicality and applicability of using ISAREG for the Horqin semiarid area.

Keywords: Model ISAREG, Horqin sandy land, simulation, validation, soil moisture.

1 Introduction

Desertification area in China mainly distributes in north and northwest, which has complex and diverse geographical ecosystem. Special position and geomorphic features result in significant differences for hydrological process between the north and south desertification area (2005). The ecosystem in North China including sandy land, desert and agricultural land, specially for the meadow and sandy dunes take shape of stripe distribution, forming meadow- transition area- sandy dunes multi-

* National Natural Science Foundation of China, No. 51139002; No.51069005; Inner Mongolia Agricultural University innovation Team Building Program Cold-Arid Region Water Resources Utilization, No. NDTD2010-6; Inner Mongolia Scientific and Technology Bureau No. 20090516; Project of the Ministry of Science and Technology of China, No. 2010DFA71460.

D. Li and Y. Chen (Eds.): CCTA 2012, Part II, IFIP AICT 393, pp. 421–429, 2013.

cascade ecological system. Soil water evaporation plant transpiration are the most important parts in soil water balance, the soil water evaporation reduced soil water content, crops has influent by water stress under arid condition. Concerns have been focus on the study of eco-hydrological process in arid and semiarid region (Yang, 2006).

2 Materials and Method

2.1 Experimental Station

The 9.71 km2 study area is located in the south of Horqin sandy land (Fig 1), Horqin sandy land is 51700 km^2 and it's situated between the Inner Mongolian Plateau and the Northeast China Plain. The area has an elevation ranging from 88.5m to 308.4m above mean sea level (Duan, 2011). The study area was chosen because it's located in the typical desertification area -- the Horqin Sandy area(Fig.1), with the dune-cascade transition zone-meadow geomorphic feature making up the unique ecosystem, which contains 54.5% sand dunes, 26.6% meadow lands, 10.4% cultivation area and 5.2% lake area (Duan 2011). Lots of observations have been done in this area from year of 2004 to support research on eco-hydrological system.

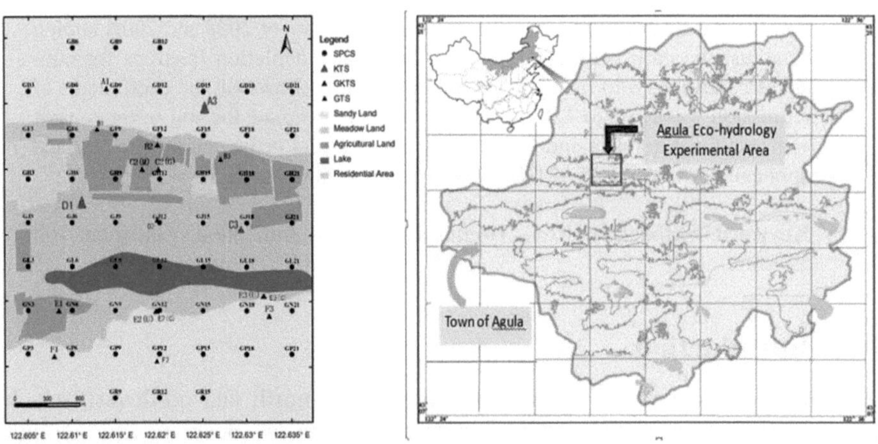

Fig. 1. The location and landscape features of the study area, soil physicochemical sites and hydrologic and meteorological stations

Table 1. Characteristics of the measurement sites for A3 and D1

Site NO.	Land Feature	Soil Texture	Type of Vegetation	Elevation
A3	Mobile dune	Sand	*Artemisia halodendron*	199.8
D1	Meadow	Sandy loam	*Leymus chinensis*	188.5

The study area has a temperate continental monsoon climate, with the average annual precipitation vary from 350mm to 550mm over the period of year 1957 to 2009, 70% of which is concentrated from June to August., with the average annual evaporation between 1700mm to 2000mm. Temperature goes along with the seasonal trend, with the average annual value is 5.8°C, and the minimum in January (-15.1°C) and maximum in July (23.8°C) (Wang 2008). The relative humidity varying from 50% to 60%, which has the lower value in summer and higher value in winter. The average annual wind speed in the area In the study area, it has mobile dune, fixed dune, semifixed dune that correspond with sand, sandy loam as the texture in the sandy sites, and the meadow sites with the sandy loam. The distribution of vegetation follows the zonality of landform. The sandy dunes are covered mainly by Artemisia halodendron L., Salix gordejevii Chang& Skvortsov, and Populus, the type of natural vegetation on meadow sites are Leymus chinensis Tzvelev, Phragmites australis Trin., and/or Ixeris chinensis Nakai.

Table 2. Soil characteristic of the site A3 and D1

Site NO.	Depth (m)	Bulk density (g/cm^3)	Field capacity (% volume)	Wilting point (% volume)
	0.00-0.10	1.70	0.26	0.10
D1	0.10-0.40	1.66	0.28	0.09
	0.40-1.00	1.57	0.38	0.23
	0.00-0.80	1.52	0.18	0.04
A3	0.80-1.60	1.57	0.17	0.03
	1.60-2.20	1.55	0.20	0.03

2.2 Instrumentation and Data Collection

The Agula eco-hydrological experimental station has been established since 2003, 16 sites has been chosen for typical points, including two key typical sites (KTS) and ten general key sites(GKTS) and four groundwater observed sites(GTS). Continuous observation has been taken of all 16 sites since 2004. Measurements are shown as below:

a. The soil physical and chemical characteristic parameters in the top 30 cm layer, including soil water content for every layer, soil classification, soil bulk density, the electric conductive(EC) ,soil pH value, soil organic content;
b. the groundwater depth of the 16 sites;
c. automatic meteorological station and monitoring of environmental gradient were set on the sites of A3 and C3(Fig .1), measurements including wind speed, humidity, temperature, of 2m, 3m and 10 m, the soil water content, soil temperature and soil salinity of every soil layer. Also total radiation, photosynthetically active radiation, sunlight hour, rainfall.

d. The soil water content measured by the time domain reflectometer (TDR) and neutron probe.
e. The crop information about the crop height, root depth, and development stages.

The characteristic of the soil for sites have been chosen for simulation showed in Table.1 and Table. 2.

2.3 Model ISAREG

The Win ISAREG model was used to performing simulations of soil water balance, which is the integration of two different models, the EVAP56-for computing reference evapotranspiration, and the ISAREG part as to describe(2003, Pereira et al). The model and its calibration and validation for Portuguese conditions were presented by Teixeira and Pereira (1992).

Input data for the model ISAREG

(1) Meteorological data: precipitation, Pe(mm) , wind speed (m/s) and minimum relative humidity, RHmin(%), the calculation module EVA56 following the FAO-Penman Monteith method was used for computing ETo (Allen et al,1994).
(2) Crop data: including the dates of crop growth stages and crop coefficients Kc, soil water depletion fractions for non-stress conditions p, root depth Zr (m). the information of the sites A3 and D1 has been shown in Table 3.
(3) Soil data: according to the soil profile the respective depth for each soil layer, d(m), and the soil water characteristic parameters as field capacity and wilting point for computing the total evaporable water(TEW) and Readily evaporable water (REW).
(4) Irrigation characteristic data: for the case there is no irrigation during the crop growth stages, concentrating on the Groundwater Contribution data, according to the proposed value by Liu (Liu, 2006) to choose parameters, input the water table depth of the chosen days as well.

Table 3. Crop information from field experiments utilized in the model simulation for Artemisia and Leymus

Crop Stages		Artemisia		Leymus	
		Date	Root depth(m)	Date	Root depth(m)
stage A	Initial period	5/1	1.0	5/5	0.5
Stage B	Start of vegetative growth	5/15	1.0	5/20	0.5
Stage C	Full cover	6/10	1.1	6/15	0.5
Stage E	Ripening	10/1	1.1	10/01	0.5
Stage F	End season	11/1	1.1	11/1	0.5

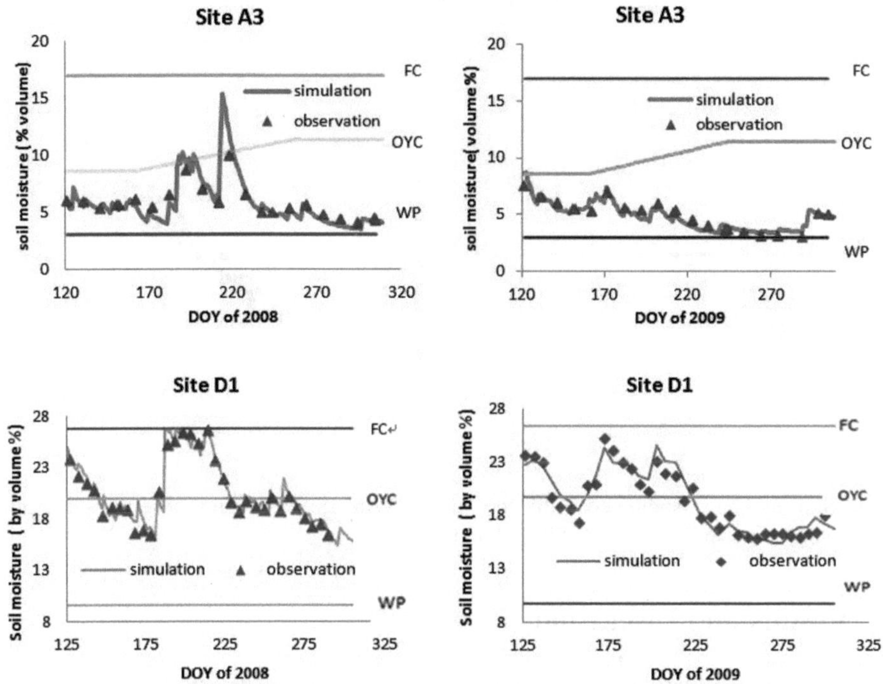

Fig. 2. Observed and simulated. soil moisture for site A3(Artemisia) and D1 (Leymus) of year 2008 and 2009 curves *F*c and *W*p represent soil moisture at field capacity and at wilting point

• The root mean square error (same units as O_i), which characterizes the variance of the errors.

$$\mathrm{RMSE} = \left(\frac{\sum_{i=1}^{n} (P_i - O_i)^2}{n} \right)^{0.5} \tag{1}$$

• The average absolute error (same units as O_i), which expresses the size of estimation errors in alternative to RMSE.

$$AAE = \frac{1}{n} \sum_{i=1}^{n} \left| O_i - P_i \right| \tag{2}$$

• The average relative error (%) that indicates the size of errors in relative terms.

$$AAE = \frac{100}{n}\sum_{i=1}^{n}\left|\frac{O_i - P_i}{O_i}\right| \tag{3}$$

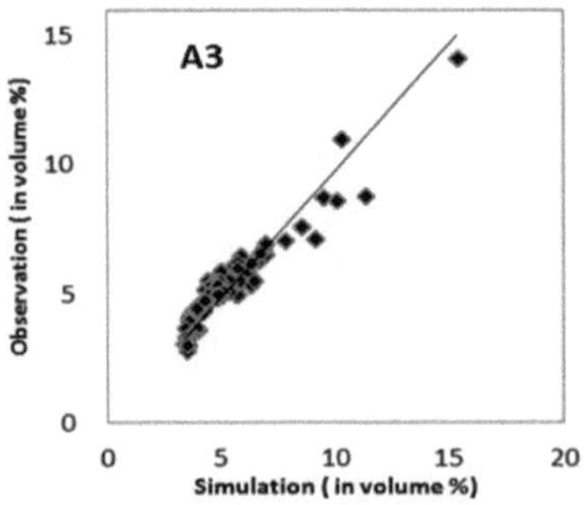

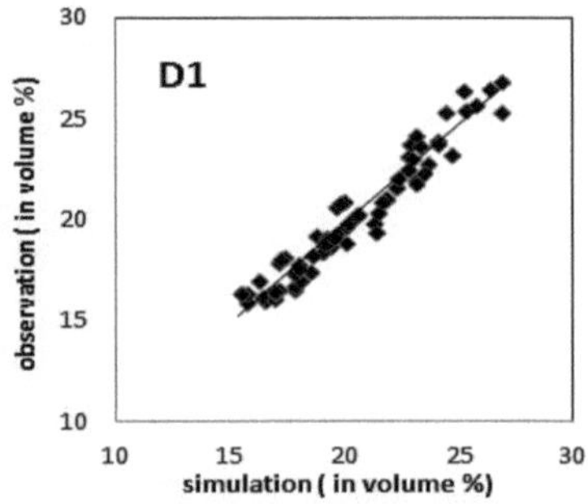

Fig. 3. Soil moisture simulated with the model Win ISAREG and observed at Agula eco-hydrology experimental station. Site A3 and Site D1

3 Results and Discussion

Calibration and validation of model for soil moisture, model ISAREG calibrated using 2008 data collected from Agula experimental sites, the crop coefficients Kc and soil water depletion fraction p were adjusted for both two sites A3 and D1, the simulation of soil moisture along the development periods were compared with the observation(Fig 2). Seasonal changes of the soil moisture of simulations comparing with observations were shown (Fig. 2 and Fig. 3) for sandy site A3 with Artemisia and meadow site D1 with leymus relative to 2008 and 2009. The calibrated crop parameters Kc and p are presented in Table 4.

The first value of Kc_{ini} of A3(0.5) and D1 (0.3)was taken from Allen et al(1998)and then adjust for the actual condition., as for Artemisia there is no irrigation during the whole period, during the initial period, evapotranspiration mainly come from soil evaporation, as the low frequency of rainfall, Kc_{ini} is low. For the site D1, leymus starts to grow quickly, Kc-0.3, which obtained in the model was the same as to the 0.30 suggested by Allen et al (1998). The midseason for Artemisia is obtained as 1.10, when the leymus is lower as 0.9, and between the value (0.85-1.05) proposed by Allen et al (1998). This is explained by the soil characteristic and the relative humidity, site D1 is close to the lake, with higher humidity.

Kc_{end} was adjusted as 0.5 for leymus site and 0.4 for Artemisia site, water supply was applied only for small amount of rainfall during the last season, due to the dry soil surface, the value was lower than proposed in FAO.

The soil water depletion fractions for non-stress conditions *p* is relative to the climate, as the desiccation result in higher Etc, p for both sites was adopted as 0.5 (A3) and 0.6(D1), which are lower than value in the FAO table (Allen et al, 1998). The initial value for *p* used in this study was taken from the guideline of FAO (Allen et al., 1998). Because of the condition of semiarid area, with the high frequency of strong wind, the Kc_{mid} should be higher, but as the crop wasn't seed ordered but natural growth, especially for A3 the smaller surface coverage.

The coefficients of regression through the origin, b is very close to 1, when the coefficient of determination R^2, for both D1 and A3 are acceptable (Table. 4).

The root mean square error(RMSE) for calibration and validation is 0.006 and 0.004 of site A3, and 0.008 and 0.010 of site D1,which indicate that the estimation errors are small.

The average of absolute errors (AAE) of soil moisture is low, less than 1% in volume (Table 4). Results show that the AAE is acceptable for prediction.

The average relative error of estimates, ARE (Table 4) confirm the analysis above,for A3, it is 9.10 and 8.90, not exceeding 10%, for D1 it's 2.88 and 4.42, lower errors are expected. Results showed that the model is appropriate for this area.

Table 4. Quality of model simulation of soil moisture for data sets utilized to calibration and validation

Parameters	Site NO.	Regression coefficient b	determination coefficient R^2	RMSE (m^3/m^3)	AAE (m^3/m^3)	ARE(%)
Model calibration	A3	0.98	0.96	0.006	0.005	9.10
	D1	0.98	0.91	0.008	0.006	2.88
Model validation	A3	1.00	0.89	0.004	0.004	8.90
	D1	0.99	0.89	0.010	0.009	4.42

4 Conclusions

This study is based on collecting data from the typical semiarid area- Agula eco-hydrological experimental station. Calibration and validation for Model ISAREG was performed using soil water content observations. By using continuous observations data of 2008 and 2009 comparing with simulations using ISAREG, calibration and validation for crop parameter K_c was analysis, results showed the Model ISAREG has accurately simulated the variability of soil water content of Artemisia for sandy dunes and Leymus for meadow area. Also with Model ISAREG, the K_c has been estimated for both Artemisia and Leymus of different development stages in the semiarid area.

The Kc_{ini}, Kc_{mid} and Kc_{end} for Artemisia are 0.50, 1.10 and 0.60, for Leymus it's 0.30, 0.90 and 0.50. It can be concluded that the model for prediction in this area is available and acceptable.

Acknowledgements. This study was financially supported by the National Natural Science Foundation of China under contract #51069005, and Inner Mongolia Scientific and Technology Bureau under contract #20090516. With my sincere thanks to Professor Tingxi Liu, Prof. Luis Pereira, Paula Perareds and colleagues in IMAU and CEER for supporting and guidance. Special thanks to the people Dakang Liu, Guanli Wang, limin Duan, who made the contribution for establishment of the experimental station and collected data.

References

Allen, R.G., Smith, M., Pereira, L.S., Perrier, A.: An update for the calculation of the reference evapotranspiration. ICID Bull. 43(2), 35–92 (1994)

Allen, R.G., Smith, M., Pruitt, W.O., Pereira, L.S.: Modifications to the FAO crop coefficient approach. In: Camp, C.R., Sadler, E.J., Yoder, R.E. (eds.) Proc. Int. Conf. on Evapotranspiration and Irrigation Scheduling, San Antonio, pp. 124–132. ASAE, St. Joseph (1996)

Allen, R.G., Pereira, L.S., Raes, D., Smith, M.: Crop Evapotranspiration. Guidelines for Computing Crop Water Requirements. Irrigation and Drainage Paper 56. FAO, Rome (1998)

Liu, Y., Teixeira, J.L., Zhang, H.J., Pereira, L.S.: Model validation and crop coefficients for irrigation scheduling in the North China Plain. Agric. Water Manage. 36, 233–246 (1998)

Liu, Y., Pereira, L.S., Fernando, R.M.: Fluxes through the bottom boundary of the root zone in silty soils: Parametric approaches to estimate groundwater contribution and percolation. Agric. Water Manage. 84, 27–40 (2006)

Teixeira, J.L., Pereira, L.S.: ISAREG, an irrigation scheduling simulation model. ICID Bull. 41(2), 29–48 (1992)

Pereira, L.S., van den Broek, B., Kabat, P., Allen, R.G. (eds.): Crop–Water Simulation Models in Practice, p. 332. Wageningen Press, Wageningen (1995)

Chen, F., Cai, H., Wang, J., et al.: Estimation of evapotranspiration and crop coefficients of winter wheat and summer maize in Yangling zone. Transactions of the Chinese Society of Agriculture Engineering 22(5), 191–193 (2006) (in Chinese with English abstract)

Liu, Y., Pereira, L.S.: Validation of FAO methods for estimating crop coefficients. Transactions of the Chinese Society of Agriculture Engineering 16(5), 26–30 (2000) (in Chinese with English abstract)

Doorenbos, J., Pruitt, W.O.: Guidelines for Predicting Crop Water Requirements. Irrigation and Drainage Paper 24, Food and Agriculture Organization of the United Nations, Rome, Italy (1997)

Wang, G.: Simulation analysis for water transforming based on field test for GSPAC system in dune–meadow–dune area in Horqin Sand, Ph. D. thesis, Inner Mongolia Agricultural University, 117 p. (2008)

Duan, L., Liu, T., Wang, X., et al.: Spatio-temporal variations in soil moisture and physicochemical properties of a typical semiarid sand-meadow-desert landscape as influenced by land use. Hydrol. Earth Syst. Sci. 15, 1865–1877 (2011)

The Classification Method of Multi-spectral Remote Sensing Images Based on Self-adaptive Minimum Distance Adjustment

Junhua Liu[1,2], Chengming Zhang[2,3,*], and Shujing Wan[3]

[1] Chinese Academy of Surveying and Mapping, Beijing, China
[2] Shandong Agricultural Science Colleague, Jinan, China
[3] School of Information Science &Engineering, Shandong Agricultural University, Taian, China

Abstract. The phenomenon of "Same Object with Different Spectra" in the issue of multi-spectral remote sensing images land use classification makes major effects on improving accuracy. The paper based on the analysis of modeling on classification problems, proposed a method based on minimum distance self-adaptive adjustment to realize the split of cluster centers and solved the problem of identified scope intersection leading to improving the accuracy in the classifying methods difficultly. By experiments compared with the traditional methods, it can improve classification accuracy about 4% and the results prove the validity of this method.

1 Introduction

Multi-spectral remote sensing images are the main data source in the land use applications. Studying the appropriate classification algorithm according to the fundamental characteristics of the multi-spectral remote sensing images is the primary means to obtain high-precision land-use information.

In the 1980s, it was mainly used statistical pattern recognition methods to classify remote sensing images by computer [1]. Since the 1990s, it emerged a large number of remote sensing image classification methods, such as the artificial intelligence classification, combination of remote sensing and GIS, object-oriented classification, composite classification and so on which achieved better results [2]. Dixon et al. took the support vector machine method to do land use classification of TM images that obtained the best classification accuracy [3].

Wardlow et al. used the decision tree method to classify crop in the central Plains of United States, which gained the overall classification accuracy of better than 80% [4]. Taochao et al. [5] came up with a high-resolution remote sensing image classification method based on probabilistic platen semantic model for the phenomenon of "same spectrum with different objects" in high-resolution remote sensing image. Li Gang

* Corresponding author.

D. Li and Y. Chen (Eds.): CCTA 2012, Part II, IFIP AICT 393, pp. 430–437, 2013.

proposed an uncertainty classification technology based on high-dimensional cloud model and RBF neural network modified [6].

It wasn't satisfied with the traditional remote sensing image classification methods in the aspects of automation, intelligent, and classification accuracy [7]; although the technology based on neural network has the advantages of strong fault tolerance and adaptability, it existed structure selected difficult, too fast of local convergence and study process difficulty controlling in the practical application, which resulted that was not wholly better than the traditional technologies; the method based on decision tree was not ideal in the treatment of the boundary [8,9].

For the fundamental characteristics of multi-spectral remote sensing images, with comprehensive considering the speed and accuracy of land use classification, the paper proposed a classification method of introducing self-adaptive minimum distance adjustment in order to better meet the needs of the actual production.

2 Self-adaptive Minimum Distance Adjustment Classification Method Principle

The existing remote sensing classification methods research more on how to improve the formation of the cluster center, but in judgment the attribution problem of the pixel to be classified it usually adopts the way that calculation the distance of pixel to be classified to the cluster center first, then taking it into the category with minimum distance, which results in the difficult to improve the classification accuracy.

2.1 Remote Sensing Image Classification Problem Modeling

Under the ideal conditions, the eigenvector of pixels representing similar surface features in remote sensing images will distribute in the same feature space regions; however, as different surface features with different eigenvectors, they should distribute in different surface features, which means that any kind of feature in images has only one gray value scope in any band for a remote sensing image to be classifies with b bands and for some feature c, it can obtain the value set of the pixels in every band, as Figure 1 shown:

To the band i of class c, if use li as the lower bound of the gray value and hi as the upper, in the band i, the value of its center mi is: $m_i = \frac{(h_i + l_i)}{2}$ and based on it, it can define m=(m1,m2,…,mb)T as the center feature vector of the class which can describe the adaptive changes of the cluster center better. Use ri as the permissible error radius surrounded, by the analysis above, the value of ri is |hi-li|/2 and it can name all the vectors made by maximum permissible error of all the bands as effective radius vector. For pixel x to be classified in class c, compare the distance of gray value of each band to

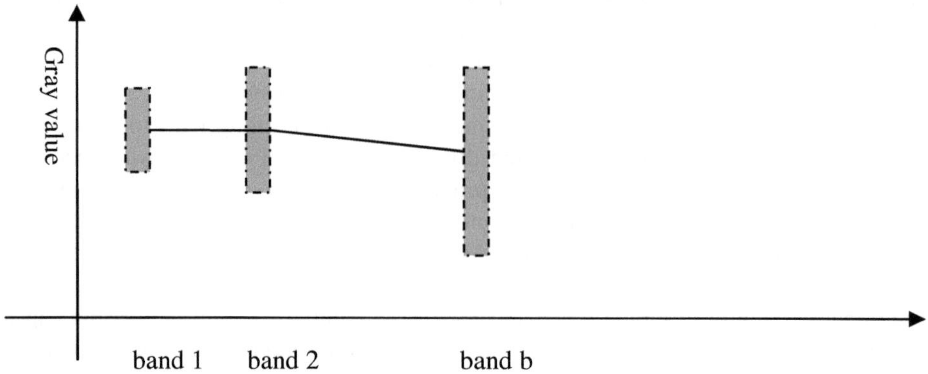

Fig. 1. The schematic diagram of spectral range under the ideal conditions

the surrounding component of the center feature vector, and if all the distances don't exceed the surrounding permissible radius, take it into class c.

2.2 The Processing Method of "Same Object with Different Spectra" Problem

The "Same Object with Different Spectra" phenomenon for a variety of reasons makes the error value in some bands or all bands too large, which lead to the feature vector of the class distributing too scattering in the feature space and difficult to attribute to same cluster as excepted. To better illustrate this problem, now explain it with an example of a two-band image classification, assuming that the image will be divided into two classes A, B and the pixels distribution of each class are shown in Figure 2.

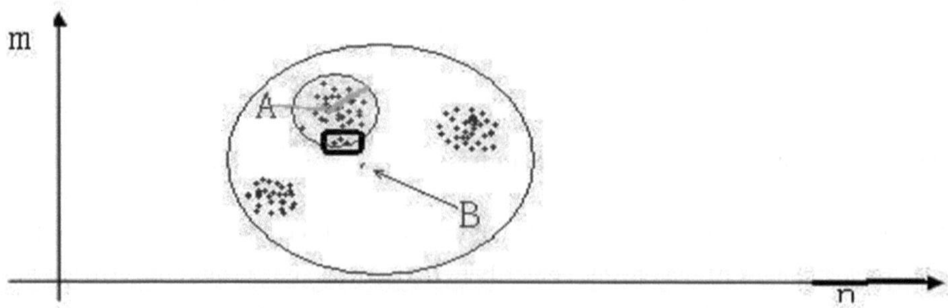

Fig. 2. The sketch map of "Same Object with Different Spectra" resulting in classification error

In Figure 2, for the pixel of class B distributing in two regions, the distance of cluster centers generated with two regions is too large leading to the range identified (blue circles) too large. As a more extreme case, the area of another class A is surrounded in the scope identified by class B and the part surrounded by an orange rectangle should be divided into pixels of class A but to class B.

From the analysis above, we can see that the root cause of classification accuracy reducing is too large of cluster centers and identified radius leading to the identified scope of two cluster centers intersect. For this, it can split cluster centers by adaptive minimum distance adjustment to eliminate intersection of identification range which can improve the classification accuracy effectively.

3 The Method Realization of Self-adaptive Minimum Distance Adjustment Classification

3.1 The Process of Self-adaptive Minimum Distance Adjustment

Based on the second analysis, it gives the method of self-adaptive minimum distance adjustment: doing the intersecting judgment to the cluster centers after the end of the sample study. If there are the intersecting parts generated on the basis of the two cluster centers' identification radius, it needs to split the cluster centers. The split steps are:

(1) Calculate the cluster centers generated one to another to determine whether they intersect. If intersect, process according to (2).

(2) Split the bigger one of identified scope in the two cluster centers. Select any sample from the classes to be split, choose a value of less than the biggest identified distance as threshold randomly, divide the samples into different subsets, study separately for each resulting in each cluster center and estimate with the classes not split. If there are still intersecting with the cluster center of some subset, select a less threshold and split again.

3.2 Algorithm Design

The algorithm uses K-Means algorithm to split the cluster center into two subsets each time and do sample set division in the way of binary tree. Each class generates a binary tree named as the subset tree of the class. A node of binary tree is corresponding to a sphere. Record each radius of sphere centre and the surrounding sample subset. The center of the sphere is defined as the center of the sample subset in this node and the radius is the maximum of Euclidean distance from the sample point on that node to the center of the sphere. The sphere obtained by the subdivision of a node's subset is said in two child nodes of it.

The classification steps of adaptive minimum distance classifier towards points to be classified i are shown in Figure 3, the specific steps:

(1) Find out the distance D of the point to subset tree corresponded by each class. The distance D(T,P) of unclassified point P to the subset tree is defined as:

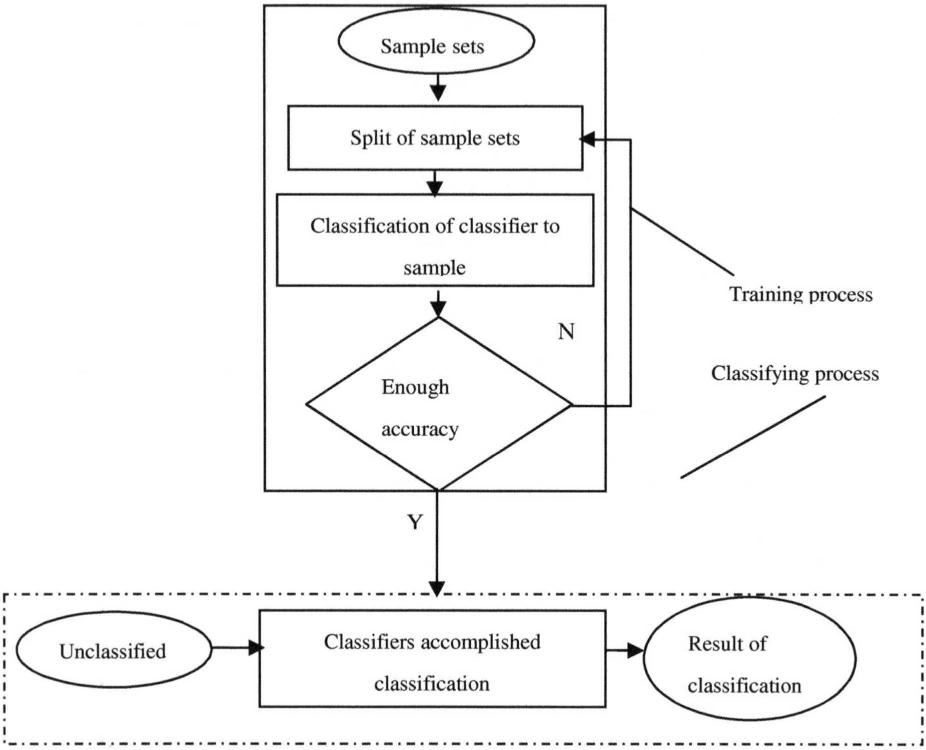

Fig. 3. The flowchart of self-adaptive minimum distance classification algorithm

A. If the centre of sphere Euclidean distance d corresponds to the root node of P to T is greater than twice sphere radius of the node, ignore all globules subdivided by the node and let D = d;

B. If the root of T is already the leaf node, let D=d;

C. If it is not satisfied with A and B, definite recursively as the less distance D1 and D2 of P to the left and right sub-tree T1 and T2 of T.

(2) Give the points to be classified to the class of a subset tree with minimum distance to D.

With using binary tree search in the distance calculation, there are a part of nodes joined in finding out the distance D to reduce the number of joining in the real calculation which guarantees that calculated amount will not increase much under the condition of sample number or sphere subdivided increasing.

Distance calculation using the only part of the nodes participate in the strike of the distance D, can effectively reduce the actual participation of computing d, the number does not guarantee the number of samples increases, or broken down to get the sphere increased computation does not increase.

4 Experiment and Discussion

The paper selected Xueye basin of Laiwu City, Shandong Province as the study area. It selected three different kinds land use categories of water, forest and bare soil to classify and compare.

4.1 The Experimental Results and Analysis of Different Classification Methods in XueYe Basin

The data sources used in the experiment are ETM+ images OF 30m with a total of eight bands and the size of 1500 pixels × 1200 pixels acquired in May 31, 2007.

4.1.1 The Classification of Common Ways

From the classification results of figure 4, it can be seen that with using the minimum distance method, the number of pixels to be classified increased significantly, the phenomenon of mixed classification with forest is more serious and the misclassification of the water affected by shadow was existence. K-nearest neighbor method can distinguish water and bare soil well, but recognize forest not well with classifying a part as water. Although the maximum likelihood classification method can classify water well, it exits serious mistake phenomenon of classifying the forest and bare soil into the kind to be classified. K-nearest neighbor method can distinguish water and bare soil well, but it cannot recognize forest commendably with classifying a part to water. BP neural network has better effects to classify each kind, but a part of vegetation still not distinguishes. The classification accuracy is shown in Table 1.

Fig. 4. The classification results of XueYe Reservoir

Table 1. Classification accuracy of common ways

Classification	water	forest	Bare soil	Average
Maximum likelihood	85.6	84.1	86.5	85.4
Minimum distance	87.7	86.8	85.3	86.5

Table 1. *(Continued)*

KNN （K=19）	90.2	85.7	85.1	87.0
BP natural network	87.0	85.2	85.9	86.0

4.1.2 The Process and Result of Self-adaptive Minimum Distance Classification

While using the self-adaptive minimum distance classification, the experimental results are shown in the Figure 5.

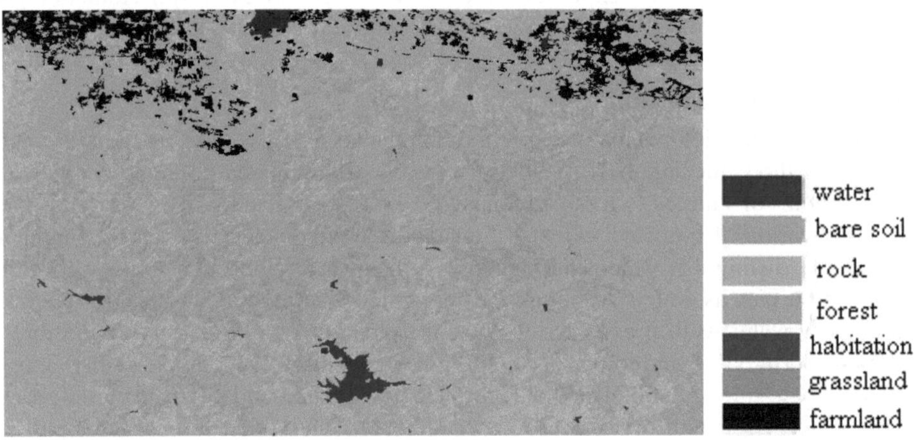

Fig. 5. The extraction results of XueYe reservoir

The classification accuracy of different training samples and category schema is shown in Table 2.

Table 2. The classification accuracy of different training samples and category schema

Water	Forest	Bare soil	Average
91.3	92.8	89.2	91.1

Contrasting the classification results of different methods, it showed that using the method of paper to classify, the pixel in the results obviously reduced and the accuracy increase 1.06% in average.

Acknowledgments. This study has been Funded by (1) Basic Research Fund of Chinese Academy of Surveying and Mapping under Grant (Contract Number: 7771109), (2) The Research Award Fund for Outstanding Young and Mid-career

Scientists in Shandong (Contract Number: BS2011DX031), (3) Technology and Evolving Countries Program in Shandong (2012GGB01093), (4) Basic Research Fund of Chinese Academy of Surveying and Mapping under Grant (Contract Number: 7771211). Among them, (1) and (4) are the basic business founding of CASM.

References

1. Luo, L.: The Research of Fuzzy Pattern Recognition and Decision Tree Methods Application in Remote Sensing Image Classification. Capital Normal University (2006)
2. Shi, Z., Ma, Y., Wang, Y.: Review on the Classification Methods of Land Use/Cover Based on Remote Sensing Image. Chinese Agricultural Science Bulletin 28(12), 273–278 (2012)
3. Dixon, B., Candade, N.: International Journal of Remote Sensing 29, 1185 (2008)
4. Wardlow, B.D., Egbert, S.L.: Remote Sensing of Environment 112, 1096 (2008)
5. Tao, C., Tan, Y., Peng, B.: A Probabilistic Latent Semantic Analysis Based Classification for High Resolution Remotely Sensed Imagery. Acta Geodaetica et Cartographica Sinica 40(2), 155–161 (2011)
6. Li, G., Wan, Y.-C.: Uncertainty classification method of remote sensing image based on high-dimensional cloud model and RBF neural network. Science of Surveying and Mapping 37(1), 115–118 (2012)
7. Qian, R.: The Comparative Study of Remote Sensing Images Classification Methods. Chang'an University (2007)
8. Jia, K., Li, Q., Tian, Y.: A Review of Classification Methods of Remote Sensing Imagery. Spectroscopy and Spectral Analysis 31(10), 2618–2623 (2011)
9. Guo, Y., Wang, Z., Le, X.: Minimum distance classifier ensemble based on adaptive distance metric. Journal of Computer Applications 26(7), 1703–1706 (2006)

Analysis on Agricultural E-Commerce Platform Construction in Developed Areas Based on Rural Residents' Needs – Take the Case of Beijing*

Jian Cao[1] and Yubin Wang[2,**]

[1] Beijing University of Agriculture, Beijing, China
c_jcaojian@163.com
[2] China Agricultural University, Beijing, China
wyb@cau.edu.cn

Abstract. This paper takes Beijing as an instance, launches investigation and survey on 152 rural residents, and analyzes the real demands of rural residents in the progress of e-commerce implementation in developed areas to provide government with support for the further construction in the relevant aspect. 71.71% of rural residents have ever heard about this concept, but among all the interviewees, 24.34% of them don't know what e-commerce is. There are 61.84% of rural residents who once conducted online shopping, but there are only 32.89% of rural residents having tried to sell or market their own agricultural products on the internet. Econometric model is used to find out the factors influencing rural residents to choose E-commerce. Rural residents with younger age are more willing to launch agricultural business through e-business, rural residents with higher annual income are more willing to attempt; and the ones with experiences of e-commerce online shopping are more willing to accept e-commerce of agricultural products. These suggestions are to provide with more trial opportunities and to provide rural residents with more training opportunities.

Keywords: E-Commerce Platform, Agricultural E-Commerce, Rural Residents' Needs, Influence Factors.

Agriculture is based on production, which is determined by two important factors: market and circulation. Information technology, network technology and e-commerce, such a new business pattern have injected new vitality and vigor into the circulation of agricultural products in China. The agricultural e-commerce has inspired the government and the society to promote the agriculture modernization in China to realize the great-leap-forward development of agriculture.

The agriculture modernization was put on the agenda since the 3rd Plenary Session of the 17th CPC. The informationized agriculture was increasingly valued, which brings a direction to the agriculture e-commerce. Since the development of e-commerce in the

* Supported by the people's Republic of China Ministry of agriculture in 2012 special, Horticultural crop standard park to create result analysis and evaluation(No. 20125002).

** Corresponding author.

D. Li and Y. Chen (Eds.): CCTA 2012, Part II, IFIP AICT 393, pp. 438–446, 2013.

late 20th century in China, agricultural e-commerce is no longer a new buzz word for Chinese, but its development has been limited greatly due to the immature conditions in all aspects. At present, the agriculture e-commerce is mainly headed by the government and the agricultural enterprises, while the individuals rarely go further under the present agricultural economic structure and confined by the capital condition. The government just advocates to promote the agriculture modernization by the agriculture industrialization and to optimize the agriculture industrialization by the agriculture informationization.

The agriculture informationization development in China achieved a lot during the past years. Yet, it is rarely researched on the topic what the proper information technology and e-commerce platform that the rural residents---the main users are in great need. In this article, the authors, taking 152 rural residents in Beijing as objects, analyze the real need of the rural residents in the developed rural areas in China, in order to provide a strong support for the further step related on such an issue.

1 The Development Level of Agricultural Informationization and E-commerce Both in Beijing and China

1.1 Level of Agricultural Informationization in Beijing and China

The development of agricultural e-commerce relies on agricultural informationization. As a weak industry, agriculture lags behind industry in terms of the development degree of informationization.

According to a report by CNNIC in January 2012, the number of the netizens in China has exceeded 50 million, reaching 51.3 million, and the Internet penetration has arrived at 38.3% by the end of 2011. The rural netizen number in China is about 13.6 million, which is approximately equal to one third of that of the urban on the corresponding period of last year, and the Internet penetration is about equal to one sixth. Compared with the year of 2010, the proportion of rural netizens decreased by 0.8%, its growth rate was still less than that of urban netizens. The number of the netizen who make a living on agriculture, such as farming, forestry, animal husbandry and fishery, takes about 4.0% of that of the total netizen, which means most of the rural residents in China don't surf online. The information services provided by agricultural websites can reach rural residents only if with the help of other information facilities or channels. The issue of information landing has become one of the bottlenecks restricting rural citizens from using information[1]. Beijing ranks top among other areas in China in terms of netizen popularity, its ratio is up to 70.3%, doubling that of the national level. This means that the internet surfing level of rural citizens of Beijing also ranks top nationwide.

The proportion of administrative villages with internet access has been enhanced to 91.5% (CNNIC, 2010). At present, the target of "four ones", namely "equip each village with one information service station, equip each town with one information service point, build one website for each village and launch one online program of agricultural and sideline products for each town", has been realized principally in 15 provinces.

1.2 The Current E-Commerce Situation on Agricultural Products

According to the analysis of Hu, there are seven business models of e-commerce of agricultural products in China, which can be sorted into three types: the first type are the primary ones, including catalog model, information model, medium model and virtual community model; the second type are the advanced models, including e-shop model, e-procurement model and value-chain integration model; the later type of models is the third party trading market model suitable for China's national conditions at the present stage[2].

Currently, the application of the e-commerce of agricultural products can be classified into two types, the first is service provision of supply and demand information hosted by government, represented by http://www.agri.gov.cn/ (Chinese agriculture information network hosted by China's Ministry of Agriculture), and other agriculture related websites organized by governments at all levels. Websites hosted by governments also include the styles of B2B and B2C, such as http://12582.10086.cn/ (Nongxintong), http://www.118326.com/index.html (Info Village) and http://nxw.ln.10109555.com/web/index.asp (Jinnongtong) hosted by three basic telecommunications enterprises, and other basic comprehensive information platforms that have already covered the whole country. The second type is business service provision hosted by various economic entities. These websites mainly engage in providing with the service of commercial electronicalization of business links related to the production, supply and sales of agricultural products. Their main customers are enterprises with a certain scale. B2B and B2C are the major two forms being applied[1].

From the aspect of e-commerce application, there are a futures market for wheat, cotton and bean, etc, and a spot market online dealing with agricultural products. it has enlarged the online product category that includes poultry, vegetables, local specialty, flowers, fruits, aquatic products, tea and so on, instead of the former products mainly concerning on grain, oil and fertilizer.

However, from a research by CNNIC, the first five online applications of the rural netizens are online music, online news, online game, searching engine and instant messaging. The application of online games tops among the five and only 17.6% rural netizens has experienced online shopping. Therefore, it can be concluded that few rural netizens take a good advantage of the current agricultural e-commerce platform in China.

2 The Recognition of the Rural Citizens on E-Commerce in Beijing

2.1 General Information of the Survey

This survey is distributed to the suburb, including Haidian district (rural and urban fringe zones), Fangshan District, Tongzhou District, Shunyi District, Changping

District, Daxing District, Huairou District and Pinggu District, and the exurban areas, including Yanqing County and Miyun County. There are 200 questionnaires totally, of which 152 are valid. There are 66.45% rural residents in plain areas and 33.55% in mountain areas and semi-mountain areas .

The following table 1 shows the basic character of the sample in this survey.

Table 1. The generalization of the survey

Item		Ratio (%)
gender	male	42.76
	female	57.24
age	below 20	6.58
	20-30	35.53
	30-40	20.39
	40-50	26.32
	50-60	7.24
	Over 60	3.95
education	university/college and above	34.87
	high school	24.34
	middle school	27.63
	primary school	10.53
	below primary school	2.63
job	the 1st industry	55.92
	the 3rd industry[1]	44.08

Data Source: original survey, similarly hereinafter.

2.2 The Information Facilities in Rural Areas

The survey shows 88.16% of the rural residents have their own computer and 80.92% can surf online. That means 9.7% of the rural residents have not possessed the skills to surf online. Among those who surf online, 74.34% prefer to be online at home. This indicates the basic network building in Beijing is excellent, which will play a good role in promoting e-commerce.

Viewing from the internet access, at present, there are three ways including DUN (dial-up networking), ADSL and broadband. 79.53% of rural residents can log on the internet through broadband, showing that the internet speed is guaranteed to lay a foundation for the e-business.

With the popularity of internet, the channels for famers to get information have undergone great changes; internet has become the major information source of rural

[1] The 3rd industry mainly refers to the leisure travel and service industry.

residents, and there are 44.73% of rural residents choosing internet to gain information. But through our survey, we find that only 30 interviewees (accounted for 19.74% among all the interviewees) often surf on the relevant agriculture websites to gain information, showing that at present rural residents still haven't formed the habit of gaining information from professional agriculture websites, that is to say the first primary phase of e-commerce of agricultural products hasn't been universal among rural residents currently.

2.3 The Recognition of the Rural Residents on E-Commerce

The concept of "e-commerce" has been advocated in recent years that 71.71% rural residents have heard as the survey shows. However, there are different understandings. Among all the interviewees, 24.34% of them don't know what e-commerce is, 21.71% of them think e-commerce just means issuing relevant sales information on the internet, only 26.31% of them believe that e-commerce indicates "sales on the internet".

As to the question what will benefit the agricultural products by e-commerce, 46.05% rural residents believe they can learn more information and 28.95% think e-commerce will help them sell their products with good price.

Although rural residents can't figure out what is e-business, some of them have started practicing. There are 61.84% of rural residents who once conducted online shopping, but there are only 32.89% of rural residents having tried to sell or market their own agricultural products on the internet. Meanwhile, the passion among rural residents to sell through the model of e-commerce is not relatively high, accounting for less than 40% among all the rural residents.

3 The Factors to the Rural Residents to Do E-Commerce

Through this survey, we have found that rural residents' awareness of e-commerce is still vague at present. As a result, scene description to rural residents is needed before the research on the prerequisites influencing on rural residents when they are choosing the operation model of e-commerce. First, investigators explain the basic concept of e-commerce to rural residents in straightaway language, then describe to them the various advantages of e-commerce comparing to the traditional sales mode. After the situation description, conduct the investigation about whether rural residents are willing to use agricultural e-commerce platform as the main operation form.

In this survey, there are three options to the question of "Are you willing to sell your agricultural products or to get various commercial opportunities through e-business?", "Yes", "No" and "I don't care". In the data processing, "Yes" is defined as "1," "No" and "I don't care " are defined as "0". Therefore, the explained variable in researching on the e-commerce demand of rural residents is (0, 1); this paper chooses Logistic model for regression analysis.

When y takes a certain value (such as y = 1), the relation between the probability of occurrence (p) and some factors (x) is as following:

$$P\ (y=1/x),\ \text{ie. } p=f(x) \tag{1}$$

P (probability) is valued between 0~1.

The Logistics model is used with a grouped data to predict the relation between a number of x and a dependent variable with different values, thus, showing the reliable relationship.

The General Equation is as followed:

$$y = \log it(p) = \beta_0 + \beta_1 x_1 + \cdots \beta_n x_n \tag{2}$$

3.1 Sample Selection

Obviously, e-commerce is based on internet access, therefore, when discussing whether rural residents will choose e-commerce or not, we only select a subsample with 121 interviewees who have mastered the skills of surfing on the internet from the sample with152 interviewees.

3.2 Variable Selection

In general, the individual character, the related facilities and so on are what will be taken into consideration of the action of the rural residents. Therefore, we will focus on the following four factors in this article:

a. The individual character of the rural residents: Age, education, job and income of very month are taken as the dependent variables.

b. The situation of information infrastructure in rural areas: The distance to the county and the way to surf online as the dependent variables represent the situation of information infrastructure in rural areas.

c. The internet usage of the rural residents: Three options “Occasional”, “often” and “every day” are taken to represent the online frequency.

d. The recognition on e-commerce of the rural residents: The question “have you ever shopped online” is indicated whether they know e-commerce, as it is hard for them to tell the concept of e-commerce.

Table 2. Basic Information of Variables

Variable Name	Variable Abbreviation	Variable Explanation	Variable Statistics
Dependent Variable	Y		Willing to use d-business
Age	AGE	Physical Age	33.44
Education	EDU	Represented by number of years one receiving education	12.49

Table 2. *(Continued)*

Industry	IND	Rural residents engaging in breeding and cultivation are set as 1, rural residents engaging in other tertiary industries are set as 0 (leisure and sightseeing industry is also regarded as tertiary industry)	28.10% are engaging in primary industry
Annual Household Income	INC		27851.24
Distance between home to County Town	DIS	Unit: Kilometer	24.79
Internet Connection Mode 1	MODE1	DUN=1, Others=0	4.13%, UNI
Internet Connection Mode 2	MODE2	ADSL=1, others=0	12.40%, ADSL
Internet-using Frequency 1	FRE1	Occasionally=1, others=0	33.07%, occasionally connecting to the internet
Internet-using Frequency 2	FRE2	Often=1, Others=0	13.22%, often connecting to the internet
Have you ever bought anything through e-business?	BUY	Yes=1, No=0	74.38% once bought something through e-business

3.3 Presumption to Be Testified

a. The one who is younger and with higher education will be more likely to do e-commerce.
b. The one who has the e-commerce experience will be more likely to do ecommerce with the agricultural products.
c. The development of the internet will be a factor to affect e-commerce.
d. The one with a higher frequency will be more likely to do e-commerce.

3.4 Assessment and Results

We use EVIEWS7.0 to assess the model in this article. We find there is a great relation between the age and the education. Therefore, we eliminate the variable of education to make an assessment. The relative result is as following:

Table 3. The result of the assessment

Variable	Coefficient	Std. Error	z-Statistic	Prob.
AGE	-0.052831	0.017224	-3.067221	0.0022
IND	0.400549	0.489629	0.818066	0.4133
INC	2.49E-05	1.40E-05	1.782153	0.0747
DIS	-0.005285	0.007459	-0.708533	0.4786
MODE1	0.424002	0.964057	0.439810	0.6601
MODE2	-0.567187	0.596467	-0.950911	0.3416
FRE1	0.460137	0.501401	0.917703	0.3588
FRE2	0.061557	0.611150	0.100724	0.9198
BUY	1.194978	0.447902	2.667942	0.0076

Viewing from the estimation result, below the significance level of 10%, there are three options including age, annual household income and the past experience of online shopping that show significance. That is to say rural residents with younger age are more willing to launch agricultural business through e-business, rural residents with higher annual income are more willing to attempt; and rural residents with experiences of e-commerce online shopping are more willing to accept e-commerce of agricultural products.Variables including internet connection mode and online frequency indicate insignificance.

4 Conclusion and Further Discussion

In developed areas (eg. Beijing), all the villages and towns have built up their own websites principally, the degree of convenience for rural residents to connect to the internet has been increased greatly; furthermore, most peasant households have had their own computers, mainly realizing the target of surging on the internet at home for rural residents. Under such background, the personal characteristic of a farmer, such as the age and income status, is the dominant factor influencing one's decision to launch online business of agricultural products or not.

Meanwhile, it is more acceptable for newly emerging things after some trying experience; thus appropriate increase of trial opportunities can be beneficial.

Through analysis, we can find that the internet speed doesn't significantly influence the application of e-commerce of agricultural products; therefore, hardware is not the main restrictive factor.

Based on the analysis above, we would like to propose some suggestions to the construction of e-commerce of agricultural products in China:

1. More attempts to do e-commerce with free charge should be provided by the platform.

With the attempts to do e-commerce, the rural residents will be encouraged and inspired, which will make a solid foundation for further development of the e-commerce on agricultural products.

2. More training should be provided to the rural residents.

In the survey, 51.25% of the rural residents think what they need most is the training on e-commerce. It is the greatest investment in the aspect of the software after the platform as the hardware is finished.

References

1. Tianshi, H.: Study on E-commerce Business Model for Chinese Agricultural Products. Chinese Academy of Agricultural Sciences (2005)
2. Yanyan, H.: Study on the Countermeasures of Developing Electronic Business of Agricultural Produce——A Case Study of Vegetables in Shouguang. Chinese Academy of Agricultural Sciences (2008)

Real Time Detection of Soil Moisture in Winter Jujube Orchard Based on NIR Spectroscopy

Xiaofei An[1], Minzan Li[1,*], Lihua Zheng[1], Yumeng Liu[1], and Yajing Zhang[2]

[1] Key Laboratory of Modern Precision Agriculture System Integration Research, China Agricultural University, Ministry of Education, Beijing 100083, China
[2] Editorial office of Transaction of CSAM, CAAMS, Beijing, China
pact@cau.edu.cn

Abstract. The measurement and control of soil moisture are the key technologies of precision agriculture. In order to real-time detect soil moisture content faster and more accurately, a portable soil moisture sensor based on NIR spectroscopy was developed. With the sixty soil samples collected from a winter jujube orchard, a linear regression model was established. The determination coefficients of the calibration (R_c^2) and validation (R_v^2) reached 0.88 and 0.92, respectively. The model passed *F*-test and *t*-test and showed robust. Subsequently, two spatial distribution maps of soil moisture were generated based on the data obtained by the portable soil moisture detector and the data obtained by oven drying method, respectively. Finally, the correlation between these two maps was investigated by using the software of Surfer 8.0. The zones of dry and wet soil could be distinguished easily in both maps. The results of the study showed that the developed detector was practical.

Keywords: Soil moisture, Real time detection, Soil sensor, Near infrared reflectance spectroscopy.

1 Introduction

Soil moisture content is one of the most critical soil components to plant growth and land management, especially in dryland [1]. It is usually measured by the oven drying method with a quite high precision. On the other hand, the oven drying method is a difficult, costly and time consuming procedure, so that it is necessary to develop a real time soil moisture detector for the precision agriculture practice.

The near-infrared (NIR) reflectance spectroscopy is a promising non-contact measurement method available to provide rapid information about some soil physical and chemical properties, such as soil organic matter content, soil total nitrogen, potassium, phosphorous, pH, soil moisture and particle size et al. When understanding the interaction between the NIR incident light and the soil properties, those physical and chemical parameters can be estimated based on Chemometrics.

* Corresponding author.

D. Li and Y. Chen (Eds.): CCTA 2012, Part II, IFIP AICT 393, pp. 447–455, 2013.

Many scholars such as Shibusawa, Christy, Xiao W [2-19] have researched on estimation of soil moisture based on NIR spectroscopy in recent years. Mouazen [2] established a soil moisture forecasting model with correlation coefficient of 0.982 in the laboratory. The model method was partial least squares regression (PLSR) with 306.5~1710.9 nm. But in the field measurement, the correlation coefficient was just 0.75 and the maximum error of measurement was above 2.5%. Hummel[3] analyzed relationship between soil moisture and near infrared spectroscopy, and a model was established by the multiple linear regression (MLR) method. Peng Y.K. [4] adopted the NIR spectra (Neotec51A) to forecast soil moisture, organic matter and total nitrogen of Loess Plateau and also obtained several estimations models. Shibusawa et al. reported a development of a real-time NIR spectroscopy measurement system for soil moisture[8]. The light illumination and reflection fibre were attached to a spectrophotometer with300 -1700 nm VIS and NIR light source.

Most of above researches established forecasting models in the laboratory conditions, and the real-time detection of soil moisture has not been realized with higher practical applicability. Thus, we decided to develop a portable and reliable soil moisture sensor for the real-time management. A comparison between thematic maps of soil moisture obtained by the developed detector and oven drying method respectively was conducted. And the field experiments were executed in a winter jujube orchard to test the performance of the detector.

2 Materials and Methods

The winter jujube orchard was located at Beijing suburb, with the area of 1000 m^2. Cultivated variety was Zhanhua winter jujube planted as south-north row direction. The range of experiment zone was E116.15268541°~116.15269492° and N 40.030982° ~ 40.031201°. The soil texture was typical loam in north China.

The detector was designed with an optical unit and a control unit. Fig.1 shows the overall structure of the soil moisture detector, and Fig.2 shows the prototype of the detector. The optical unit included a near infrared lamp-house at 1450 nm, a shared lamp-house drive circuit, a Y-type optical fiber shared by incidence and reflectance, a probe and an InGaAs photoelectric sensor. The control unit included an amplifier circuit, a filter circuit, an A/D circuit, a LCD display circuit and a storage circuit with U-disk. Light from near infrared lamp-house was adopted in the detector instead of the sunlight.

When the detector worked, it was pushed into soil under the surface of 30 cm firstly. Then the optical signal at single wavelength was transferred from the 1450 nm near-infrared lamp-house to the surface of the target soil via the incidence optical fiber. The reflected light from the soil surface was acquired and transferred to the InGaAs photoelectric sensor, where the optical signal was converted to the electrical signal. Subsequently, the obtained electrical signal was processed and stored temporarily by an 89S52 MCU. Finally, the calculated soil moisture content was displayed on a LCD and stored in a U-disk at the same time.

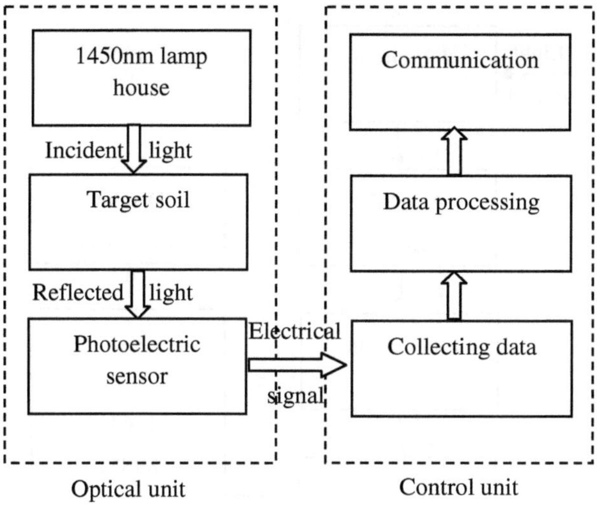

Fig. 1. Overall structure

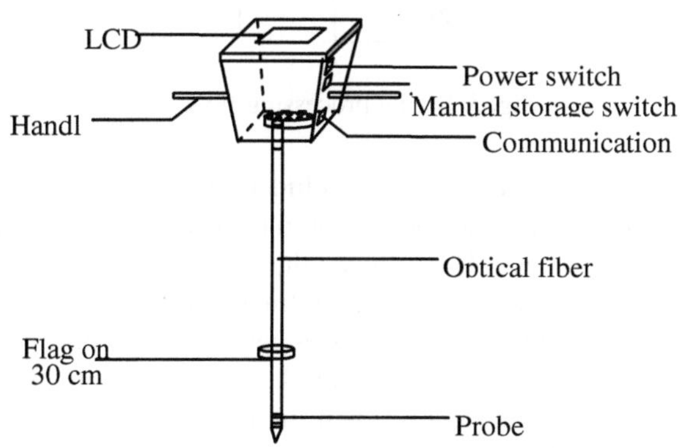

Fig. 2. Prototype of the detector

Fig.3 shows the signal processing flow in the control unit. After the I-U converting circuit, the optical signal I_0 was converted to the electrical signal U_0 by the photoelectric sensor. The electrical signal U_0 was a weak signal. It was necessary to design an amplifier circuit and a filter circuit before A/D circuit.

The processed signal U_3 was transmitted to the MCU 89 S52. In the MCU 89S52, the data were displayed with LCD and at the same time stored in the U disk. And finally the data could also be transmitted to the PC by a serial port.

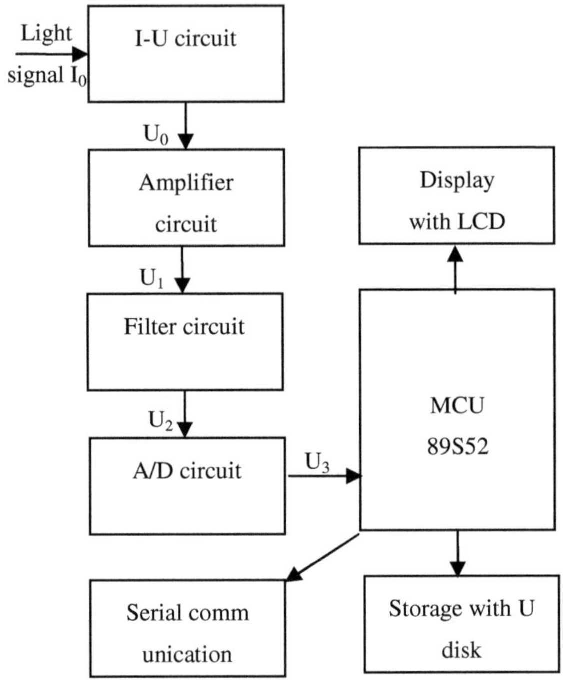

Fig. 3. Signal processing flow

In this detector several modules were included in the software, main program, subroutine of A/D conversion, subroutine of data processing, subroutine of LCD display and subroutine of storage. Fig.4 shows the flow chart of the main program.

Absorbance was selected as the spectral parameter. In order to get the absorbance of soil samples, a standard whiteboard was designed. First, the probe of the developed soil moisture detector was put on the standard whiteboard and the output voltage V_i of the instrument was measured. And then the probe was pushed into soil sample and the output voltage V_i' of the instrument was measured. According to the Equation (1), (2) and (3), the reflectance r_i , absorbance A_i and average absorbance A_{iavg} of every soil sample were calculated.

$$r_i = V_i' / V_i \times 100\% \tag{1}$$

$$A_i = \lg(1/r_i) \tag{2}$$

$$A_{iavg} = \sum_{n=1}^{6} A_i \Big/ 6 \tag{3}$$

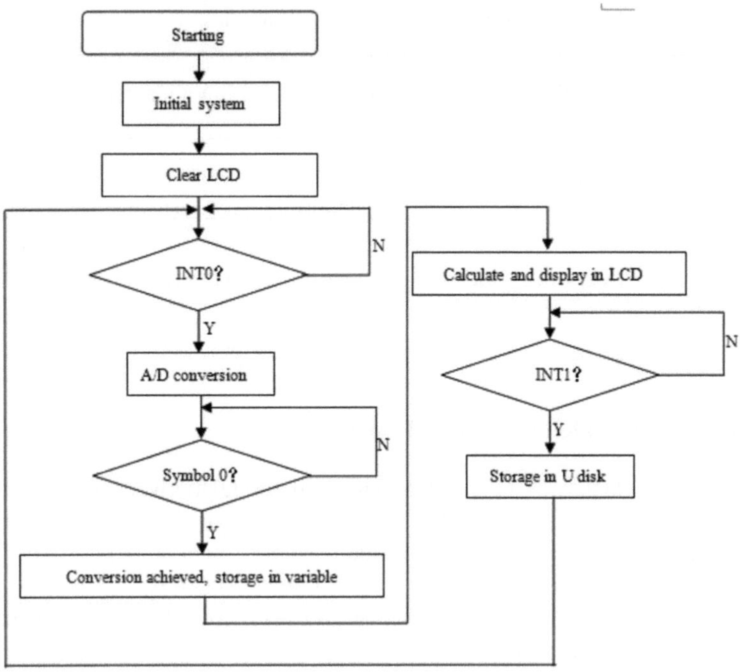

Fig. 4. Flow chart of main program

3 Results and Discussion

3.1 Data Preprocessing

Soil moisture contents of all soil samples were also measured by the oven drying method in the laboratory condition. Soil samples were subjected to oven drying at 60°C for 24 hours, and soil moisture content was obtained through the Equation (4).

$$w = \frac{m_1 - m_0}{m_0} \times 100\% \tag{4}$$

where: w is soil moisture content, m_1 is original soil weight, m_0 is soil weight after oven drying.

Statistic characteristics of soil moisture contents in the winter jujube orchard are shown in Table 1. All soil samples were divided into two groups: calibration group (including 40 soil samples) and validation group (including 20 soil samples). In the calibration group, the selected soil moisture range of 3.15%~14.19% covered the most acquired plant and organic activities during the winter jujube season. In the validation group, the range was 4.54%~12.09%, and the average was 8.03% and the standard deviation was 2.09%.

Table 1. Statistics of soil samples

Parameter	Calibration group (40)			Validation group (20)		
	Range	Average	Standard deviation	Range	Average	Standard deviation
Soil moisture (%)	3.15~14.19	8.79	2.41	4.54~12.09	8.03	2.09

3.2 Estimation Modeling of Soil Moisture

The object of this task was to build a statistical model between the absorbance of 1450nm and soil moisture content. There were several statistical modeling techniques for the proper calibration performance, such as the linear and non-linear multiple regression analyses and the principal component analysis (PCA).

In this work, the forecasting model of soil moisture content was established by the linear regression method with the average absorbance spectral data. Fig.5 shows the comparison between measured data and predicted data of soil moisture. The forecasting model is shown in Equation (5). The determination coefficient of the calibration (R_c^2) was 0.88, and the determination coefficient of the validation (R_v^2) was 0.92. The model passed the F-test and t-test, which illustrated that both the model and the parameters were significant.

$$y = 35.74 \times x_{1450} - 35.37 \tag{5}$$

where: y represents the dependent variable of soil moisture content; x_{1450} is average absorbance of soil at the wavelength of 1450 nm.

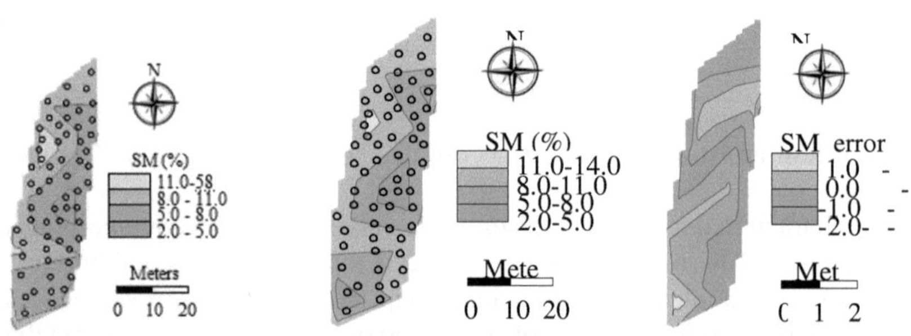

(a) Predicted SM distribution (b) Measured SM distribution (c) SM error distribution

Fig. 5. Thematic maps of soil moisture

3.3 Discussion

Use of infrared emitting diode of 1450 nm could make it possible to increase the frequency of sampling at a faster rate. It would also make the unit run on low power, compact and light, so that it would require little space and mechanical support.

Soil moisture content could be obtained by the soil moisture sensor. A linear regression model could satisfy the basic needs. The determination coefficient of the calibration (R_c^2) and validation (R_v^2) were 0.88 and 0.92 respectively. Root mean square error of calibration (RMSEC), root mean square error of prediction (RMSEP) and residual prediction deviation (RPD) were 0.94, 0.85 and 3.26 respectively. From the thematic maps of soil moisture, the zones of dry and wet soil could be distinguished easily. Both the results indicated that the sensor could meet the requirement of precision agriculture.

In order to increase the accuracy of the NIR measurement, other influencing parameters such as soil texture and density should be taken into account. Both the texture and density affected the results seriously. Extending the sensor application for all soil textures requires incorporating extra laboratory measurements of spectra taken from soil samples of another soil texture.

4 Conclusions

Soil moisture content is one of the most critical soil components to plant growth and land management, especially in dryland. In this paper, a rapid method to measure soil moisture was proposed. According to the result and discussion, the following conclusions could be obtained.

(1) A portable soil moisture detector was developed based on NIR spectroscopy. In the detector, LED of 1450 nm, *Y*-type fiber and circuits were adopted to insure the sensor stable and reliable enough. It also made the unit run on low power, compact and light so that it would require little space and mechanical support.

(2) The field experiment showed that the instrument could work stably. The established model showed that there were high correlation coefficient between average absorbance of 1450 nm and soil moisture content, with R_c^2 and R_v^2 of 0.88, 0.92 respectively. Similar spatial patterns of soil moisture measured with the developed detector and oven drying method were observed. Both above results indicated that the detector could satisfy the basic agriculture requirements.

(3) In order to improve the accuracy of soil moisture detection, soil texture should be paid more attention in the next research step.

Acknowledgements. This study was supported by Chinese National Science and Technology Support Program (2011BAD21B01), NSFC projects (61134011, and 31071330), and 863 programs (2011AA100703, 2011AA100704).

References

1. Shang, S.P., Wang, Y.L., Hong, A.J.: Establishment of calibration equation for soil moisture content by NIRS. Chinese Journal of Spectroscopy Laboratory 28(4), 1604–1607 (2011)
2. Mouazen, A.M., Baerdemaeker, J.D., Ramon, H.: Towards development of on-line soil moisture content sensor using a fiber-type NIR spectroscopy. Soil & Tillage Research 80(1), 171–183 (2005)
3. Hummel, J.W., Sudduth, K.A., Hollinger, S.E.: Soil moisture and organic matter prediction of surface and subsurface soils using an NIR soil sensor. Computers and Electronics in Agriculture 32(2), 149–165 (2001)
4. Peng, Y.K., Zhang, J.X., He, X.S.: Analysis of soil moisture organic matter and total nitrogen content in loess in china with near infrared spectroscopy. Acta Pedologica Sinica 35(4), 553–559 (1998)
5. Christy, C.D.: Real-time measurement of soil attributes using on-the-go near infrared reflectance spectroscopy. Computer and Electronics in Agriculture 61, 10–19 (2008)
6. Gashaw, A., Shane, M.W.: Development of a prototype infrared reflectance moisture meter for milled peat. Computer and Electronics in Agriculture 28, 1–14 (2000)
7. Xiao, W., Li, X.Y., Li, P.W.: Adaptability of the model for soil moisture measurement based on near-infrared spectroscopy. Transactions of the Chinese Society of Agricultural Engineering 25(3), 33–36 (2009)
8. Shibusawa, S., Made, I., Anom, S.W.: Site-specific crop response to temporal trend of soil variability determined by the real-time soil spectrophotometer. In: Proceedings of the Joint European Conference of ECPA-ECPLE, pp. 639–643. Wageningen Academic Publishers, Berlin (2003)
9. Liang, X.Y., Li, X.Y., Lei, T.W.: Study of sample temperature compensation in the measurement of soil moisture content. Measurement 44(10), 2200–2204 (2011)
10. Liang, X.Y., Li, X.Y., Wang, W.: NIR Spectroscopy Based on DWT and LS-SVM for Prediction of Soil Moisture. In: Proceedings of the 2010 International Conference on Electrical and Control Engineering (ICECE 2010), pp. 5874–5876 (2010)
11. Minasny, R., Alex, B.M., Veronique, B.M.: Removing the effect of soil moisture from NIR diffuse reflectance spectra for the prediction of soil organic carbon. Geoderma 167, 118–124 (2011)
12. Pan, L.: Development of soil organic matter fast determination instrument based on spectroscopy. China Agricultural University (2007)
13. Li, M.Z., Pan, L., Zheng, L.H., An, X.F.: Development of a portable SOM detector based on NIR diffuse reflection. Spectroscopy and Spectral Analysis 30(4), 1146–1150 (2010)
14. Li, M.Z., Han, D.H., Wang, X.: Spectral analysis and application. Science Press, Beijing (2006)
15. Chen, H.H., Li, X., Ding, W.X.: Twelve Kinds of Gridding Methods of Surfer 8.0 in Isoline Drawing. Chinese Journal of Engineering Geophysics 4(1), 52–57 (2007)
16. Li, X.Y., Xiao, W., Li, P.W.: Near-infrared spectral detection of soil moisture based on feature extraction of FFT. Transaction of the Chinese Society for Agricultural Machinery 40(5), 64–67 (2009)
17. Liang, X.Y., Li, X.Y.: The Research on Sample Temperature Compensation for Soil Moisture Content Based on Multi-Wavelength and LS-SVM. Chinese Journal of Sensors and Actuators 24(8), 1228–1232 (2011)

18. Krapez, J.-C., Olioso: A combination of temperature, vegetation indexes and albedo, as obtained by airborne hyperspectral remote sensing, for the evaluation of soil moisture. Qirt Journal 8(2), 187–200 (2011)
19. Song, T., Bao, Y.D., He, Y.: Research on the Method for Rapid Detection of Soil Moisture Content Using Spectral Data. Spectroscopy and Spectral Analysis 29(3), 675–677 (2009)

Correlations between Nitrogen Content and Multispectral Image of Greenhouse Cucumber Grown in Different Nitrogen Level

Wei Yang, Nick Sigrimis, Minzan Li, Hong Sun, and Lihua Zheng,

Key Laboratory of Modern Precision Agriculture System Integration Research, Ministry of Education, China Agricultural University, Beijing, China
limz@ cau.edu.cn

Abstract. Short fruit cucumber was chosen as object. Different nutrition conditions were applied to cultivate plants.CCD camera with special filters was used to gather images. Correlation analysis results shows that four indexes, CNDVI, GNDVI, NDGI and NDVI, had much higher correlation with nitrogen content of cucumber leaves. LS-SVM algorithm was proposed to construct models between vegetation indexes and nitrogen content. The results verified that the method of CCD camera with filters to get multispectral images could be used for analyzing nitrogen content of cucumber leaves.

Keywords: CCD camera, vegetation indexes, multispectral image, LS-SVM.

1 Introduction

The short fruit cucumber imported recent years is a kind of new variety of cucumber with better taste. It is favored by consumers and has enormous market potential. The short fruit cucumber can grow all year round using soilless culture in greenhouse, and the nitrogen content is the most important factor to cucumber's quality and quantity. The traditional measurement method of nitrogen content is destructive and time-consuming. When cucumber is in deficient nitrogen state, there will be long serials of changes in color, thickness and morphological characteristics of cucumber leaves. It will make light reflectance characteristic (reflectance intensity, color, texture and so on) change. Thus it is important to conduct research on nondestructive measurement of nitrogen content based on reflection spectra and multispectral images. Scholars both in China and abroad have carried out many researches on this subject based on multispectral images. However, complicated algorithm is the main problem of most multispectral cameras used for analysis of plant multispectral image, and besides high cost it also has very strict requirements of condition for taking photos and choosing object. In this paper, we tried to provide a new approach using a CCD camera with special filters to take spectral images. The demands for the camera were easy to operate, low cost, and wide application.

D. Li and Y. Chen (Eds.): CCTA 2012, Part II, IFIP AICT 393, pp. 456–463, 2013.

2 Material and Methods

The research chose short fruit cucumber as object. In order to do research in different nutrition level, different nutrition conditions were applied to cultivate plants. Since soilless cultivation technology can make sure samples are not affected by other factors like soil, it was used in planting cucumbers. All cucumbers were divided into four groups in four different nitrogen content conditions, and each group had 10 cucumbers. The standard nitrogen content (N) was set as 0.067 kg/L. The first group was planted in double standard nitrogen content condition (2N); the second group was in standard nitrogen content condition (N); the third group was in half standard nitrogen content condition (0.5N); and the last group was in one fourth of standard nitrogen content condition (0.2N).

The research chose a normal CCD camera with filters to get multispectral images. The central wavelengths of the filters were 670 nm at red band and 780 nm at near infrared (NIR) band with much more information, low correlation between reflectance data, and good separability. The CCD/CMOS used by digital camera can induce infrared light. In order to avoid color shift produced by infrared ray, digital camera manufacturers often put an Infrared Cut Filter (ICF) in front of CCD to shut off infrared ray. When moving away ICF and adding special filters in front of camera, the lens can get multispectral images at specific wavelengths. Figure1 (a) (Figure2 (a)) and Figure3 (a) (Figure4 (a)) were the examples of the images taken by the camera.

3 Result and Discussion

In order to get characteristic parameters from multispectral image, we needed to do preprocessing of images to decrease the influence of soil background and random factors as much as possible. The preprocessing of multispectral images mainly contained digital image denoising, image enhancement and image segmentation. Spectral information can be decomposing into background noise (normally belongs to low frequency), component information (frequency) and random noise (high frequency). After using Wavelet Packet Transform to spectra, every part of spectra could be totally separated in frequency domain, and then all noise could be easily filtered out.

Steps of the wavelet transform to image denoising were as following. The first was choosing suitable basis functions and decomposition layers (N) to decompose image. The second was quantifying high frequency coefficient and getting rid of high-frequency component of images. The final was, filtering images and eliminating noise according to multi-resolving coefficients of the wavelet transform and reconstructing of wavelet. The 1-5 lay's image decomposing results were compared based on three basic functions such as bior3.1, coif3, db4. The three-lay image decomposing method based on bior3.1 was chosen to process cucumber multispectral images. Images from wavelet transform are showed in Fig. 1 (b), Fig. 2 (b), Fig. 3 (b) and Fig. 4 (b). From figures we found that the noise of the images had been effectively removed. After wavelet transform, Genetic Algorithm and watershed algorithm were used to separate leaves from background, respectively.

(a) (b) (c)

(a) Original near infrared image (b) Image with wavelet transform (c) Near infrared binary image

Fig. 1. Near infrared binary image separated by Genetic Algorithm

(a) (b) (c)

(a) Original near infrared image (b) Image with wavelet transform (c) Near infrared binary image using watershed algorithm

Fig. 2. Near infrared binary image separated by watershed algorithm

Figure1 shows the near infrared binary images separated by Genetic Algorithm. Fig. 1(a) are two original near infrared images, Fig. 1(b) are images with wavelet transform, Fig. 1(c) shows the near infrared binary image. Fig. 2 shows the near infrared binary image separated by watershed algorithm. Fig. 2 (a) shows two original near infrared images, Fig. 2(b) are images with wavelet transform Fig. 2(c) shows near infrared binary images. Comparing binary images with two different methods,

(a) (b) (c)

(a) Original red image (b)Image with wavelet transform (c)Red binary image using Genetic Algorithm

Fig. 3. Red binary image separated by Genetic Algorithm

(a) (b) (c)

(a) Original red image (b) Image with wavelet transform (c) Red binary image using watershed algorithm

Fig. 4. Red binary image separated by watershed algorithm

similar good results were obtained. When processing images, any of these two methods was available. When using these two methods on red images, we also obtained good results, and both of them could be used in image processing.

After filtering and segmentation to multispectral image, the average gray value of NIR, R and G wavelength were got from multispectral image of cucumber leaves.

And then vegetation index could be calculated according to the reflectance of NIR, R and G of cucumber leaves. All vegetation indexes we used in research are shown in Table1.

Table 1. Vegetation value based on multispectral image

Name	Formula
DVI	$NIR_{IA}-R_{IA}$
GDVI	$NIR_{IA}-G_{IA}$
RVI	NIR_{IA}/R_{IA}
GRVI	NIR_{IA}/G_{IA}
NDVI	$(NIR_{IA}-R_{IA})/(NIR_{IA}+R_{IA})$
SAVI	$\frac{NIR_{IA}-R_{IA}}{NIR_{IA}+R_{IA}+L}(1+L) \quad (L=0.5)$
GNDVI	$(NIR_{IA}-G_{IA})/(NIR_{IA}+G_{IA})$
NDGI	$(G_{IA}-R_{IA})/(G_{IA}+R_{IA})$
CNDVI	$(2NIR_{IA}-R_{IA}-G_{IA})/(NIR_{IA}+R_{IA}+G_{IA})$

Analyzing correlation coefficient between nitrogen content of cucumber leaves and all kinds of vegetation index in different growth, the result shows in Table2.

Table 2. The correlation analysis between nitrogen content and multispectral images of cucumber

Index	Teen				Early flowering				Fruit			
	2N	N	0.5N	.2NN	2N	N	0.5N	0.2N	2N	N	0.5N	0.2N
NIR_{IA}	0.36	0.29	0.24	.20	0.50	0.54	0.36	0.36	0.50	0.52	0.36	0.36
DVI	0.35	0.37	0.40	.43	0.26	0.32	0.34	0.32	0.23	0.34	0.32	0.32
GDVI	0.07	0.05	0.05	.05	0.14	0.10	0.06	0.05	0.15	0.12	0.06	0.06
RVI	0.26	0.24	0.24	.24	0.32	0.30	0.25	0.25	0.33	0.32	0.28	0.28
GRVI	0.39	0.37	0.37	.36	0.43	0.42	0.38	0.36	0.44	0.40	0.40	0.38
NDVI	0.56	0.54	0.54	.54	0.59	0.49	0.55	0.56	0.52	0.53	0.56	0.49
SAVI	0.50	0.49	0.48	.46	0.63	0.63	0.60	0.59	0.65	0.63	0.55	0.61
GNDVI	0.53	0.52	0.52	.52	0.66	0.66	0.63	0.60	0.68	0.69	0.59	0.58
NDGI	0.65	0.65	0.64	.64	0.68	0.68	0.66	0.65	0.69	0.65	0.64	0.63
CNDVI	0.67	0.66	0.66	.65	0.70	0.70	0.67	0.65	0.72	0.73	0.70	0.68

The results from correlation analysis showed that the correlation between nitrogen content of cucumber leaves and spectral characteristics of the multispectral images was close in teen of cucumbers in different nitrogen levels. The correlation between

CNDVI and nitrogen content was higher than others and reached to 0.67. All of them, the correlation coefficients between GNDVI, NDGI, NDVI and nitrogen content were all higher than 0.50. In early flowering stage, the correlations between nitrogen content of cucumber leaves and spectral characteristics of the multispectral images in different nitrogen levels were close. In fruit stage, the correlations between nitrogen content of cucumber leaves and spectral characteristics of the multispectral images in different nitrogen levels existed remarkable difference, and the correlation of 2N level and N level was higher than the correlation of 0.2N level. The results suggested that vegetation index based on multispectral image had been used to reflect nitrogen content of cucumbers in different growth. CNDVI, GNDVI, NDVI and NDGI these four indexes have better ability in nitrogen content detection than other indexes.

After image processing, reflectance of cucumber leaves could be calculated from image, and then vegetation index of cucumber leaves could be obtained from the calculation of reflectance of three wavelengths. The vegetation indexes were important land vegetation ecosystem status parameters and have been found wide application in agriculture, environment and ecological environment. According to chief content of the research, four indexes, including CNDVI、GNDVI、NDGI and NDV, were chosen, and then building the correlation model between these four vegetation indexes and nitrogen content of cucumber leaves. Least squares support vector machines (LS-SVM) model was proposed to construct model in the paper.

Support Vector Machine (SVM) was a new machine learning method based on Statistical Learning Theory (SLT). The data space was mapped to high dimension feature space with nonlinear Kernel function, so that it could be used to process Higher-Order Nonlinearity classification and regression. It could overcome the shortcomings of ANN BP Networks, like long training time, easy to fall in local optima and limited recognizing range. SVM was trained with LIBSVM algorithm and radial base kernel function. Radial base kernel function had fewer parameter and value limit, and reduced the complexity of model and improved training speed. Cross-validation was used to determinate related parameter, and ample data were divided into training part and testing part.

In modeling, four kinds of vegetation indexes were chosen as input, and the nitrogen content of cucumber leaves was set as output. SVM was used to fit and forecast the nitrogen content of cucumber and built the correlation model. Fig. 5 shows the forecasting model of nitrogen content of cucumber leaves based on SVM, calibration R^2 was 0.92 and verification R^2 was 0.69. The result showed high correlation between vegetation index and nitrogen content of cucumber leaves. This method provided feasibility of collecting and testing growth information of cucumber.

4 Conclusions

The research chose a CCD camera with special filters as multispectral image gathering platform to gather images of cucumber leaves in different growth. After Wavelet De-noising and Image Segmentation, we got reflectance of cucumber leaves, and vegetation indexes of cucumber leave samples. Correlation analysis was made

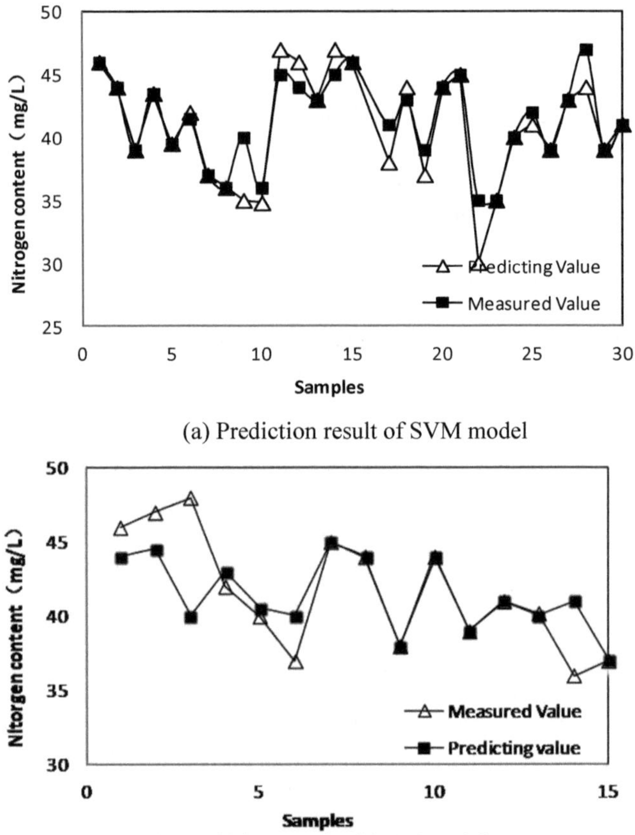

(a) Prediction result of SVM model

(b) Verifying result of SVM model

Fig. 5. SVM prediction result of nitrogen content of cucumber leaves

between vegetation indexes and nitrogen content of cucumber leaves, and the results shows that four indexes, CNDVI, GNDVI, NDGI and NDVI, had much higher correlation with nitrogen content of cucumber leaves. Least squares support vector machines (LS-SVM) algorithm was proposed to construct models between vegetation indexes and nitrogen content. The correlation coefficient between vegetation indexes and nitrogen content of cucumber leaves was 0.92, and verifying coefficient was 0.69. These results had ideal estimations, and the vegetation indexes were comprehensive reflection of nitrogen content of cucumber.

Acknowledgments. This study was supported by Chinese National Science and Technology Support Program (2011BAD21B01), Chinese National Programs for High Technology Research and Development Research Fund (2011AA100703, 2011AA100704), Doctoral Program of Higher Education of China (20110008130006, 20100008110030).

References

1. Fukagawa, T., Noguchi, N., Ishi, K.: Paddy field application of monitoring system for crop status using multi-spectral imaging sensor. Journal of Hokkaido Branch of the Japanese Society of Agricultural Machinery, 37–42 (2003)
2. Qin, X., Hen, L.: Wavelet Transform and Its Application in Spectral Analysis. Spectroscopy and Spectral Analysis 20(06), 892–897 (2000)
3. Tian, G., Yuan, H., Liu, H.: The Application of Wavelet Transform in Near Infrared Spectroscopy. Spectroscopy and Spectral Analysis 23(06), 1111–1114 (2003)
4. Jongschaa, R.E.E.: Run-time calibration of simulation models by integrating remote sensing estimates of leaf area index and canopy nitrogen. Europe Journal Agronomy 24, 316–324 (2006)
5. Tang, Q., Li, S., Wang, K.: Monitoring Canopy Nitrogen Status in Winter Wheat of Growth Anaphase with Hyper spectral Remote Sensing. Spectroscopy and Spectral Analysis 11, 3061–3066 (2010)
6. Richardson, A.J., Wiegand, C.L.: Distinguishing vegetation from soil background information. Photogram. Eng. Remote Sensing 43(12), 1541–1552 (1977)
7. Li, H., Zhang, L., Chen, X.: Image analysis method in application of digital image on diagnosing wheat nitrogen status. Chinese Journal of Eco-Agriculture 1, 155–159 (2001)
8. Guo, J., Chen, L., Yang, Y.: Effect of Nitrogen Nutrition Diagnosis on Cucumber in Greenhouse. Journal of Changjiang Vegetables 22, 60–63 (2010)
9. Li, Y., Zhu, Y., Cao, W.: Characterizing Canopy Hyper spectral and Multispectral Reflectance under Different N-application Conditions in Wheat. Journal of Triticeae Crops 2, 103–108 (2006)
10. Li, J., Dong, Z., Zhu, J.: Present Application and Outlook for Method of Nitrogen Nutrition Diagnosis. Journal of Shihezi University (Natural Science) 7(1), 80–83 (2003)
11. Wang, K., Shen, Z., Wang, R.: Effects of Nitrogen Nutrition on the Spectral Reflectance Characteristics of Rice Leaf and Canopy. Journal of Zhejiang Agricultural University 24(1), 93–97 (1998)
12. Van Henten, E.J., Hemming, J., van Tuijl, B.A.J., et al.: An autonomous robot for harvesting cucumbers in greenhouses. Autonomous Robots 13(3), 241–258 (2002)
13. Van Henten, E.J., van Tuijl, B.A.J., Hoogaker, G.J., et al.: An autonomous robot for de-leafing cucumber plants grown in a high-wire cultivation system. Bio Stems Engineering 94(3), 317–323 (2006)
14. Van Henten, E.J., Dijk, G., Kuypers, M.C., et al.: IFAC modeling and control in agriculture. In: Horticulture and Post Harvesting Processing, Wageningen, Netherlands, pp. 39–44 (2000)

A New Approach for Fast Calculation of Sloped Terrace Earthwork Based on GIS in the Hilly Regions

Qingchun Zhang

Land Consolidation and Rehabilitation Center, the Ministry of Land and Resources, Beijing Normal University

Abstract. Due to the undulating terrain, earthwork estimation was very difficult in the planning and design of land consolidation projects in the hilly areas. A new approach was proposed to estimate the earthwork of sloped terraces according to map features of planning and design of land consolidation in this paper. It was achieved by applying digital elevation model (DEM) theory and statistical function of the slope with GIS software according to the derived formula. A land consolidation project was chosen as an empirical research example to use the proposed approach in Yangxin County, Hubei Province. Study found that the use of GIS for earthwork estimation of sloped terrace is feasible, the earthwork of different slope range could be estimated quickly according to sloped terrace section (ridge height, ridge slope and the field surface slope after down slope) after getting the area of different slope range using GIS. The approach is simple and fast, with strong practical characteristics. It provided a new idea for earthwork estimation of sloped terrace in hilly area, and improved the efficiency of project planning and design.

Keywords: land consolidation, sloped terrace, land leveling.

1 Introduction

Terraces were the stepped or sloping section farmlands that were built along the contour in in the mountain areas or hilly areas. Terraces can change the terrain slope, store rainwater, increase soil moisture, and prevent soil erosion, which can achieve water conservation, soil conservation, fertilizer purposes. So terraces have significant ecological and economic benefits to improve agricultural production conditions, promote the steady development and comprehensive management of grain production. "Soil and Water Conservation Law of People's Republic of China" stipulates that the sloping fields below 25 °slopes generally can build terraces to grow crops, and the slope above 25 °should be planted trees and grass. Terraces can be divided into stepped terraces and wave-like terraces according to the section type, and stepped terraces which can be divided into horizontal terraces, sloping terraces, etc.

Land consolidation is effective measures to achieve dynamic balance of total arable land. It is also an important means to improve land quality, promote land use intensively, and protect the ecological environment. Land consolidation is extremely

D. Li and Y. Chen (Eds.): CCTA 2012, Part II, IFIP AICT 393, pp. 464–471, 2013.

important in promoting rural modernization and sustainable economic development. The earthwork amount of land leveling is directly related to the investment of land leveling in land consolidation (Liu etal, 2003). As the large amount of land leveling when sloping fields being built into level terrace, thus, by the level of investment restrictions, many land consolidation projects have carried down slope design, organized into sloping terraces in hilly areas. In hilly areas, due to topography, it is difficult to measure the amount of earthwork (Dong etal, 2004). Currently, some scholars carried out some research for the calculation method of level terrace earthwork. A level terrace earthwork method was proposed through the application of digital elevation model (DEM) theory and GIS software in hilly areas (Liu etal, 2007).The application of "3S" technologies in land leveling has also been explored in depth (Jin etal, 2004; Lu etal, 2003; Chang etal, 2008;Lu etal, 2002)."Comprehensive regulation technical specifications of soil and water conservation -Sloping land Regulation technologies of People's Republic of China " (GB/T16453.1-2008) gives the formula calculation method and calculation of level terrace, but on the sloping terrace does not make provision. It is important that quickly and accurately measuring the amount of earthwork to carry out planning and design, control total investment and allocate the capital. Earthwork calculations for sloping terraces are not the same way, the current method of calculation earthwork mainly included longitudinal method, typical plots method, and direct measuring and calculation method (Fan etal, 2010). These methods are complex, accuracy, and have certain limitations. Therefore, the measure software and methods with a broad user base, simple operation, making full use of existing data sources and estimation results precision were urgently called for in earthwork calculation of slope terrace for land consolidation.

The earthwork calculation method of slope-style terraced is relatively less currently. A fast and accurate earthwork estimation method of slope-style terraced was proposed in this paper, taking a land consolidation project in Sanxi town, Yangxin County, Hubei province as an example. The earthwork was calculated through the application of digital elevation model (DEM) theory and the statistical capabilities of GIS software according to the characteristics of land consolidation planning and design maps, which can provide a reference for earthwork estimation of slope-style terraced in hilly areas.

2 Study Area

A low hilly land consolidation project in Sanxi Town, Yangxin County, Hubei Province was taken as an example to introduce the earthwork calculation steps by using this method in this study. Project area is located in east longitude 115 ° 48'22 "~ 115 ° 49'13", latitude 29 ° 50'34 "~ 29 ° 50'57", and its construction area is 36.22hm^2. Project area has significant low hilly landscape features with undulating terrain, slope degree between 10 ° ~ 15 ° and the highest elevation of 75.27m, the lowest 31.56m.There are four sections needed to be terraced (Figure 1), with acreage of 2.61hm^2, 12.02hm^2, 1.21 hm^2 and 4.74 hm^2 respectively.

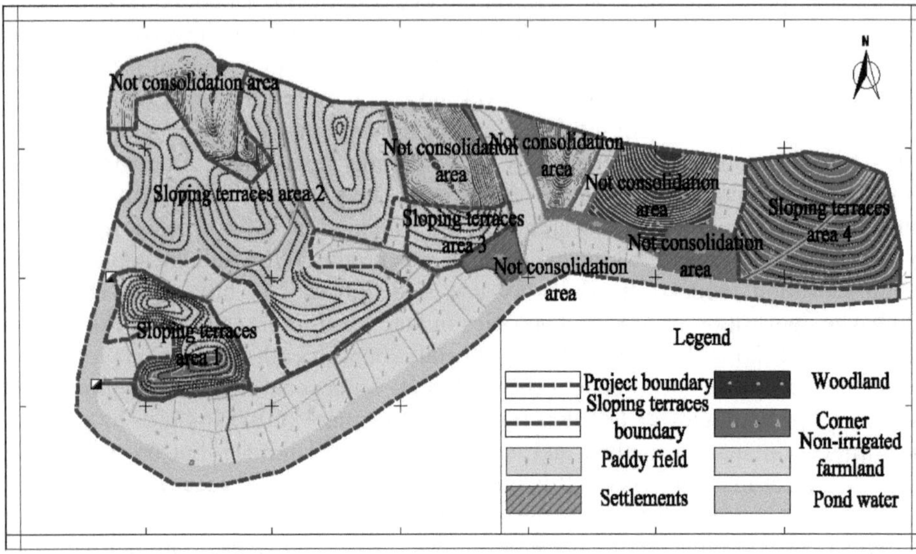

Fig. 1. Land consolidation project area in Sanxi town, Yangxi County, Hubei Province

3 Calculation Principle of Sloped Terrace Earthwork

3.1 Design Elements of Sloping Terrace Section

Sloping terrace section elements include the original ground slope, the field surface slope after down slope, ridge height, ridge outside slope, net width of field surface, gross width of field surface, land occupation width of ridge and ramp width of field surface (Zheng etal, 2009; Fan etal, 2008). A typical cross-section of slope-style terraces were shown in Figure 2.

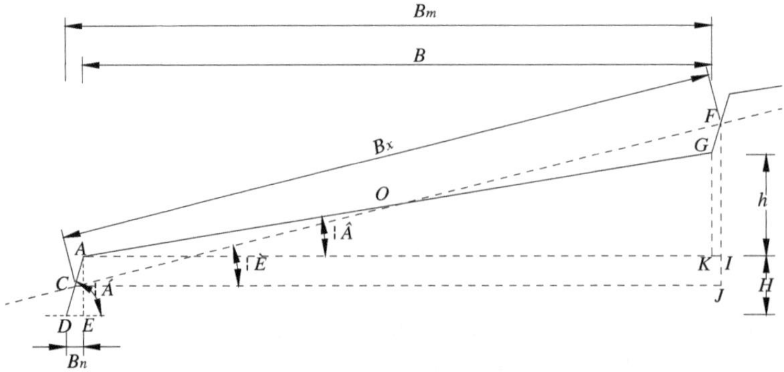

Fig. 2. Section design of sloped terrace

θ——The original ground slope, (°) ; β——The field surface slope after down slope, (°) ; α——Ridge slope, (°) ; H——ridge height, (m) ; h——Slope height of Field surface, (m) ; H'——Table height, (m) , $H'=H+h$; B ——Clear width of field surface , (m) ; B_n ——Land Occupation width of ridge, (m) ; B_m——Gross width of field surface, (m) ; B_x——Ramp width of field surface, (m) 。

3.2 Earthwork Calculation of Sloped Terrace

Calculation Principle of Sloping Terraced Earthwork

According to sloping terrace section design, per unit area (1 hm^2) sloping terrace earthwork can be calculated by:

$$V = S_{OAC} \times L \tag{1}$$

Where V is dig (fill) earthwork of per unit area (1 hm^2), m^3; S_{OAC} is cross-section area of triangular ΔOAC , m^2; L is the terraces length of per unit area (1 hm^2) , m.

Cross-Section Area of Triangular ΔOAC

$$\begin{aligned} S_{OAC} &= \frac{1}{2}\times L_{OC}\times L_{AC}\times\sin\angle ACO = \frac{1}{2}\times\frac{H}{2\sin\alpha}\times\frac{\tan\alpha-\tan\beta}{2\cos\theta\times\tan\alpha\times(\tan\theta-\tan\beta)}\times H\times\sin(\alpha-\theta) \\ &= \frac{1}{8}\times H^2\times\frac{(\tan\alpha-\tan\beta)\times\sin(\alpha-\theta)}{\sin\alpha\times\cos\theta\times\tan\alpha\times(\tan\theta-\tan\beta)} \end{aligned} \tag{2}$$

The Terraces Length of Per Unit Area (1 hm^2)

$$L = \frac{10000}{B} = \frac{10000\times\tan\varepsilon\times(\tan\theta-\tan\beta)}{(\tan\alpha-\tan\theta)\times H} \tag{3}$$

The Relationship between Table Height and Ridge Height

$$H' = H+h = H+\frac{\tan\beta\times(\tan\alpha-\tan\theta)}{\tan\alpha\times(\tan\theta-\tan\beta)}\times H = \frac{\tan\theta\times(\tan\alpha-\tan\beta)}{\tan\alpha\times(\tan\theta-\tan\beta)}\times H \tag{4}$$

The Calculation Formula of Sloping Terraced Earthwork

$$\begin{aligned} V &= S_{OAC}\times L = \frac{1}{8}\times H^2\times\frac{(\tan\alpha-\tan\beta)\times\sin(\alpha-\theta)}{\sin\alpha\times\cos\theta\times\tan\alpha\times(\tan\theta-\tan\beta)}\times\frac{10000\times\tan\varepsilon\times(\tan\theta-\tan\beta)}{(\tan\alpha-\tan\theta)\times H} \\ &= 1250\times\frac{\tan\alpha-\tan\beta}{\tan\alpha}\times H = 1250\times\frac{\tan\theta-\tan\beta}{\tan\theta}\times H' \end{aligned} \tag{5}$$

The results can be found by formula (5): Sloping terraced earthwork is closely related to ridge slope (α), the original field surface slope (θ), the field surface slope after down slope (β), ridge height (H), and table height (H'). Just make sure the parameters of the three sloping terrace section(α 、 β and H or θ 、 β and H') can be formed to determine the amount of earth sloping terrace. In soil characteristics and other factors determined conditions, ground slope is a key factor to determine the parameters α , β and H .Therefore, if the slope distribution of project area is determined, slope-style terraced earthwork calculation is easy.

4 Fast Calculation Approach of Sloped Terrace Earthwork Based on GIS in the Hilly Regions

GIS has slope statistical functions, which provide feasibility for the use of GIS for slope-style terraced earthwork calculation. To estimate the amount of earthwork by

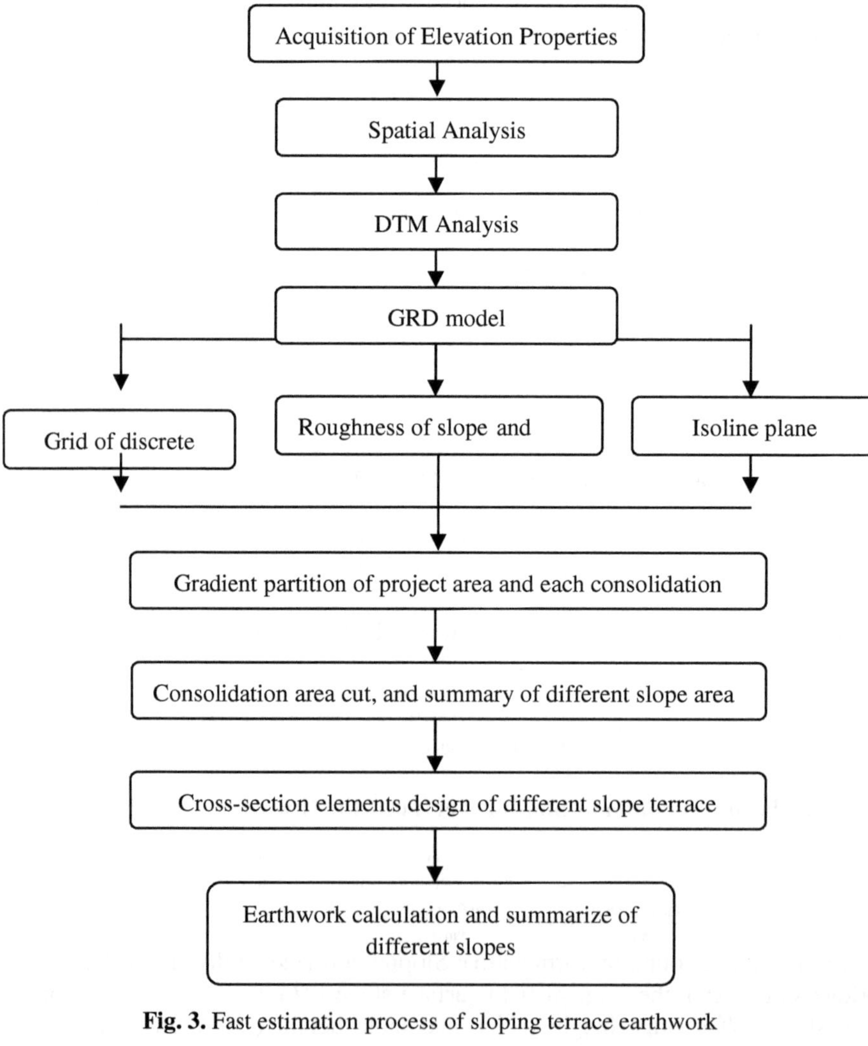

Fig. 3. Fast estimation process of sloping terrace earthwork

application of GIS technology is a manifestation of GIS software spatial analysis functions. Applying GIS software to measure the amount of earthwork needs the elevation data, including Contour lines, elevation points, etc. General project areas have the latest digital measurement of topographic maps, which provide the basic data for GIS software applications. The basic steps f rapid estimation of sloping terraces earthwork proposed by the use MAPGIS in this study were shown in Figure 3.

5 Case Studies

According to the basic steps of quick estimation of sloping terraces earthwork by the use of MAPGIS, the land consolidation project area in Sanxi town, Yangxin county, Hubei province was partitioned by slope (Figure 4).The range of different slope range were designed to reduce slope (table 1), then into calculation formula of sloping terraced earthwork, the amount of land consolidation were rapidly measured and summarized (table 1).

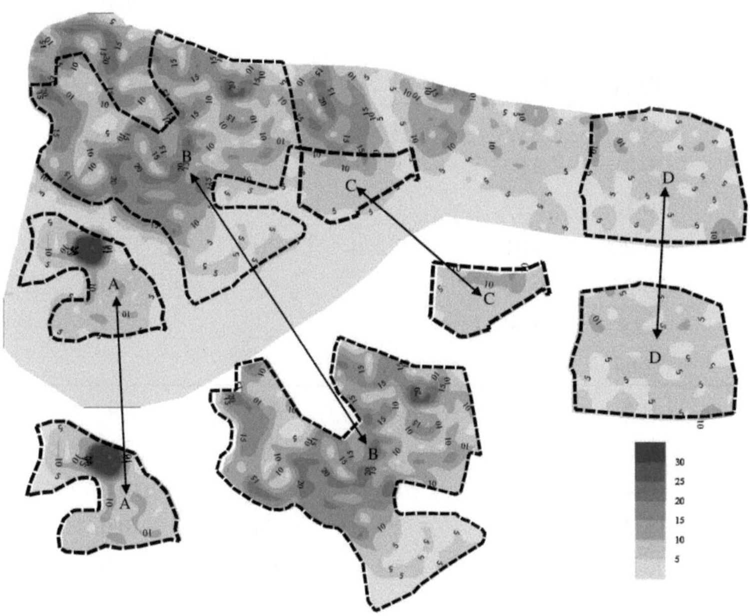

Fig. 4 Slope map of land leveling area

Table 1. Earthwork calculation of sloping terrace land leveling Font sizes of headings.

Consolidating region	Slope	Area	Ridge height	Ridge slope	Field surface slope after descending slope	Earthwork
	°	hm^2	m	°	°	m^3

Table 1. *(Continued)*

Consolidating region 1	0-5	0.34	0.8	85	0.0	341.65
	5-10	1.57	1.2	80	5.0	2322.57
	10-15	0.36	1.8	75	10.0	772.84
	15-20	0.06	2.2	70	10.0	148.89
	20-25	0.05	2.4	65	10.0	133.33
	>25	0.23				0.00
	Total	2.61				3719.29
Consolidating region 2	0-5	1.59	0.8	85	0.0	1590.04
	5-10	3.36	1.2	80	5.0	4966.74
	10-15	3.23	1.8	75	10.0	6929.53
	15-20	3.02	2.2	70	10.0	7776.79
	20-25	0.76	2.4	65	10.0	2085.71
	>25	0.06				0.00
	Total	12.02				23348.81
Consolidating region 3	0-5	0.04	0.8	85	0.0	40.40
	5-10	0.97	1.2	80	5.0	1431.40
	10-15	0.19	1.8	75	10.0	405.93
	15-20	0.01	2.2	70	10.0	23.41
	20-25	0.00	2.4	65	10.0	0.00
	>25	0.00				0.00
	Total	1.21				1901.14
Consolidating region 4	0-5	1.17	0.8	85	0.0	1167.78
	5-10	3.39	1.2	80	5.0	5013.83
	10-15	0.17	1.8	75	10.0	371.60
	15-20	0.00	2.2	70	10.0	0.96
	20-25	0.00	2.4	65	10.0	0.00
	>25	0.00				0.00
	Total	4.74				6554.17

6 Conclusions

This study proposes a new calculation method of terraced earthwork in hilly area, and this method is applied to the actual project area. Through research the following conclusions can be reached:

(1) Sloping terraced earthwork is closely related to ridge slope (α), the original field surface slope (θ), the field surface slope after down slope (β), ridge height (H), and table height (H').Just make sure the parameters of the three sloping terrace section(α 、 β and H or θ 、 β and H') can be formed to determine the amount of earth sloping terrace. Use of GIS for slope-style terraced earthwork calculation is feasible.

(2) Earthwork calculation principles of the proposed method are clear, and easily measured. The earthwork of different slope range could be estimated quickly according to sloped terrace section (ridge height, ridge slope and the field surface slope after down slope) after getting the area of different slope range using GIS. The workload of earthwork calculation was greatly reduced.

(3) The approach is simple and fast, with strong practical characteristics. It provides a new idea for earthwork estimation of sloped terrace in hilly area, and improves the efficiency of project planning and design.

References

1. Liu, C., Du, L.: Application of Arcview for calculating earthwork in land consolidation project. Transactions of the CSAE 19(2), 224–227 (2003)
2. Dong, J., Zhang, A.: Practice in land consolidation in hilly regions of south china: case study of shangsi county, guangxi zhuang autonomous region. China Land Science 18(2), 53–55 (2004)
3. Liu, N., Yun, W., Lei, T.: Method for fast estimation of terrace earthwork in the hilly regions and its application. Transactions of the CSAE 23(4), 47–51 (2007)
4. Jin, X., Li, H., Zhou, Y., et al.: Assistant decision making for DEM-based land readjustment project in hills and mountainous areas. China Land Science 18(1), 39–44 (2004)
5. Lu, C., Wu, C., Wang, F., et al.: Application of GIS technology in earthwork calculation and allocation of farmland consolidation. Transactions of the CSAE 19(6), 289–292 (2003)
6. Chang, L., Zhang, W., Li, C., et al.: Research of some problems in land consolidation based on GIS technology. Journal of Southwest Agricultural University (Social Science Edition) 6(3), 9–12 (2008)
7. Bao, H., Wu, C., Ye, Y., et al.: Dimension design of farmland and application of GPS-GIS-RS technology to land consolidation. Transactions of the CSAE 18(1), 167–172 (2002)
8. Fan, Y., Luo, Y., Wei, C.: On level terrace engineering design of slope land in the hilly and mountainous, southwestern China. Journal of Mountain Science 28(5), 560–565 (2010)
9. Zheng, C., Qiu, D., Tian, S.: A Formula Deduced for the Calculation of Earthwork Volume of Sloped Terrace in Hilly Area. Journal of Southwest University (Natural Science Edition) 31(1), 173–178 (2009)
10. Zhou, W., Liu, X., Yang, C.: Study on land consolidation status in the hilly and mountainous region of southwest China. Journal of Southwest China Normal University (Natural Science Edition) 33(2), 104–109 (2008)

The Estimation of Tree Height Based on LiDAR Data and QuickBird Imagery

Wei Su[1,2], Rui Liu[2], Ting Liu[2], Jianxi Huang[2], Xiaodong Zhang[2], and Junming Liu[2]

[1] Key Laboratory of Agri-informatics,
Ministry of Agriculture, Beijing, 100081, P.R. China
[2] College of Information and Electrical Engineering,
China Agricultural University, Beijing 100083 P.R. China

Abstract. The estimation of tree height is advanced following the development of LiDAR technique. The estimation model of tree height considering suppressed trees is developed in order to extract tree height accurately using LiDAR data. Filtered LiDAR data and Quickbird imagery are segmented using watershed segmentation method based on mathematical morphology to get the boundary of trees. And the highest point in each canopy object is used to estimate tree height. Weibull distribution is used to estimate height distribution of the suppressed trees. The experiment results indicate that the watershed segmentation method based on mathematical morphology is an effective method to extract the boundary of trees. And the R^2 between the tree height estimated using estimation model of tree height considering suppressed trees and the tree height measured by field work is 0.93.

Keywords: estimation of tree height, LiDAR, watershed segmentation, Weibull function.

1 Introduction

LiDAR (Light Detection And Ranging) is a remote sensing technology that can be used to extract detailed 3D structure information. LiDAR is an active system which emits high frequency bursts of laser pulses. So they are able to penetrate relatively dense vegetation canopies, and can be used to get the 3D structure information of trees even for suppressed trees.

There are many researchers explore how to estimate tree height using LiDAR data in recent years, most of them do this using CHM (Canopy Height Model). This CHM is the difference between DSM (Digital Surface Model) and DTM (Digital Terrain Model), which is the height of non-ground objects such as trees, buildings, vehicles etc.[1-5] Liu [6] *et al.* develop smoothed CHM to estimate tree height. Firstly, they smooth CHM using Gaussian smoothing method. Then, they develop a double tangent crown edge recognition (DTCER) algorithm to detect the canopy boundary, and the local maximum search algorithm with fixed or variable window was used to recognize crown tops. Persson *et al.* create DCM (Digital Canopy Model) from subtracting DTM from DSM, then which is smoothed with different scales. A parabolic surface was fitted

D. Li and Y. Chen (Eds.): CCTA 2012, Part II, IFIP AICT 393, pp. 472–482, 2013.

to the elevation data to determine which scale to choose for different parts of the image, and the height and crown diameter are estimated for the identified trees [7]. Popescu *et al.* use regression models and cross-validation to estimate tree height based on LiDAR data. They found that estimating tree height for deciduous plots gave superior results without calibrating the search window size based on forest type [8]. Zimble *et al.* search potential canopy top points using a moving radius circular search window with 0.9m radius, and adjoining pixels that are higher than 85% of their neighbors are combined into clumps that are assumed to include the peaks of tree crowns[9]. A common ground of all of these researches is that there should be some prior knowledge about their study area which is the basis of determining initial parameters used in model. Therefore, there are some difficulties in extending those methods to other areas.

We develop an estimation model of tree height considering suppressed trees in this study. The watershed segmentation algorithm is used to segment tree canopy objects, the highest point in each tree canopy object is searched using a moving window size, and the height of suppressed trees is estimated using Weibull function. The rest of the paper is organised as follows: Section 2 describes the study area and data source. Section 3 presents the methodology of the estimation model of tree height considering suppressed trees. And the rresults, discussion and conclusions are presented in Section 4 and Section 5.

2 Study Area and Data Sources

The study area is an urban area (Kuala Lumpur City Centre, from 101°40'46" to 101°43'25"E, 3°8'00" to 3°10'39"N) located midway along the west coast of Peninsular Malaysia, at the confluence of the Klang and Gombak rivers. This region is characterized by rolling topography and flourishing vegetation, and the height of high trees range from 20m to 30m. A subset area covered with dense high trees in Kuala Lumpur City Centre is selected as study area in this study. And the QuickBird imagery and CHM image produced using LiDAR data of this subset are as Figure 1 (a) and (b) respectively, which are resample to 1m spatial resolution.

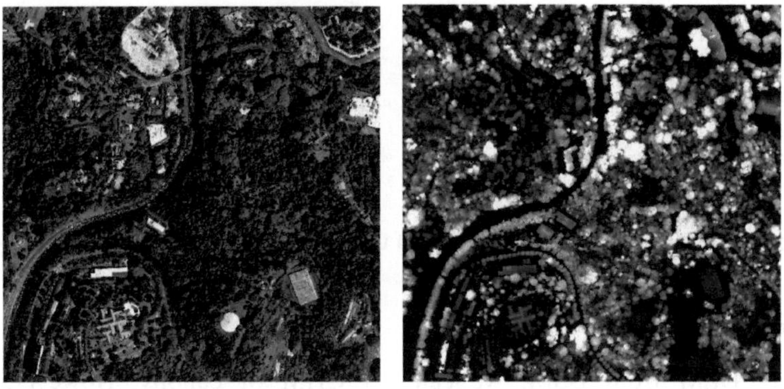

Fig. 1. Study area (subset of Kuala Lumpur City Centre, Malaysia). (a) Quickbird image, (b) CHM image

3 Methodology

The developed estimation model of tree height considering suppressed trees works on CHM and Quickbird image. The ground objects such as trees, buildings, roads, are segmented based on CHM image using watershed segmentation based on mathematical morphology method. The spectral features coupling with height information of image objects are used to identify canopy objects. So the boundary of canopy are acquired. Searching the highest points in each object, we can estimate the height of trees. At last, we use Weibull function to simulate and estimate the height of suppressed trees. And there are 4 components in this algorithm: CHM image smoothing, image segmentation using watershed segmentation method based on mathematical morphology, extracting the tree vertex points and estimating the height of canopy, estimating the height of suppressed trees using Weibull function. This flow chart of the whole algorithm is as follows:

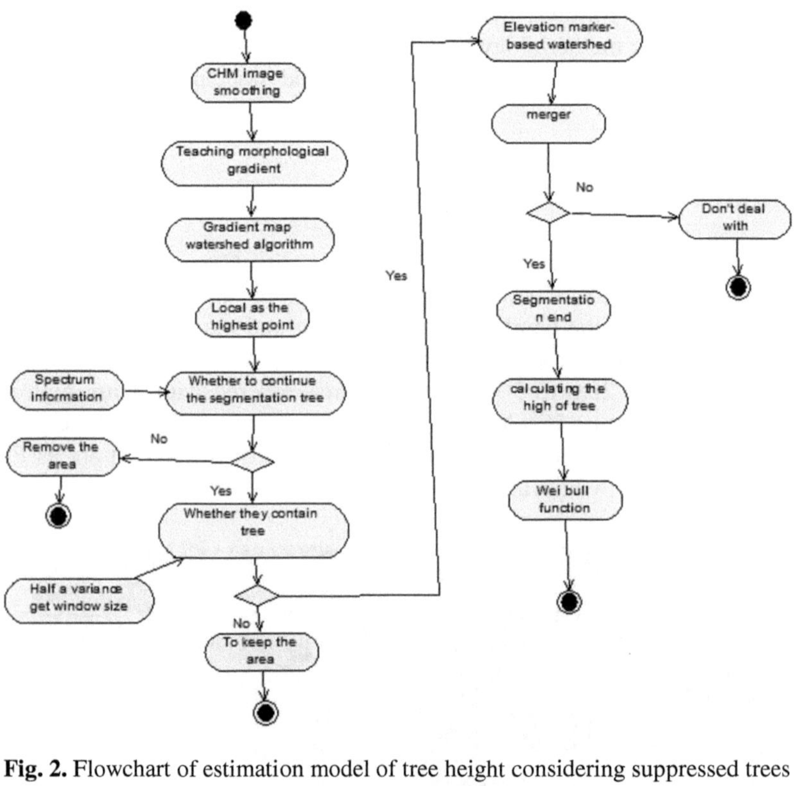

Fig. 2. Flowchart of estimation model of tree height considering suppressed trees

3.1 Image Smoothing

Subtracting DEM from DSM, we can get CHM of canopy. However, for the irregularity of LiDAR data and the twigs of the canopy surface, there will be false tree vertex if we use unsmoothed CHM image directly to identify the tree vertex and crown boundary. So we smooth the CHM image in order to remove these noise before identifying tree vertex and crown boundary[10]. However, this smoothing will weaken the difference between tree vertex and around pixels, which will decrease the ability to identify the tree vertices. So he Gaussian filter method is used to smooth CHM image. The filtering kernel of Gaussian filter is as follows:

$$g(x,y,\sigma) = \frac{1}{2\pi\sigma^2} e^{-(\frac{x^2+y^2}{2\sigma^2})} \tag{1}$$

Where, x and y are the distant to the center of smoothing nuclear, σ is used to adjust the parameters of the Gaussian function. Figure 2 shows a schematic diagram of a 3×3 filtering kernel of Gaussian smoothing algorithm.

g(-1,1, σ)	g(0,1, σ)	g(1,1, σ)
g(-1,0, σ)	g(0,0, σ)	g(1,0, σ)
g(-1,-1, σ)	g(0,-1, σ)	g(1,-1, σ)

Fig. 3. 3×3 filtering kernel of Gaussian smoothing algorithm

Moving this filtering kernel, we smooth CHM image. For the pixel p(x,y) located in the centre of filtering kernel, its DN value is computed as:

$$p(X,Y) = \sum_{i=-d/2+0.5}^{d/2-0.5} \sum_{j=-d/2+0.5}^{d/2-0.5} p(X+i, Y+j) \times g(x+i, y+j, \sigma) \tag{2}$$

X and *Y* are coordinates of the pixel in the image, d is the size of the filtering template which is an odd number such as 3, 5, 7 etc., *i* and *j* are the distant to the studied pixel.

3.2 Watershed Segmentation Based on Mathematical Morphology

3.2.1 Principle of Watershed Segmentation

The watershed segmentation algorithm is a robust image segmentation tool, which can be used to extract interested objects in the image effectively. Vincent's simulate immersion algorithm is used in this study [12]. The pixels in gradient image are classified as four types during the watershed segmentation: marked pixel, waterline pixel, unlabelled pixel and classified pixel. The watershed is finished when all pixels are classified pixels or waterline pixels. There are two steps in Vincent's method: sorting and immersion.

(1) Sorting: ascending the pixel value in line with the DN value of pixel, we can get a sorted pixel matrix. In the process of gradually submerged, we can call the

pixels which need to be processed. But we needn't process all pixels in every loop, which can speed up the calculation.

(2) Immersion: this immersion is performed in ascending order to call each pixel based on sorted gradient image. Starting from the minimum DN value in the whole gradient image, immersion is done one pixel by one pixel. This loop queue is used to extend labeled catchment basins. Based on the distance to the neighborhood, we label finished watershed.

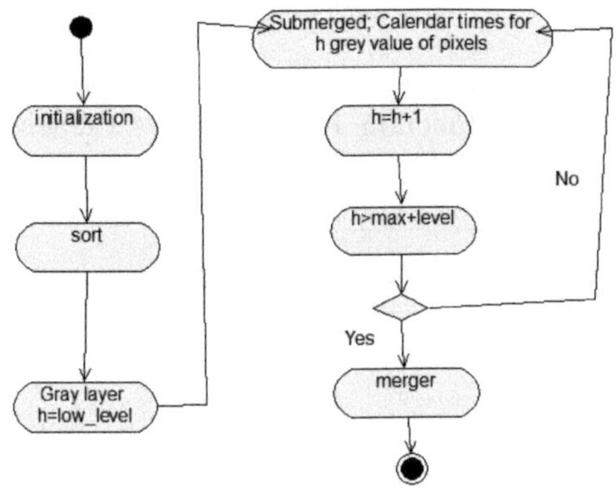

Fig. 4. Flowchart of Vincent's segmentation algorithm

The study area is a tropical rain forest area, where there are dense and high tropical trees. This results in a difficulty in canopy segmentation because there is overlapping between one canopy and another. Watershed segmentation method is an effective way to extract canopy objects because which can take advantage of the height difference between trees fully. In addition, the watershed segmentation method is a method without any parameters, which is suitable in many situations. So we can get an ideal image segmentation result without prior knowledge. However, there will be a minor gradient change if there are two trees which is located near and the canopy top is relatively flat. In this case, it is difficult to identify these tree objects using watershed segmentation based on gradient. Fortunately, watershed based on the distribution of height is used to do further segmentation. The watershed segmentation algorithm based on the distribution of height is as follows: searching the maximum points in each objects produced from previous segmentation using fixed window, the further segmentation will be done if there are two or more maximum points in one objects.

Waterline pixel in the watershed segmentation is the intersection of two or more tree objects. The classification of these points is done by the inclination angle between this point and crown vertex which the studied point is possible belonged to. And the studied point should be classified as the tree objects which have a larger inclination angle. For an example in Figure 4, there is an intersection between tree A and tree B. r1 and r2 are the inclination angles between this intersection and tree A and tree B. And this point should be classified as tree A, otherwise, classified as tree B.

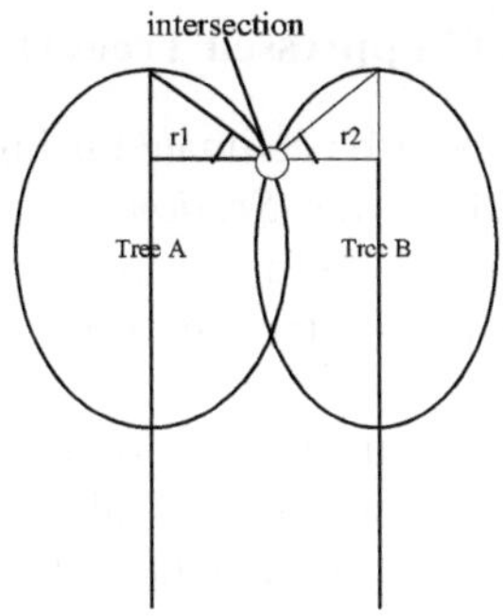

Fig. 5. Schematic diagram of ownership of the intersection point

3.2.2 Merging of over-Segmentation

There is over-segmentation in watershed segmentation sometimes, especially when there are larger branches highlighted in studied objects. So we should merge these over-segmented image objects. The Full Lambda Schedule algorithm of Robinson *et al.* is used in this study[13], which is as follows:

$$t_{i,j} = \frac{|o_i| \times |o_j|}{|o_i| + |o|_j} \times \|u_i - u_j\|^2 / \text{length}(\alpha(o_i, o_j)) \tag{8}$$

$|o_i|$ and $|o_j|$ are the area of image object i and j respectively, u_i and u_j are the mean DN value of image objects i and j, $\|u_i - u_j\|^2$ is the Euclidean distance between the image object i and the image object j, length$(\alpha(o_i, o_j))$ is the length of the public side of the image object i and the image object j. $t_{i,j}$ is computed using Formula (8), we will merge image object i and image object j when $t_{i,j}$ is less than threshold.

3.3 Extracting Tree Vertex

The segmented results using watershed image segmentation method is only the candidate tree objects, which are not necessarily the real tree objects. So we must identify if or not every classified tree objects is tree in the next step. And the NDVI produced from the spectral bands of Quickbird image is used to do this judgment process. The mean NDVI is used to classify the tree objects. The objects with mean NDVI value less than the threshold will be classified as non-tree, which will be deleted; otherwise, which should be classified as tree.

The highest point in an object is classified as tree vertex after this classification, these are used to estimate tree height. We use a 5×5 moving window to calculate the tree height if there more than on maximum value in one image object, and the mean DN value in this 5×5 window acts as the tree height in this case. If the maximum DN value is still equivalent in 5×5 window, we will extend window size until we can get the unique maximum DN value.

3.4 Estimating the Height of Suppressed Trees Using Weibull Function

The height of suppressed trees can't be estimated using those existing methods [14], including the method described in this *Section 3.3*. So we should develop some method to do this. Weibull function is proposed by Weibull, who is the Swedish physicist, which is used widely in forestry measuring and research for the flexibility of its form. Weibull function is

LiDAR usually only can find the trees in the dominant, and the distribution of the underlying wood is difficult to obtain [14], to make use of the Weibull function for modeling and prediction and extract the height distribution of the lower wood.The Weibull function is proposed by the Swedish physicist Weibull, because is widely used in forestry research, it is the main method for the prediction of the height distribution and the diameter distribution [15]. The three-parameter Weibull function of the density function f (x) and distribution function F(x):

$$f(x)=\begin{cases} \frac{c}{b}(\frac{x-a}{b})^{c-1}\times e(\frac{x-a}{b})^{c} & x\geq a \\ 0 & x<a \end{cases} \tag{9}$$

$$F(x)=1-e^{-(\frac{x-a}{b})^{c}} \tag{10}$$

a is a location parameter, b is a scaling function, c is a shape parameter.

In the application of using the Weibull function to simulate the trees high, a is the height of minimum spanning tree height, so that a becomes a set value .At this point, the three-parameter Weibull function is converted to two-parameter Weibull function used here are a set worthy of the two-parameter Weibull function. Because the determination of the laser radar, lidar accuracy combined with mutual shadowing between trees, and stacked cover, trees and undergrowth of wood is difficult to be identified, measured by the tree is not up to the minimum spanning tree, so only is the measured number of large trees, trees number is unknown, the corresponding can not be used to complete the Weibull function to the left truncated Weibull function to simulate the distribution of trees, get the value of b and c, then a complete weibull function simulate the distribution of the trees. Left truncated Weibull function of the density function f (x) as follows:

$$f(x)=\begin{cases} \frac{c}{b}(\frac{x}{b})^{c-1}\times e((\frac{t}{b})^{c}-(\frac{x}{b})^{c}) & x>t \\ 0 & x<t \end{cases} \tag{11}$$

t is the measured height of the laser radar.

Weibull function parameter estimation There are several ways, the most common and accurate maximum likelihood estimates the parameters of the weibull function, 11 b and c, values, and then substituted into equation (10) to get a high degree of less than t probability F (t).

4 Results and Analysis

Using the algorithms discussed in section 2 to segment LIDAR data of the two pilot areas, and the results as shown below.

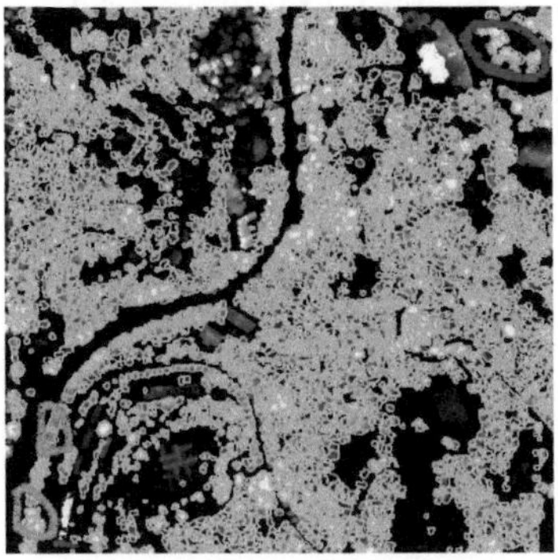

Fig. 6. Segmentation results ((a) segmentation result of test area one (b) segmentation result of test area two)

We can be seen from Figure 6 that trees are basically of two pilot areas have been extracted, and the border in line with the actual situation, some spread between the artificial construction of the trees are extracted. However, due to the spectrum appear synonym spectrum with the spectrum of foreign bodies, and some buildings were classified as trees, some trees are not extracted.

NDVI can remove a large number of non-vegetation information. Can be seen from Figure 5, trees and buildings are well separated. But can be found in Figure 5 (a): the top right of the red marked man-made structures has also been judged as trees, the average NDVI of these areas is 0.4, but the bottom left of the green marked regional NDVI trees region NDVI only 0.36. Staggered addition of artificial buildings and trees, mixed distribution, but when the building height to a certain extent, the shadow of the building will be obscured to some trees, then trees NDVI values also fall, leading to an average of some trees image object NDVI dropped to less than the threshold, thus false positives for non-trees objects. Artificially built-up areas in the second image, the trees did not get to some of the buildings shadow, but in the forest part of the overall NDVI values are relatively high, so the shadow of the forest part of the less affected. In order to retain more information of the crown, set low NDVI, the first image of NDVI threshold is relatively high with 0.35, while the second image of 0.3.

Image segmentation and image objects of the merger after the completion of the judgment, the process to complete the Full Lambda the Schedule method, this method taking into account the spectral information of the Quickbird image object, area information and public side length of the adjacent area. Spectrum and size of the difference is proportional to the object with two images, side length and area with the adjacent edges of the two image objects is inversely proportional to the differences. This method is more reasonable to merge some of the region has a long common edge, but because of the high spectral similarity of the two image trees region, when the trees were over-segmentation, some of this should be a whole tree is divided into multiple regions ,but adjacent region area does not necessarily have similar spectral information, if it belongs to the edge of the area from the whole partition, it would be more similar to the edge of the area and other adjacent tree spectrum rather than it should be subordinate the tree itself. And because of the generally over-segmentation regions of various shapes, so the shape parameter and the height distribution of the information is difficult to join .So we use spectral information, but the merger of the two adjacent regions, the spectral information is too close to the merger, only two adjacent regional spectral information over a certain range when the merger. In this paper, spectral information is also used NDVI values.

Extracted from the tree height and the measured points were compared, the results are as follows:

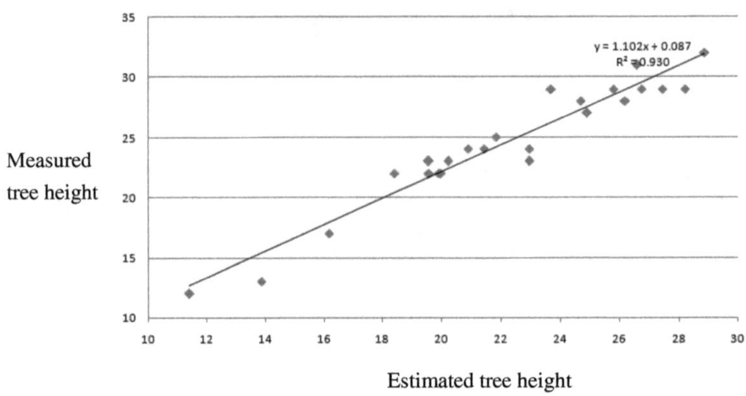

Fig. 7. Tree height fitting results

We can see in Figure 7 that using LiDAR measured tree height and measured height has a high correlation, which is close to the 1:1 ratio. Measured tree height = 1.12 × the LiDAR tree height +0.087, R2 = 0.93.

Simulated trees using Weibull function, and predict the two images to detect the elevation distribution of the trees greater than 5m tree. 4057 and 1284, respectively, on the simulation results of the trees greater than 5m R20.96 and 0.90, predicted less than 5m of trees 26 and 12. As the first image is mainly forest areas ,so using the tree height distribution measured in the first image as a test ,set to detect minimum tree height is 15m, left-truncated Weibull function simulation to obtain between 5-15mthe tree height distribution, the results show that the 3702 tree greater than 15m 5-15m of the tree 443, the distribution of more than 5m tree obtained by the Weibull function and to detect the distribution of greater than 5m tree compared to the R2 is 0.93. The above three Weibull function simulation through the Kolmogorov - Smirnov test can account for that using left truncation Weibull function can be effective simulation and forecasting detected and undetected trees.

5 Conclusions

This paper combined the use of LiDAR data with Quickbird images and constructed the model of finding tree height which take into account the height of the trees of lower wood .This method relative to the maximum search approach, be able to very good use of the spatial variation of the height distribution, rational use of trees and between trees existing gap in the watershed segmentation process, bifurcation identification of trees overlap or tree good effect; After the first watershed segmentation using watershed segmentation on a high level and some smooth crown can be split effectively, and using the Full Lambda the Schedule algorithm can effectively merge the over-segmentation area, so not only guarantee the adequacy of the segmentation to reduce over-segmentation; Combination of LiDAR data and Quickbird images can effectively distinguish between forest and other surface features; Using the Weibull function to simulate the distribution of tree height and predict the distribution of the trees. This method of height information extraction can obtain high extraction efficiency in the region of dense forest and artificial buildings and trees interlaced distribution. Proposed in this study to consider the underlying wood to a high degree of tree height to find the height model extraction process does not require too much prior knowledge, application scalability. Of course, the proposed algorithm there is need for further improvements:

1. In the process of tree height extraction, the threshold of NDVI and canopy size need to manually set and the threshold size would affect the results, the NDVI impact on the distinction between forest and other surface features, and crown size affects the crown dense region segmentation results.
2. In the process of canopy object mergering, each region is relatively small and are independent, the use of the consistency of every region of the spectrum to the merger will bring some errors. If you can use the multi-echo or echo image of the laser radar, canopy and lower operating directly on point clouds, merged correctly.

Acknowledgements. This research is funded by Agricultural Research Elitist Fund of Ministry of Agriculture, P. R. China, Open-end Research Fund of Key Laboratory of Agri-informatics, Ministry of Agriculture, P. R. China[2012003], and National Key Technology R & D Program of the Ministry of Science and Technology [2012BAD20B0103]. We would like to thank the reviewers for valuable comments and constructive advice.

References

1. Xiao, X.W.: China's forest resource inventory. China Forestry Publishing House, Beijing (2005)
2. Andersen, H.E., Reutenbush, S.E., Mcgaughey, R.J.: A rigorous assessment of tree height measurements obtained using airborne LIDAR and conventional field methods. Canadian Journal of Remote Sensing 32(5), 355–366 (2006)

3. David, R.S., Nancy, F.G.: LiDAR measurement of sagebrush steppe vegetation heights. Remote Sensing of Environment 102, 135–145 (2006)
4. Husch, B., Miller, C., Beers, T.: Forest mensuration, 2nd edn. Ronald Press Company, New York (1972)
5. Hans, O.Ø., Erik, N., Ole, M.B.: Classifying species of individual trees by intensity and structure features derived from airborne laser scanner data. Remote Sensing of Environment 113, 1163–1174 (2009)
6. Liu, Q.W.: Study on the Estimation Method of Forest Parameters Using Airbone LiDAR, Beijing (2009)
7. Persson, A., Holmgren, J., Soderman, U.: Detecting and measuring individual trees using an airborne laser scanner. Photogrammetric Engineering & Remote Sensing 68(9), 925–932 (2002)
8. Sorin, C.P., Randolph, H.W.: Seeing the trees in the forest: using lidar and multispectral data fusion with local filtering and variable window size for estimating tree height. Photogrammetric Engineering & Remote Sensing 70(5), 589–604 (2004)
9. Daniell, A.Z., David, L.E., George, C.: Characterizing vertical forest structure using small-footprint airborne LiDAR. Remote Sensing of Environment 87(2), 171–182 (2003)
10. Tomas, B., Timothy, A.W., Rick, E.L.: Detection and analysis of individual leaf-off tree crowns in small footprint, high sampling density lidar data from the eastern deciduous forest in North America. Remote Sensing of Environment 85, 290–303 (2003)
11. Solberg, S., Maesset, E., Bollandsas, O.M.: Single tree segmentation using airborne laser scanner data in a structurally heterogeneous spruce forest. Photogrammetric Engineering & Remote Sensing 72(12), 1369–1378 (2006)
12. Vincent, L., Soille, P.: Watersheds in digital spaces: An efficient algorithm based on immersion simulations. IEEE Transactions on Pattern Analysis and Machine Intelligence 13(6), 583–598 (1991)
13. Robinson, D.J., Redding, N.J.: Implementation of a fast algorithm for segmenting SAR imagery. Scientific and Technical Report 1(1), 4–5 (2002)
14. Maltamoa, M., Eerikainen, K., Pitkanene, J.: Estimation of timber volume and stem density based on scanning laser altimetry and expected tree size distribution functions. Remote Sensing of Environment 90, 319–330 (2004)
15. Meng, X.Y.: A study of relation between D and H-Distribution by using the Weibull funtion. Journal of Beijing Forestry University 1(10), 40–48 (1988)

Author Index

MIX
Papier aus verantwortungsvollen Quellen
Paper from responsible sources
FSC® C105338

FSC
www.fsc.org

If you have any concerns about our products,
you can contact us on
ProductSafety@springernature.com

In case Publisher is established outside the EU,
the EU authorized representative is:
Springer Nature Customer Service Center GmbH
Europaplatz 3, 69115 Heidelberg, Germany

Printed by Libri Plureos GmbH
in Hamburg, Germany